INFORMATION GRAPHICS

A Comprehensive Illustrated Reference

ROBERT L. HARRIS

Management Graphics
Atlanta, GA

Oxford University Press
New York   Oxford

Oxford University Press

Oxford   New York

Athens   Auckland   Bangkok   Bogotá   Buenos Aires   Calcutta
Cape Town   Chennai   Dar es Salaam   Delhi   Florence   Hong Kong   Istanbul
Karachi   Kuala Lumpur   Madrid   Melbourne   Mexico City   Mumbai
Nairobi   Paris   São Paulo   Singapore   Taipei   Tokyo   Toronto   Warsaw

and associated companies in
Berlin   Ibadan

Copyright © 1999 by Robert L. Harris

Published by Oxford University Press, Inc.
198 Madison Avenue, New York, New York 10016

Oxford is a registered trademark of Oxford University Press.

First published in 1996 by Management Graphics, Atlanta, Georgia.

First issued as an Oxford University Press paperback in 1999.

This publication is designed to provide accurate, illustrative, and authoritative information in regard to the subject
matter covered. It is sold with the understanding that neither the author nor the publisher is engaged in rendering
legal, accounting, investment, or other professional services. If expert assistance is required, the services of a compe-
tent professional person should be sought.

Neither the author nor the publisher assumes any responsibility for the uses made of the material in this book or for
decisions based on its use and makes no warranties regarding the contents of this product or its fitness for any partic
ular purpose.

All data used in the examples in this book are fictitious. Numeric values in the example may or may not be similar to
those occurring in actual applications.

Acknowledgment and thanks are extended to Glenda Jo Fox Hughes, C. Dwight Tabor, Jr., and Merwyn L. Elliott for
reviewing portions of the material in this book.

Library of Congress Cataloging-in-Publication Data
 Harris, Robert L., 1930–
Information graphics : a comprehensive illustrated reference / Robert L. Harris.
p.   cm.
Includes bibliographical references.
ISBN 0-19-513532-6
1. Graphic methods Encyclopedias.   2. Charts, diagrams, etc., Encyclopedias.
3. Computer graphics.   I. Title.
QA90.H287   1999
001.4'226'03—dc21   99-30920

9

Printed in the United States of America
on acid-free paper

Major focus of book

This book addresses charts, graphs, maps, diagrams, and tables used in all areas;
however, its major focus is on their uses for operational purposes.

To many people, information graphics are the images frequently used in presentations at
formal meetings or the stylized charts and graphs used in newspapers and magazines.
Many are used for these purposes; however, for every chart, graph, map, diagram, or
table used in a presentation or publication, there are thousands that are utilized for what
are called operational purposes. Information graphics for operational purposes are used
by millions of people on a daily basis for such things as improving their efficiency and
effectiveness, improving quality, solving problems, planning, teaching, training,
monitoring processes, studying the geographic distribution of data, looking for trends and
relationships, reviewing the status of projects, developing ideas, writing reports,
analyzing census data, studying sales results, and tracking home finances. With the need
to cope with increased amounts of data and at the same time increase quality and
productivity, charts, graphs, and maps are being used more and more in operational
situations. Fortunately, as a result of developments in computer equipment and software,
most of the popular charts and graphs used on a daily basis can be generated rapidly,
easily, and with little or no special training.

This single volume is the first book to focus on the needs of the millions of individuals
who currently use or would like to use information graphics for their operational needs. It
is of particular value to those who might generate their charts and graphs on computers.
The book brings together the necessary information on charts, graphs, maps, diagrams,
and tables so individuals can rapidly discover the wide range of information graphics that
is available, how they are being used, how to construct them, and how to interpret them.

Organization and major contents

The contents of this book are organized for ease of use. Towards this goal:

- Entries are alphabetized using the letter by letter system in which spaces, hyphens, commas, etc., between words in the major headings are ignored.
- Although the major headings of Chart, Graph, Map, Diagram, and Table are each listed in their proper alphabetical location, the individual graphics and features that make up these major categories also have their own headings in the master alphabetical listing.
- Specific applications are described and illustrated to relate the theoretical to the practical.
- Entries are accompanied by one or more graphic examples (frequently annotated) that complement the written descriptions.
- Terminology applicable to specific graphics are explained in the text, shown on the example, or both.
- General terminology such as Variable, Fill, Legend, Matrix, Polygon, Plane, Coordinate, etc., are discussed under their individual headings.
- For the convenience of the reader, some information is repeated under multiple headings.

The following additional features have been incorporated.

Four major types of cross-referencing

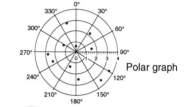

Polar graph

- When an information graphic has a single name but might be classified several different ways, the major write-up is included under one of the headings and cross-referenced under the others. For example, the main write-up for polar graph (right) is under Polar Graph but it is referenced under Graph, Circular Graph, and Point Graph.

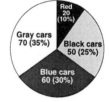

Sometimes referred to as:
Pie chart
Cake chart
Circle diagram
Circle graph
Circular percentage chart
Divide circle
Sectogram
Sector chart
Segmented chart

- When an information graphic is commonly referred to by several different names, the major write-up is shown under only one of the headings. The other names are included at their proper alphabetical locations along with a short description, an example, and a reference as to where the major write-up is listed. For example, various users refer to the chart at the right by nine different names. The major description is included under Pie Chart, with only thumbnail descriptions under the other eight headings.

- In cases where the same name is used to describe entirely different information graphics, each situation is handled on an individual basis. An example is shown below in which each of the graphics is sometimes referred to as a bar chart. In this particular case, the information graphics are different enough that each is given its own write-up. Each is cross-referenced appropriately, although all are not referenced to one another. Superscripts are assigned in cases where the exact same name is the one most frequently used for multiple graphics.

Column graph

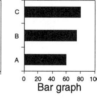
Bar graph

Bar chart[1]

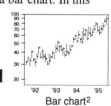
Bar chart[2]

- When a topic is related to an entry or it would be beneficial for the reader to be aware of a related topic, the related topic is noted in the write-up.

Meaningful groups and families of information graphics

With so many different information graphics used in such diverse applications, it is sometimes helpful to group them into families or categories. For example, it is useful to know which graphs are used to study data distribution or to look for correlations. It is helpful to know which graphs present percent-of-the-whole data most efficiently or what types of graphs are used to determine probabilities. In the area of maps, it is useful to know that most maps that are used as charts fall into six major categories, the four major ones being statistical, descriptive, flow, and topographic. The sections that discuss the various groupings of information graphics are in addition to the sections that discuss the specifics of the various graphics.

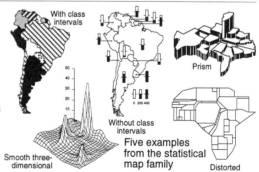

With class intervals

Prism

Without class intervals

Smooth three-dimensional

Five examples from the statistical map family

Distorted

Key construction details

Most information graphics software programs have construction details designed into them. For instance, the initial decisions are made largely by the software manufacturer regarding line size, tick marks, grid lines, scales, type of text, etc. Many programs give the operator the option of changing these, and as people become more proficient they often generate their own unique graphics. If the graphic is being made by hand, all the construction decisions must be made by the person generating the information graphic. Sections of this book are devoted to a discussion of each of the major construction features.

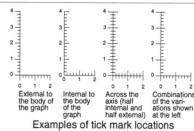

External to the body of the graph | Internal to the body of the graph | Across the axis (half internal and half external) | Combinations of the variations shown at the left

Examples of tick mark locations

Organization and major contents (continued)

Design features that might mislead the viewer

Certain methods of presenting data have been found to frequently mislead the viewer. Many times these methods are used because the person making the graphic is unaware of the hazard or does not know an alternative. These misleading design features, such as broken scales and perspective views, are discussed under their individual headings such as Scales and Perspective Projection, as well as under certain specific types of graphs and maps.

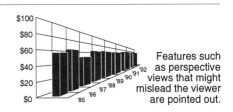

Features such as perspective views that might mislead the viewer are pointed out.

The most up-to-date developments in information graphics

Many advances have been made in the area of information graphics, both as a result of creative individuals such as W.S. Cleveland, E.R. Tufte, and J.W. Tukey as well as many excellent software developers. In some cases an entirely new information graphic has been invented, such as the box graph. In other cases it might be a component, such as a framed rectangle symbol, or a concept, such as the data-ink ratio. Because previously there has been no vehicle to bring these developments to the attention of the vast majority of users, many of the new designs and techniques are largely underutilized.

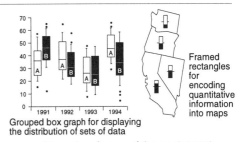

Grouped box graph for displaying the distribution of sets of data

Framed rectangles for encoding quantitative information into maps

Examples of some of the most recent developments in information graphics

Information graphics available as a result of new software

There are a number of information graphics that have been around for many years, but because there was no efficient way to generate them, they have not been widely used. With the development of powerful desktop computer software, these graphical tools are now economically available to anyone interested. These charts are discussed in the context of all of the other charts with no special category assigned to them. Three examples are shown here.

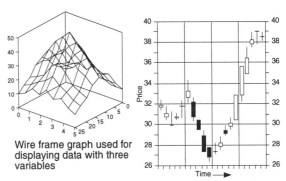

Wire frame graph used for displaying data with three variables

Candlestick chart used for recording the price of stocks, commodities, etc.

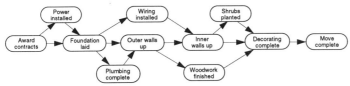

PERT chart used for planning and tracking major programs

Information graphics used in many different fields

In some cases information graphics developed in one field can be directly applied in other fields. In other cases, a slight modification might make the graphic useful, or in still other cases, a specific information graphic may not work but the idea of how the chart elements are used might trigger a completely new chart design. One of the purposes for including application-specific information graphics is to serve as a catalyst in the transfer of graphic ideas from one field to another.

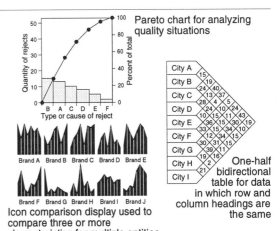

Pareto chart for analyzing quality situations

One-half bidirectional table for data in which row and column headings are the same

Icon comparison display used to compare three or more characteristics for multiple entities

Interrelationships of complex information graphics

In most cases a brief explanation plus an example is all that is required for readers to understand how a chart or graph is constructed and functions. In a few cases it is not obvious how a particular graph or map is generated or how two or more graphs or maps relate. In these cases a more detailed explanation is sometimes given, as shown at the right. Taking the time to study these more detailed explanations is not necessary for an understanding of the basic graphs or maps. Such explanations can be skipped without detracting from the main content of the section.

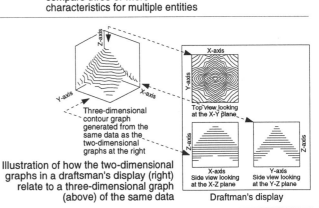

Three-dimensional contour graph generated from the same data as the two-dimensional graphs at the right

Top view looking at the X-Y plane

Side view looking at the X-Z plane

Side view looking at the Y-Z plane

Draftman's display

Illustration of how the two-dimensional graphs in a draftsman's display (right) relate to a three-dimensional graph (above) of the same data

Abscissa	The distance a data point is located from zero along the horizontal axis of a graph.
Abscissa Axis	A technical name for the horizontal or X-axis on a two-dimensional graph.

Abstract Graph

Sometimes referred to as a diagram. An abstract graph typically has no scales, tick marks, or grid lines and is used primarily to investigate, illustrate, and convey ideas and concepts rather than specific values. A break-even graph, shown at right, is one of the better-known examples of an abstract graph. This graph illustrates some of the basic concepts associated with sales, costs, and profits without using numeric values.

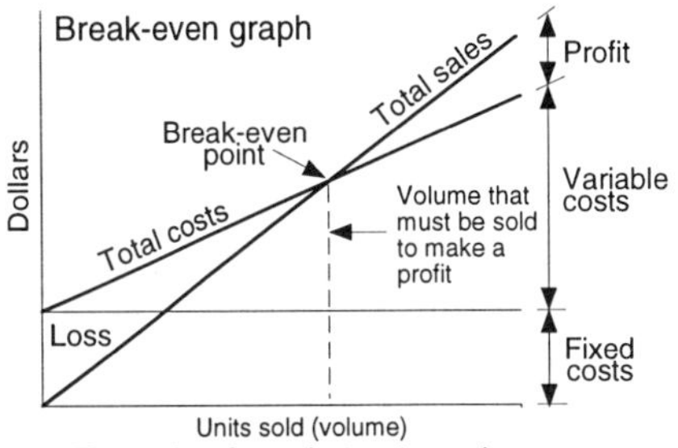

Example of an abstract graph

Abstract Map

A map that conveys a focused but limited amount of specific information such as location, direction, relationships, time, or information by area with little concern for the presence or accuracy of information outside its focus. For instance, in the example at the right, the distances on the map show equal driving times between locations. Most of the actual ground distances are distorted. Abstract maps generally convey one or at most a few sets of facts, ideas, concepts, etc. Many distorted maps, strip maps, cartograms, and diagrammatic maps qualify as abstract maps.

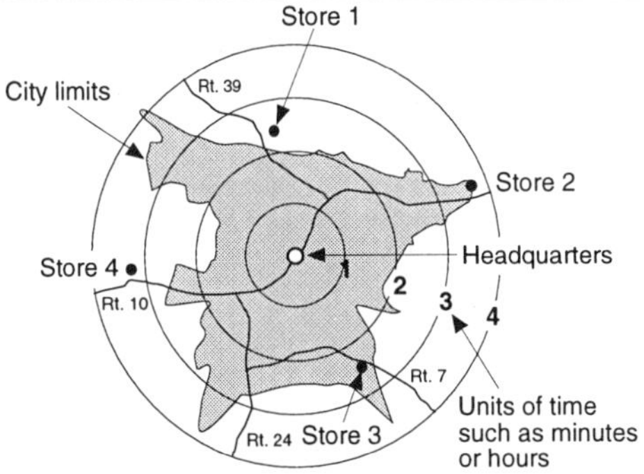

An abstract map indicating points of equal driving time from a central location, taking into consideration type of roads, terrain, congestion, etc. Most of the actual ground distances are distorted.

Age and Sex Pyramid

Also referred to as a population pyramid. A specific application of a pyramid graph or two-way histogram for studying a population by age and sex. Age intervals are plotted on the vertical axis and the number of males and females in each age interval on the horizontal axis. See Population Pyramid.

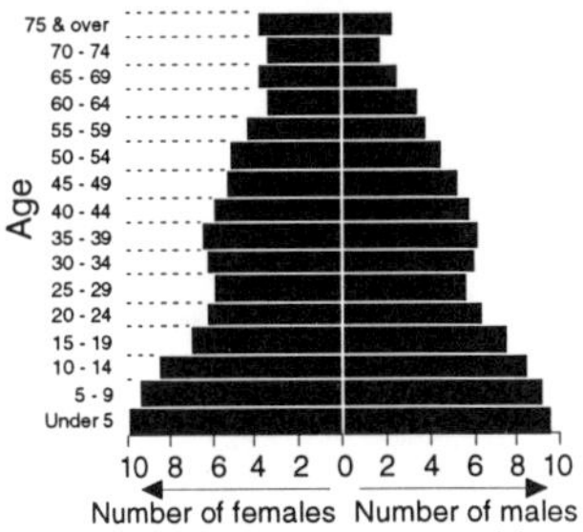

Age and sex pyramid

Alignment Graph

Sometimes referred to as a nomograph, nomogram, or calculation graph. An alignment graph is designed to solve an equation involving three or more variables. Such graphs consist of three or more scales arranged so that a straight line crossing all the scales intersects the scales at values that satisfy the equation. The graph at the right satisfies the equation $A + B = C$ (e.g., 8 on scale A plus 5 on scale B equals 13 on scale C). See Nomograph.

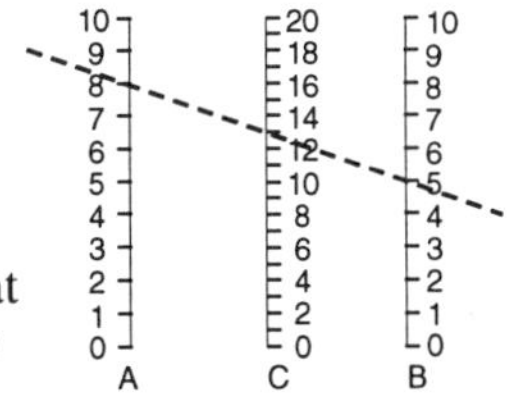

Alignment graph

Alignment of Text

A term used to describe how text is formatted. There are five major types of alignment: the four shown below plus alignment on the decimal point. See Text.

Aligned flush left	Aligned flush right	Centered	Justified
Sometimes referred to as justified left and/or ragged right. When text is aligned left, the left edges of all lines are even. The right edge of each line is determined by the full words, partial words, and spacings in the line.	Sometimes referred to as justified right and/or ragged left. When text is aligned right, the right edges of all lines are even. The left edge of each line is determined by the full words, partial words, and spacings in the line.	Sometimes referred to as ragged right and left. Uniform spacings are used between words and letters. Excess space is distributed equally at the two ends of each line. Generally used only with a few lines of text.	Sometimes referred to as flush right and left. Text that is justified normally has the right and left edges aligned. This is accomplished by varying the space sizes between words and letters and the frequent use of hyphens.

Amount Scale

Sometimes referred to as a quantitative, numeric, value, or interval scale. An amount scale consists of numbers organized in an ordered sequence with meaningful and uniform spacing between them. Quantitative variables are typically measured along amount scales. See Scale.

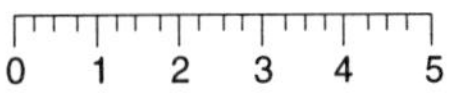

Example of a linear amount scale. Other types of amount scales include logarithmic, probability, and power.

Analytical Chart/Graph	The term refers to the way a chart or graph is used rather than a specific format. As the name implies, analytical charts are used to analyze information rather than for such purposes as reference, monitoring, presenting, entertainment, or advertising. An analytical chart might be a simple scatter graph or a complex conceptual diagram. Such charts are often helpful when studying information for relationships, similarities, correlations, patterns, trends, etc.

Analytical Table

Sometimes referred to as a statistical or summary table. This type of table is used primarily for analyzing information, as opposed to referencing, scheduling, advertising, etc. There are always two or more variables and the data in the body are normally numeric. Information is generally arranged, ranked, or sorted to make relationships, trends, comparisons, distributions, or anomalies stand out.

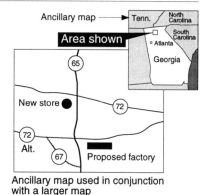

Number of students with given combination of scores											
Total	3	8	13	22	21	20	14	10	4	3	1	119
100%										1		1
90%									1	1	1	3
80%						2	4	3	2	1		12
70%				1	3	4	3	4	1			16
60%			1	4	5	6	4	2				22
50%			1	6	6	4	2	1				20
40%	1	2	4	6	3	2	1					19
30%	1	2	4	2	2	1						12
20%	1	3	2	3	2	1						12
10%		1	1									2
0%												

(Score on test A on vertical axis; Score on test B on horizontal axis: 0% 10% 20% 30% 40% 50% 60% 70% 80% 90% 100% Total)

Example of an analytical table

Ancillary Map

Sometimes referred to as an inset or supplementary map. If viewers are not familiar with the area shown on a detailed map, a small ancillary map is often used to help orient them. For example, the larger map at right shows a hypothetical area in northern Georgia. If shown only the detailed map, most people would not know what state it was located in, whether it was in the northern, southern, eastern, or western portion of the state, and what was close to it that they might recognize. An ancillary map, shown in the upper right-hand corner of the main map, supplies answers to most of these questions.

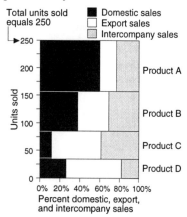

Ancillary map used in conjunction with a larger map

Angular (Polar) Scale

Angular scales are used with circular graphs. They typically are arranged with equal distances and equal numbers of degrees between the major tick marks (major values). The number of degrees is normally an even fraction of 360 degrees, such as 1/24 (15°), 1/12 (30°), 1/6 (60°), or 1/4 (90°). The labels might be shown as degrees or radians. See Polar Graph.

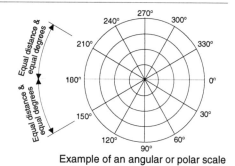

Example of an angular or polar scale

Area Band Graph — Sometimes referred to as a band, range, or silhouette graph. See Range Symbols and Graphs.

Area Bar Graph

In an area bar graph, the widths of the bars are proportional to some measure or characteristic of the data element(s) represented by the bars. For example, if the bars are displaying the profitability of various product lines, the width of the bars might indicate what percentage of the total sales the various products represent. Bar widths can be displayed along the vertical axis in terms of percents (left) or units (right). The concept can be used with a simple bar graph (left) or a stacked bar graph (right). See Bar Graph.

Simple area bar graph showing the profitability of each product on the horizontal axis with the width of the bars proportional to the percent the product line represents of total sales.

(left chart: Percent of total sales vs. Percent profitability, Products A, B, C, D)

(right chart: Total units sold equals 250; Domestic sales, Export sales, Intercompany sales; Units sold; Percent domestic, export, and intercompany sales; Products A, B, C, D)

Unit scale on vertical axis

Area Chart

Sometimes referred to as a proportional area chart. A variation of a proportional chart used for communicating differences in size, quantity, value, etc., by drawing data graphics in the same proportions as the things they represent. For example, if the sales of A are twice as big as the sales of B, the area of the data graphic for A would be twice as big as the area of

The area of column A is twice that of column B

The area of circle A is twice that of circle B

Two variations of area charts showing the relative size or value of two things

the data graphic for B. Area charts may be used to display actual values (left) or percents-of-the-whole (right). Pie charts are the most widely used variation for presenting percent-of-the-whole data. Area charts can also be used to compare the

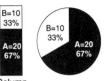

Column chart   Pie chart

Two variations of area charts used to present percent-of-the-whole data

same thing at multiple points in time (right). • Since it is difficult to accurately estimate values using area charts, the actual values are generally shown on or adjacent to the data graphics. • The areas of data graphics, which can be any shape (right and below), are not meant to convey exact data but simply to give the viewer some visual indication of the relative sizes of the items they represent.

Area charts used to depict changes in the value of something, such as coffee consumption, over time.

48%
38%
14%

35   28 ⟨10⟩

Irregularly shaped area charts

Area charts are used almost exclusively for communication purposes. They are seldom used for analytical purposes. See Proportional Chart and Pie Chart.

Area Column Graph

A variation of column graphs in which the width of the columns have significance. In the area column graph, the widths of the columns are proportional to some measure or characteristic of the data elements represented by the columns. For example, if the columns are displaying the profitability of various product lines, the width of the columns might

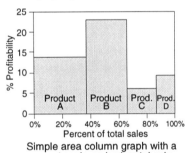

Simple area column graph with a percent scale on horizontal axis.

indicate what percentage of total sales the various products represent. Column widths can be displayed in terms of percents (left) or units (right). There may be a scale on the horizontal axis or a scale may be provided in a legend. In still other cases, the

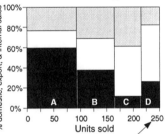

Total units sold of A, B, C, & D equals 250

Stacked area column graph with a unit scale on the horizontal axis

values might be noted directly on the graph. If the values along the horizontal axis are cumulative, the columns are generally joined so there are no unaccounted-for spaces. For

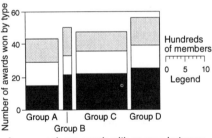

An area column graph with spaces between columns. In this type of graph the values along the horizontal axis are not cumulative, the total is unimportant, and the scale is shown in a legend.

example, if all of the columns add up to a certain value, such as 100%, or if the sum of all of the values along the horizontal axis is important, no spaces are left between columns. Spaces can be used between the columns if the values are not cumulative, the total is unimportant, and a scale is part of the legend (example at left). Sometimes the widths and heights of the columns are related such that as one varies, the other varies also. For example, in a histogram, if the width of a column is increased to encompass a broader class interval, the height of the column is adjusted accordingly (example at right). In these cases the area of the column is a truer measure of the thing being represented than either the height or width. The concept of using the width of columns to convey additional information is generally applied only to simple and stacked column graphs, including 100% stacked graphs. See Column Graph.

The histogram is probably the most widely used application of an area column graph. In a histogram, as the width of a column is expanded to cover a broader class interval, the height is adjusted accordingly.

Area Graph

An area graph can be considered as much a process or technique as a basic graph type since area graphs are generated by filling the areas between lines generated on other types of graphs. For instance, when only one data series is involved, filling the area between the data curve and the horizontal axis, as shown at right, converts the line graph to an area graph. Area graphs are

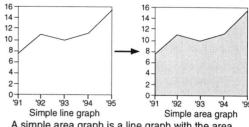

Simple line graph   Simple area graph
A simple area graph is a line graph with the area filled between the data curve and the horizontal axis.

generally not used to convey specific values. Instead, they are most frequently used to show trends and relationships, to identify and/or add emphasis to specific information by virtue of the boldness of the shading or color, or to show parts-of-the-whole.

Curves used with area graphs

As with line graphs, there are three basic types of curves used to generate area graphs – segmented, stepped, and smooth. An area graph is sometimes referred to by the

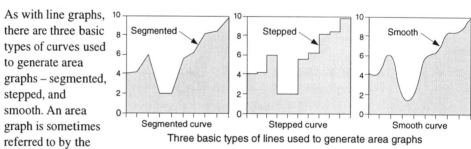

Segmented curve   Stepped curve   Smooth curve
Three basic types of lines used to generate area graphs

type of line used to generate the graph. For example, an area graph generated by a stepped curve might be called a stepped area graph. The line and/or symbols representing the data points may or may not be shown.

Simple area graphs - (area graphs with a single data series)

Sometimes referred to as a filled line, surface graph, or silhouetted graph. A simple area graph typically has a quantitative scale on the vertical axis and a category, quantitative, or sequence scale on the horizontal axis. Examples of all three are shown at the right. With the category scale, a data point is located directly above every label

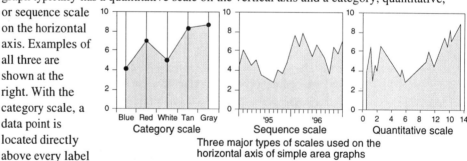

Category scale   Sequence scale   Quantitative scale
Three major types of scales used on the horizontal axis of simple area graphs

(category) on the horizontal scale. All categories must be shown on the scale for the viewer to properly interpret the graph. With sequence and quantitative scales, the data points may or may not be located above the labels. In fact, many sequence and quantitative graphs can be interpreted even if only the beginning and ending values are labeled. Positive and negative values can be plotted on all quantitative scales as shown below.

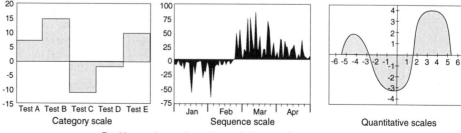

Category scale   Sequence scale   Quantitative scales
Positive and negative values plotted on simple area graphs with the three major types of scales on the horizontal axes

Area Graph (continued)

Area graphs with multiple data series

The examples below indicate the major variations possible when plotting multiple data series on an area graph. Line and surface graphs are included to illustrate the relationships between the three types of graphs. The same data is plotted on all of the examples. The stacked variation of area graphs are the most widely used. The two-dimensional, grouped area graph is used much less frequently and can cause misinterpretation since many people automatically associate the stacked variation with area graphs. • On a three-axis graph, when the area between curves is filled, as opposed to the areas between the curves and the axis, the graph is generally referred to as a surface graph. See Surface Graph [1].

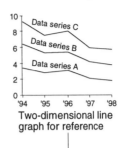

Two-dimensional line graph for reference

Two-dimensional area graphs

Three-dimensional line, area, and surface graphs

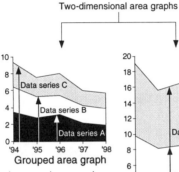

Grouped area graph

In grouped area graphs, the bottoms of all the data series touch the horizontal axis. Thus the top of each filled area represents the actual values of a single data series. If any of the data series intersect or if the data graphics are not arranged in the proper sequence, portions or all of one or more data series may be hidden.

• Grouped area graphs are not often used, largely because many people assume that area graphs are always stacked. In fact some software programs do not offer grouped area graphs as an option.

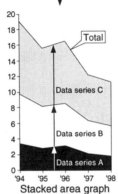

Stacked area graph

In stacked area graphs – probably the most widely used variation of area graphs – each data series is added to the one below it. Only the bottom data series touches the horizontal axis. The top of the upper data series represents the total of all the data series plotted. Stacked area graphs are sometimes referred to as multiple-strata, stratum, strata, divided area, layer, subdivided area, or subdivided surface graphs.

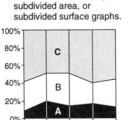

100% stacked area graph

Same as above except each data series is shown as a percent of the whole.

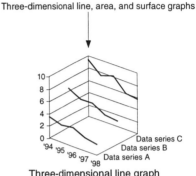

Three-dimensional line graph

When a two-dimensional line graph is converted to a three-axis, three-dimensional line graph, the data series are distributed uniformly along the third axis.

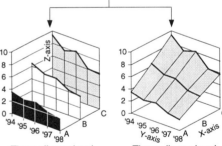

Three-dimensional grouped area graph

When the areas below the lines representing the data series are filled vertically along the Z-axis, the result is a three-dimensional, grouped area graph.

Three-dimensional surface graph

When the areas between the lines representing the data series are filled along the X and Y axes, the result is a three-dimensional surface graph.

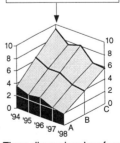

Three-dimensional surface graph with sides filled

When the areas below the lines representing the data series on the exposed edges of the graph are filled, the resulting surfaces are referred to as sides or skirts. The sides are the equivalent of two-dimensional area graphs for the data series on the edges.

Grouped area graph – (multiple data series all referenced from the same zero axis)

Sometimes referred to as an overlapped or multiple area graph. In grouped area graphs the data graphics representing each of the multiple data series are placed one behind the other. With grouped area graphs, the values on the vertical axis for all data series are referenced from a common zero baseline axis, generally the horizontal axis. As a result, the values for each data series can be read directly from the graph, as with grouped line graphs. A comparison of a grouped line graph and a grouped area graph is shown at right. Grouped area graphs sometimes cause confusion because there is no easy way to differentiate between a grouped and stacked area graph. In addition, many people think of area graphs as always being stacked. In grouped area graphs, the shaded areas between curves represent differences, not actual values of the data series.

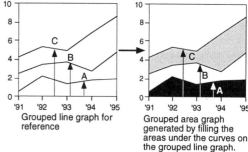

Grouped line graph for reference

Grouped area graph generated by filling the areas under the curves on the grouped line graph.

Comparison of a grouped line graph and a grouped area graph

Intersecting data series

When data series intersect on a grouped area graph, there is no sequence of data graphics that will assure full visibility of the curves of all data series. The examples below show techniques that are sometimes used to convey information about hidden curves.

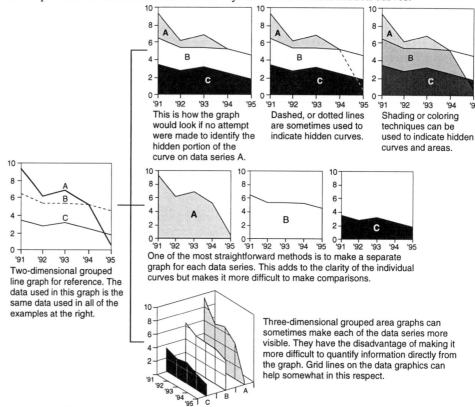

This is how the graph would look if no attempt were made to identify the hidden portion of the curve on data series A.

Dashed, or dotted lines are sometimes used to indicate hidden curves.

Shading or coloring techniques can be used to indicate hidden curves and areas.

Two-dimensional grouped line graph for reference. The data used in this graph is the same data used in all of the examples at the right.

One of the most straightforward methods is to make a separate graph for each data series. This adds to the clarity of the individual curves but makes it more difficult to make comparisons.

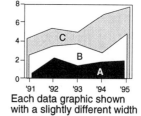

Three-dimensional grouped area graphs can sometimes make each of the data series more visible. They have the disadvantage of making it more difficult to quantify information directly from the graph. Grid lines on the data graphics can help somewhat in this respect.

Examples of techniques used to overcome the problem of hidden data curves.

Methods for reducing the confusion associated with grouped area graphs

Grouped area graphs sometimes cause confusion because the viewer cannot determine whether the areas for the data series extend down to the zero axis. Shown below are two methods for partially overcoming this problem by slightly exposing the data graphics in the back, thus letting the viewer see that all data graphics extend down to the zero axis.

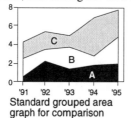

Standard grouped area graph for comparison

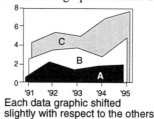

Each data graphic shifted slightly with respect to the others

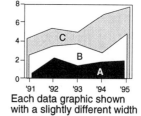

Each data graphic shown with a slightly different width

Grouped area graph (continued)
——————**Variations of data and scales** ——————

Shown below are examples of five major variations of scales and data used with grouped area graphs. These variations are of most interest/concern when the graphs are being generated on a computer.

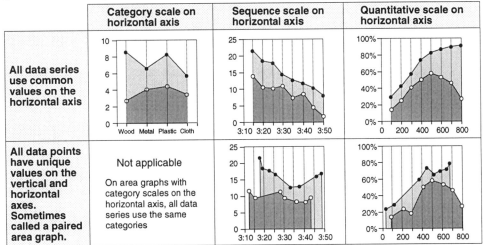

	Category scale on horizontal axis	Sequence scale on horizontal axis	Quantitative scale on horizontal axis
All data series use common values on the horizontal axis			
All data points have unique values on the vertical and horizontal axes. Sometimes called a paired area graph.	Not applicable On area graphs with category scales on the horizontal axis, all data series use the same categories		

Five variations of types of data and types of scales used on grouped area graphs

——————**Three-dimensional grouped area graphs** ——————

The examples at right demonstrate how the width of the data graphics representing the data series on three-dimensional graphs can be varied to improve readability and/or appearance. In addition to

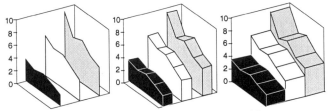

Examples of various widths of data graphics

being a potential solution for hidden data, grouped three-dimensional area graphs are sometimes used to give additional insight into complex data series. In the example below, the combination of the two-dimensional line graph and the three-dimensional area graph help the viewer more rapidly become oriented as to how the three data series relate.

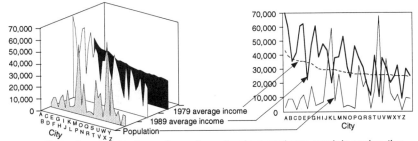

Illustration showing how a three-dimensional grouped area graph in conjunction with a two-dimensional line graph can sometimes help orient the viewer.

——————**Negative values with grouped area graphs** ——————

Grouped area graphs can handle negative values somewhat better than stacked area graphs but they still have the problem of all or portions of data curves being hidden by the data series towards the front. Shown below are two- and three-dimensional sample graphs of two data series, one of which has negative values. The same data is also shown on a grouped line graph at the left for reference.

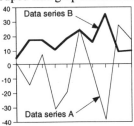

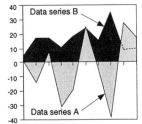

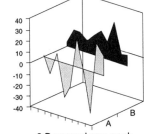

| Grouped line graph for reference | 2-D grouped area graph | 3-D grouped area graph |

Examples of negative values plotted on grouped area graphs.

Stacked area graph – (multiple data series positioned on top of one another)

Sometimes referred to as a layer, strata, multiple-strata, stratum, divided area, subdivided area, or subdivided surface graph. The stacked area graph is the most widely used variation of area graph. It is similar to a linked stacked column graph. For comparison, examples of a linked stacked column graph and two types of stacked area graphs are shown below. In all three graphs, the data series are stacked on top of one another such that the tops of the columns and the top curve on the area graphs represent the total of all the data series shown on the graph. In each example, only the bottom data series is referenced from or touches the horizontal axis. Each of the other data series on the graph are referenced from the data series immediately below it. One can think of a stacked area graph as a stacked column graph with the areas between the columns filled.

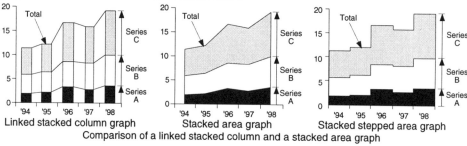

Comparison of a linked stacked column and a stacked area graph

In an area graph with multiple data series, but no indication as to what type it is, it is fairly safe to assume it is a stacked area graph. The stacked area graph is often used to show relative sizes of the components that make up the whole. For example, stacked area graphs are frequently used to show how many dollars of sales each product line contributes to the overall dollar sales of a company and how the overall sales dollars have trended. Such a graph also gives a rough idea as to how the sales of the individual product lines have performed over time. When approximations are all that are needed, stacked area graphs are usually adequate. When accuracy is desired, this type of graph is generally not used, particularly when the values fluctuate significantly and/or the slopes of the curves are steep. For instance, the top graph in the box at the right gives an indication as to what data series A and the sum of data series A and B look like, but gives little or no insight into the performance of data series B by itself (bottom of box at right). In this example, reversing the relative positions of A and B in the stacked graph would not improve the situation. In other cases, reversing the relative positions of the components can significantly improve the readability of the information – as shown in the example below.

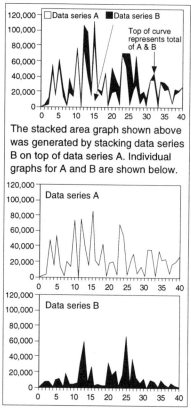

The stacked area graph shown above was generated by stacking data series B on top of data series A. Individual graphs for A and B are shown below.

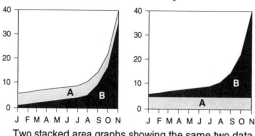

Two stacked area graphs showing the same two data series. In the graph on the left, A is stacked on top of B. In the graph on the right, B is stacked on top of A, illustrating the major impact the relative position of the data series can sometimes have.

Sometimes the most important data series is placed on the bottom. In other cases the selection is based on which variable is the most stable or what will make the overall graph most meaningful. There are no restrictions as to the relative positions of the data series on the stacked area graph. The vertical scale of a stacked area graph typically is linear, starts at zero (no shortening of the scale), is continuous (no breaks), incorporates only positive values, and has an upper value equal to or greater than the largest value plotted on the graph.

Stacked area graph (continued)

100% stacked area graph

An important variation of the stacked area graph is the 100% stacked area graph which displays percent-of-the-whole data. With this type of graph, instead of plotting the actual values for each data series, the percents that the values represent of the total of all the data series are plotted. For example, if the total sales for a company were $16 million and the sales for a product line in a given year were $4 million, 25% would be plotted for that product line instead of $4 million. In this type of graph, the sum of all the components, which is represented by the top curve, is always 100%. An example of a 100% stacked graph is shown here along with examples of grouped and stacked area graphs using the same data for comparison.

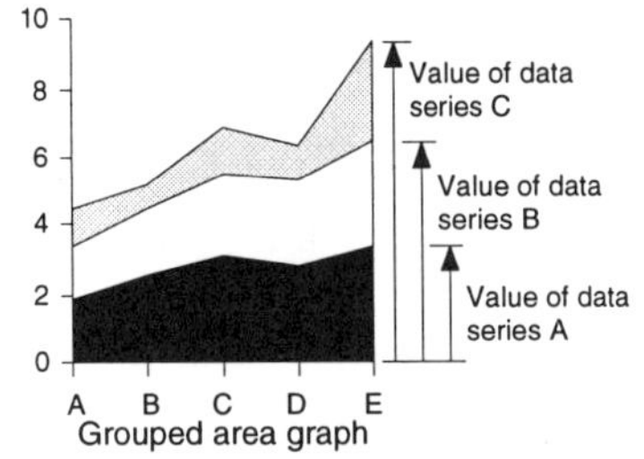

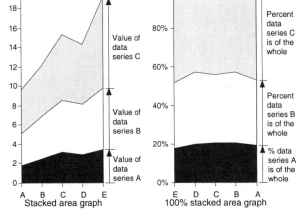

Comparison of grouped, stacked, and 100% stacked area graphs

Negative values with stacked area graphs

Technically, both positive and negative values can be plotted on the vertical axis of a stacked area graph. Practically, it is seldom done because the data graphics that result can be very difficult to interpret. The examples below show the combining of two simple data series into a stacked area graph. Only one of the series has negative numbers. With effort, one can interpret the resulting graphs at the right; however, with a different type of graph the information could be interpreted much easier and faster.

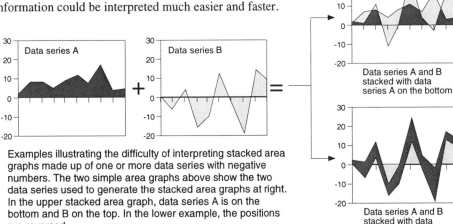

Examples illustrating the difficulty of interpreting stacked area graphs made up of one or more data series with negative numbers. The two simple area graphs above show the two data series used to generate the stacked area graphs at right. In the upper stacked area graph, data series A is on the bottom and B on the top. In the lower example, the positions are reversed.

Three-dimensional stacked area graph

Two- and three-axis stacked area graphs might be displayed as three-dimensional (right). The two-axis variation is used primarily for its aesthetic value. Depth may or may not be added to the data graphics in the three-axis graph. There is no limit on the number of variables that can be stacked in such graphs; however, more than three segments generally becomes confusing.

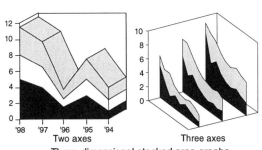

Three-dimensional stacked area graphs

15

Range and high-low type area graphs

Range or band graph

When the area between the lines representing the upper and lower values on a graph is filled in, the result is called a range, silhouette, area band, or band graph.

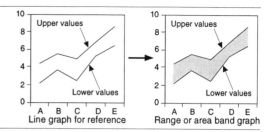

Line graph for reference · Range or area band graph

High-low graph

When a line representing some variation of midvalues is shown in conjunction with upper and lower values, the result is sometimes called a high-low graph. The upper and lower values might represent maximum and minimum, plus and minus 1, 2, or 3 standard deviations, confidence limits, etc. The midvalues might represent averages, medians, fitted curves, etc.

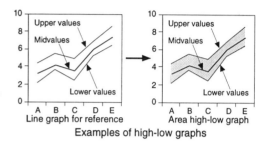

Line graph for reference · Area high-low graph
Examples of high-low graphs

Envelope

A specialized variation of a high-low graph is used to record stock prices. The upper and lower ends of the symbols designate the highest and lowest prices of the stock for the period represented. The midvalues are generally the closing or average price for the period. In this example, the area between the high and low values is filled. The filled area is sometimes referred to as an envelope.

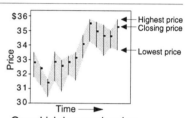

Open-high-low graph or bar chart with the range of prices indicated by the shaded area

Error bars

Another way of showing a range or set of upper and lower values on an area graph is to incorporate symbols such as error bars into the area graph itself. With this technique, as well as most of the others, one can designate both an upper and lower limit, or just one or the other. For example, if the graph displays expected values and values that are not to be exceeded, only the upper half of the symbols might be displayed, as shown in the example at right.

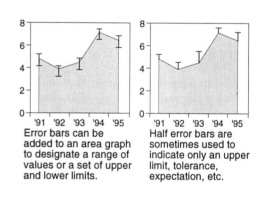

Error bars can be added to an area graph to designate a range of values or a set of upper and lower limits.

Half error bars are sometimes used to indicate only an upper limit, tolerance, expectation, etc.

Difference and deviation area graph

Area graphs are occasionally used to show differences or deviations. There are four major variations for this application. In a gross deviation graph, one color or pattern indicates positive values or deviations, and another color indicates negative values or deviations. A net deviation graph plots the difference between the two. A third type plots both gross and net deviations on the same graph. The cumulative net variation graph plots a continuous sum of the net deviations.

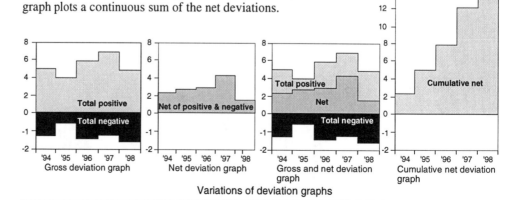

Gross deviation graph · Net deviation graph · Gross and net deviation graph · Cumulative net deviation graph
Variations of deviation graphs

Stepped area graph

Stepped area graphs are sometimes used to differentiate various data series more clearly, to highlight differences or comparisons rather than trends, and to emphasize the nature of the data such as abrupt changes and then constant values – for example, prices, interest rates, plant capacity, etc., that change abruptly and then remain level for a period of time. The example at right clearly illustrates when the price changed, by how much, and how long each price was in effect.

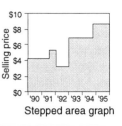

Stepped area graph

———— **Width of horizontal portions of curve** ————

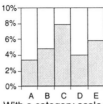

With a category scale on the horizontal axis, the widths of the horizontal portions of the curve are as wide as the space allocated to each category.

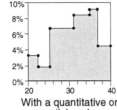

With a quantitative or sequential scale on the horizontal axis, the horizontal portion of the curve extends from data point to data point.

When a category scale is used on the X-axis, the horizontal portions of a stepped area curve are as wide as the space allotted to each category. When a sequential or quantitative scale is used on the X-axis, the horizontal portions of the curve extend from point to point regardless of how near or far apart the points are. Plot symbols may or may not be used to designate the data points. Examples are shown at the left.

———— **Location of plotting symbols** ————

The locations of plotting symbols on stepped area graphs are sometimes used to convey additional information about the data. For example, symbols located in the centers of the horizontal portions of a curve may indicate the values are averages over a period of time. Symbols at the left or right of the horizontal portions may emphasize that actions or processes start or stop at those points, causing the curve to move to a new level. If the symbols are on the left ends of the horizontal portions of the curve, the graph is sometimes referred to as left stepped graph. If the symbols are on the right ends we have a right stepped graph, and in the center, a center stepped graph

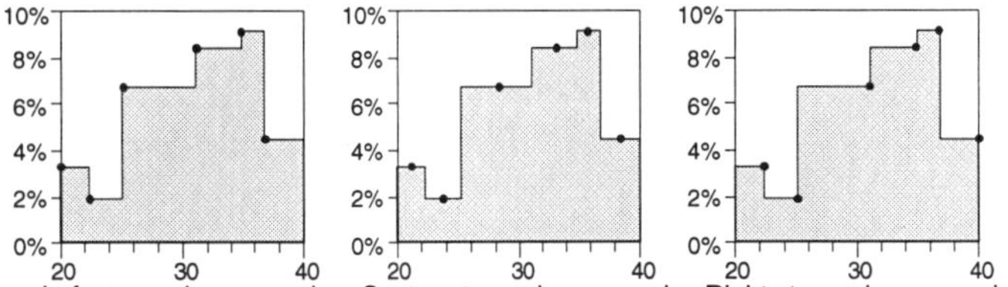

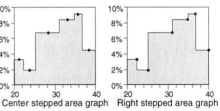

Left stepped area graph   Center stepped area graph   Right stepped area graph

Examples of left, center, and right stepped area graphs

———— **Stepped versus segmented curve** ————

Depending on the nature of the data and what one is attempting to accomplish with the graph, an area stepped graph can sometimes highlight differences and similarities more dramatically that an area graph using segmented lines. In the illustration at the right, both graphs are of the grouped type and both have the same data plotted on them. The segmented version on the left tends to give a better impression as to how the two series compare, while the stepped graph accentuates the differences and similarities.

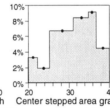

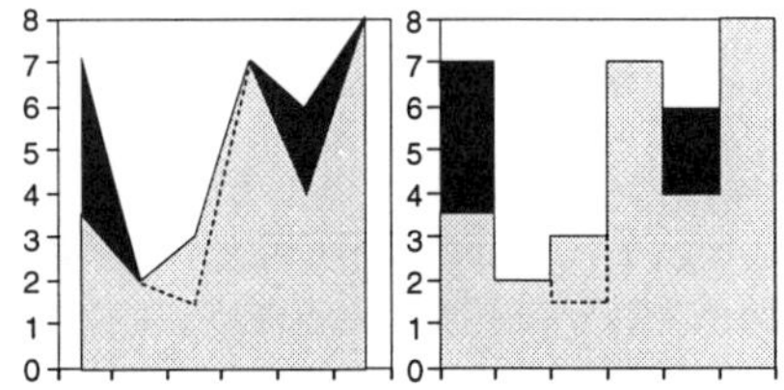

Area graph displaying two data series using segmented line curves

Area graph displaying two data series using stepped line curves

A comparison of grouped area graphs using segmented and stepped line curves. The same data is plotted on both graphs.

Orientation of area graphs

The curves on area graphs almost always progress from left to right. Although it is technically possible to have them run up and down, it is only occasionally done, probably because area graphs are so firmly associated with time series – and time series typically progress from left to right. The examples at right display the same data plotted both ways. Generally, viewers feel most comfortable with the example on the far right.

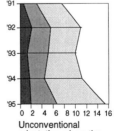

Unconventional orientation where the areas run up and down

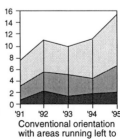

Conventional orientation with areas running left to right

Comparison of area graphs in which the areas have horizontal and vertical orientations

Other graphs where key information is designated by filling the area between lines

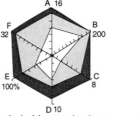

Radar/spider graph where each filled area represents a different entity

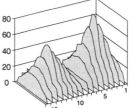

Profile graph where three-dimensional data is plotted/analyzed incrementally along the X- or Y-axis

Circular area graphs have the area within the data line filled

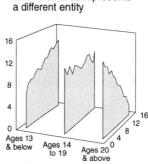

Slices graph in which three-dimensional scatter graph data is grouped into planes

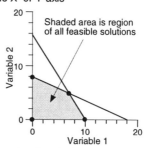

Graphs where areas are filled as part of a problem-solving process such as graphical linear programming

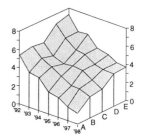

Three-dimensional surface graphs have the areas between data series filled

Variations of grid lines and drop lines on area graphs

There are a number of different ways that grid lines and drop lines can be used to assist the viewer in reading an area graph. Shown here are four of the more widely used variations. The ability to accurately determine values without cluttering the graph with grid and drop lines is one of the challenges in generating an area graph. See Grid Lines.

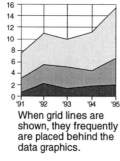

When grid lines are shown, they frequently are placed behind the data graphics.

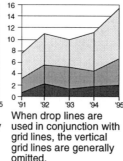

When drop lines are used in conjunction with grid lines, the vertical grid lines are generally omitted.

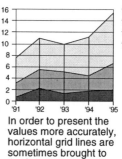

In order to present the values more accurately, horizontal grid lines are sometimes brought to the front.

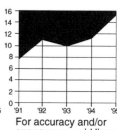

For accuracy and/or appearance, grid lines are sometimes placed only on the data graphics.

Missing or irregular data

Area graphs are often used to plot sequential data, particularly time series. It is generally recommended that there be no breaks in a time scale, since breaks can distort the pattern or trend of the data. Even though data might be irregular and in some cases missing, the time scale is generally continuous and uniform. Shown here are several alternatives for calling missing data to the viewer's attention.

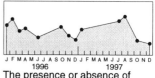

The presence or absence of plotting symbols indicates where actual data elements were used.

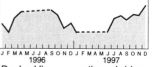

Dashed lines sometimes bridge gaps where data is missing.

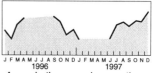

A gap in the curve is sometimes used.

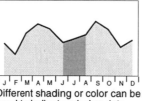

Different shading or color can be used to indicate missing data.

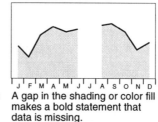

A gap in the shading or color fill makes a bold statement that data is missing.

Area Graph (continued)

Scales on area graphs

Although the actual value designated by an area graph is determined by the upper edge of the shaded or filled area, many people at first glance associate the height or area of the shaded or filled area with its value. For that reason, the following general guidelines are offered regarding the vertical quantitative scale on area graphs.

– Include zero unless it is clearly identified otherwise.

– Locate zero where the horizontal axis crosses the vertical axis.

– Have positive values increase upward and negative values downward.

– Make scales continuous (no breaks) unless clearly identified otherwise.

– The upper scale value should be larger than any value plotted on the graph.

– The lower scale value should be smaller than any value plotted on the graph.

– Use only linear scales (i.e., don't use nonlinear scales such as logarithmic).

There are exceptions to these guidelines.

– With range and difference graphs, the scale sometimes does not start at zero because the shaded or filled area does not extend to the zero axis.

– With simple and overlapped area graphs, the scales are sometimes broken, expanded, and/or cropped to make changes and differences more visible. Several methods for accomplishing this are shown below.

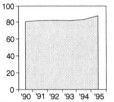 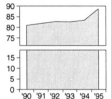 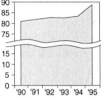 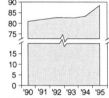

For reference, the scale starts at zero on this example. Because of the large actual values and small differences, the variations from year to year are difficult to estimate.

These three examples expand the scale and then take a section out of the middle (sometimes referred to as a scale break) to keep the size of the graph the same. If the viewer overlooks the scale break, the presence of the zero on the vertical axis can be misleading. The center example tends to make the scale break most obvious.

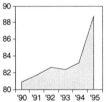 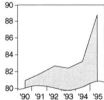 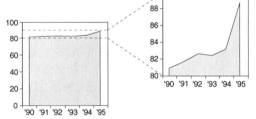

An alternative to a scale break is to eliminate the lower portion of the scale and start at some value just under the lowest value being plotted. However, the viewer may overlook the fact that the lower portion of the scale is missing and reach a conclusion based on the height or area of the data graphic, instead of the value at the top of the shaded or filled area.

Making the lower portion of the graph uneven helps call attention to the fact that the lower portion of the scale is missing.

The use of two graphs of the same data, as shown here, is sometimes employed. One graph incorporates a zero axis so the viewer can put the values in perspective. The enlarged graph (right) showing just the area with the data points lets the viewer see the detailed variations between data elements.

• Scales are generally located on the left side and bottom. There typically is only one scale on the horizontal axis. On the vertical axis the norm is one scale; however, it is not uncommon to repeat the quantitative scale on the right side, particularly if the graph is wide and grid lines are not used. Two different quantitative scales can be used but seldom are. If there are two different quantitative scales and both do not apply to all data series, care must be taken to identify which scale applies to which data series. Examples of multiple quantitative scales are shown below.

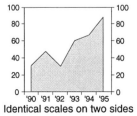 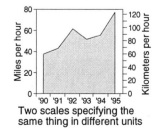 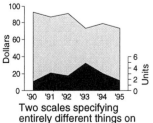

Identical scales on two sides

Two scales specifying the same thing in different units

Two scales specifying entirely different things on a grouped area graph.

Curve fitting on area graphs

Curve fitting can be applied to area graphs, just as to line graphs. There are several different types of curves that may be fitted, such as linear, polynomial, or exponential. Examples of linear and polynomial fitted curves are shown at the right. See Curve Fitting.

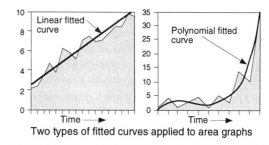

Two types of fitted curves applied to area graphs

Displaying actual values on area graphs

When it is advantageous to have the viewer know exact values , the values are sometimes included on the graph. Three common variations are shown below.

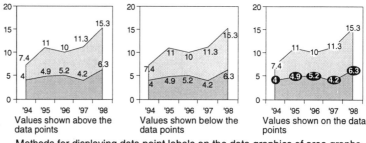

Methods for displaying data point labels on the data graphics of area graphs

Curve, line, and symbol variations

Variations and combinations of types of lines and symbols are sometimes used to distinguish data series from one another, emphasize certain aspects of the graph, encode additional information, or improve the appearance of the graph. Four examples of ways of displaying lines and symbols are shown below.

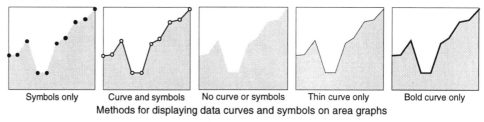

Methods for displaying data curves and symbols on area graphs

Variations of the frame and ends of the data areas

Sometimes when no grid lines are shown it is difficult for the viewer to determine whether the top portion of an area graph is background or another data series.
 – One way of avoiding this is to omit the top and right side of the graph frame.
  – Another method is to leave a space between the start and finish (left and right sides) of the data graphic and the graph frame.
   – A third method incorporates both of the first two options.

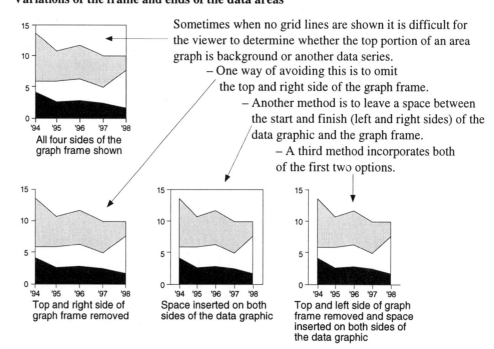

Area Graph (continued)	**Area graph icons**

Miniature (1" square or less) area graphs, frequently without titles, labels, tick marks, or grid lines, are sometimes called icons, profiles, or symbols. Area graph icons can be used as symbols on graphs, charts, and maps or grouped together to compare multiple entities with regards to three or more variables. For instance, the example below compares ten desktop computers with regards to eight different characteristics. A template, as shown in the legend, is used when generating the icons to assure that the same characteristic is always in the same location on every icon and that the same scale is used for all icons. A copy of the template is generally shown in a legend adjacent to the display. Since values and units of measure can vary significantly plus the fact that some variables might be qualitative while others are quantitative, a common scale of 0 to 1.0 is generally used for all variables. The highest value or rating in each variable is assigned the value of 1.0. The lowest value or rating is assigned some value less than 1.0 (often zero). Other than legibility, there is no limit to the number of variables that can be included in each icon. There also is no limit as to how many icons can be included in the same display. The data graphic may or may not be filled and drop lines may or may not be used. Icons used in this way are sometimes referred to as icon comparison displays. See Icon and Icon Comparison Display.

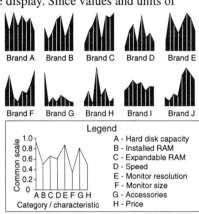

Example of an icon comparison display

Arithmetic Graph	An arithmetic graph is one that has linear scales on one or more of its quantitative axes.

Arithmetic Scale	Many times referred to as a linear scale. An arithmetic scale is arranged such that the distance and value between any two major tick marks (major values) is always the same for a given scale. For example, the distance between any two major tick marks might be one inch and represent ten units of measure (e.g., 10 gallons, 10 houses, etc.) The distance between any two minor tick marks might be one-tenth of an inch and represent one unit of measure (e.g., 1 gallon, 1 house, etc.). On an arithmetic scale, these two criteria would apply at any point along the entire length of the scale, as in the example below.

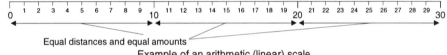

Example of an arithmetic (linear) scale.

Arrow Diagram	A term sometimes used to refer to one or both of the major graphical techniques for planning and tracking large programs and projects. One of the techniques is called PERT and the other is called Critical Path Method (CPM). Both techniques are discussed under their individual headings. The example below gives an indication as to why the term "arrow diagram" was coined. See PERT Chart and Critical Path Method.

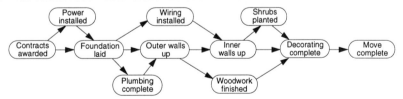

Basic PERT chart showing activities (arrows) and events (nodes)

Attribute Data Control Chart	A large number of quality control charts are categorized as attribute data charts because of the type of data plotted on them. Attribute data is generally discrete data that can be counted but not measured. Typical examples are quantity of acceptable parts, percent that pass a no-go gauge, number of scratches, defects per unit, etc. See Control Charts.

Attribute Information	Attribute information is the term given to the characteristic data added to a base map to generate all statistical maps and many descriptive maps. For example, a base map might show the boundaries of all counties in a state. Attribute data would be encoded onto the map to show such information as the population of each county, the number of houses in each county, the crops raised, the dollar sales, the crime rates, etc. Attribute information tends to be variable, while information on the base map tends to be more constant. See Base Map.

This section discusses axes as used on graphs. Axes for maps are discussed under Map. An axis is a line that serves many functions including providing the reference from which coordinates are measured, orienting the graph and all of the elements on it, acting as a vehicle along which tick marks and scales are displayed, and forming a frame around the graph. Axes are referred to by several different terms. The major ones are discussed here.

Zero base line axis for quantitative scales

Zero base line axes on rectangular graphs:
- Serve as reference lines from which quantitative values are measured in order to establish the proper location of data points
- Define the origin of the graph
- Form the traditional two-dimensional quadrants (example at right) and three-dimensional octrants
- Form the basic planes for three-dimensional analysis (see Plane)
- Run perpendicular to quantitative axes
- Exist whether or not they are actually drawn on a graph
- In three-dimensional graphs, the zero base line axes are actually planes formed by the two axes that are perpendicular to a quantitative axis. For example, the X and Y base line axes form the plane from which quantitative values are measured along the Z-axis.

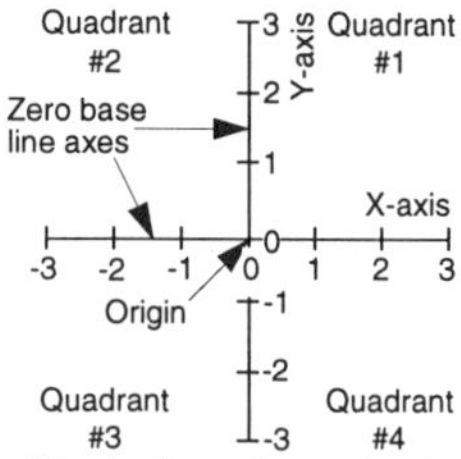

Illustration of quadrants

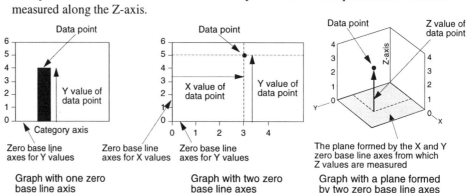

Graph with one zero base line axis

Graph with two zero base line axes

Graph with a plane formed by two zero base line axes

The plane formed by the X and Y zero base line axes from which Z values are measured

—— **Location of zero base line axis on graph** ——

The zero base line for the Y-axis might be located at the bottom of the graph, at the top, or somewhere in between. If the zero for the Y-axis scale is outside of the graph, the zero base line axis will also be outside of the graph and therefore not shown. The same principle apply to the X- and Z-axes. For illustration, four of the more common situations for the Y-axis zero base line are shown below.

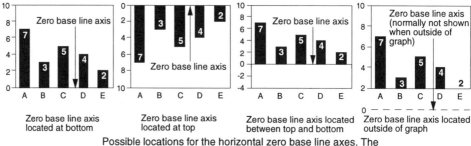

Zero base line axis located at bottom

Zero base line axis located at top

Zero base line axis located between top and bottom

Zero base line axis located outside of graph

Possible locations for the horizontal zero base line axes. The same principles also apply to the vertical zero base line axis.

—— **Relationship of zero base line axis and label axis** ——

Although many times the zero base line axis serves as the label axis also, this is not a requirement. The zero base line axis is independent of the location of the labels, as illustrated in the three examples at the right, all of which display the same data. In these examples, there are three variations of label axes but the zero base line axis remains unchanged.

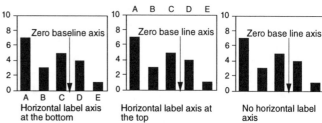

Horizontal label axis at the bottom

Horizontal label axis at the top

No horizontal label axis

The zero base line axis is independent of the label axis. Three variations of horizontal label axes are shown. In each case the zero base line axis remains located at the bottom of the graph.

Scale or label axes

An axis is sometimes referred to by the type of scale it has on it. For example, if an axis has a quantitative scale on it, the axis might be referred to as a quantitative axis. Alternate terms such as value, interval, numeric, or amount might be used instead of quantitative. Other scale terms commonly used are category and sequential, or variations of these two (see Scales). Examples of the three major classifications of scales are shown below.

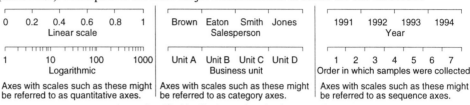

Axes with scales such as these might be referred to as quantitative axes. | Axes with scales such as these might be referred to as category axes. | Axes with scales such as these might be referred to as sequence axes.

Axes classified by the type of scale shown on them

In some situations, instead of referring to an axis by the specific type of scale on it, such as quantitative axis, category axis, or sequence axis, a broader term such as scale axis or label axis is sometimes used. • Lines on which scales are displayed (normally the lines bounding the graph) are sometimes referred to as scale lines. Scale axes or scale lines on rectangular graphs are almost always parallel to their zero base line axis.

Letter designations for axes

Use of the letters X, Y, and Z to designate axes is widely accepted. In two-dimensional graphs it is almost universal that Y designates the vertical axis and X the horizontal axis. In three-axis graphs, Z is widely accepted as the designation for the vertical axis. Which of the other two axes of a three-axis graph are designated X and Y varies depending on the application, the software, and/or the individual preparing the graph.

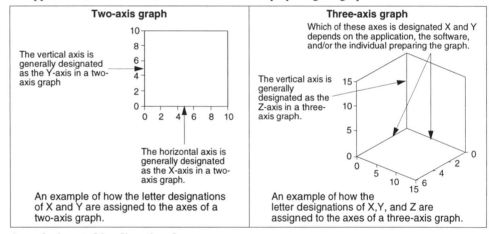

An example of how the letter designations of X and Y are assigned to the axes of a two-axis graph. | An example of how the letter designations of X, Y, and Z are assigned to the axes of a three-axis graph.

Axes designated by direction drawn

In rectangular graphs the axes are sometimes referred to by the direction in which they are drawn. In one- and two-axis graphs it is common to refer to axes as vertical and horizontal. In three-axis graphs, the term vertical axis is also widely used; however, the term horizontal axis is seldom used since most three-dimensional graphs are shown tilted and consequently there are generally no horizontal axes.

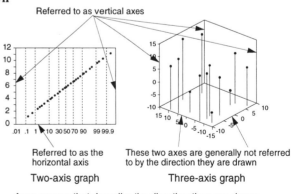

Two-axis graph | Three-axis graph

Axes names that describe the direction they are drawn

Abscissa and ordinate axes

In two-dimensional graphs the vertical or Y-axis is sometimes referred to as the ordinate axis and the horizontal or X-axis as the abscissa axis.

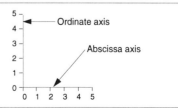

Axes designated by position on graph

– Axes are sometimes referenced by their position on the graph. On two-dimensional graphs it is common to refer to the upper and lower horizontal axes or the right and left vertical axes as shown in the example at the right.

– On three-dimensional graphs it is common to refer to the left and right vertical, or Z axes and the upper (top) and lower (bottom) X and Y axes. On three-dimensional graphs, caution should be exercised when referring to the X and Y axes as right or left, since they reverse their relative positions from top to bottom. In the example at right, the upper X label axis is to the right of the upper Y label axis, while at the bottom, the relative positions are reversed.

– With the increasing ability to rotate graphs 360° in almost any direction, the use of terms such as vertical, horizontal, top, bottom, right, and left become less and less precise.

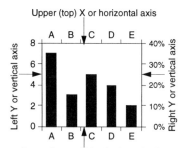

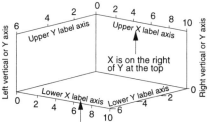

Two-dimensional graph illustrating how axes are sometimes named based on location

Three-dimensional graph illustrating how the relative positions of the X and Y label axes reverse from top to bottom

Axes referred to by the title or units of measure of the scale

It is not uncommon for an axis to be referred to by the title or units of measure of the scale on the axis. For example, if a graph shows sales in units along the horizontal axis and dollars of profit along the vertical axis, it is not unusual to have the horizontal axis referred to as the units or sales axis and the vertical axis as the dollars or profit axis.

Reference axis

Sometimes referred to as a reference line or shifted reference axis. If it is desirable to see fluctuations around a given value as well as the actual values of the data, a reference axis can be used, as shown below. In such graphs, the data points are plotted from the zero base line axis according to their actual values. Short drop lines, bars, or columns are then drawn to a reference axis. When the additional axis is used in this manner, the graph functions as a combination of a standard graph showing actual values and a deviation graph showing differences between the actual values and some reference value.

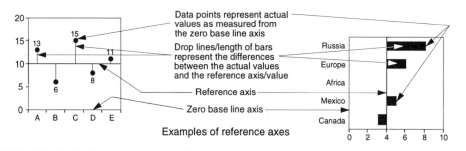

Examples of reference axes

Axis versus frame

Some individuals and software developers refer to all lines around a graph as axes. Others call them scale lines or frame lines and still others use a combination of all of them. The differences in terminology apply whether or not the lines have tick marks and/or labels.

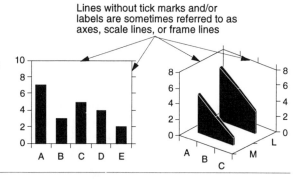

Axis, Graph (continued)

Number of scale axes or scale lines on rectangular graphs

There are no limitations on the number or location of scale axes, label axes, scale lines, frame lines, etc., that can be used with a given graph. Several variations of three-dimensional graph configurations are illustrated below utilizing different numbers and locations of axes/lines. The number and location of the axes is generally a function of such things as type and amount of data, purpose of the graph, software used, and observations one is attempting to make.

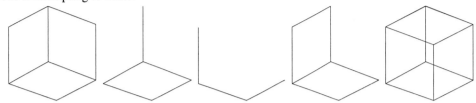

Examples of a few of the many possible combinations of axes, scale lines, or frame lines that might be used with three-dimensional rectangular graphs

Circular graph axes

Lines with quantitative scales radiating from the centers of circular graphs are generally referred to as radial axes. The radial axis at the zero point on the circular axis is a zero base line axis (sometimes referred to as polar axis when degrees are measured on the circular scale). The line around the circumference of the circle is referred to as the circular axis or polar axis. There are two variations of three-dimensional circular graphs. One, referred to as cylindrical, uses a vertical Z-axis as the third axis along which values are measured. The other variation, referred to as spherical, has a Z-axis for reference purposes but also incorporates a second circular axis along which the third variable is measured in degrees. Examples of axes on four variations of circular graphs are shown below.

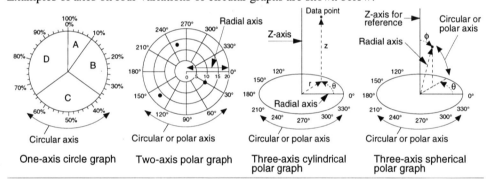

| One-axis circle graph | Two-axis polar graph | Three-axis cylindrical polar graph | Three-axis spherical polar graph |

Trilinear graph axes

The three primary axes of a trilinear graph are the three lines that run from the vertexes to the bases opposite them (altitudes). The primary axes form right angles with the bases. The three lines that form the outside of the triangle are generally used as scale or label axes. The scale values on the label axes are projections of the values measured on the primary axes. The label axes are generally referred to by their position such as right, left, bottom (also referred to as horizontal), their title, or X,Y, and Z. There are no guidelines as to which letter should be assigned to which axis.

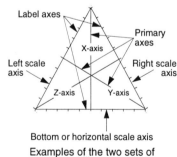

Examples of the two sets of axes on a trilinear graph

Axis Scatter Graph

A specialized scatter graph designed to assist in the interpretation of three-dimensional data. One of the primary functions of this type of graph is to display patterns of data points; therefore, scales are many times omitted. Sides are also omitted and the floor is frequently positioned somewhere within the cloud of data points, many times at the vertical midpoint. The location of the vertical axis is chosen so as to highlight particular characteristics of specific data. Axis scatter graphs are especially helpful in conjunction with the ability to rotate the graph.

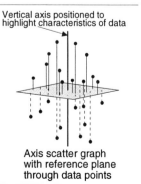

Axis scatter graph with reference plane through data points

Axonometric Projection (View)

When a graph or map with data on three axes is shown in three-dimensional form, an axonometric projection or view is generally used. In this type of view, the graph, map, or other object is tilted and rotated so the viewer sees the top or bottom and two sides as shown at the right. Two-axis graphs are sometimes shown in axonometric view, as illustrated at the left, but the special view yields no additional information over a standard two-dimensional or oblique view. • Although not a necessity, depth is frequently added to data graphics for category type data to improve appearance. (illustration at the right)

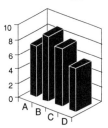

Axonometric view of a two-axis graph. Sometimes referred to as three-dimensional.

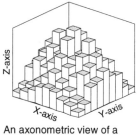

An axonometric view of a three-dimensional graph with data plotted on three axes

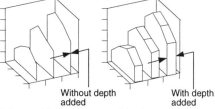

Without depth added   With depth added

Data graphics on axonometric views may or may not have depth added. They frequently do.

• When axonometric views are used for analytical purposes, the direction of rotation can have an effect on the readability of the data. If the high points on a graph or map are concentrated on one side or if it is important for the

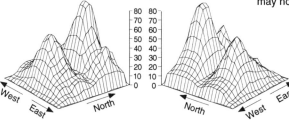

Two maps of the same data with two different orientations. The one on the left is most frequently used.

viewer to see particular data, the chart may need to be rotated in a direction that will best display the most important characteristics. For instance, the two examples at left show maps of the same data with the same degrees of tilt and rotation. The difference between the two is that they are rotated in opposite directions. More features of the smaller peak on the east side are visible in the example on the left than in the example on the right. • With three-axis charts, the amount of tilt and rotation can also have an effect on the readability of the chart. For example, when the values on the Z-axis are relatively small, a slight amount of tilt might be adequate. When there are many high peaks or values relatively close together, a greater amount of tilt is advisable so the data graphics in the front do not obscure the ones in the back. Examples of several different combinations of tilt and rotation are shown below. Users of these charts should note that when the degrees of rotation and/or tilt approach zero or a multiple of 90°, some of the advantages of an axonometric view are lost. Varying either the tilt or the rotation without varying both will result in charts that are difficult to interpret. • An isometric projection is a variation of axonometric projection in which the angles between each pair of axes is 120°. See example below.

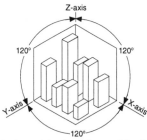

An isometric view of something is an axonometric view in which the angles between each pair of axes is 120°

See Projection[2] for a comparison of graphs and maps generated using oblique and axonometric projections.

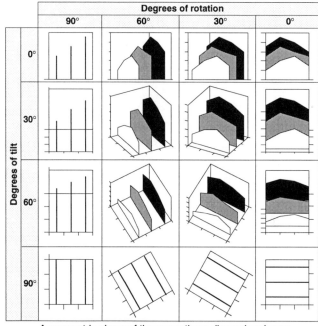

Axonometric views of the same three-dimensional area graph with varying amounts of rotation and tilt
(Degrees of rotation and tilt are based on an arbitrary set of reference angles. A different set of reference angles would result in different numbers of degrees but the appearance of the graphs would be the same.)

Background	Background is the material shown behind the major elements of a chart. It might be the area behind and around a graph, map, diagram, or table, or the area on a larger chart on which one or more other information graphics are mounted or superimposed. Reasons for using backgrounds include improving the appearance, reducing eye strain on the viewer, subtly conveying additional information, enhancing the material being presented, setting a mood, or helping to orient the viewer. There is a large variety of backgrounds to choose from including transparent, white, solid colors or fills, gradient colors or fills, pictures, patterns, and text. See examples at right. 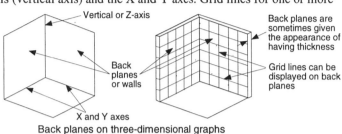
Back Plane	Sometimes referred to as a wall. Back planes are the vertical planes on three-dimensional graphs formed by the Z-axis (vertical axis) and the X and Y axes. Grid lines for one or more of the axes are many times displayed on the back planes. Back planes are sometimes given the appearance of having thickness for aesthetic purposes. 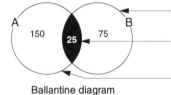 Back planes on three-dimensional graphs
Ballantine Diagram	Also referred to as a Venn diagram or set diagram. Ballantine diagrams are used to describe the relationship between two or more sets of things or bits of information by the relative positioning of geometric shapes (normally circles). See Venn Diagram.
Band Graph	Sometimes called a silhouetted, area band, or range graph. A band graph displays spreads, ranges, or differences. Often the midvalue is displayed within the band as shown at right. An example of a midvalue might be the average water pressure each day. In that case the upper and lower boundaries of the band represent the high and low pressure readings for the same days. The band may or may not be filled. When midvalues are shown, the graph is sometimes called a high-low graph. 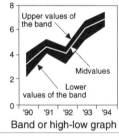 Band or high-low graph
Bank	Sometimes called cycle, deck, tier, or phase. The major interval on a logarithmic scale.
Bar and Symbol Graph	Sometimes a series of symbols are shown on a bar graph to designate reference points or points of comparison for each of the bars. The reference points might designate such things as budget, plan, last year's value, or specification values. By means of the symbols, the viewer can easily make comparisons between the actual values (as shown by the ends of the bars) and some other set of values designated by the symbols. Any symbol that crisply identifies the reference value can be used. Bar and symbol graphs typically have only one data series represented by horizontal rectangles. Multiple sets of symbols can be used on the same graph to indicate two or more sets of reference values. Symbols are seldom used with grouped or stacked type graphs. 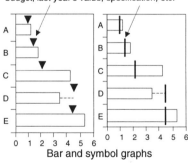 Bar and symbol graphs
Bar Chart[1]	A variation of the proportional area chart family. In a bar chart, the size of each bar is proportional to the value it represents. Bar charts do not have scales, grid lines, or tick marks. The value that each bar represents is shown on or adjacent to the data graphic. The major purpose of the graphical portion of a bar chart is to visually orient the viewer to the relative sizes of the various elements of a data series. With a quantitative scale on the horizontal axis, the graphic is frequently called a bar graph. See Bar Graph and Proportional Chart. 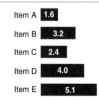 Bar chart where the sizes of the bars are proportional to the values they represent

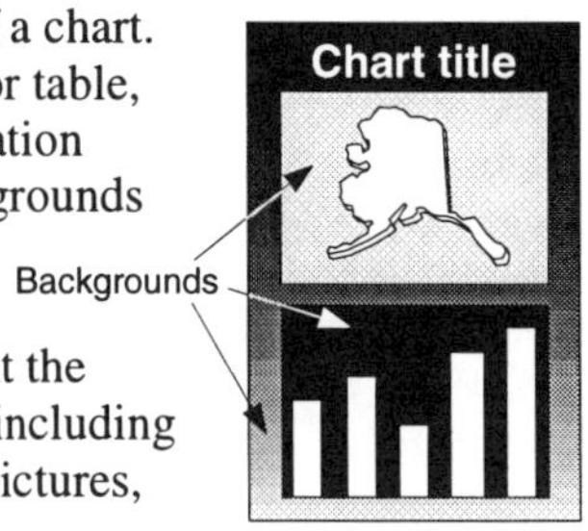

Bar Chart[2]

Sometimes referred to as a price chart, vertical line chart, open-high-low-close (OHLC) graph, high-low-close-open (HLCO) graph, high-low-close (HLC) graph, or high-low graph. Bar charts are used to record and analyze the historical selling prices of such things as stocks, bonds, and commodities. The plot might represent a single entity (e.g., General Electric, IBM, McDonald's, etc.) or the average of multiple entities (e.g., all stocks in a particular exchange, selected industrial companies, selected utilities, etc.). Price is measured along the vertical axis and uniform intervals of time progress from left to right along the horizontal axis. Time intervals range from minutes to years depending on the purpose of the chart. Days, weeks, and months are among the most frequently used. Whichever time interval is used, one symbol represents each time period. Vertical price scales can be shown on the left side, right side, or both. Because of the width of the charts, the large amount of data, and the fact that the data on the right side of the chart is usually of greatest interest, two vertical scales or a single scale on the right side are commonly used.

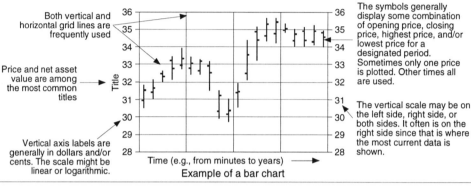

Example of a bar chart

Symbol used with bar charts

The symbol used with bar charts consists of one vertical line and one or more short horizontal lines intersecting or abutting it. The horizontal lines are sometimes referred to as ticks or markers. The key elements of the symbol are described below.

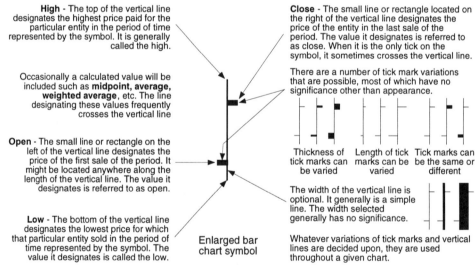

All or none of the tick marks shown here might be used with a given bar chart. Examples of the most widely used variations are shown below.

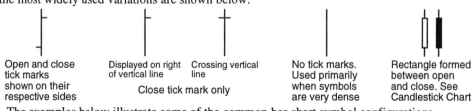

The examples below illustrate some of the common bar chart symbol configurations resulting from particular combinations of open, close, high, and low prices.

Price relationships	Close higher than open	Close lower than open	Close the same as high	Open the same as low	Open, low, and close all the same	Open the same as close	Open the same as low and close the same as high
Configuration of symbol							

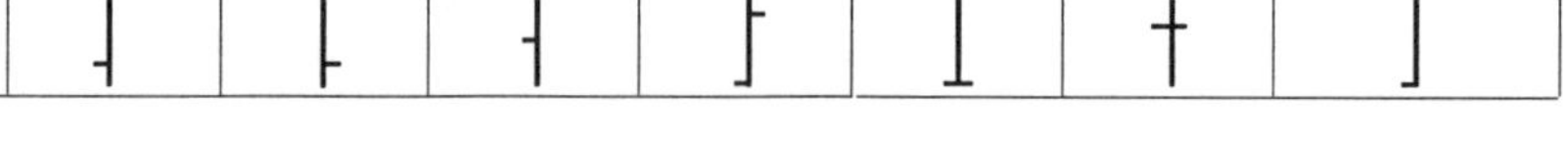

28

Data points connected with lines

In most charts the symbols are not connected by lines. In some cases, particularly where a trend is considered more important than the spread of data, lines are used to connect selected data points such as all closing values or all midvalues. When specific points, such as closing values, are connected, some or all of the other data points such as high, low, or open might be left as is, deemphaized by making them faint, or eliminated completely. Examples in which closing values have been connected by lines are shown below.

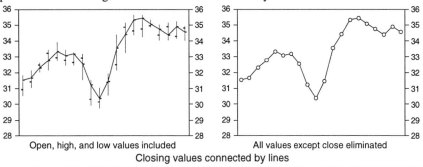

Open, high, and low values included — All values except close eliminated

Closing values connected by lines

Grid lines and frame

Whether vertical grid lines, horizontal grid lines, both, or neither will be used is largely dependent on the purpose of the chart. In many cases the viewer wants to know the exact time period associated with a particular symbol. When this occurs the vertical grid lines can be very helpful. Similarly, with three to five different data points associated with each symbol, the horizontal grid lines can become almost a necessity. On the other hand, if one is simply looking for a general trend, perhaps no grid lines are needed. Examples of various combinations of grid lines are shown below.

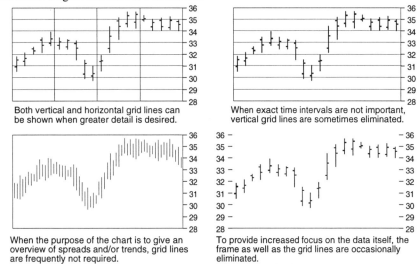

Both vertical and horizontal grid lines can be shown when greater detail is desired.

When exact time intervals are not important, vertical grid lines are sometimes eliminated.

When the purpose of the chart is to give an overview of spreads and/or trends, grid lines are frequently not required.

To provide increased focus on the data itself, the frame as well as the grid lines are occasionally eliminated.

Variations in grid line and frame combinations

Vertical grid lines generally encompass groups of time intervals. For example, if daily intervals are plotted, grid lines are generally placed between every fifth interval to group the symbols by week (Saturdays and Sundays are omitted). If weekly intervals are used, the vertical grid lines typically group the intervals by month.

Insets and notes

Insets and notes are many times added to bar charts to give the reader additional background data to interpret the information.

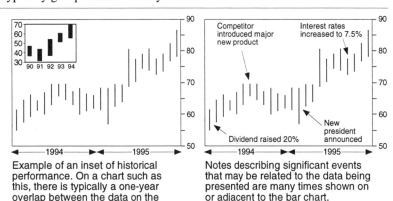

Example of an inset of historical performance. On a chart such as this, there is typically a one-year overlap between the data on the inset and the data on the main chart.

Notes describing significant events that may be related to the data being presented are many times shown on or adjacent to the bar chart.

Vertical scale

Selling price is displayed on the vertical axis with values increasing upward. Price scales are often expanded to provide greater detail. It is common for scales not to start at zero. It is very uncommon for a price scale to have a break or discontinuity in it. In most cases the units on the vertical scale are in dollars. In specialized cases the units might be in cents, points, multipliers, etc. Generally the labels represent whole numbers such as 57, 58, 59, and 60. In some cases, particularly where the spread of values from the lowest to the highest value is small, fractions or decimals might be used. Depending on the minimum size units at which the entity is traded and the customs of the particular market, fractional values may be displayed in several different ways. See examples at right.

58.5
58¹/₂
58¹⁶/₃₂
5816
58³²/₆₄
58^32

Different ways the same value ($58.50, in this example) may be displayed on bar charts.

Reasons for using logarithmic scales

Both linear and logarithmic scales are widely used on the vertical price axis. The comparisons below illustrate three major reasons logarithmic scales are used. In each example only the midpoints are plotted. The same principles apply when multiple data points are plotted for each time interval.

To make certain values more legible when the range of values is large

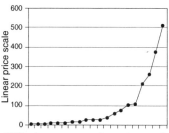

These graphs illustrate how the use of a logarithmic scale makes some values more legible. For example, in the graph on the left with a linear price scale, it is

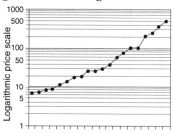

difficult to accurately estimate the values at the beginning of the chart. When the same data are plotted on the logarithmic scale at the right, the values at the beginning of the chart become more readable. Unfortunately, the readability of the values at the other end of the chart is decreased.

To make changes in growth rates more visible

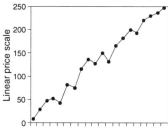

These two graphs illustrate the value of a logarithmic scale for highlighting changes in rates of growth of the price of a stock. Based on the graph on the left with the linear scale it

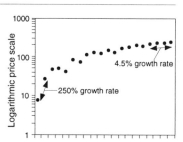

4.5% growth rate

250% growth rate

might appear that the stock has had a steady and uniform growth rate. The graph of the same data with the logarithmic scale (right) indicates that the rate of growth of the stock has steadily decreased over time. This is evidenced by the continuing decrease in the slope of the curve over time. Actual calculations show a growth rate of 250% between the first and second time intervals and a four to five percent in the last four time intervals.

To compare the growth rates of multiple entities

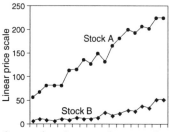

Stock A

Stock B

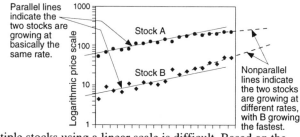

Parallel lines indicate the two stocks are growing at basically the same rate.

Stock A

Stock B

Nonparallel lines indicate the two stocks are growing at different rates, with B growing the fastest.

Comparing the growth rates of multiple stocks using a linear scale is difficult. Based on the graph on the left with the linear scale, one might be inclined to conclude that stock A is growing at a faster rate than stock B. Based on the steepness of the slopes in the graph at the right with the logarithmic scale, it can be seen that both stocks grew at about the same rate during the early periods (parallel fitted curves), but during the latter time intervals stock B grew slightly faster than stock A.

Bar Chart[2] (continued)

Horizontal scale

The horizontal axes of bar charts always have time scales. The time intervals range from minutes to years depending on the purpose of the chart. Time intervals of one day are the most widely used. Time intervals in the same chart are generally all the same. The following examples illustrate some of the unique features of bar chart time scales.

─────── **Daily stock price chart** ───────

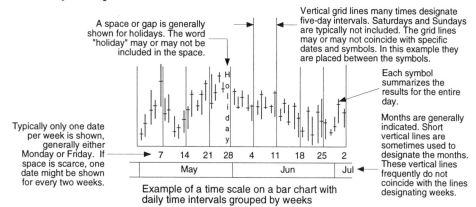

A space or gap is generally shown for holidays. The word "holiday" may or may not be included in the space.

Vertical grid lines many times designate five-day intervals. Saturdays and Sundays are typically not included. The grid lines may or may not coincide with specific dates and symbols. In this example they are placed between the symbols.

Each symbol summarizes the results for the entire day.

Typically only one date per week is shown, generally either Monday or Friday. If space is scarce, one date might be shown for every two weeks.

Months are generally indicated. Short vertical lines are sometimes used to designate the months. These vertical lines frequently do not coincide with the lines designating weeks.

Example of a time scale on a bar chart with daily time intervals grouped by weeks

The scale format shown above grouping daily symbols by week is one of the most widely used. Occasionally, daily symbols are grouped by month, as shown at the right. With this high concentration of symbols, many times only the vertical lines are displayed, or specific points such as closing are connected by lines and the other ticks are eliminated.

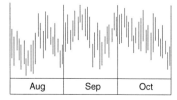

Daily time intervals grouped by month

─────── **Weekly stock price chart** ───────

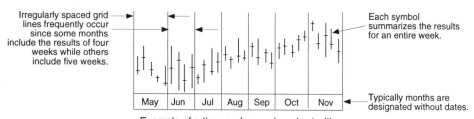

Irregularly spaced grid lines frequently occur since some months include the results of four weeks while others include five weeks.

Each symbol summarizes the results for an entire week.

Typically months are designated without dates.

Example of a time scale on a bar chart with weekly time intervals grouped by month

In this variation, vertical grid lines designate approximately one-month intervals. Whatever month a particular day of the week (generally Friday) is in determines what month the total results for that week will be shown. For example, if the first four days of a week are in March but Friday is in April, the results for the week will be shown in April. This procedure results in some months showing four weeks and others five weeks. Some charts use equally spaced vertical grid lines and leave a gap in those months with only four Fridays. Others have no gaps, resulting in irregularly spaced grid lines. The example above has irregularly spaced grid lines and no gaps. (Holidays are typically not noted in weekly graphs.) Occasionally weekly results are grouped by year, as shown at the right. With this many symbols, ticks are often omitted. When data is this dense, specific points such as closing are sometimes connected by lines and the open and close ticks eliminated.

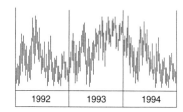

Weekly time intervals grouped by year

─────── **Monthly stock price chart** ───────

When price data is summarized by month, it is generally grouped by year as shown at the right. Generally individual months are not labeled.

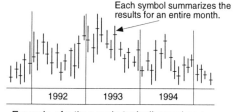

Each symbol summarizes the results for an entire month.

Example of a time scale typically used on a stock price chart with monthly time intervals

Bar Chart[2] (continued)

Graphical methods to assist in the interpretation of bar charts

Many graphical techniques have been developed to assist in analyzing price charts, particularly bar charts. Among the many reasons for such techniques are:
- To gain a deeper understanding of what has happened or is happening to prices in order to determine the ramifications or significance; and
- To attempt to project a stock's future performance so that decisions and/or recommendations might be made.

Among the things analysts look for in bar charts are the general slope of the price line and its characteristics, such as whether the slope is steep, shallow, positive, negative, changing, oscillating, etc. Sometimes they look for lines that cross. It may be the price line crossing a trend or support line, or, in other cases, it might be two construction lines crossing one another, such as two moving average curves. The examples shown below are representative of some of the more widely used graphical methods. The interpretation, value, and significance of any of these methods can only be decided by someone knowledgeable in that field.

Trend lines

The term trend line is defined in a number of different ways. One widely accepted definition is a straight line passing through major peaks or valleys of a bar chart spanning a given period of time. Trend lines are used with individual items such as a specific stock, segments of markets, or, in some cases, entire markets. Uptrend lines pass through the lowest points (valleys) of a price chart when each successive major valley is higher than the previous major valley. Downtrend lines pass through the highest points (peaks) of a price chart where each major successive peak is lower than the previous major peak. When a continuous trend line spans a long period of time it is referred to as a major trend line. Short trend lines are referred to as minor trend lines. Examples are shown below.

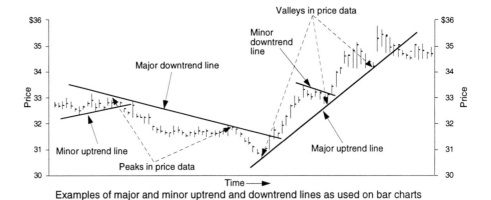

Examples of major and minor uptrend and downtrend lines as used on bar charts

Support and resistance lines

Trend lines drawn through the peaks of a price chart are sometimes referred to as resistance lines, whether or not the lines are ascending or descending. Trend lines drawn through the valleys are sometimes referred to as support lines, whether ascending or descending. Ascending lines are called bullish and descending lines bearish.

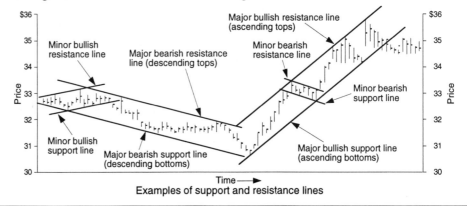

Examples of support and resistance lines

Channels

The area between essentially parallel support and resistance lines is sometimes referred to as a channel, trading zone, or trading pattern. If ascending, the channel is referred to as an up channel; if descending, a down channel. If flat (horizontal), the channel may be referred to as rectangular, trading zone, or line formation. Support or resistance lines covering long periods of time are referred to as major, shorter ones as minor.

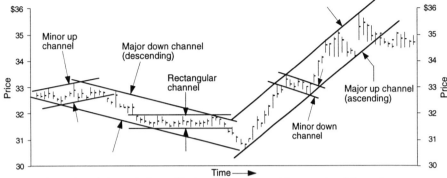

Examples of channels formed by trend lines, support lines, and resistance lines.

Formations and patterns

Certain recurring formations or patterns have been identified and in some cases given names. Many of the patterns are formed by minor trend lines. To experienced analysts, these patterns sometimes give meaningful indications. A few of the formations and patterns used with bar charts are shown below.

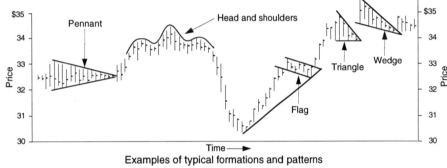

Examples of typical formations and patterns

Linear regression line

A linear regression line is sometimes described as a straight line that best represents a data series. Such a line is sometimes used for assessing the general short and/or long term trends of the data and may be applied to ascending or descending values. The slope of the regression line is normally determined using established mathematical techniques. Regression lines seldom pass through all data points and sometimes pass through none. They can be generated for any of the key price figures such as open, close, high, low, midpoint, or any combination thereof. When multiple regression lines are drawn such as high, low, and close, the lines may or may not be parallel. In addition to determining the historical slope or trend of a stock, regression lines are sometimes used to project or estimate future performance. Projections are normally differentiated from the main body of the regression line by means of dashes, colors, or shading. One of the more difficult aspects of applying regression lines to bar charts is deciding what period of time should be used, since different time periods can yield significantly different lines, as shown here.

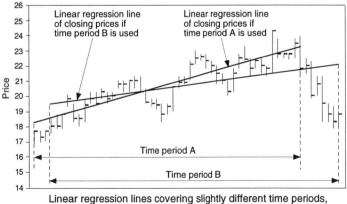

Linear regression lines covering slightly different time periods, illustrating the impact that time period selection can have on the slope of the regression line

33

Moving average curves

Moving average curves are frequently used with bar charts to smooth the data, make trends easier to see, and identify potential decision points. Moving average curves might be based on open, high, low, mid, or close values. Occasionally a combination of two or more values is used. It is not uncommon for multiple moving average curves to be shown on the same bar chart. A moving average curve smooths the data by plotting points that are the average of several time intervals. For example, on a chart that plots data on a daily bases, the value plotted for each day might be the average of that day's value plus the values for the previous 19 days. This would be called a 20-day moving average curve and would be much smoother than a curve of the actual daily data.

Variations in number of periods in average

There is no limit as to how many time intervals can be included in the average. For short-term analysis, 3 to 10 days might be used (sometimes referred to as fast moving averages). For longer-term analysis, 50 to 200 days are frequently used (sometimes referred to as medium or slow moving averages). The more periods included in the averaging process, the smoother the curve but the slower changes in direction appear.

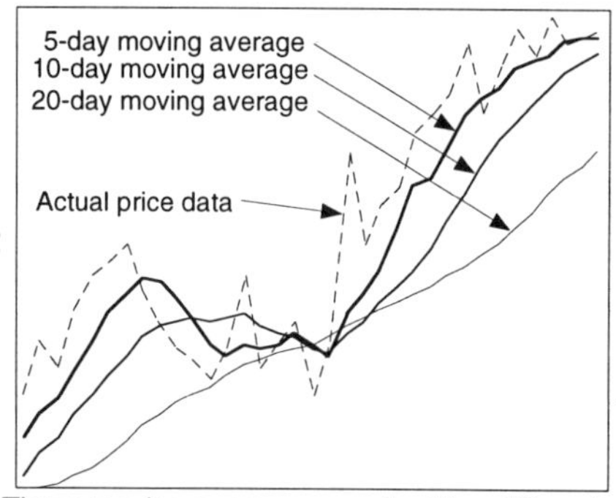

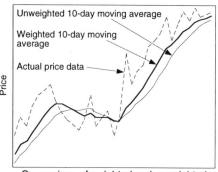

Three moving averages, each with a different number of days used in the average

Front loaded methods

Front loaded moving average curves place greater emphasis on the most current prices than on the older prices. This tends to make the curve more responsive to current fluctuations in price. Three of the methods typically used are weighted, stepped weighted, and exponential. A comparison of weighted and unweighted ten-day moving average curves is shown at right. Note that the front loaded curve more closely follows the actual data curve.

Comparison of weighted and unweighted moving average curves

Envelope or band

Sometimes envelopes or bands are formed with moving average curves. Some analysts believe that when the price curve goes beyond the envelope, actions are indicated. One way to form an envelope is to generate two or more additional curves at fixed percentages above and below a standard moving average curve. Another is to generate moving average curves for the highs and lows. (There are also methods for generating envelopes without using moving averages.)

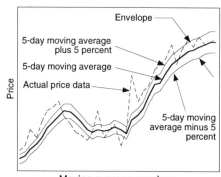

Moving average envelope

Shifted moving average curve

Occasionally one or more moving average curves are shifted or displaced horizontally. If only one displaced moving average curve is used, the analyst frequently looks where the stock price curve crosses the shifted moving average curve. In another case, the moving average curve might serve as a trend line. If two moving average curves are used, an analyst might look for the point at which the two curves cross as an indication to take some action. An example of such a situation is shown at the right.

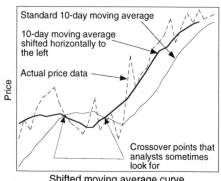

Shifted moving average curve

Bar Chart[2] (continued)

Retracement lines

Retracement lines are a series of lines that enable the viewer to see what percent of an advance or decline has been or is likely to be reversed. Retracement lines are frequently drawn horizontally between two reference prices. If a stock experiences a marked increase over a period of time, the retracement lines would be drawn between the price at which the increase began and the price where the increase ended. For example, a horizontal line might be drawn at a price that represents the new higher price, minus 38% of the difference between the prices at which the increase began and ended. In this way the viewer can visually determine if and/or when the price of the stock loses 38% of the gain it had made.

With a series of lines such as 25%, 33%, and 50%, the viewer can determine what percent retracement has taken place at any time. Lines might be shown for the full width of the graph, as tick marks, or any variation in between. Retracement lines can be used when prices are ascending or descending. Sometimes arcs are used instead of straight lines; in this case both vertical and horizontal distances are taken into consideration. Examples of horizontal retracement lines are shown at the right.

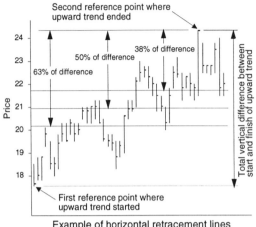

Example of horizontal retracement lines

Volume chart

Many analysts feel that the relationship between the price of a thing being sold, such as a stock, and the volume of it that is sold is important to an overall analysis. Volume charts are used to provide the second bit of information. Two variations of volume graphs are shown at the right. Technically they are not part of a bar chart; however, since they are most frequently displayed and used in conjunction with bar charts they are discussed here. Volume charts are used to record several things including shares of stock, number of

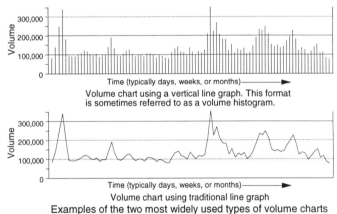

Volume chart using a vertical line graph. This format is sometimes referred to as a volume histogram.

Volume chart using traditional line graph
Examples of the two most widely used types of volume charts

contracts executed, number of contracts open at the end of trading, number of price reversals, etc. The most familiar use is the recording of the number of shares of stock sold in a given period of time. The period of time is the same as that used in the price bar chart the volume chart accompanies. Whether the volume plotted applies to an individual stock or a family of stocks depends entirely on the price bar chart. Whatever the bar chart applies to, the companion volume chart applies to the same thing. When used in conjunction with

Typical price and volume chart in which the same time scale applies to both

bar charts, the volume chart typically uses a vertical line graph format; however, occasionally a segmented line graph format is used. In both types of graph, volume is recorded on the vertical axis and time on the horizontal axis. • The example at left represents a typical combination chart showing stock price information at the top by means of a bar graph and volume information at the bottom using a vertical line graph. In this example, the two graphs share a common time scale, but each has its own vertical value scale.

Indicators

Dozens of indicators are used to analyze the data on bar charts and to predict future performance of a stock, a market, or a segment of a market. A large portion of the indicators are mathematical derivatives of information on the price and/or volume charts. Most utilize a simple line graph to display the indicator, similar to those shown below. Viewers are generally looking for such things as the trend of the indicator, when the indicator crosses a certain value, changes in direction, or cyclical behavior of the indicator.

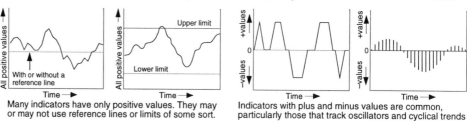

Many indicators have only positive values. They may or may not use reference lines or limits of some sort.

Indicators with plus and minus values are common, particularly those that track oscillators and cyclical trends

Examples of the types of graphs frequently used for indicators on stock price charts

Some indicators are superimposed on the stock price chart while others are juxtaposed, generally above or below. Whether superimposed or juxtaposed, the indicator graph frequently shares a common time scale with the price chart as shown at right. If the actual values of the indicator

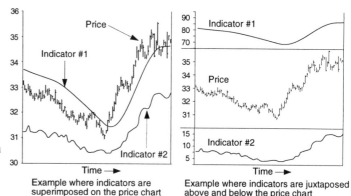

Example where indicators are superimposed on the price chart

Example where indicators are juxtaposed above and below the price chart

are important and its values and/or units are different from the bar chart, an additional vertical scale may be used. Where shape, slope, points of crossing, etc., are the significant characteristics of the indicator instead of actual values, additional scales may not be necessary.

Superimposing other curves on bar charts

When reviewing the overall performance of a company and its stock price, many curves covering different aspects of the company are sometimes superimposed onto the same bar chart with the stock price. For example, it might be of interest to see how the price of the stock relates to earnings, changes in dividends, the volume of the shares purchased, and the performance of other markets. The information included on a particular graph varies significantly depending on its purpose and the audience. Some of the most common types of information included are shown below.

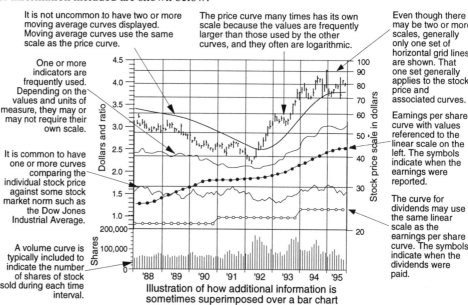

It is not uncommon to have two or more moving average curves displayed. Moving average curves use the same scale as the price curve.

The price curve many times has its own scale because the values are frequently larger than those used by the other curves, and they often are logarithmic.

Even though there may be two or more scales, generally only one set of horizontal grid lines are shown. That one set generally applies to the stock price and associated curves.

One or more indicators are frequently used. Depending on the values and units of measure, they may or may not require their own scale.

It is common to have one or more curves comparing the individual stock price against some stock market norm such as the Dow Jones Industrial Average.

A volume curve is typically included to indicate the number of shares of stock sold during each time interval.

Earnings per share curve with values referenced to the linear scale on the left. The symbols indicate when the earnings were reported.

The curve for dividends may use the same linear scale as the earnings per share curve. The symbols indicate when the dividends were paid.

Illustration of how additional information is sometimes superimposed over a bar chart

Bar Graph

Sometimes referred to as horizontal bar, horizontal column, or rotated column graph. Bar graphs are a family of charts that display quantitative information by means of a series of horizontal rectangles. Sometimes the term bar graph is used to refer to graphs with vertical rectangles. In this book, graphs with vertical rectangles are called column graphs (see column graph). Bar graphs are most frequently used to show and/or compare the values of multiple entities at a given point in time – for instance, the year-to-date sales for each salesperson or the population of the five largest cities in a country as of the end of 1995. An example is shown below.

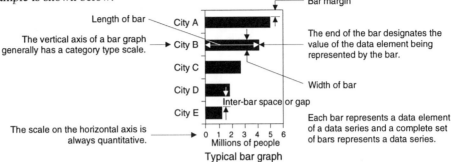

Typical bar graph

Each rectangle represents a data element in a data series and a complete set of bars represents a data series. The end of each bar is located at the value it represents. Because the ends of the bars are so pronounced, this type of graph is considered good for showing specific values. Because of the stand-alone nature of the bars, it is also well suited for representing discrete data. Bar graphs have a quantitative scale on the horizontal axis which is normally linear and typically starts at zero. Dual quantitative scales are seldom used. Bar graphs typically have a category scale on the vertical axis, with entries consisting of word descriptions such as names, events, products, organizations, etc. A sequential scale, such as a time series, can be used on the vertical axis; however, a column graph is normally used for this type of application. The few cases where a quantitative scale is used on the vertical axis are discussed below.

Simple bar graph

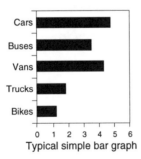

Typical simple bar graph

When a bar graph displays a single data series similar to the example at left, it is generally referred to as a simple bar graph. The bars can be any width; however, with few exceptions, bars are uniform in width throughout a given graph. The same is true for the spaces between the bars. The horizontal scale is always quantitative and both positive and negative values can be plotted as shown in the example at right. When both positive and negative values are plotted it is sometimes referred to as a deviation graph. Linear scales are almost always used on the quantitative scale. When the bars become so wide there are no spaces between them, the graph is sometimes called a joined bar graph, connected bar graph, stepped bar graph, or histogram. Two examples are shown below. In some cases a

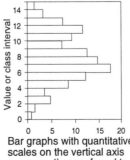

Both positive and negative values can be plotted along the horizontal axis.

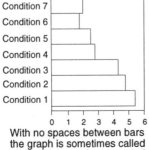

With no spaces between bars the graph is sometimes called a joined bar graph, connected bar graph, stepped bar graph, or histogram.

Bar graphs with quantitative scales on the vertical axis are many times referred to as frequency bar graphs, histograms, or area bar graphs.

joined graph is used to emphasize abrupt changes from one item or condition to another. In a few cases quantitative scales are applied to the vertical axis of a joined bar graph. When this is done the graph is sometimes referred to as a frequency bar graph, histogram, or area bar graph. Area bar graphs are discussed below. Histograms are discussed under that heading.

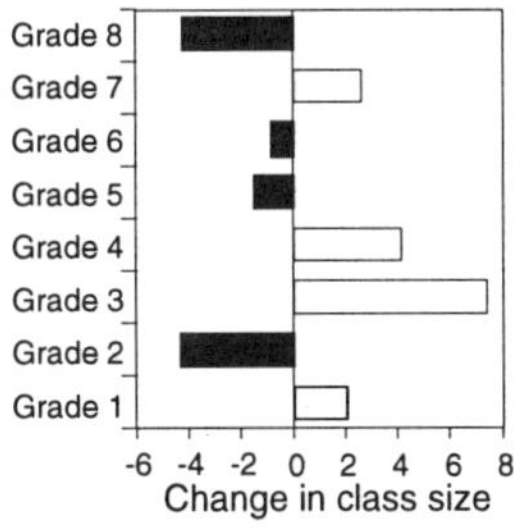

Grouped bar graph – (multiple data series all referenced from the same zero axis)

When two or more data series are plotted side-by-side on the same bar graph, it is referred to as a grouped, clustered, side-by-side, or multiple bar graph. With this type of graph, each data element of each data series is represented by a separate bar. The data series are differentiated from one another by filling the bars of each data series with a different

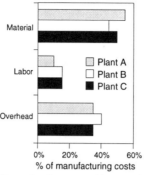

Example of a grouped bar graph.

color, shade, or pattern, generally explained in a legend (see example at left). The bars from each data series that correspond to the first category on the vertical axis are placed side-by-side to form a group or cluster in the center of the space allocated to the first category. The bars from each data series that correspond to the second category on the vertical axis are placed side-by-side in the center of the space allocated to the second category. This continues until all of the columns representing data elements are entered on the graph. To illustrate the procedure, a graph might be generated to compare the percents that material, labor, and overhead represent of the total manufacturing costs for three

plants that produce the same product. The bars that show the percent that material represents for each of the three plants would be grouped in the space on the vertical axis allocated to the material category. The bars that show the percent that labor represents for each of the three plants would be grouped in the space allocated to the labor category, The same process would be followed for overhead (see above). A space is almost always provided between each group of bars so that the data associated with each category can be easily distinguished. The spaces used in a given graph can be of any width; however, they are generally uniform throughout a given graph. • The example below describes some of the design features typically used with grouped bar graphs.

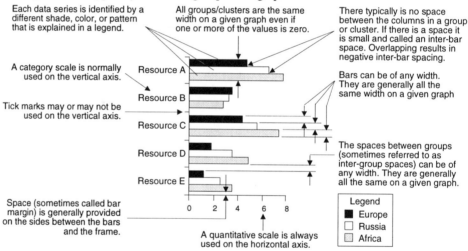

Design features typically used with grouped bar graphs

Technically, there is no limit as to the number of data series that can be plotted on a single graph. Practically, if the number goes above three or four the graph becomes confusing. Most variations of grouped bar graphs can accommodate both plus and minus values on the horizontal axis, as shown in the example at right. Three-dimensional grouped bar graphs (shown below) are sometimes used for

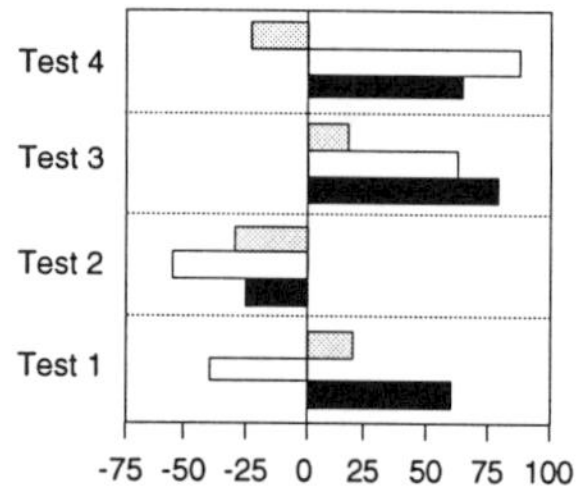

Grouped bar graph with positive and negative values

their aesthetic value. The three-axis variation is sometimes difficult to construct and interpret and therefore, a three-axis grouped column graph is frequently used instead.

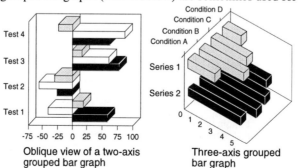

Oblique view of a two-axis grouped bar graph

Three-axis grouped bar graph

Bar Graph (continued)

Grouped bar graph (continued)
———**Overlapped grouped bar graph**———

In an attempt to make grouped bar graphs easier to read, the bars are sometimes overlapped. In this process the bars representing complete data series are shifted vertically such that the bars of each successive data series are partially hidden by the bars of the data series in front. The bars can be shifted up or down by any amount ranging from barely to completely overlapped (100%). Care should be taken as the limit of 100% overlap is approached; the graph can easily be mistaken for a stacked bar graph. The data series with the shortest bars is normally positioned in the front so as not to be hidden by the longer bars. Examples of 25, 50, 75, and 100% overlap are shown below. It should be noted that when the width of the graph stays the same, it is possible to increase the width of the bars as the percent of overlap increases.

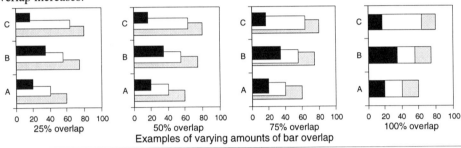

Examples of varying amounts of bar overlap

Stacked bar graph – (multiple data series positioned on top of one another)

———**Simple stacked bar graphs**———

Sometimes referred to as segmented, extended, divided, composite, or subdivided bar graphs. A simple stacked bar graph has multiple data series stacked end-to-end instead of side-by-side as in a grouped bar graph. The far right ends of the bars represent the totals of all components in the bars. The horizontal scale is always quantitative, linear, and starts at zero. The vertical scale is almost always categorical. Stacked bar graphs are used to show how a larger entity is divided into its various components and the relative effect each component has on the total entity. For example, a graph might show the domestic, export, and inter-company components of sales for each major product line, as shown at the right. Each data series is identified by a different shade, color, or pattern (explained in a legend). Depending on the nature of the data, the largest, most important, or least variable components might be placed on the bottom of the stack. There are no rigid rules regarding relative placement of the various components.

———**100% stacked** bar graph———

Sometimes called percent-of-the-whole. An important variation of the stacked bar graph is the 100% stacked bar. With this type of graph, instead of plotting the actual values for each data series, the percents the values represent of the total of all the data series are plotted. For instance, in the example shown here, the first component of the bar for product A shows the percent that domestic sales represent of the total sales of product A. The second component shows the percent that export sales represent of the total sales of product A, and the third component shows the percent that intercompany sales represent. The two examples shown here contain the same data used in the product example immediately above. Eliminating the spaces between the bars (called joined) sometimes makes relationships more visible.

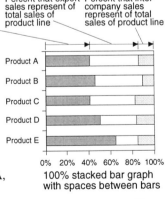

100% stacked bar graph
with spaces between bars

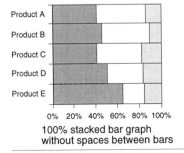

100% stacked bar graph
without spaces between bars

Stacked bar graph (continued)

Linked/connected stacked bar graph

In an attempt to make relationships easier to see, lines are sometimes drawn connecting the boundaries between the data series. The resulting graph is referred to as a linked or connected stacked bar graph. Connecting lines can be used with simple stacked or 100% stacked bar graphs.

Simple stacked bar graph — 100% stacked bar graph

Linked/connected stacked bar graphs

Three dimensional stacked bar graph

A third axis is seldom used with stacked bar graphs. Oblique views of two-axis stacked bar graphs, as shown at right, are sometimes used for their aesthetic value, though the technique generally results in some loss in accuracy of decoding.

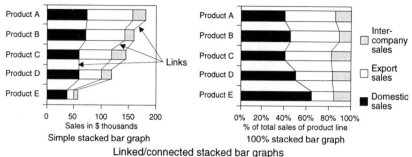

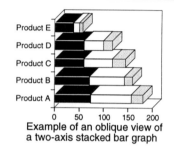

Example of an oblique view of a two-axis stacked bar graph

Negative values

Data series with all negative values can be plotted on stacked bar graphs. Data series with combinations of positive and negative values result in very confusing and misleading stacked bar graphs and therefore are rarely plotted.

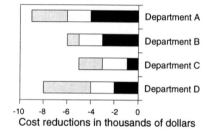

Stacked bar graph with a single bar (one-axis bar graph)

A typical stacked bar graph shows how the relative sizes of components change or differ from one situation, condition, location, etc., to another. Each bar of such a graph shows the make up of the components at a given point in time or under one set of conditions. Sometimes the interest is in just one point in time or one condition, in which case a single bar is used (left). When used in this way, the graph functions very much like a circle graph. • When a single bar is used, the quantitative scale is some-times placed directly on the top or bottom of the bar. The resulting graphic is sometimes referred to as a one-axis graph. Actual value scales, percent-of-the-whole scales, or both might be used, as shown at the right. • Sometimes the scales are eliminated and the values placed on the data graphic. Without the scale, the graphic is called a bar chart instead of a bar graph (left).

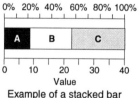

Example of a stacked bar graph with a single bar

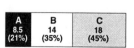

Example of a bar chart (Since it has no scales, it is not called a graph.)

Single stacked bar with a percent-of-the-whole scale

Single stacked bar with a quantitative value scale

Single stacked bar with percent-of-the-whole and quantitative value scales.

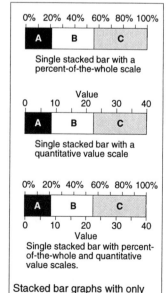

Stacked bar graphs with only one bar. Sometimes called one-axis bar graphs.

Bar Graph (continued)

Paired bar graph

Sometimes referred to as a sliding bar, opposed bar, two-way bar, or bilateral bar graph. A paired bar graph is a variation of a bar graph in which two data series are plotted on the horizontal axis. Values for one data series are measured to the right, and values for the second data series are measured to the left. Values might be expressed in either units or percents. The major purpose of the paired bar graph is to compare two or more data series with particular attention to correlations or other meaningful relationships. The units of measure and scale intervals for the left and right scales may or may not be the same. For example, the number of males and females in five different departments might be compared, in which case the units of measure (number of people) would be the same on both scales. In another case the total number of employees in five different departments might be compared with the expenses incurred in each department. Then one scale would be in numbers of people and the other in dollars of expenses. Linear quantitative scales are used on the horizontal axis, and generally all positive or all negative values are plotted. Grid lines can be used but many times are not. The bars representing the different data series can be filled with different colors, shading, or patterns to clearly differentiate them. Examples of five major types of paired bar graphs are shown below.

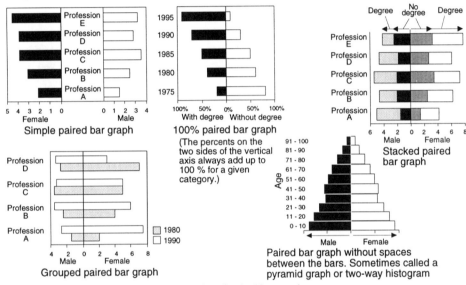

Examples of paired bar graphs

Pictorial bar graph

Sometimes referred to as a pictographs. A pictorial bar graph is a graph in which the rectangular bars have been filled with pictures, sketches, icons, etc., (below), or where pictures, sketches, icons, etc., have been substituted for the rectangular bars (right). In both variations, one or more symbols might be used in each bar. The two major reasons for using pictorial graphs are to make the graph more appealing visually and to facilitate better communications (see Pictorial Charts and Graphs).

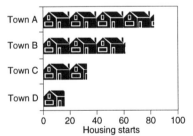

Each symbol represents a certain number of units. The symbol on the end is cropped such that the edge of the graphic is in line with the value on the scale that the bar represents.

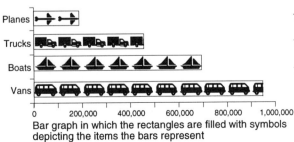

Bar graph in which the rectangles are filled with symbols depicting the items the bars represent

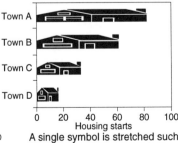

A single symbol is stretched such that the end of the symbol is in line with the value on the scale that the bar represents

Bar Graph (continued)

Range bar graph

Range bar graphs designate upper and lower boundaries by means of short bars. The boundaries might be measured or calculated values, maximum and minimums, confidence limits, 10th and 90th percentile values, etc. Additional values, referred to as inner values, are sometimes also designated. Inner values are typically the same values designated by the end of the bars in a simple bar graph. They always lie at or between the upper and lower values. Inner values many times designate such things as average and median values..

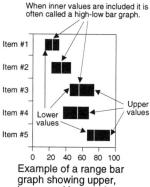

When inner values are included it is often called a high-low bar graph.

Example of a range bar graph showing upper, lower, and inner values

In a range bar graph, one end of the bar (normally the right end) designates the upper value and the other end designates the lower value. The bars for a particular range of values are generally all the same color. When an inner value is designated, it is often shown as a vertical line across the bar. For example, if the scores for a group of people are plotted for a battery of tests, the right ends of the bars might indicate the highest scores achieved on each test and the left ends the lowest scores; a line across each bar might indicate the average or median scores. See Range Symbol and Graph.

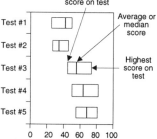

Range bar graph showing the highest, lowest, and average scores on a battery of five tests

Box type range bars

In some situations it is desirable to convey information about the distribution of the data as well as the inner, upper, and lower values. The amount of additional information and how it is communicated varies significantly. One example is shown at the right. See Box Graph for additional examples.

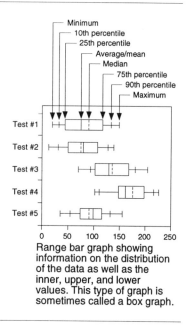

Range bar graph showing information on the distribution of the data as well as the inner, upper, and lower values. This type of graph is sometimes called a box graph.

Grouped range bars

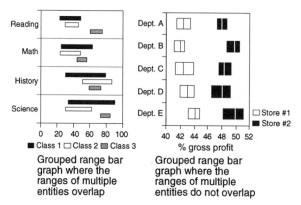

Grouped range bar graph where the ranges of multiple entities overlap

Grouped range bar graph where the ranges of multiple entities do not overlap

A grouped range bar graph format can be used to compare the ranges of multiple entities. For example, the ranges of scores of other classes might be compared with the range of scores for the class under study. When the ranges of the multiple entities overlap they are plotted side-by-side. When the ranges do not overlap, the range bars can be plotted in line with one another. Examples of both variations are shown at left.

Bar Graph (continued)

Graphs with ranges designated on bars

Solid range symbol

The ends of the bars on bar graphs many times correspond to values such as average or median. Range bars or similar type symbols are sometimes added to the end of the bars to indicate values that tend to bracket the average or median values such as high and low values, plus and minus 1, 2, or 3 standard deviations, plus and minus 90% confidence limits, etc. When this is done, the graph is sometimes referred to as a high-low bar graph or a bar graph with error bars. There are a number of different symbols used for this purpose. Examples of three of them are shown below. See Range Symbols and Graphs for other examples. High-low symbols are generally used with simple and grouped bar graphs. They are almost never used with stacked or three-dimensional bar graphs.

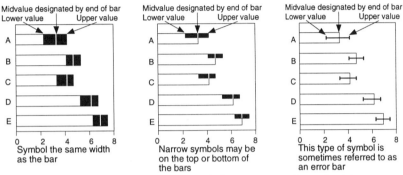

Examples of typical symbols used to indicate ranges on the ends of bars

Gradient range symbol

When crisp vertical lines are used to indicate confidence intervals or ranges in which a value probably lies, the viewer sometimes assumes there is a higher degree of certainty associated with the values than actually exists. This is true of both the ends of the bars and the symbols. To counter this, the crisp lines are sometimes replaced by graphics that are blurred. Without defining a specific number that the viewer can focus on, the blurred graphic encourages the viewer to think in terms of a range of values within which the actual value exists. In addition to designating a range instead of an exact value, some of the blurred graphics also indicate where the actual value most probably lies within the range. In this type of graph, the end of the bar may or may not designate a specific value such as average or median. Several examples are shown below. One would interpret the data graphic associated with item A in the graph on the left as indicating that with a certain degree of confidence, the actual value for A lies somewhere between two and four, with a greater probability that it is closer to four than to two. The degree of confidence that applies may or may not be noted on the graph.

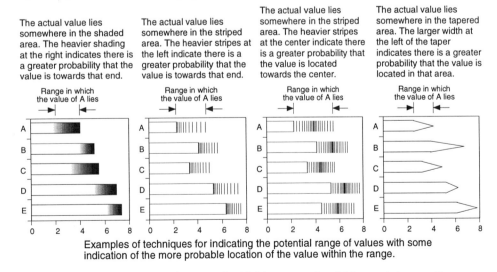

Examples of techniques for indicating the potential range of values with some indication of the more probable location of the value within the range.

The gradient technique is sometimes used with histograms in which case it is sometimes referred to as a fuzzygram. See Fuzzygram.

Difference bar graph

There are three major types of difference bar graphs – simple difference, change, and deviation. In all cases a quantitative scale is used on the horizontal axis and typically a category scale is used on the vertical axis. Each is reviewed separately below.

Simple difference bar graph

A simple difference graph compares two data series by plotting the actual values of the two data series and connecting the two values with a bar. The entire bars from zero to the values for each data series might be included; however, generally only those portions of the bars where the differences exist are shown. In some cases no effort is made to indicate which series has the largest value. In these cases the bars are generally all the same color, shade, or pattern. When it is desirable to indicate which of the data series is the largest, the data series associated with the higher and/or lower values might be noted at the ends of the bars or the information might be encoded into the data graphics. The actual numeric differences may or may not be noted on the bars. Typical information that might be plotted on a simple difference bar graph are:

  – Average scores for two different groups of students on a battery of tests
  – Male and female median incomes for different professions
  – Frequency of various diseases for two different countries

Three ways of presenting the same difference data are shown below.

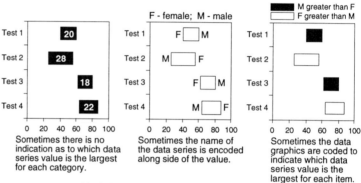

Variations of simple difference bar graphs

Change bar graph

A change bar graph compares multiple factors at two points in time or under two different sets of conditions. Actual values are plotted. The values for the first time/condition and the second time/condition for each item are connected with a bar. With this type of graph the direction of change is generally considered important; therefore, a coding system is used to indicate direction. Methods for accomplishing this are shown below. Since neither left to right or right to left is always favorable (desirable) or unfavorable (undesirable), an additional coding system is sometimes used to designate whether the change is favorable or not. The numeric value of the change may or may not be shown on the data graphics. Typical information plotted on a change graph includes:

– An athlete's performance on several tests before and after training
– A person's response time to different types of stimuli before and after ingesting drugs
– The price of a stock at the opening and closing of trading over various periods of time

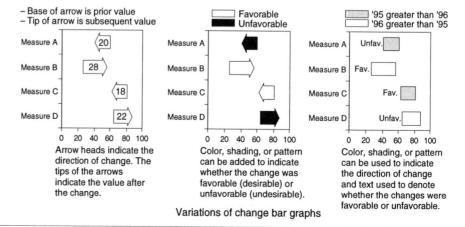

Variations of change bar graphs

Difference bar graph (continued)

Deviation bar graph

A deviation graph focuses almost entirely on the differences between a data series and a reference or another data series. Frequently the actual values of the data series are not plotted. Instead, the difference or deviation values are plotted against a zero axis with bars connecting the deviation values and the zero axis. For example, if plant growths with various fertilizers are being compared, only the difference in growth with and without fertilizer might be plotted. If the plant growth with the fertilizer is greater than without, the difference is shown as a positive value. If the plant growth with the fertilizer is less than without, the value is shown as a negative number. If the growth is the same with and without the fertilizer, a zero is shown since there is no difference or deviation from the reference – which is no fertilizer. • Sometimes the actual data consists of two data series such as cash in versus cash out, or the number of stocks that increased in price versus the number that decreased in price. When this type of actual data is plotted it is sometimes referred to as a gross deviation graph. When the differences between the two data series on a gross deviation graph are plotted, it is called a net deviation graph. Examples of deviation graphs resulting from both types of actual data are shown below.

Deviation when comparing a data series against a reference

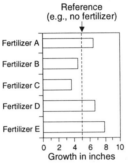

Standard bar graph of actual growth of plants using various fertilizers, with the reference of no fertilizer

The graph on the left displays the actual data with a vertical reference line indicating the reference or control value of the plants without fertilizer. The graph on the right, which is a deviation graph, plots only the amounts by which the plants using the fertilizer differed from the reference or control value of the plants without fertilizer.

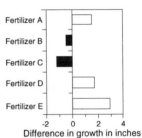

This deviation graph shows only the differences in growth between the plants using the various fertilizers and the reference of the plants without fertilizer.

Gross deviation when comparing two data series

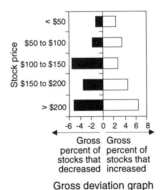

Gross deviation graph

The graphs on the left and right both show the percent of stocks in various price categories that went up and down during a given time period. The only difference between the two is the location of the labels. When the total positive and negative values are shown, as they are here, the graph is sometimes referred to as a gross deviation or split bar graph.

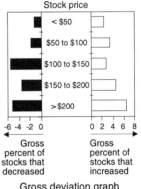

Gross deviation graph

Net deviation when comparing two data series

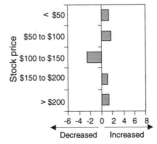

Net percent of stocks that changed price (i.e., difference between the percent of stocks that increased and the percent that decreased)

The graph on the left is referred to as a deviation or net deviation graph. The values shown are equal to the algebraic sum of the gross values shown above. The graph at the right combines the gross and net information into a single graph.

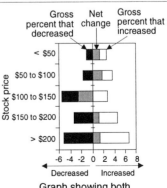

Graph showing both gross and net deviations

Area bar graph

Area bar graphs are the only variation of bar graphs in which the width of the bars has significance. In the area bar graph the widths of the bars are proportional to some measure or characteristic of the data element represented by the bars. For example, if the bars are comparing the average salary of different groups of people along the horizontal axis, the width of the bars might indicate how many people are in each group. If the bars are displaying the profitability of various product lines, the width

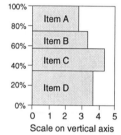

Simple area bar graph with a percent scale on vertical axis

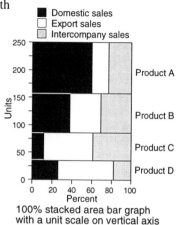

100% stacked area bar graph with a unit scale on vertical axis

of the bars might indicate what percentage of the total sales the various products represent. Bar widths can be displayed along the vertical axis in terms of percents (left) or units (right). There might be a scale on the vertical axis or provided in a legend. In still other cases the values might be noted directly on the graph. Examples of all three variations are shown below.

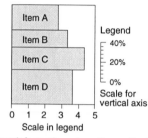

Scale on vertical axis

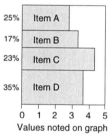

Scale in legend

Values noted on graph

Examples of how width of bar information along the vertical axis can be communicated

If the values along the vertical axis are cumulative, the bars are generally joined so there are no unaccounted-for spaces. For example, if all of the values add up to a certain figure such as 100%, or if the sum of all of the values along the vertical axis is important (e.g., the total number of units sold), no spaces are left between bars. If the values are not cumulative, the total is unimportant, and the scale is part of the legend, spaces can be used between the bars without affecting the accuracy of the graph. An example is shown at the left. • Sometimes area bar graphs are categorized based on the following two criteria:

1. The widths of the bars are basically descriptive with little or no relationship to the length of the bars. For example, the width might denote the population of cities and length might designate the number of violent crimes. In this case the actual area of the columns have little or no significance.

Area bar graph with spaces between bars. In this type of graph the values along the vertical axis are not cumulative, the total is unimportant, and the scale is shown in a legend.

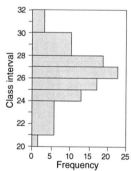

The histogram is an example of an area bar graph where the width and length of the bar are related. As the width of a bar is expanded to cover a broader class interval, the length of the bar is changed accordingly.

2. The widths and lengths of the bars are related such that as one varies, the other varies correspondingly. For example, in a histogram, if the width of a bar is increased to encompass a broader class interval, the length of the bar must be changed accordingly. In these cases the area of the bar can be as or more meaningful with regards to the data element it represents as either the length or the width of the bar.

• In some cases the width of the bar is only meant to convey approximate or relative values and no scale is shown. The concept of using the width of columns to convey additional information is generally not applied to grouped bar graphs.

Bar Graph (continued)

Circular bar graph

A circular bar graph is the equivalent of a rectangular bar graph wrapped into a circle. The horizontal axis of the rectangular graph becomes the circular or value axis of the circular graph, and the vertical axis of the rectangular graph becomes the radial or category axis of the circular graph. An example of a circular bar graph and its associated terminology is shown below. In a typical circular bar graph, the bars representing the various categories are uniformly spaced along the radial or category axis. The ends of the bars designate the values they represent. The lengths of the bars have no significance and can be misleading since the lengths are not proportional to the

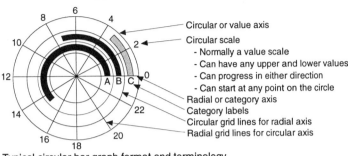

Typical circular bar graph format and terminology

values they represent. The value scale can have any upper and lower value. It can progress in either direction and start at any point around the circle. Although simple circular bar graphs are generally used, the same technique can be applied to more complicated variations such as stacked, grouped, paired, 100%, etc. Occasionally a time series scale is shown on the circumference and the graph is used to display a repetitive schedule, as shown at the right. When this is done, it is sometimes referred to as a scheduling or circular Gantt chart. Circular bar graphs are generally used for their aesthetic value. Functionally, they offer little or no advantage over the rectangular bar graph.

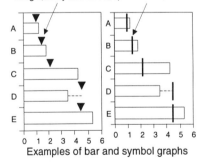

A specialized circular bar graph with a time series scale shown on the circumference. Sometimes called a scheduling or circular Gantt chart.

Bar and symbol graph

Sometimes a series of symbols are shown on a bar graph to designate reference points or points of comparison for each of the bars. The reference points might designate such things as budget, plan, last year's value, or specification value. By means of the symbols the viewer can make comparisons between the actual values as designated by the ends of the bars and some other set of values indicated by the symbols. Any symbol that crisply identifies the reference value can be used. Two examples are shown at the right. Bar and symbol graphs typically have only one data series represented by horizontal rectangles. Multiple sets of symbols can be used on the same graph to indicate two or more sets of reference values. The bar and symbol technique is seldom used with grouped or stacked graphs.

Symbols typically designate individual reference or comparison values for each bar. The reference values might be such things as budget, last years value, specification, etc.

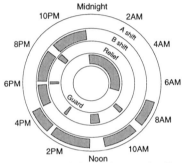

Examples of bar and symbol graphs

Progressive bar graph

Sometimes referred to as stepped, step-by-step, or staggered bar graph. A progressive bar graph is the equivalent of a stacked bar graph with only one bar and with the individual segments of that bar displaced vertically from their adjacent segments. The segments can be displaced up or down and by any amount. They are generally shifted slightly more than the width of the bar. Whatever amount and direction of displacement are selected, they are used throughout the graph. Displacing the segments adds visibility and emphasis to the individual segments while still maintaining the concept that the segments all add up to the whole.

Progressive bar graph (continued)

In the example at the right, the major overhead expenses for a hypothetical company are shown. The bottom scale enables the viewer to determine cumulative and overall expense dollars. The upper scale enables the viewer to determine cumulative percents. The actual dollar value and/or percent that each expense represents of the total may or may not be shown on each data graphic. As shown in the next example,

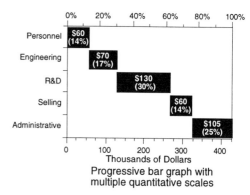

Progressive bar graph with multiple quantitative scales

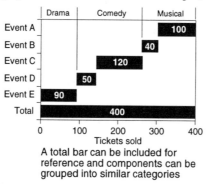

A total bar can be included for reference and components can be grouped into similar categories

an additional bar representing the total of all of the individual segments is sometimes included. In addition, the example at left illustrates how segments can be grouped to convey additional information to the viewer. Time and activity charts are sometimes considered variations of a progressive bar graphs where the intervals on the quantitative horizontal scale are units of time such as days, weeks, or months. With this type of chart the lengths of the bars correspond to the lengths of time required to complete each subprogram of a larger project. The left end of the bar is positioned at the number of units of time from the start of the overall project that the particular subprogram is to begin. With such a graph, the viewer is able to determine when each subprogram will begin, how many units of time it will take to complete each sub-program, and the cumulative units of time consumed at any point in the project.

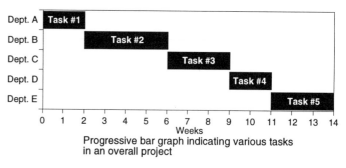

Progressive bar graph indicating various tasks in an overall project

Mosaic graph

A mosaic graph is a combination of 100% stacked column graphs and 100% stacked bar graphs. The major purpose of such graphs is to display a system of interrelated values in such a way that groupings and relative sizes of the various elements can be seen graphically. For instance, in the example below, 50% of the sales dollars go into the cost to manufacture. Of the costs to manufacture, about 55% goes into materials and about 36% of the material dollars go into raw materials. See Mosaic Graph.

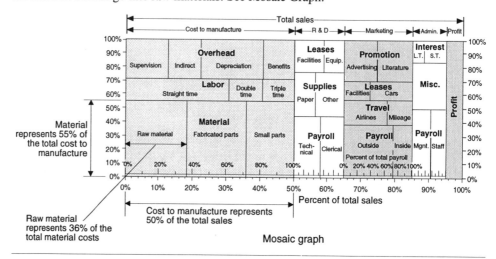

Mosaic graph

Bar Graph (continued)

Curve fitting on bar graphs

Although seldom done, curves can be fitted to bar graphs. Most data plotted on bar graphs is not continuous and many times adjacent data points have little or no relationship to one another. Since curve fitting is generally applied to continuous data, it is not applicable to most bar graphs. In the few cases where it is done, there is a quantitative scale on the vertical axis and the assumption is made that the data points are located in the center of the end of each bar.

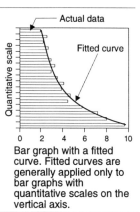

Bar graph with a fitted curve. Fitted curves are generally applied only to bar graphs with quantitative scales on the vertical axis.

Location of axis from which values are referenced

By displaying data against a reference other than zero, additional information can be displayed on a single graph. One widely used technique is to place bars between the actual values and some meaningful value or set of values such as budget, forecast, goal, industry average, etc. In this way the graph communicates three values at the same time:

    – Actual values
    – Reference values
    – Difference between actual value and the reference value

Three examples are shown below. The first one is for reference and is a conventional bar graph with the bars between zero and the actual values. The second example uses the same non-zero reference value for each category, and the third uses a unique reference value for each category. The same data is plotted in all three examples.

──────**Standard bar graph for reference**──────

This example is for reference only. It is a standard bar graph with bars drawn from the zero axis to the actual values represented by the bars.

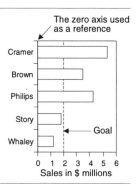

──────**Common reference value for all categories**──────

In some cases the axis from which the bars are drawn is shifted horizontally away from zero. When this is done, the ends of the bars still correspond to the actual values as measured from the zero base line axis; however, it might be either the right or left end of the bar that is positioned over the actual value. For example, in plotting the sales performance of each salesperson against a $2 million sales goal for each salesperson, the reference axis might be shifted to correspond to that $2 million goal as shown at the right. In this way, the bars for salespersons who exceeded goal extend to the right, and the actual values are designated by the right ends of the bars. The bars for the salespersons who were below goal extend to the left, with actual values designated by the left ends of the bars.

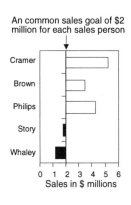

An common sales goal of $2 million for each sales person

──────**Unique reference value for each category**──────

If each salesperson had a different goal, the reference line would be stepped as shown at the right. As in the example above where the goals were all the same, the bars for those who exceeded their goal extend to the right, and the bars for below goal extend to the left.

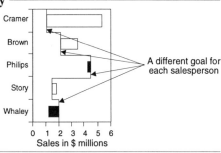

A different goal for each salesperson

Labels and supplemental information

One of the advantages of bar graphs is that they accommodate long labels and supplemental information better than almost any other type of graph. Several examples are shown below.

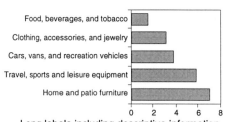

Long labels including descriptive information can be accommodated on either the left or right side of a bar graph.

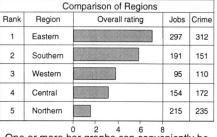

Comparison of Regions				
Rank	Region	Overall rating	Jobs	Crime
1	Eastern		297	312
2	Southern		191	151
3	Western		95	110
4	Central		154	172
5	Northern		215	235

One or more bar graphs can conveniently be included in a table of data.

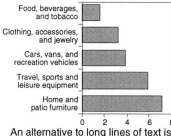

An alternative to long lines of text is to shorten the lengths and use two or more lines for each label.

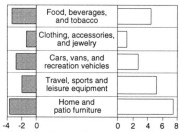

In paired bar and certain deviation graphs, the labels can sometimes be positioned in the center.

Bar fill

When different variations of the same thing are plotted on a simple bar graph, the bars typically all have the same fill to convey the idea that variations of the same thing are being plotted, not a series of different things.

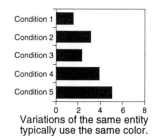

Variations of the same entity typically use the same color.

When distinctly different items are plotted on a simple bar graph, all the bars might have the same fill, or each one might have a different fill to emphasize the differences.

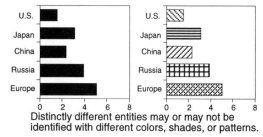

Distinctly different entities may or may not be identified with different colors, shades, or patterns.

When positive and negative values are plotted, the two types of bars might have the same or different fills. The unfavorable values are sometimes colored red, whether they are positive or negative.

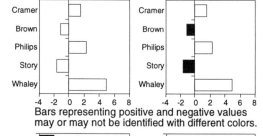

Bars representing positive and negative values may or may not be identified with different colors.

In grouped and stacked bar graphs, all bars or segments of bars that represent the same thing (the same data series) typically are the same color, shade, or pattern.

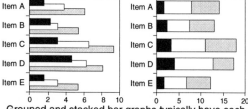

Grouped and stacked bar graphs typically have each data series identified with a different color or shade.

Bar Graph (continued)

Scales

Although in most cases the actual value designated by a bar is determined by the location of the end of the bar, many people associate the length or area of the bar with its value. As long as the scale is linear, starts at zero, is continuous, and the bars are the same width, this presents no problem. When any of these conditions are changed, the potential exists that the graph will be misinterpreted. For this reason the following general guidelines are offered regarding quantitative scales on bar graphs.

- Have all scales include zero unless it is clearly identified otherwise.
- Locate zero where the vertical axis crosses the horizontal axis.
- Have positive values increase to the right and negative values to the left.
- Make scales continuous (no breaks).
- The upper scale value should be larger than any value plotted.
- Use only linear scales (e.g., avoid nonlinear scales such as logarithmic).
- Except for area bar graphs, make all bars the same width.

There are exceptions to these guidelines.

- With range and difference graphs the scale sometimes does not start at zero since the bars do not extend to the zero axis. With these types of graphs, the lower scale value should be lower than the lowest value being plotted.
- With simple and grouped bar graphs the scales are sometimes broken and/or expanded to make changes and differences from bar to bar more visible. Several methods for accomplishing this are shown below.

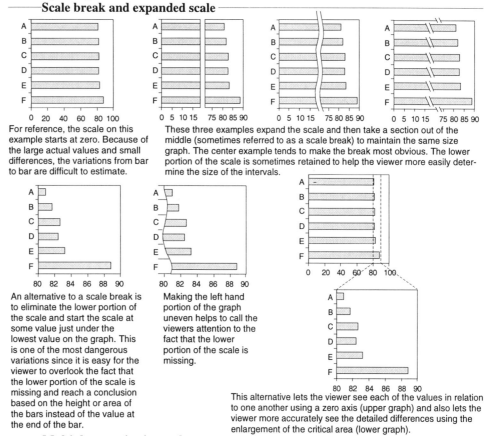

─── **Scale break and expanded scale** ───

For reference, the scale on this example starts at zero. Because of the large actual values and small differences, the variations from bar to bar are difficult to estimate.

These three examples expand the scale and then take a section out of the middle (sometimes referred to as a scale break) to maintain the same size graph. The center example tends to make the break most obvious. The lower portion of the scale is sometimes retained to help the viewer more easily determine the size of the intervals.

An alternative to a scale break is to eliminate the lower portion of the scale and start the scale at some value just under the lowest value on the graph. This is one of the most dangerous variations since it is easy for the viewer to overlook the fact that the lower portion of the scale is missing and reach a conclusion based on the height or area of the bars instead of the value at the end of the bar.

Making the left hand portion of the graph uneven helps to call the viewers attention to the fact that the lower portion of the scale is missing.

This alternative lets the viewer see each of the values in relation to one another using a zero axis (upper graph) and also lets the viewer more accurately see the detailed differences using the enlargement of the critical area (lower graph).

─── **Multiple quantitative scales** ───

Scales are generally located on the left side and bottom. Typically there is only one scale on both the vertical and horizontal axis. Occasionally the identical quantitative scale will be repeated on the top, particularly if the graph is tall. Two different quantitative scales can be used but seldom are. If there are two different quantitative scales and both do not apply to all bars, care must be taken to identify which scale applies to which bars if it is not obvious. Examples of multiple quantitative scales are shown at right.

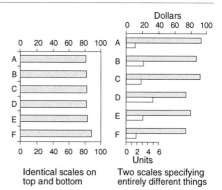

Identical scales on top and bottom

Two scales specifying entirely different things

Grid lines

Grid lines are generally displayed behind the data graphics. Occasionally the grid lines will be shown on the bars themselves or drop lines will be used. Examples of all three variations are shown at the right. See Grid Lines.

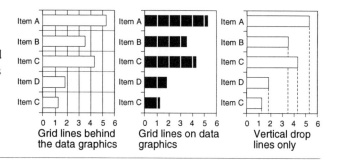

Values on graph

When it is desirable to convey exact values, they are sometimes placed directly on the data graphics, as shown below. Generally only the value represented by the bar is shown; however, it is possible to include additional values – for example, the percent-of-the-whole that each value represents, the percent or ratio the value is to a given reference value, the amount the value deviates from a standard, etc. An example of multiple values is shown at the right.

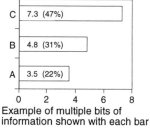

Example of multiple bits of information shown with each bar

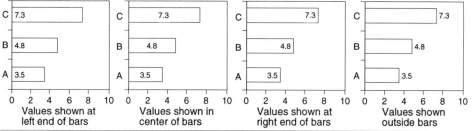

Frames

Because of the boldness of the bars, the frame around a bar graph can sometimes be reduced significantly without detracting from the value of the graph. This is especially true if the values are shown on the data graphics. Four variations are shown below, ranging from a full frame on the left to no frame on the right.

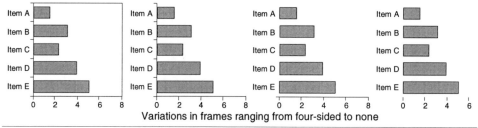

Variations in frames ranging from four-sided to none

Bar chart

For presentation purposes, the quantitative scale is sometimes eliminated as well as the frame. When this is done the values are generally noted on or adjacent to the data graphics. Without a quantitative scale, the resulting configuration is no longer a graph, but a proportional area chart in which the size/areas of the data graphics are proportional to the data elements they represent. See Proportional Chart.

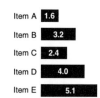

Proportional area bar chart with no quantitative scale

Column graph

Sometimes the term "bar graph" is used to refer to a graph with vertical rectangles similar to the example at the right. This type of graph is frequently called a column graph and in this book it is classified as such. See Column Graph.

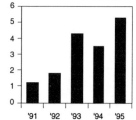

This type of graph is sometimes referred to as a bar graph. In this book it is classified as a column graph.

Base Line	Sometimes referred to as a reference or datum. The base line is a line, plane, or point on a chart, graph, or map from which other values or information are compared, referenced, or measured. In many cases the base line is zero. Other values are also used such as budget, prior performance, industry average, an arbitrary elevation, sea level, etc.	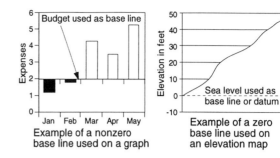

Example of a nonzero base line used on a graph

Example of a zero base line used on an elevation map

Base Map

A basic map on which attribute data is superimposed to form most statistical and descriptive maps. As a general rule, the amount of information on a base map is kept to a minimum so as not to detract from the attribute information conveying the major theme of the map. An illustration combining attribute data and base map to form a statistical map is shown below. See Map.

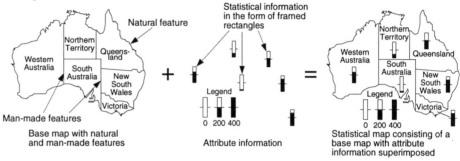

An illustration of how attribute data is superimposed on a base map to form a statistical map

Bathtub Curve

A bathtub curve is so named because of its distinctive shape of higher values at the beginning and end with relatively constant, low values in between. Time is typically plotted on the horizontal axis, with zero at the left and time increasing to the right. A bathtub curve is sometimes used to illustrate how the failure rates of certain products change with time. When used in this way, time zero is considered the time at which each product first goes into service. Uniform periods of time are measured along the horizontal axis and the percent of failures for each period of time are plotted on the vertical axis. The higher failure rates at the left of the graph occur shortly after the products are put into service. After these early failures, the rate

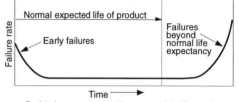

Bathtub curve, sometimes used to illustrate one type of product failure rate profile

typically drops to a low level where it stays for the rest of the products' expected life. If the products continue to be used beyond their normal life expectancy, the failure rate begins to rise again as the products wear out. Because of the relatively short period of time over which the early failures occur, versus the long periods of time for the other phases, the time scale in the early periods is sometime expanded to improve the readability of the curve.

Bathymetric Map and Bathymetric Contour Lines

A bathymetric map is a map that portrays the form of the land (floor or bed) below the surface of a body of water. Looked at another way, a bathymetric map indicates the depth of the water. This is generally accomplished by means of a series of isolines that connect points on the surface of the floor or bed that are at equal depths below the surface of the water. These are frequently called bathymetric contour lines or isobaths.

Bell-Shaped Curve	A normal distribution curve is many times referred to as a bell-shaped curve because of the similarity between the shape of the curve and a bell. See Normal Distribution Curve.	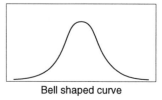

Bell shaped curve

| **Belt Chart** | A variation of a stacked pie chart that enables the viewer to look concurrently at the distribution of interrelated data in a number of different ways. By doing this, patterns and relationships sometimes are easier to note. Using the example shown below, the following are representative of the types of observations that can be made from a belt chart. |

- 64% of the work force is female and 36% male
- Of the 64% females, 48% are nonexempt and 16% exempt
- Of the 36% males, 15% are nonexempt and 21% male
- 64% of all employees work in the office and 36% in the field
- 63% of the employees are nonexempt and 37% are exempt
- Of the 63% nonexempt, 48% are female and 15% are male
- Of the 37% exempt, 16% are female and 21% male
- Of the 63% nonexempt, 51% work in the office and 12% in the field
- Of the 37% exempt, 13% work in the office and 24% in the field

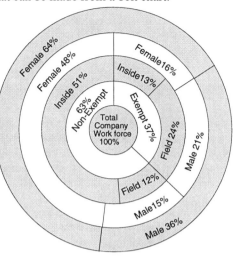

Example of belt chart

Bidirectional Table

Sometimes referred to as a matrix. When the same headings are used for both the rows and columns of a table, the table is sometimes referred to as a bidirectional table. Such tables are frequently used to indicate distances between locations. In a full bidirectional table, as shown on the left below, the data is repeated in the upper and lower triangles. In some cases different units of measure, such as miles and kilometers, are used in the two halves to enhance the value of the table. In other cases the duplicate values are eliminated and the table is referred to as a half bidirectional table. Two examples of half bidirectional tables are shown on the right below.

	City A	City B	City C	City D	City E	City F	City G	City H	City I
City A		15	19	40	37	5	24	43	19
City B	15		24	13	4	10	11	30	10
City C	19	24		28	24	15	15	34	15
City D	40	13	28		10	36	15	34	15
City E	37	4	24	10		33	12	31	11
City F	5	10	15	36	33		30	39	16
City G	24	11	15	15	12	30		19	2
City H	43	30	34	34	31	39	19		21
City I	19	10	15	15	11	16	2	21	

Full bidirectional table

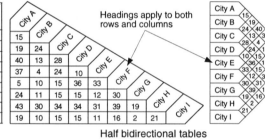

Half bidirectional tables

Bilateral Bar Graph

Sometimes referred to as an opposed, paired, sliding, or two-way bar graph. This type of graph is a variation of a bar graph in which positive values are measured both right and left from a zero on the horizontal axis. The major purpose of such a graph is to compare two or more data series with particular attention to correlations or other meaningful relationships. See Bar Graph.

Bin

Sometimes referred to as a class, class interval, group interval, or cell. Bins are the groupings into which a set of data is divided for purposes of generating any one of several data distribution graphs or maps. An example of a graph using bins is shown at the right. See Histogram and Frequency Polygon for a discussion of bins used with graphs and Statistical Map for bins used with maps.

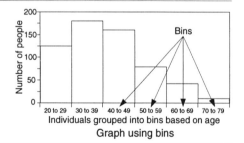

Graph using bins

Binary Decision Diagram

If each decision point on a decision chart allows only one of two decisions, such as yes or no, the chart is sometimes referred to as a binary decision diagram. See Decision Diagram.

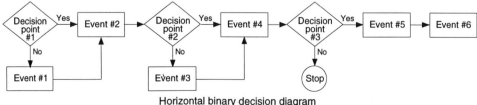

Horizontal binary decision diagram

Bivariate	A subcategory of multivariate. A chart, graph, map, group of data, etc., with two variables.

Block Chart

A variation of a unit chart. A block chart is a chart used for representing quantities of things by making the number of geometric symbols displayed proportional to the quantity of things being represented. For example, 50 cars might be depicted by displaying five symbols, where each symbol represents 10 cars. Block charts might be two- or three-dimensional. See Unit Chart.

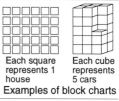

Each square represents 1 house

Each cube represents 5 cars

Examples of block charts

Block Diagram

A variation of a conceptual diagram. A block diagram uses geometric figures and symbols (referred to as blocks) to lay out things schematically such as systems, networks, concepts, circuits, procedures, structures, etc. Block diagrams are used for such things as planning, development, communication, and organization of thoughts. For example, a block diagram for starting a company might have five blocks representing five key elements. Those elements might be developing the idea or concept; raising capital; assembling a management team; acquiring facilities, product, and a work force; and initiating the business. If these five elements are enclosed in rectangles, as shown at the right, the resulting graphic might be called

Develop idea or concept

Assemble management team

Raise capital

Acquire facilities, product and work force

Initiate business

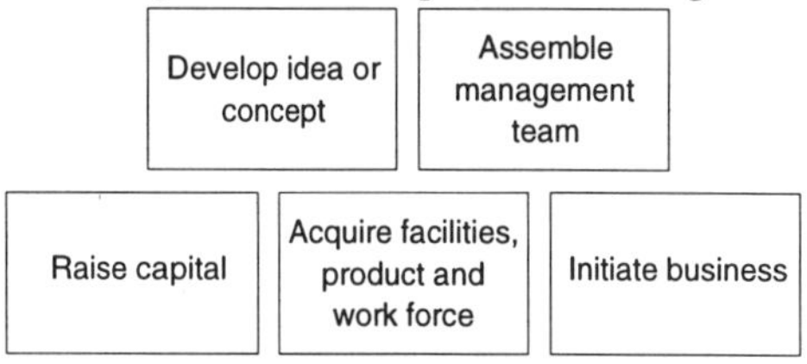

Block diagram indicating the five major elements in starting a business

a simplified block diagram (even though the blocks are not connected). When used for planning purposes, a group of people might shuffle these blocks around until they agree on the best sequence for a particular situation. Once agreement is reached, the blocks might be arranged in their proper sequence and connected by arrows as shown below.

Develop idea or concept → Raise capital → Assemble management team → Acquire facilities, product, and work force → Initiate business

Block diagram with blocks connected by arrows

When used in this way, block diagrams function as an aid in the planning, development, and communicating processes by providing a graphical means of addressing major elements of a project without getting involved in too much detail. The blocks on a diagram may have different shapes to indicate type of function, facility, organization, etc., and may be connected with one or more lines or arrows, as shown in the example at the left. When

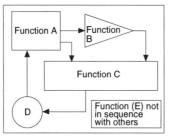

Block diagram using multiple geometric shapes to encode additional information

the purpose of the block diagram is simply to indicate such things as relative physical location or the existence of marginally related ideas, or information, there may be no interconnecting lines (below). Block diagrams generally address the broad overview of a subject. When they become detailed descriptions of the subject, they are many times referred to as flow charts, process charts, or procedural charts.

Engine compartment | Passenger compartment | Luggage compartment

Block diagram of a typical automobile showing the location of major elements

Block Diagram Map

When three-dimensional maps represent portions of a larger entity, they are sometimes referred to as block diagram maps (examples below). Originally hachures and outlines were used to generate such maps. Today most are of the fishnet or shaded relief type. The sides of the block sometimes display geological information or details on man-made structures such as tunnels, sewers, shelters, etc.

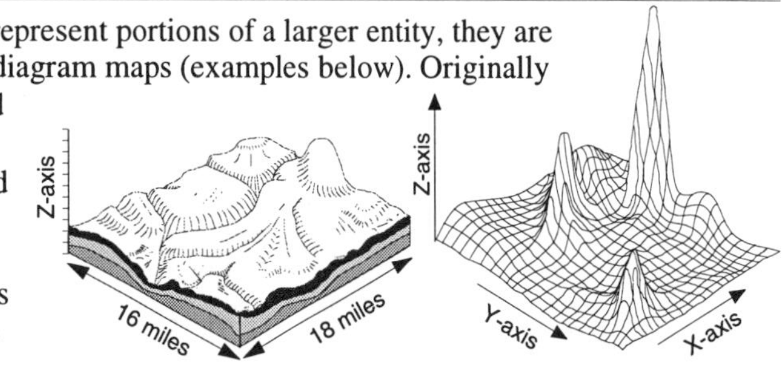

Block diagram maps

Block Map

Sometimes referred to as a stepped relief, stepped, or prism map. In this type of statistical map, areas are elevated in proportion to the values they represent. For example, the height of the various states might be proportional to the grain produced per acre in each of the states. The more grain produced per acre, the taller the prism for that state. See Statistical Map.

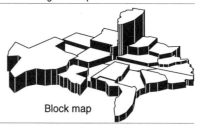

Block map

55

Blot Map

Sometimes referred to as a patch map. Blot maps are a variation of descriptive maps on which such things as natural resources, types of soil, agricultural usage, wetlands, etc., are identified with filled areas. The size and shape of the filled areas generally approximate the entities they represent. The meaning of the fills is explained in a legend.

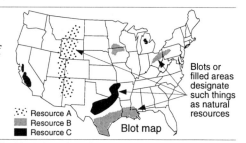

Blots or filled areas designate such things as natural resources

Resource A
Resource B
Resource C

Blot map

Boldface

Sometimes abbreviated as bold. A heavier, darker version of a standard or plain typeface, generally used to add emphasis. Sometimes an entire paragraph or page will be done with boldface type. **In other cases, such as here, a sentence will be bolded.** In still other cases a single **word** or a few words will be put in boldface. **O**ccasionally single letters are bolded, such as the first letter of the first word in a paragraph. Boldface can apply to any type size, plain or *italicized* .

Entire sentence in boldface

Single word in boldface

Single letter in boldface

Italics in boldface

Border[1]

In many situations the terms border and frame are used interchangeably with regards to charts and graphs. Some of the major items the terms are applied to include:

- Title border/frame
- Chart border/frame
- Map border/frame (sometimes referred to as neat line)
- Table border/frame
- Graph border/frame (sometimes referred to as grid border, box, rectangle, plot area border, scale line, label axis, scale axis, or axis)
- Data graphic border/frame
- The area outside of a graph, map, table, or diagram. is sometimes called border or margin
- Legend border/frame
- Note border/frame

Also see Frame.

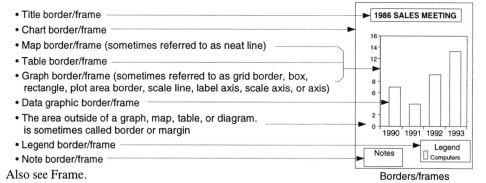

Borders/frames

Border[2]
and
Border Line

On maps, the term border sometimes has a definition that encompasses more than just one line. When used in this way, the term border includes the neat line, the border line (a second line outside of the neat line) and the material between the two lines, including any labels.

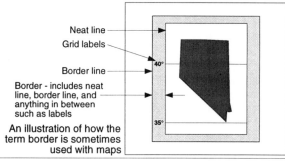

Neat line
Grid labels
Border line
Border - includes neat line, border line, and anything in between such as labels

An illustration of how the term border is sometimes used with maps

Border Plot

Sometimes referred to as a marginal frequency distribution graph. Border plots are used for displaying the distribution of data along one axis of a two-axis graph by putting a plot of the data in the margin/border of the graph. The technique, normally used with scatter graphs, can be applied to one or both axes. The graphic displays in the borders might be one of a number of different types including one-axis point graphs, stripes, boxes, or histograms. An example is shown at the right. See Marginal Frequency Distribution Graph.

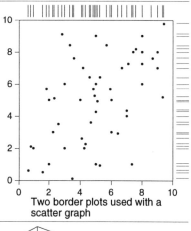

Two border plots used with a scatter graph

Bottom

Sometimes referred to as a floor or platform. The term bottom is used to identify the lower plane of a three-dimensional graph.

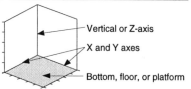

Vertical or Z-axis
X and Y axes
Bottom, floor, or platform

Bottom on a three-dimensional graph

Box Graph or Box Diagram or Box Plot or Box-and-Whisker Plot and Box Symbol

An individual box symbol summarizes the distribution of data within a data set. For example, if the times it takes 25 people to do a task are measured, only the average of the measurements would typically be shown. By using a box symbol, in addition to the average value, a significant amount of other information about the distribution of the measurements can also be encoded, as shown at right. A box symbol consists of a rectangle (box) that generally has a line extending from both ends (whiskers). The ends of the rectangle generally designate the 25th and 75th percentiles of the data set. The ends of the lines projecting from the rectangle generally designate the 5th and 95th or the 10th and 90th percentiles. Lines across the rectangle indicate the average and/or median values. Data points above and below the outer percentiles (e.g., 5th and 95th) are generally indicated by individual data points. Based on information of this type, one can frequently make the following observations.

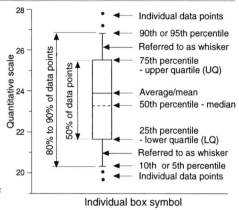

Individual box symbol

– What the key values are such as average, median, 75th percentile, etc.
– Whether there are outliers (unusual data points) and what their values are
– Whether the data is symmetrical and how tightly the data is grouped
– Whether the data is skewed and if so which direction
– What values do 50% and 90% of the data points lie within (i.e., 25% to 75% and 5% to 95%, respectively)

The box plot is a relatively new type of graph that is widely used. Wide usage plus efforts to simplify its construction and interpretation have resulted in many modifications to the original design. Consequently, it is sometimes unclear what the various elements of the plot represent without an accompanying explanation. Except for very technical works, many of the original definitions of the various indicators on box symbols are seldom used. (An explanation of the original terminology is shown at the end of this section.) The examples below illustrate some of the many variations that have been developed.

Key measures indicated	Example of symbol
Average, 10th, 25th, 75th and 90th percentiles	10th, 90th
Average, 5th, 25th, 75th and 95th percentiles	5th, 95th
Average, 25th percentile, 75th percentile, minimum and maximum	Min., Max.
Average, median, 25th percentile, 75th percentile, plus one standard deviation, and minus one standard deviation	-1 SD, +1 SD
Median, plus and minus one standard deviation, minimum and maximum	Min., -1 SD, +1 SD, Max.
Average, median, 10th, 25th, 75th, and 90th percentiles, plus markers indicating plus and minus one, two, and three standard deviations	-3SD, -2SD, -1 SD, 10th, 90th, +1 SD, +2 SD, +3 SD
Average, 5th, 25th, 75th, and 95th percentiles, plus individual points beyond the 5th and 95th percentiles	5th, 95th
Average, 5th, 25th, 75th, and 95th percentiles. Sometimes referred to as a percentile graph	5th, 95th

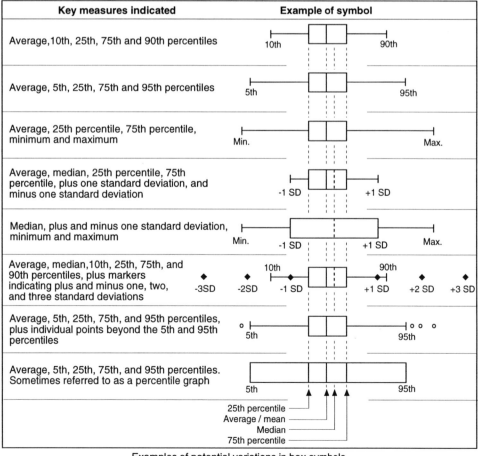

Examples of potential variations in box symbols

Box Graph
or
Box Diagram
or
Box Plot
or
Box-and-Whisker Plot
and
Box Symbol
(continued)

General

Orientation

Box symbols can run vertically or horizontally. When used individually, they typically have a single axis.

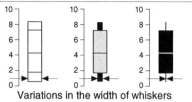

Fill of box and width of whiskers

The rectangle of a box symbol might be unfilled or filled with a color or pattern. The whiskers can be of any width but never exceed the width of the box. The widths of all whiskers are the same in any given graph.

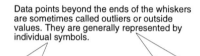

Variations in the width of whiskers

Symbols at ends of whiskers

The short perpendicular lines at the ends of the whiskers (sometimes called ticks) can be replaced by other symbols or eliminated. In some cases the short perpendicular lines are retained and the whiskers are eliminated. In still other cases the short perpendicular lines become the ends of the rectangle.

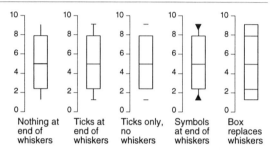

Nothing at end of whiskers | Ticks at end of whiskers | Ticks only, no whiskers | Symbols at end of whiskers | Box replaces whiskers

Unusual data points

Data elements above and below the high and low percentiles represented by the ends of the whiskers are shown as individual data points. Generally they all use the same symbol. Occasionally different symbols are used to indicate different ranges of distances from the average, median, or ends of the box.

Data points beyond the ends of the whiskers are sometimes called outliers or outside values. They are generally represented by individual symbols.

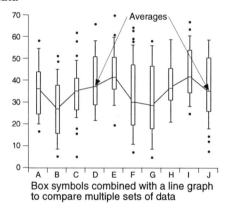

Extreme data points (sometimes called far outside values) are occasionally denoted by a different type of symbol. What is considered an extreme value depends on the application.

Box graphs used to compare multiple sets of data

In addition to analyzing a single set of data, box symbols can be used to analyze multiple data sets on a graph, as shown at the right. Typically, data such as this would be analyzed by plotting the average of each data set and perhaps connecting the points with a line. By superimposing a set of box symbols over the point graph or line graph of the averages, one can see not only how the averages compare but also how distributions and unusual data points compare.

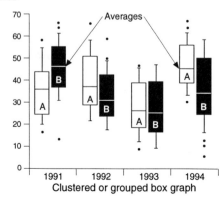

Box symbols combined with a line graph to compare multiple sets of data

Box graphs used to compare multiple data series

Sometimes called a grouped or clustered box graph. Box graphs can be used to compare multiple data series similar to the grouping of columns in a grouped column graph. In the example at the right, data series A and B are compared over a four-year period. With this type of graph, one can make comparisons and see changes in the distribution of the data elements in each data set in addition to making observations about the averages.

Clustered or grouped box graph

Box Graph
or
Box Diagram
or
Box Plot
or
Box-and-Whisker Plot
and
Box Symbol
(continued)

Information encoded by width of boxes

Normally the widths of all the rectangles in a box graph are the same and have no significance. Occasionally, the widths are varied to encode additional information, as shown at right. Box widths might be proportional to such things as the size of the sample represented by the box, the size of the entire family of data from which the sample was taken, market share, or relative importance. Either quantitative or qualitative characteristics might be encoded.

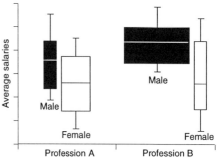

Grouped box graph in which the box widths are proportional to the size of the family from which the data was taken

Confidence intervals

Confidence intervals can be indicated by means of notches or shading, as shown at the right. The two ends of the notch or shading indicate the two limits of the confidence interval. This communicates to the viewer that, although, due to such things as sample size, the average value shown may not be the average of the entire family from which the sample was taken, there is a 90%

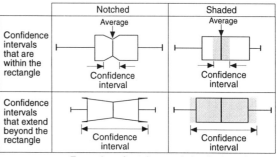

Examples of notches and shading used to designate confidence intervals in box plots

or 95% probability that the overall average lies somewhere within the confidence interval. The actual size of confidence intervals depends on the nature of the data, the size of the

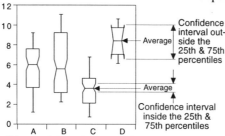

If the confidence intervals of any two box plots do not overlap, the averages of those data sets are said to have a significant difference.

sample, and the method used for calculating the limits. When confidence intervals are used on a box graph one can determine whether there is a significant difference between the averages of multiple data sets (left). If the confidence intervals for two boxes do not overlap, the differences in the averages of data represented by the boxes is generally considered significant. Confidence intervals and variable width boxes can be used in conjunction with one another.

Box graph and point graph combined

When additional detail on the distribution of the data is desired, the actual data points are sometimes superimposed over the box symbol as shown at the right. In this way the viewer can see the summary information of the box plot as well as the distribution and clustering of the actual data points. Either dots or stripes can be used in this application.

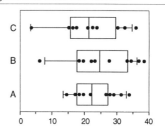

Illustration of how actual data points can be superimposed over box symbols

Original box plot designations

This diagram illustrates how key points were designated on the original box symbols.

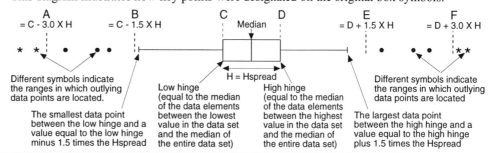

Break-Even Graph

A break-even graph is a special application of a grouped line graph. It is used to estimate when the total sales of a company equals the total costs of the company: the break-even point. The same concept and graph can also be applied to a product, a type of service, or any facet of a business where sales and fixed and variable costs can be identified and the variable costs calculated on a per unit basis. One of the main purposes in establishing the break-even point is to determine the volume of business a company must do to begin making a profit. In theory the graph is valuable and easy to use and understand. In practice it is not easy to apply with a high degree of accuracy due to the difficulty of accurately estimating and allocating each of the values. • Units (e.g., pieces, volume, time) are generally shown on the horizontal axis. Total sales dollars are plotted as one of the variables on the vertical axis. Total sales typically include such things as product mix, various discount schedules, returns, etc. The other variable plotted is total costs, both direct and indirect. The categories of fixed and variable may differ from application to application. Where the sales and total cost line intersect is considered the break-even point, beyond which the company, product, etc., being studied theoretically begins making a profit. In a theoretical break-even graph, the lines are generally drawn straight. In practice these lines might be straight, curved, or stepped. Two examples are shown at the right. The upper one is the most widely used version of the graph, though both yield the same result.

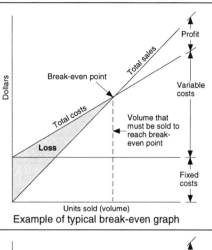
Example of typical break-even graph

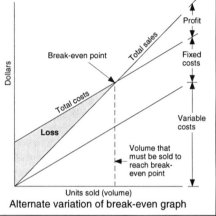
Alternate variation of break-even graph

Broken Line Graph

Sometimes referred to as a segmented line, fever, thermometer, or zigzag graph. A broken line graph is a variation of a line graph in which the data points are connected by straight lines as shown at the right.

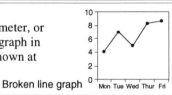
Broken line graph

Brushing

Sometimes referred to as querying. A technique used to relate data points, in a graph or map on a computer screen, with the corresponding entries in the spreadsheet from which the graphic was generated. When there are many data points on a graph or map, it is sometimes difficult to identify each of them individually and/or determine such things as similarities or relationships. With a computer and the proper software, this type of information can be obtained, even when there are large amounts of data. There are a number of different ways the process works, depending on the software. A few examples:

– With a special cursor, point to a specific data point or location and information regarding it appears on the computer screen beside the entity, or the entry is highlighted in the spreadsheet from which the graphic was generated.
– Enclose a group of data points on a graph or an area on a map with a special tool, and ask a question about what you have enclosed. For example, specific data about every data point in the enclosure; whether there is a hospital located in the geographic area; whether there are any companies included in the data points or area with more than 5,000 employees; etc. The information might appear on the map or graph, or be shown on a separate screen.
– Enclose a group of data points or an area with a special tool and have the computer generate a new graph or map using only the selected data points or area.
– Highlight specific entries in the spreadsheet, and the data points representing those entries are identified on the graph or map.
– Select two points on a map and the distance between the points appears on the screen.
– If a series of data points are plotted using two variables, all data points with a particular third variable can be highlighted, or only those data points with a third variable less than or greater than a particular value might be identified.

Bubble Graph

A bubble graph is a variation of a point or line graph where the data points (dots) have been replaced by circles (bubbles). The major advantage of a bubble graph versus a point or line graph is the ability to encode one or more additional variables by means of the bubble symbol. Bubble graphs might be two- or three-dimensional, as shown in the examples at the right.

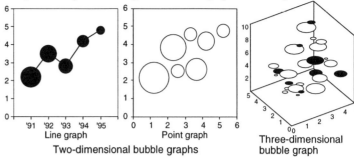

Two-dimensional bubble graphs — Line graph — Point graph

Three-dimensional bubble graph

Bubbles representing quantitative variables

When the variable is quantitative, either the diameter or area of each bubble is proportional to the value it represents. The two alternatives give somewhat different visual appearances, as shown in the comparison at the left. To enable the viewer to decode the quantitative information, a legend is provided, the values are shown on the bubbles, or the circles are coded and a cross reference is included. Examples of legends are shown at the right. • All positive or negative values can be encoded by means of bubbles – but not mixtures of the two.

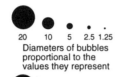

20 10 5 2.5 1.25
Diameters of bubbles proportional to the values they represent

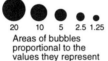

20 10 5 2.5 1.25
Areas of bubbles proportional to the values they represent

Comparison of bubbles bases on diameters and areas

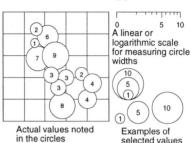

Actual values noted in the circles

A linear or logarithmic scale for measuring circle widths

Examples of selected values shown

Various ways the values of the bubbles might be indicated

Bubbles representing qualitative variables

When bubbles are used to convey qualitative or ordinal information, only a fixed number of different sizes of circles is generally required. In the example at the right, each bubble represents the sales/expense information for a different sales person. The sizes of the bubbles indicate what region the sales person is assigned to. Since there are four regions, there are four different bubble sizes. The same information could be encoded using four different colors of bubbles.

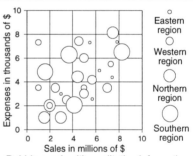

Eastern region
Western region
Northern region
Southern region

Bubble graph with qualitative information encoded into bubbles

Bubbles representing multiple variables

Multiple variables can be encoded by using pie charts as the bubbles. The pie charts might be all the same size or still another variable can be encoded by varying the sizes of the pie charts similar to the techniques discussed above.

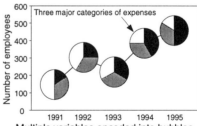

Three major categories of expenses

Multiple variables encoded into bubbles

General

When opaque bubbles are used, there is the risk of some of the circles being hidden. One solution is to make the bubbles transparent as shown at right.
• Transparent bubbles with dots at their centers are sometimes used in conjunction with grid lines to enable the viewer to more accurately determine the exact location of the data point. An example is shown at the left.

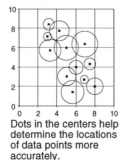

Dots in the centers help determine the locations of data points more accurately.

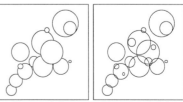

Opaque circles — Transparent circles

Transparent bubbles are sometimes used to assure that all bubbles are visible.

Buffer Map

A buffer is a region or zone around a particular point, line, or area that is selected for special attention or because it has particular characteristics. For example, if a new store is to be built, an analysis might be done to see how many potential customers live in a five-mile radius of the proposed new site (called a ring buffer). If one is studying crime along a highway, a buffer might be established to collect data on crimes committed one mile on either side of the highway (called a corridor buffer). In still other cases, the buffer zone might be irregularly shaped and defined by natural and/or man-made features, such as express ways, ravines, political boundaries, etc. See example at right.

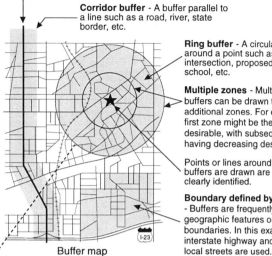

Line around which corridor buffer is drawn

Corridor buffer - A buffer parallel to a line such as a road, river, state border, etc.

Ring buffer - A circular buffer around a point such as an intersection, proposed store site, school, etc.

Multiple zones - Multiple concentric buffers can be drawn to establish additional zones. For example, the first zone might be the most desirable, with subsequent zones having decreasing desirability.

Points or lines around which buffers are drawn are normally clearly identified.

Boundary defined by map features - Buffers are frequently defined by geographic features or political boundaries. In this example an interstate highway and a series of local streets are used.

Buffer map

Multiple buffers can be used in conjunction with one another. There is no limit as to how many buffers can be used in the same map.

Buffers can be displayed within a larger map, as shown above, or as a stand-alone map, as shown at the left. It is common for buffers to be used inside of buffers. For example, if someone wants to open a liquor store within a one-mile radius of a given point they might draw a circle with a radius of one mile. If there is a restriction that liquor stores can not be located within one block of a school, smaller buffer zones might be drawn around each school, as shown at the left. • Buffers do not have to be symmetrical, can be shaded or colored, and sometimes include dimensions and statistical data on the map.

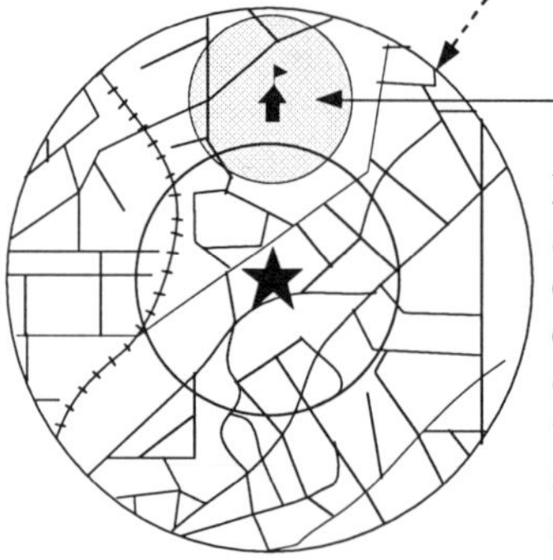

Stand alone buffer map with a smaller buffer zone drawn within it

---

Build Chart

Sometimes referred to as a reveal chart. One of a series of charts used to develop or present an overall message, idea, or concept. For example, if used in a presentation where only text is used to discuss a series of points, the first chart of the series would have only the first point on it. The next chart would have points one and two on it. The next chart would have points one, two, and three on it. This would continue until the subject was fully explained and all points had appeared. A similar technique can be used with graphs, maps, diagrams, etc.

Sales Strategy	Sales Strategy	Sales Strategy
✓ More sales calls	✓ More sales calls ✓ Bigger incentives	✓ More sales calls ✓ Bigger incentives ✓ More backup
Build chart #1	Build chart #2	Build chart #3

Build charts in which each chart includes the material from the previous chart, plus one or more additional points

---

Bullet

The circle, star, dot, or other symbol used before a sentence, phrase, or word to:

Examples of bullets. Bullets are a variation of dingbats.

- Clearly designate it as a separate item
- Encode additional information (e.g., time, optional, new, etc.)
- Assist in noting different levels of information (indents)
- Help the viewer identify transitions in the material

Bullet Text Chart

A variation of a text chart in which symbols (bullets) highlight the key thoughts or phrases. Dots are most widely used; however, any symbol is acceptable. In addition to highlighting individual points, different symbols can be used to encode information into the chart or to indicate different levels of indentation. Bullet charts are used extensively in presentations.

Picnic Agenda	Picnic Agenda	Picnic Agenda
• Adult activities • Cookout • Races • Table games • Youth activities	o Adult activities o Cookout • Races • Table games • Youth activities o Morning activity ● Afternoon activity	• Adult activities ◇Bingo ◇Dancing • Cookout • Races ◇Bag • Table games • Youth activities ◇Swimming
Example where the same symbol is used for the entire chart	Example where information is encoded into the symbols	Example where different symbols are used for indents

Variations of bullet charts

Business Graphs	Business graphs tend to fall into three major categories: those used for analysis and planning; those used for monitoring and controlling; and those used to communicate, inform, and instruct. A few examples of the first two categories are shown below. The standard graphs such as point, line, area, column, and bar are not included in the examples since they are used extensively in all areas.

- **Analyzing and planning** - A large percent the graphs used for these purposes are of the relational and comparative types. For example, at key times such as planning, the start or finish of major programs, periodic reviews, etc., many organizations analyze such things as how their performance compares with competition, how one department compares with another, how the causes of rejects compare, whether progress has been made, or how contributions relate to campaign expenses.

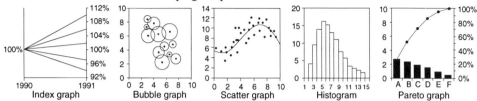

| Index graph | Bubble graph | Scatter graph | Histogram | Pareto graph |

- **Monitoring and controlling** - Most of these are of the time series type and are used to track things against plans, forecasts, expectations, previous performance, or specifications that may have been established during a previous analysis or planning phase.

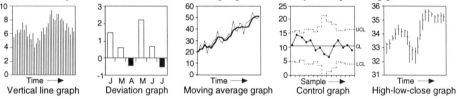

| Vertical line graph | Deviation graph | Moving average graph | Control graph | High-low-close graph |

- **Communicating, informing, and instructing** - There are no special types of graphs in this category, only variations of graphs from other categories with special emphasis on ease of understanding, focused message, and aesthetics. Included in this category are graphs used in presentations, sometimes referred to as presentation graphs. Although used by all disciplines, presentation graphs are frequently classified as business graphs.

Business Matrix

The basic grid of a business matrix typically consists of two to five rows and columns. The vertical and horizontal axes display variables which are often subjective, such as breadth of product line, industry attractiveness, or relative quality. There generally are a minimum of labels; simple terms such as low, medium, and high or relative values such as one to five are frequently used. Once the labels are designated, cells may be assigned terms or phrases that describe the things plotted (e.g., stars, dogs, cash cows, achievers, people-oriented) or suggest an action that might be taken (e.g., invest heavily, divest, analyze in more detail, put on fast track, deemphasize). The things being studied are generally represented by words, a simple symbol (dots and circles are most common) or small pie charts (for designating multiple variables). The sizes of the circles and pie charts might be uniform or proportional in size to some characteristic of the thing the circle represents. When circles and pie charts are used, the centers of the symbols designate the actual locations of the entities represented. Typical applications of business matrices are analyzing product potentials, planning acquisitions and divestitures, developing market strategies, competitive analyses, comparing financial performances, prioritizing projects, evaluating candidates, analyzing human characteristics, etc.

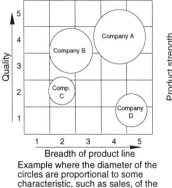

Example where the diameter of the circles are proportional to some characteristic, such as sales, of the companies they represent.

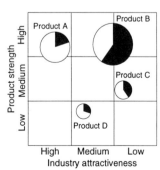

Example where two variables are encoded by means of pie charts. A third variable is encoded by the size of the pie charts.

Example where labels are shown in each quadrant to describe or categorize entities in that quadrant. In this example it is salespersons.

Three of the many variations possible with business matrixes

| **Cake Chart** | Sometimes referred to as a pie chart, divided circle, circular percentage chart, sector chart, circle diagram, sectogram, circle graph, or segmented chart. A cake chart consists of a circle divided into wedge-shaped segments. Each segment is the same percent of the total circle as the data element it represents is of the sum of all the data elements in its data set. The purpose of this type of chart is to show the relative sizes or values of components to one another and to the whole. See Pie Chart. |  Cake chart |

Calculation Chart/Tree

A calculation chart illustrates graphically how a given value is calculated. Instead of using typical mathematical symbols and notations, a calculation chart generally uses only words, boxes, and lines. Shown below is a chart for calculating return on assets, a widely used application of this type of chart.

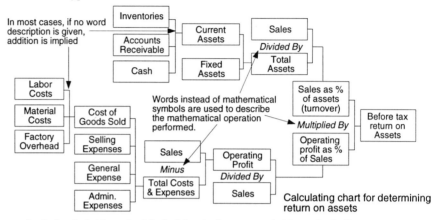

Calculating chart for determining return on assets

When a calculation is laid out in this fashion it does several things, including:
– Makes the calculation more understandable for those not familiar with mathematics
– Shows the elements that go into the final result
– Helps in explaining and trouble shooting variations in the final result
– Defines the areas, departments, indicators, etc., that impact the end result
– Shows by the use of multiple charts, how the same result can be arrived at by using different calculation methods.

The complexity of a calculation chart depends on its intended purpose. If, for example, the purpose of the above chart was to show the simplest way to calculate return on assets, the diagram might have had only three boxes – one each for operating profit, total assets, and

Simplest form of calculation chart for calculating return on assets.

return on assets (as shown at the left). If, on the other hand, its purpose was to show more of the elements that affect return on assets, it might have broken each of the categories down even further. Fixed assets might have been broken into land, facilities, equipment, and vehicles. Inventories might have been itemized as raw material, work-in-process, and finished goods. In still other cases the purpose of the chart might be to display the actual dollar value of each element to give the viewer a feel for the actual size of each element and its impact on the ultimate value. Various shaped enclosures, different colors, different type styles, etc., might be used to encode additional information such as the organization responsible, over or under budget, internally or externally controlled, value increasing or decreasing, etc. The diagrams might run horizontally or vertically and use arrows or plain lines. Data included may be actual values, planned values, specification values, or combinations of different types of data such as planned and actual values.

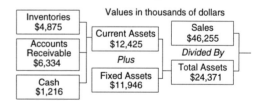

Example indicating how actual dollar amounts can be included in boxes along with headings

Calculation Graph

Sometimes referred to as a nomograph, nomogram, or alignment graph. A calculation graph is designed to solve an equation involving three or more variables. Graphs consist of three or more scales arranged so that a straight line crossing all the scales intersects the scales at values that satisfy an equation. The graph at the right satisfies the equation $A + B = C$ (e.g., 8 on scale A plus 5 on scale B equals 13 on scale C). See Nomograph.

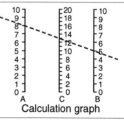

Calculation graph

Candlestick Chart

Sometimes referred to as a Japanese candlestick chart. Candlestick charts are used to record and analyze the selling prices of stocks, bonds, commodities, etc., with particular attention to indications of price reversals. They are similar to the bar charts used in the field of investment; the two are sometimes used in conjunction with one another. The chart might represent the stock price of a single entity (General Electric, IBM, McDonald's) or the average of multiple entities (all stocks in a particular exchange, selected industrial companies, selected utilities). Price is measured along the vertical axis, and uniform intervals of time progress along the horizontal axis from left to right. Time intervals range from minutes to years. Days, weeks, and months are among the most frequently used. One symbol represents each time interval. Vertical price scales can be shown on the left side, right side, or both. Because of the width of the charts, the fact that most charts include large amounts of data, and the fact that usually the greatest interest is in data on the right side of the chart, two vertical scales or a single scale on the right side are commonly used.

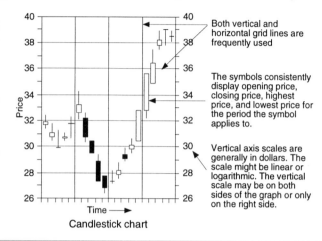

Both vertical and horizontal grid lines are frequently used

The symbols consistently display opening price, closing price, highest price, and lowest price for the period the symbol applies to.

Vertical axis scales are generally in dollars. The scale might be linear or logarithmic. The vertical scale may be on both sides of the graph or only on the right side.

Candlestick chart

Symbols used with candlestick charts

The basic candlestick symbol consists of a rectangle with a straight line extending from the top and bottom. If the closing price is lower than the opening price, the rectangle is one color, generally black (sometimes referred to as filled). If the closing price is higher than the opening price, the rectangle is another color, generally white (sometimes referred to as unfilled). A description of each part of the candlestick symbol is noted below. Definitions are based on tracking a particular stock but apply to any thing that might be plotted.

Closing price lower than opening price

Closing price higher than opening price

High - The top of the vertical line designates the highest price for which that particular stock sold for in that period of time. It is generally called the high. When there is no line, the high is designated by the top of the rectangle

Open - The top of the rectangle designates the opening value if the rectangle is filled

Close - The top of the rectangle designates the closing value if the rectangle is unfilled.

The filled body of the symbol is a strong graphical indicator that the closing price of the stock during that period was lower than the opening price.

Referred to as upper shadow or wick

Referred to as real body. In some cases two different colors are used instead of black and white. Typically the width of the rectangle has no significance. Occasionally, however, it is proportional to the volume of shares sold during that period.

The unfilled body of the symbol is a clear graphical indicator that the closing price of the stock during that period was higher than the opening price.

Referred to as lower shadow or wick

Close - The bottom of the rectangle designates the closing value if the rectangle is filled.

Low - The bottom of the vertical line designates the lowest price for which that particular stock sold for in that period of time. It is generally called the low. When there is no line, the low is designated by the bottom of the rectangle.

Open - The bottom of the rectangle designates the opening value if the rectangle is unfilled.

Shown below is a comparison of candlestick symbols with bar chart symbols for the same price relationships. Many individual candlestick symbols have names, as shown here. Most bar chart symbols do not have individual names.

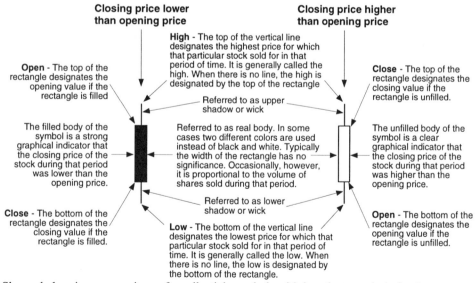

Price relationship	Close higher than open	Close lower than open	Close the same as high	Open the same as low	Open, low, and close all the same	Open the same as close	Open the same as low and close the same as high
Bar symbol configuration							
Candlestick configuration and name associated with it	Spinning top when real body is small		Shaved head	Shaved bottom	Grave stone	Doji	Close cut

A comparison of bar and candlestick symbols for similar sets of data

Comparison of bar and candlestick charts

Although bar and candlestick symbols both have unique features, each communicates the same basic four bits of data: open, close, low, and high. Bar charts tend to focus on highs and lows and changes in closing prices over time. Candlestick charts tends to make the relationships between opening and closing prices stand out more clearly and focus more on individual or groups of symbols. For purposes of comparison, the same data is plotted below on a candlestick chart and a bar chart .

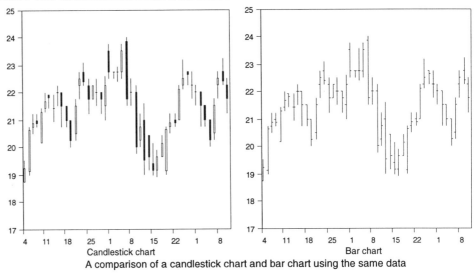

A comparison of a candlestick chart and bar chart using the same data

Formations and patterns

Patterns of symbols are used more frequently with candlestick charts than with bar charts and in some cases are considered as leading indicators of future price trends, particularly price reversals. Sometimes the patterns are the same in candlestick and bar charts even though they have different names. For example, a head-and-shoulders pattern on a bar chart is called a three-Buddha-top on a candlestick chart. In other cases the patterns are significantly different. In some situations, the location of the pattern with regards to surrounding data must be known to determine the name and significance of a symbol or a group of symbols. For example, a particular symbol might be called a hanging man symbol if located at the top of an uptrend, or a hammer if the same symbol is located at the bottom of a downtrend. A few of the patterns used with candlestick charts are shown below.

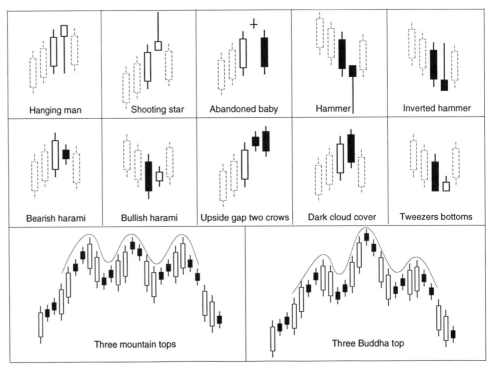

Examples of patterns used for analyzing candlestick charts

| **Candlestick Chart** (continued) | **Graphical methods to assist in the interpretation of candlestick charts** |

Since candlestick charts contain the same data as bar charts, most of the techniques used with bar charts can also be used with candlestick charts. For example, regression lines can be constructed, indicators can be superimposed or juxtaposed, and moving averages, trend lines, resistance lines, channels, etc., can be drawn. Shown at the right is a candlestick chart with a resistance line, several trend lines, and a channel to illustrated the point.

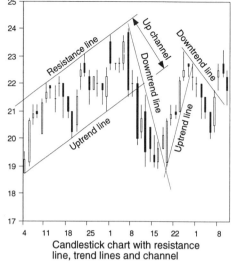

Candlestick chart with resistance line, trend lines and channel

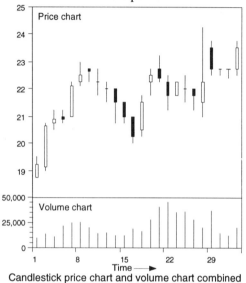

Candlestick price chart and volume chart combined

Scales, gird lines, insets, frames, and volume charts are handled on candlestick charts the same as they are on bar charts. An example of a volume chart used in conjunction with a candlestick chart is shown at the left. See Bar Chart ·

Caption

The term caption is sometimes used interchangeably with the terms title or label in all types of charts.

Cartesian Coordinates

Sometimes referred to as rectangular coordinates. Cartesian coordinates are the values by which data points are located on rectangular graphs. On two-dimensional graphs, two coordinates or numbers define the location of each data point. The two values are written one one behind the other with a comma between (e.g., 2,4 ; 5,1; 4,-3). The first number indicates how far the point is located from zero on the horizontal or X-axis. The second number indicates how far the point is located from zero on the vertical or Y-axis. Both positive and negative values can be used. With three-dimensional graphs the same procedure is used, except that there are three coordinates to designate the location along all three axes. The dashed lines in the example at the right are for illustrative purposes only. In an actual graph such lines would typically not be used.

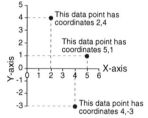

Examples of points located by means of Cartesian coordinates

Cartogram

In the broadest sense, cartogram refers to any map containing statistical data. In practice it usually refers to an abstract map, generally with a single purpose or theme. Such maps normally are simplified and shown in diagrammatic form to present a specific type of information. For example, if the key purpose of a subway map is to match terminal numbers with the towns in which they are located, unnecessary information such as miles between stations, the direction the train travels from point to point, etc., is not shown. Distances, sizes, or shapes are frequently distorted to make the key message easier to understand. Distorted maps, strip maps, and diagrammatic maps are sometimes classified as cartograms.

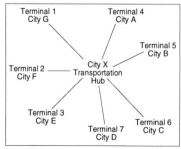

Cartogram matching city names with terminal numbers and showing general locations for a subway system

Cartoon Faces	Sometimes called faces or Chernoff faces. Cartoon faces are icons that are occasionally used as symbols to encode three or more variables into charts, graphs, maps, or icon comparison displays. The variables are encoded by assigning values, characteristics, etc., to variations in facial features. See Chernoff Faces.	

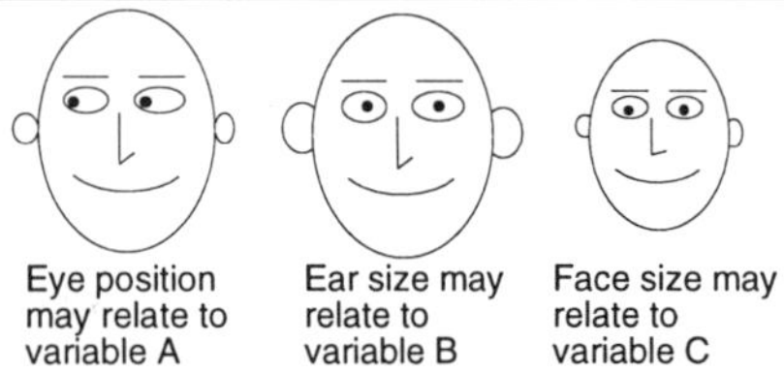

Eye position may relate to variable A Ear size may relate to variable B Face size may relate to variable C

Example of how facial features might be used to encode variable information

Casement Display

A casement display is a group of two or more two-dimensional graphs, typically shown side-by-side, that show the distribution of data points on multiple planes or slices passed through a three-dimensional scatter graph. The purpose of casement displays is to assist in the analysis of data with three variables by summarizing the data incrementally along one of the variables/axes. An example of a casement display is shown below. In addition, two graphs are shown at the right to illustrate how the casement display might have been arrived at graphically. In practice, the two preliminary graphical steps are not used. Instead, the information for plotting the casement display is determined mathematically and the casement display plotted directly. See Slice Graph.

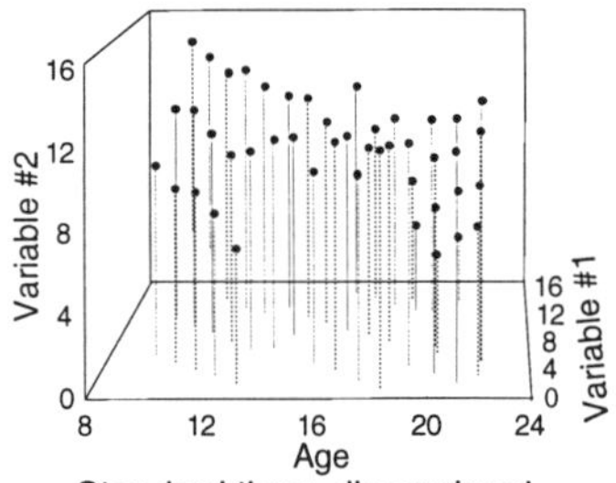

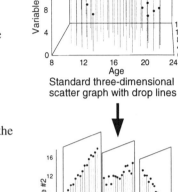

Standard three-dimensional scatter graph with drop lines

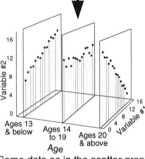

Same data as in the scatter graph above, except that the data points are condensed into three planes along the age axis

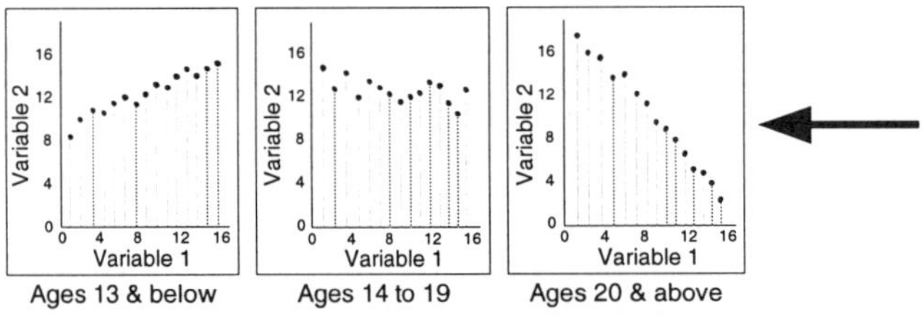

Casement display in which the three slices in the graph at the right have been turned parallel to the plane of the paper and placed side-by-side

Cash Flow Graph

Sometimes referred to as a product life cycle graph. A cash flow graph is a graph that projects and/or monitors key financial figures over the life cycle of a project, program, investment, etc. One of the key functions of this type of graph is to show the relative timing of key activities and their resulting financial impact. For instance, in the example shown here, it can be seen that this new hypothetical product does not become profitable until the end of the introduction phase, and the cash flow does not become positive (more money coming in than going out) until more than halfway into the growth phase. A companion graph may use the same time scale but show cumulative values for each of the items being tracked. The names and the lengths of the phases along the horizontal axis, as well as the specific items plotted, can differ depending on the specific application. In many cases, because of the relative sizes of the values, separate scales are required for some of the data series.

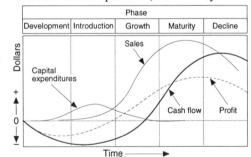

Cash flow graph for projecting and/or tracking key financial figures through the phases of a program

Category Axis

An axis that has a category scale on it is sometimes called a category axis.

Category Graph

A graph that has one or more category scales is sometimes called a category graph.

Category Scale

Sometimes referred to as a qualitative or nominal scale. This scale consists of a series of words and/or numbers that name, identify, or describe people, places, things, or events. The items on the scale do not have to be in any particular order. When numbers are used on a category scale, they are for identification purposes only, since category scales are not quantitative. Each word or number defines a distinct category which contains one or more entities. In other words, a category might refer to one man or to many men. See Scale.

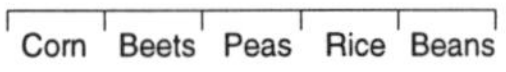

Corn Beets Peas Rice Beans

Category scale. Sometimes called qualitative or nominal scale.

Cause-and-Effect (CE) Diagram

Sometimes referred to as an Ishikawa Diagram, fishbone diagram, or characteristic diagram. The purpose of the cause-and-effect diagram is to provide a method for systematically reviewing all factors that might have an effect on or contribute to a given situation such as a quality problem or a cost reduction program. This is accomplished by assigning each potential contributor to a line or arrow (sometimes referred to as branch, bone, or category line) and then arranging the lines or arrows in a hierarchial fashion in meaningful clusters. For example, there might be four major categories of causes that are potentially contributing to a given problem. These categories might be equipment, people, parts, and processes. Each of these major categories would be assigned an arrow leading into a bold horizontal arrow (spine) that points towards a word description of the problem or objective. Next, a series of arrows are added indicating what secondary causes might be contributing to the primary causes. The arrows for these secondary causes lead into one of the four arrows representing the major categories. A third set of causes and arrows might then be added indicating what tertiary causes have an effect on the secondary causes. An example is shown below.

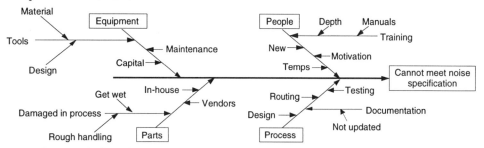

Cause-and-effect diagram using arrows where solving a quality problem is the objective

There are multiple ways to lay out a cause-and-effect diagram for exactly the same problem or objective. For example, in the diagram above, people and parts are two of the major categories. In an alternate diagram (below) the major departments or functions might be the major categories and people and parts would be secondary or tertiary factors in each of the categories. • In generating cause-and-effect diagrams, it is generally more important that all of the factors are shown and considered rather than how the diagram is constructed.

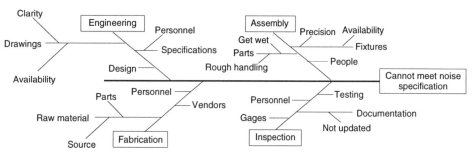

Same objective as above, except clusters have been arranged differently and without arrows

Cause-and-effect diagrams are basically qualitative tools. They sometimes are made partially quantitative by assigning specification values to the branches where a specification exists and giving some indication as to how well the specification is being met. By means of color, symbols, enclosures, and line size, a considerable amount of information can be encoded such as priorities, probability of having a significant effect, assignments, factors needing additional study, and factors for which solid quantitative information exists. Cause-and-effect diagrams are almost always constructed horizontally. Examples of terminology used with cause-and-effect diagrams are shown below.

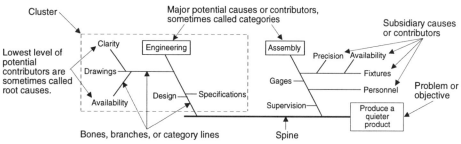

Terminology sometimes used with cause-and-effect diagrams

		Ceiling	

Ceiling

Sometimes referred to as a top. The term ceiling is used to identify the upper plane of a three-dimensional graph.

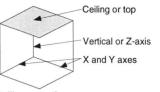

Ceiling on a three-dimensional graph

Cell[1]

The area of a table or chart defined by the intersection of a row and a column.

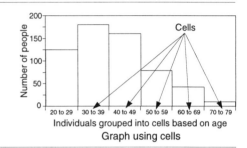

	1990	1991	1992	1993
Model A	19	18	15	13
Model B	2	5	6	4
Model C	5	6.5	7	6
Model D	8	10	9	12
Model E	13	17	12	12.5

A cell is the area formed by the intersection of a column and a row (shaded gray in the example).

Cell[2]

Sometimes referred to as a class, class interval, group interval, or bin. Cells are the groupings into which a set of data is divided for purposes of generating any of several data distribution graphs or maps. An example of a graph using cells is shown at the right. See Histogram and Frequency Polygon for a discussion of cells used with graphs. See Statistical Map for use with maps.

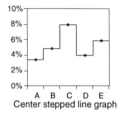

Individuals grouped into cells based on age
Graph using cells

Center Stepped Line Graph

A variation of stepped line graph in which the plot symbols are located in the center of the horizontal portions of the curve, as opposed to being located at the right, left, or at both ends. See also Line Graph.

Center stepped line graph

Change Graph

A variation of a difference graph that compares multiple factors at two points in time or under two different sets of conditions. Actual values are plotted. The numeric value of the change may or may not be shown on the data graphics. For example, if the values before and after a change were 40 and 60 respectively, the data graphic would extend from 40 to 60. The numeric change of 20 may or may not be noted on the graph. • With this type of graph, the direction of change or difference is generally considered important; therefore, a coding system is used to indicate the direction of change. Three ways of accomplishing this using both bar and column type graphs are shown below. • Since no particular direction is always favorable (desirable) or unfavorable (undesirable), an additional coding system is sometimes used to designate whether the change is favorable or not. Examples of information that might be plotted on a change graph are:

    – Changes in sales forecasts from one month to the next
    – Changes in unemployment from one period to the next
    – Changes in key business indicators from one period to another

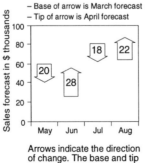

Arrows indicate the direction of change. The base and tip of the arrows indicate the prior and subsequent values, respectively.

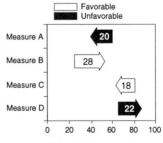

Color, shading, or patterns can be added to indicate whether the change was favorable (desirable) or unfavorable (undesirable).

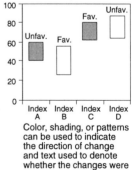

Color, shading, or patterns can be used to indicate the direction of change and text used to denote whether the changes were favorable or unfavorable.

Variations of change graphs

Characteristic Diagram

Sometimes referred to as a cause-and-effect diagram. See Cause-and-Effect Diagram.

Chart[1]

Sometimes referred to as an information graphic. A chart is a vehicle for consolidating and displaying information for purposes such as analysis, planning, monitoring, communicating, etc. Previously, charts were tangible things such as single sheets of paper, display boards, or flip charts. Today, many charts are generated and displayed electronically so that no hard copy ever exists. This book discusses charts without regard to the medium used. Instead, it discusses them in terms of content, function, format, etc. • There are five major categories of charts: graphs, maps, diagrams, tables, and other (those charts that do not fit into one of the other four categories). Each major category is broken into multiple subcategories, some of which are noted in the table below. • All individual information graphics can be included in multiple subcategories depending on the criteria that is used, such as shape, format, function, type of scales, type of data displayed, usage, number of axes, etc. For example, a widely used graph for plotting the distribution of data elements in a data set might be called as a histogram, data distribution graph, joined column graph, column graph, two-axis graph, two-dimensional graph, rectangular graph, quantitative graph, graph, or chart. All 10 terms are perfectly correct.

		Major Categories of Charts				
		Graph (plot)	Map	Diagram	Table	Other Charts
				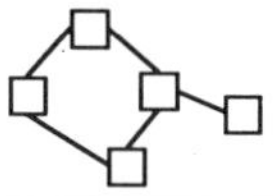	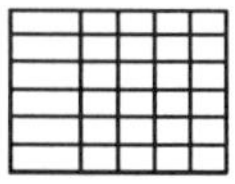	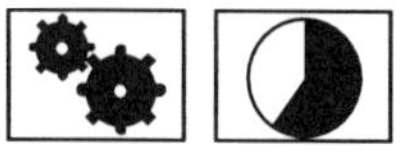
	Key features	Quantitative patterns and comparisons	Spatial and directional relationships	Nonquantitative interrelationships	Preciseness of information and ease of reference	Differs depending on specific chart
Primary Function		Examples are shown below				
Quantitative	**Shows patterns and/or relationships of quantitative data at point in time**	Area graph Bar graph Circle graph Column graph Line graph Nomograph Polar graph Radar graph Scatter graph Trilinear graph	Contour map Demographic map Distorted map Elevation map Prism map Shaded map Smooth statistical map Weather map		Analytical table Bidirectional table General table Quantitative table Reference table Spreadsheet	Icon comparison display Pie chart Proportional chart Ranking chart Unit chart Venn diagram
Quantitative	**Shows patterns and/or relationships of quantitative data over time**	Area graph Candlestick chart Cash flow graph Column graph Control chart Index graph Line graph Run chart			Analytical table General table Quantitative table Reference table Spreadsheet	Comparative chart Icon comparison display Point & figure chart Proportional chart Stair chart Unit chart
Quantitative & Nonquantitative	**Shows how/where things are distributed or located**	100% graph Border plot Box plot Histogram Pareto graph Population pyramid Probability graph Quantile graph	Block map Blot or patch map Dot density map Geological map Pin map Profile map Topographic map Weather map	Block diagram Network diagram Voroni diagram	Analytical table Frequency table General table Percent table Quantitative table Reference table Spreadsheet	Business matrix Conceptual chart Floor plan
Nonquantitative	**Relates time and activities**			CPM chart PERT chart Time line	Calendar Time table	Explanatory chart Gantt chart Loading chart Milestone chart Scheduling chart
Nonquantitative	**Shows how nonquantitative things are organized, arranged, interconnected or interrelated**		Network map Ray map Road map Strip map Thematic map	Schematic: Block diagram Cause & effect diag. Conceptual diagram Flow chart Network diagram Organization chart Relational diagram Venn diagram	General table Pictorial table Reference table	Conceptual chart Dendrogram Distribution channel chart Exploded diagram Gantt & milestone charts Minimum spanning Process chart Structure diagram
Nonquantitative	**Shows how nonquantitative things proceed**		Flow map Weather map	Schematic: Conceptual diagram Decision chart Flow chart PERT & CPM charts Process chart Tree chart	Analytical table General table Reference table Spreadsheet	Conceptual chart Gantt & milestone charts Illustration chart Process chart Vector chart
Nonquantitative	**Shows how nonquantitative things evolve or work**			Schematic: Conceptual diagram Flow chart Process chart	Analysis table General table Pictorial table Reference table	Conceptual chart Cross-section Exploded diagram Illustration chart Pictorial chart
Nonquantitative	**Shows how to do things**			Calculation chart "How to" diagram Procedural diagram		Calculation chart How-to chart Illustration chart Pictorial instruction

Chart[2]	A map used for navigation either by water (hydrographic map) or by air (aeronautical map).

Chart Element

Any graphic ingredient of a chart might be called a chart element. Included in the definition are major elements such as graphs, maps, and diagrams as well as minor elements such as notes, arrows, images, and symbols. Sometimes a smaller element included within a larger element (a symbol on a map or a legend on a graph) will be considered a map element or a graph element, and the larger graphic, (the map or graph) considered the chart element.

Chartjunk

Sometimes called clutter. Chartjunk consists of the things included in charts and graphs that are not essential to understanding the chart or that might detract from the main purpose of the chart. Examples are unnecessary grid lines, excessive tick marks, redundant data, material added for cosmetic purposes, etc. What qualifies as chartjunk is sometimes quite subjective. The two graphs at the right show the same data with and without what some might consider chartjunk.

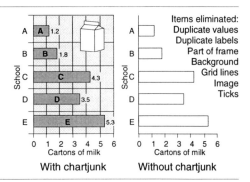

Chernoff Faces

Sometimes called faces or cartoon faces. Chernoff faces provide a graphical technique for encoding multivariate information (generally three or more variables) into the facial features of small icons so that the viewer gets an overview of the data based on facial

Eye position might relate to variable A
Ear size might relate to variable B
Face size might relate to variable C
Examples of how information is encoded by faces

expressions. For example, the position of the eyes might be proportional to one variable (e.g., looking to the left indicates a large number, to the right a small number, and straight ahead an average number). The size of the ears might be proportional to another variable (e.g., big ears, 100,000; little ears, 1,000). Quantitative and nonquantitative information can be encoded. Chernoff faces may be used as stand-alone images, in matrixes called icon comparison displays, or in conjunction with another type of chart such as a map or graph. For example, the financial performance of four companies might be compared by showing four faces, one for each company. Each company's financial data would be encoded into the features of the face representing that company. The four faces could be used by themselves or as symbols on a graph (right). The map at

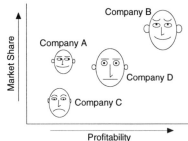

Chernoff faces as symbols on a graph

the left shows another variation in which faces represent data for seven different states. • Information can be encoded into the faces randomly, in which case the decoding process can be tedious. If; however, during the encoding process, care is taken to assure that positive values are associated with the more pleasant facial features, the viewer can get a general feel for each entity based on how happy or sad the expression is.

Chernoff faces used to encode information onto a map

Choropleth Map

Sometimes referred to as a shaded, crosshatched, or textured map. A choropleth map is a variation of a statistical map that displays area data by means of shading, color, or patterns. The areas (sometimes called areal units) might be countries, states, territories, counties, zip codes, trading areas, etc. The data is generally in terms of ratios, percents or rates as opposed to absolute units. For example, incomes would typically be given in terms of dollars per capita or dollars per household as opposed to total dollars for the area. Data is often organized into class intervals as shown in the example at the right. See Statistical Map.

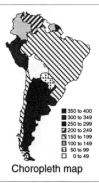

Choropleth map

Chronological Scale	Sometimes referred to as a time series scale. See Time Series Scale.

Circle Diagram **or** **Circle Graph**[1]	Sometimes referred to as a pie chart, cake chart, divided circle, circular percentage chart, sector chart, sectogram, or segmented chart. A circle diagram, a variation of a proportional area chart, consists of a circle divided into wedge shaped segments. Each segment is the same percent of the total circle as the data element it represents is of the sum of all the data elements in its data set. The purpose of this type of chart is to show the relative sizes of components to one another and to the whole. See Pie Chart.

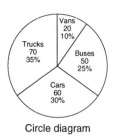

Circle diagram

Circle Graph[2]	Sometimes referred to as a circular percentage graph. A circle graph looks and functions like a pie chart. The only difference is that the circle graph has a scale around the circumference which classifies it as a graph. A circle graph has no radial scale and therefore is considered a one-axis graph. Typical characteristics of the circular scale are: – It is almost always linear. – The units of measure can be anything but usually are percent. – It can have any lower and upper value. Typically the lower value is zero and the upper value is 100%. – Zero and 100% are frequently located at the top of the circle. – Values can proceed in either direction; however, they typically increase clockwise. – Tick marks can be internal or external; external is typical. Many of the observations that apply to pie charts also apply to circle graphs. See Pie Chart.

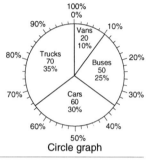

Circle graph

Circular Area Graph	A circular area graph is the equivalent of a rectangular area graph wrapped into a circle. The horizontal axis of the rectangular graph becomes the circular axis of the circular graph and the vertical axis of the rectangular graph becomes the radius axis of the circular graph. Sequential, particularly time series scales are frequently used on the circular axis. The quantitative value being measured is shown on the radius or value axis. The example at the right represents a typical application of a circular area graph in which average electric power usage is recorded for a 24-hour period.

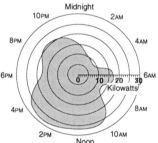

Circular area graph showing average use of electric power over the course of a day

──────**Major elements of a circular area graph**──────

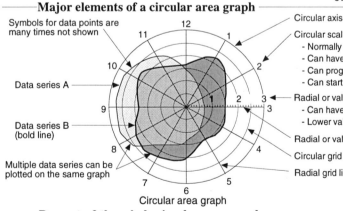

Circular area graph

──────**Percent-of-the-whole circular area graph**──────

A circular area graph can be used as a percent-of-the-whole or 100% graph. When used in this manner, the viewer can see how the relationships between the elements of a data series change over time – for example, the relative percent that each product line contributes to the overall sales of a company as the seasons change. An illustration of such an application is shown at the right.

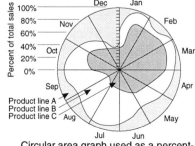

Circular area graph used as a percent-of-the-whole or 100% graph

Circular Bar Graph

A circular bar graph is the equivalent of a rectangular bar graph wrapped into a circle. The horizontal axis of the rectangular graph becomes the circular or value axis of the circular graph, and the vertical axis of the rectangular graph becomes the radial or category axis of the circular graph. An example is shown at the right. In a typical circular bar graph, the bars representing the various categories are uniformly spaced along the radius or category axis. The ends of the bars designate the

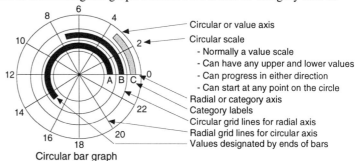

Circular or value axis
Circular scale
 - Normally a value scale
 - Can have any upper and lower values
 - Can progress in either direction
 - Can start at any point on the circle
Radial or category axis
Category labels
Circular grid lines for radial axis
Radial grid lines for circular axis
Values designated by ends of bars

Circular bar graph

values they represent. The lengths of the bars have no significance and can be misleading, since bars representing the same numerical value can be different lengths depending on where they are located radially. • The value scale on a circular bar graph can have any upper value. The lower value is typically zero. The scale can progress in either direction and start at any point around the circle. Although simple circular bar graphs are generally the only type used, the same concept can be applied to more complicated variations such as stacked, grouped, and 100%. • Occasionally a time series is shown on the circumference and the graph is used to display a repetitive type schedule, as shown at the left. This configuration might be called a scheduling chart. • Circular bar graphs are generally used for their aesthetic value. Functionally they offer little or no advantage over rectangular bar graphs.

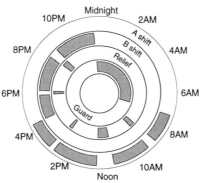

A specialized circular bar graph with a time series shown on the circumference is sometimes called a scheduling or circular Gantt chart.

Circular Column Graph

Sometimes referred to as a star, radial line, or radial column graph. A circular column graph is the equivalent of a rectangular column graph wrapped into a circle. The horizontal axis of the rectangular graph becomes the circular axis of the circular graph, and the vertical axis of the rectangular graph becomes the radial or value axis of the circular graph. An example is shown at the right. The value scale can have any upper and lower values. It typically has its lower value at the center and its upper

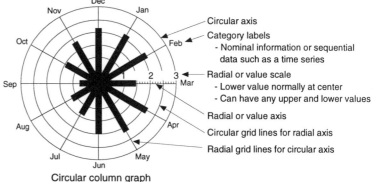

Circular axis
Category labels
 - Nominal information or sequential data such as a time series
Radial or value scale
 - Lower value normally at center
 - Can have any upper and lower values
Radial or value axis
Circular grid lines for radial axis
Radial grid lines for circular axis

Circular column graph

value at the circumference. The value scale might be linear or logarithmic, but is normally linear. • Category or sequence scales are most frequently used on the circular axis. Circular scales can start at any place on the circle and proceed in either direction. • A single data series is most frequently plotted. More complex variations such as stacked, grouped, range, etc., are occasionally used. Two examples are shown at the right. • When the columns become so narrow that they resemble lines, the graph is sometimes called a radial line graph.

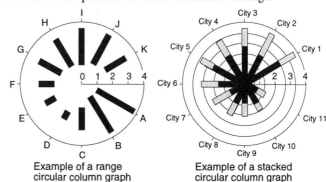

Example of a range circular column graph

Example of a stacked circular column graph

Circular Graphs

Circular graphs are used for a number of different reasons, depending on the type of data being plotted, the purpose of the graph, and the type of information to be communicated. Following is a list of some features of circular graphs. All features are not applicable to all types of circular graphs.

- Directional information can be encoded more easily by means of vectors or a circular axis with a scale of degrees.
- When recurring data is plotted on a circular graph, the continuity of information is not interrupted as it would be between the right- and left-hand edges of a rectangular graph.
- When recurring data is plotted, circular graphs make it easier to compare the values at a given time in each cycle.
- In some cases the circular graphs are more compact.
- In some cases viewers find it easier to relate to a circle as representing the "whole".
- Circular graphs more clearly indicate the repetitive nature of certain data.
- Circular graphs sometimes have a more interesting appearance.

On the negative side, some people find it more difficult to relate to and understand circular graphs; they generally are more difficult to generate; and they sometimes require more space. The examples shown below are representative of the major types of circular graphs. Details for each are listed under their individual headings.

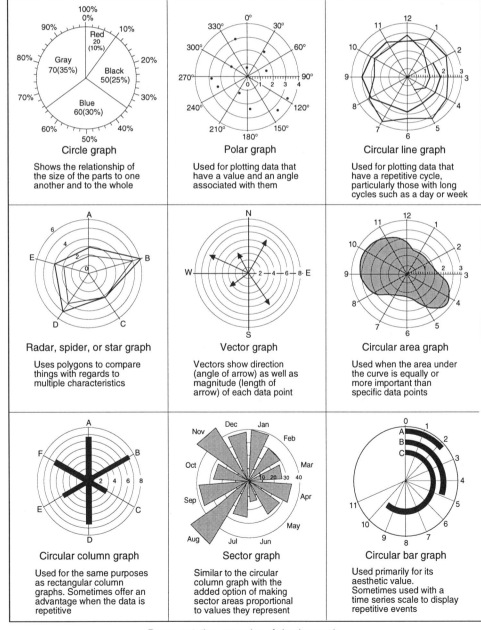

Circle graph

Shows the relationship of the size of the parts to one another and to the whole

Polar graph

Used for plotting data that have a value and an angle associated with them

Circular line graph

Used for plotting data that have a repetitive cycle, particularly those with long cycles such as a day or week

Radar, spider, or star graph

Uses polygons to compare things with regards to multiple characteristics

Vector graph

Vectors show direction (angle of arrow) as well as magnitude (length of arrow) of each data point

Circular area graph

Used when the area under the curve is equally or more important than specific data points

Circular column graph

Used for the same purposes as rectangular column graphs. Sometimes offer an advantage when the data is repetitive

Sector graph

Similar to the circular column graph with the added option of making sector areas proportional to values they represent

Circular bar graph

Used primarily for its aesthetic value. Sometimes used with a time series scale to display repetitive events

Representative examples of circular graphs

Circular Line Graph

A circular line graph is the equivalent of a rectangular line graph wrapped into a circle. The horizontal axis of the rectangular graph becomes the circular axis of the circular graph, and the vertical axis of the rectangular graph becomes the radial axis of the circular graph. Sequential scales, particularly time series, are often used on the circular axis. The quantitative value being measured is shown on the radial or value axis. The example at the right represents a typical application of a circular line graph in which pollution levels are recorded monthly.

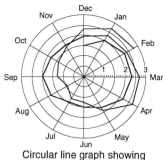

Circular line graph showing monthly pollution levels

Major elements of a circular line graph

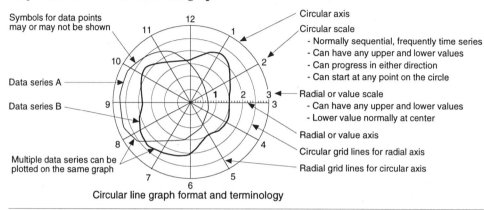

Circular line graph format and terminology

Superimposing repetitive information

It is common when plotting repetitive data to have successive periods or cycles plotted over top of one another. If, for instance, the temperature is recorded every two hours for four days, all four days of data might be superimposed over each other. By doing this, it is sometimes easier to spot similarities and dissimilarities at the same time during each cycle. In the example at the right, the data graphic clearly shows that at 4 PM each day, the temperature was exactly the same, while at 10PM the spread was about four degrees.

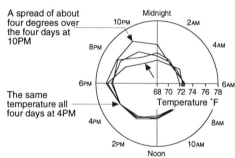

Four days of temperatures recorded on a circular line graph

Circular Organization Chart

Sometimes referred to as a radial organization chart. A circular organization chart is typically a hierarchical organization chart that has the highest ranking person or entity in the center and lower ranking persons or entities arranged in descending order as their names, positions, etc., appear further from the center of the chart. The names, descriptions, etc., of the lowest ranking entities form the circumference of the circular chart. See Organization Chart.

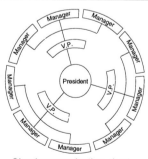

Circular organization chart

Circular Percentage Chart

Sometimes referred to as a pie chart, cake chart, divided circle, sector chart, circle diagram, sectogram, circle graph, or segmented chart. A circular percentage chart, a variation of a proportional area chart, consists of a circle divided into wedge-shaped segments. Each segment is the same percent of the total circle as the data element it represents is of the sum of all the data elements in its data set. The purpose of this type of chart is to show the relative sizes of components to one another and to the whole. See Pie Chart.

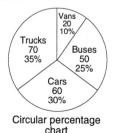

Circular percentage chart

Circular Percentage Graph

Sometimes called a circle graph. A circular percentage graph looks and functions like a pie chart, except that it has a quantitative scale around the circumference that ranges from zero to 100%. See Circle Graph [2].

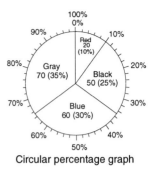

Circular percentage graph

Class or Class Interval

Sometimes referred to as a bin, group interval, or cell. When a large set of data is systematically divided into a limited number of groups, these groups are many times called classes or class intervals. Class intervals are frequently used for analyzing the distribution of data. For example, one might want to make a graph of the distribution of ages in an organization with 600 people. Instead of plotting all 600 ages, class intervals could be set up, the number of people in each class interval determined, and that set of numbers plotted. If the youngest person in the

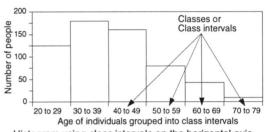

Histogram using class intervals on the horizontal axis.

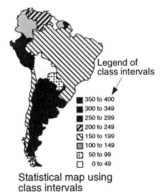

Statistical map using class intervals

group was 22, the first class interval might be 20 to 29, the next 30 to 39, the next 40 to 49 and so on until a class interval that included the oldest person in the group was established. The number of people in each class interval would be counted and the data (in this example six data points) plotted. The graph might look like the example above. Such a graph is generally called a histogram. See Histogram. • A similar technique is sometimes used when preparing data for statistical maps, as shown at the left. Class intervals for use with maps are sometimes determined differently than for graphs; see Statistical Map.

Classless Map

Sometimes referred to as a no class or unclassed map. A map that does not use class intervals to encode statistical information. See Statistical Map.

Clock Graph

A circular graph with a time scale on the circular axis that progresses in the clockwise direction. The type of graph might be line, point, bar, column, or area. The circular line graph as shown at the right is a typical example.

Clock graph

Cloud

The term cloud is sometimes used when referring to the cluster of data points on a scatter graph. The term can be applied to two- or three-dimensional graphs. It is most frequently used with three-dimensional graphs.

Sometimes referred to as clouds

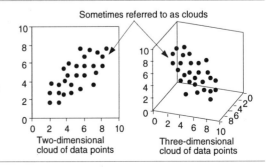

Two-dimensional cloud of data points

Three-dimensional cloud of data points

77

Clustered Bar Graph	Sometimes called a grouped, multiple, or side-by-side bar graph. A bar graph on which two or more data series are plotted side-by-side to form clusters. The bars for a given data series are always in the same position in each cluster throughout a given graph. Each data series typically is a different color, shade, or pattern. See Bar Graph.	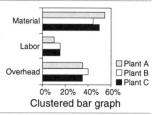 Clustered bar graph
Clustered Box Graph	Sometimes referred to as a grouped box graph. A clustered box graph is a box graph on which two or more data series are plotted side-by-side. Each data series typically has a different color, shade, or pattern. See Box Graph.	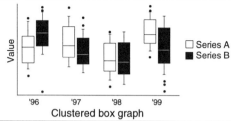 Clustered box graph
Clustered Column Graph	Sometimes called a grouped, multiple, or side-by-side column graph. A column graph on which two or more data series are plotted side-by-side to form clusters. The columns for a given data series are always in the same position in each cluster throughout a given graph. Each data series typically is a different color, shade or pattern. See Column Graph.	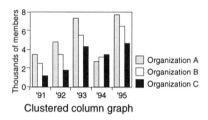 Clustered column graph
Clustered/Grouped Graphs	Using groups or clusters of two or more graphs on the same or related subject often offers synergistic advantages over the use of a single graph. See Graph.	
Cluster Map	Sometimes referred to as a tree diagram, linkage tree, or dendrogram. A cluster map is a graphical means of organizing information for the purpose of establishing groupings and/or categorizing individual elements. For example, one might study different groups of consumers to determine how they perceive a product, or a cluster map might be used to look for relationships in a group of archeological specimens. See Dendrogram.	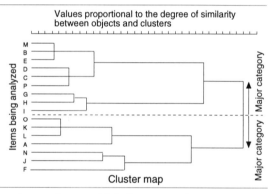 Cluster map
Coffee Grinder	A slang expression for certain vertical flow charts where multiple inputs are shown at the top of the chart and only one or at most a few outputs are shown at the bottom.	
Color	Color is a powerful tool to use with information graphics. In many of the examples throughout this book, color could be substituted where shades of gray are used to show varying values or where differently shaped symbols or line patterns are used to identify different data series. In some cases entirely different colors are used. In other cases shades of the same color are used for the same purpose. Noted below are some of the many potential applications of color with charts and graphs.	

Potential applications for color in information graphics

- Differentiate various data elements such as multiple data series on a graph, types of natural resources on a map, types of paths on a schematic, etc.
- Encode areas of equal value such as filled contour graphs and quantitative maps.
- Alert the viewer when a predetermined condition occurs such as values going negative, measurements exceeding limits, or unfavorable results of any kind.
- Provide emphasis to key elements on all types of charts.
- Identify particular values, things, places, actions, etc.
- Indicate all items, values, organizations, etc., that are the same or similar by making them all the same color.
- Signify changes in directions, time periods, trends, responsibilities, type of condition, etc.
- Improve the appearance of charts. This might involve any or all of the chart elements such as data graphics, backgrounds, borders, illustrations, etc.
- Get and hold the viewer's attention.
- Improve the viewer's retention of the information presented on the chart.

Color (continued)

- Encode information into text – for example, the use of red for unfavorable values, color coding labels for easy identification with their data graphics, differentiating miles and kilometers on maps.
- Distinguish lines such as grid lines and reference lines from data lines.
- By means of gradations, indicate gradual transitions from one set of conditions to another or emphasize the direction of flow of activities or data on diagrams and graphs.
- By means of faint backgrounds, organize information and identify such things as time periods, area of responsibility, new or old, desirable or undesirable, etc.
- Give more meaning and impact to pictures, illustrations, and images.
- Color-code similar information throughout a series of charts and graphs so the viewer can easily follow a thread of related information or ideas.
- By means of stripes of color, assist the viewer in visually following long columns or rows of data.
- On maps, differentiate such things as land, water, vegetation, etc.
- Flag end-of-month or end-of-year summaries by using different color text for lines on entire charts, graphs, or tables.

———— **Terminology** ————

The following terms are frequently used when specifying or describing colors.

- **Hue** - This is the name of the color such as red, yellow, blue, green, violet, etc.
- **Saturation or chroma** - These terms refer to the vividness or intensity of the color. The higher the saturation, the brighter, more vivid the color. The lower the saturation, the duller the color. Very weak or low saturation colors approach a neutral gray.
- **Value, lightness, or brightness** - Used when describing how light or dark a color or hue is. This is sometimes broken down into **shade** or darkness which is the result of adding black to the basic color, and **tint** which is the result of adding white to the basic color.

Column

The term column is used to identify at least three major items in the area of information graphics.

1. The vertical rectangle used with graphs, maps, and diagrams.

2. The vertical grouping of words used with text, as found in newspapers.

3. The vertical arrangements of information used in tables.

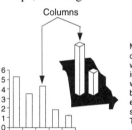

Columns

Many times large bodies of text, particularly on wide pages, are broken into multiple columns which are easier to read because the reader's eyes do not have to keep shifting back and forth. This paragraph has been

arranged in two columns for illustrative purposes only. In an actual application this small, the text might be presented in a single column. Text arranged into columns may or may not be justified left as these are.

Columns

	1990	1991	1992
Model A	19	18	15
Model B	2	5	6
Model C	5	6.5	7
Model D	8	10	9
Model E	13	17	12

Column and Symbol Graph

Sometimes a series of symbols are shown on a column graph to designate reference points or points of comparison for each of the columns. The reference points might designate such things as budget, plan, last year's value, or specification value. Symbols enable the viewer to make comparisons between the actual value, as shown by the top of the columns, and some other set of values designated by the symbols. Any symbol that crisply identifies the reference value can be used. More than one set of symbols can be used on the same graph; however, only one set of rectangles is typically used.

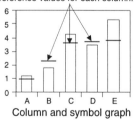

Symbols typically designate individual reference values for each column.

Column and symbol graph

Column Chart

Column charts are members of the proportional chart family. In a column chart, the size or area of each column is proportional to the value it represents. Column charts do not have scales, grid lines, or tick marks. The value they represent is typically shown on or adjacent to the data graphic. The major purpose of the graphical portion of a column chart is to visually orient the viewer to the relative sizes of the various elements of a data series. Actual values are shown on or adjacent to the data graphics. With a quantitative scale on the vertical axis, the graphic is often called a column graph.

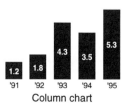

Column chart

Column Graph

Sometimes referred to as a bar, vertical bar, or rotated bar graph. Column graphs are a family of graphs that display quantitative information by means of a series of vertical rectangles. Column graphs are frequently used to compare multiple entities or to show how one or more entities vary over time. An example is shown below.

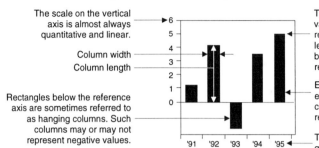

The scale on the vertical axis is almost always quantitative and linear.

Column width
Column length

Rectangles below the reference axis are sometimes referred to as hanging columns. Such columns may or may not represent negative values.

The top of the column designates the value of the data element being represented by the column. The length of the column may or may not be proportional to the value it represents.

Each column represents a data element of a data series; a complete set of columns represents a data series.

The scale on the horizontal axis is generally sequential or categorical.

Column graph and terminology

Each column represents a data element, and a complete set of columns represents a data series. The top of each column is located at the value it represents. The length and/or area of the columns may or may not be proportional to the values they represents. Because the tops of a columns are so pronounced, this type of graph is one of the best for showing specific values. Because of the stand-alone nature of the columns, it is also well suited for representing discrete data. Column graphs normally have one linear quantitative scale on the vertical axis and generally have a category or sequence scale (e.g., time series) on the horizontal axis. The few exceptions to this are discussed below.

Simple column graph

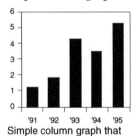

Simple column graph that displays a single data series

When a column graph displays a single data series similar to the example at the left, it is referred to as a simple column graph. The columns can be of any width; however, with a few exceptions (to be reviewed later), columns are uniform in width throughout a given graph. The same is true for the spaces between the columns. The vertical scale is always quantitative, and both positive and negative values can be plotted on it, as shown in the example at the right. Linear scales are almost always used. When both positive and negative values are plotted, the result is sometimes called an over-under or deviation graph. On a typical column graph, the horizontal axis has either a category or sequence (e.g., time series) scale.

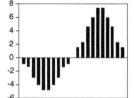

Both positive and negative values can be plotted on the vertical quantitative scale.

Vertical line graph

When columns are so narrow they approximate vertical lines, as shown at the right, the graph is frequently referred to as spike, needle, vertical line, or point graph with drop lines. With this type of graph, quantitative data can be plotted on the horizontal axis, since each line (column) can be assigned to a specific value. Vertical line graphs are reviewed under Line Graph.

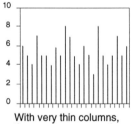

With very thin columns, the graphs are sometimes called vertical line graphs.

Joined column graph

When the columns become so wide there is no space between them, as shown at the right, the graph is sometimes called a joined column graph, connected column graph, stepped column graph, or histogram. A quantitative scale is sometimes used on the horizontal axis, in which case the column widths indicate the spread or interval of values on the horizontal axis that the value on the vertical axis applies to. Histograms and area column graphs are examples of this type of graph. Area column graphs are discussed later in this section. See also Histogram.

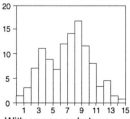

With no spaces between columns, the graph is sometimes called a histogram or a joined, connected, or stepped column graph.

Column Graph (continued)

Grouped column graph – (multiple data series all referenced from the same zero axis)

When two or more data series are plotted side-by-side on the same column graph, the result is referred to as a clustered, grouped, multiple, or side-by-side column graph. Grouped column graphs enable the user to:

    – Compare multiple items at various points in time

    – Show how relationships between multiple items change with time

    – Look for correlations or meaningful relationships between multiple data series

    – Condense onto one graph what otherwise would required multiple graphs

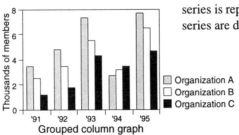

Grouped column graph

• With this type of graph, each data element of each data series is represented by a separate column. The data series are differentiated from one another by assigning a different color, shade, or pattern to each data series (left). The columns from each data series that correspond to the first interval on the horizontal axis are placed side-by-side to form a group or cluster in the center of the first interval. The columns from each data series that correspond to the second interval on the horizontal axis are placed side-by-side in the center of the second interval. This placement continues until all of the columns representing data elements are entered on the graph. To illustrate the procedure, if a comparison was made of the number of members in three different organizations for the years 1991 to 1995, columns representing the number of members for each organization in 1991 would be grouped together in the first interval along the horizontal axis. Columns representing the number of members for each organization in 1992 would be grouped in the second interval, and so forth through 1995. A space is almost always provided between each group of columns so that the·data associated with each interval can be easily distinguished. The spaces used in a given graph can be of any width. They are generally uniform throughout the graph. The example below shows some of the design features and terminology used with grouped column graphs.

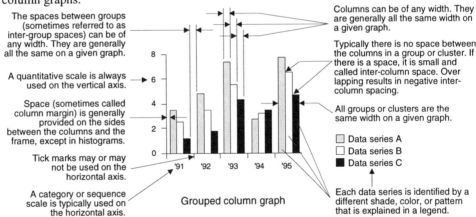

Grouped column graph

Technically there is no limit as to the number of data series that can be plotted on a single graph. Practically, if the number goes above three or four, the graph can become confusing. Most variations of grouped column graphs can accommodate both plus and minus values on the vertical axis, as shown in the examples below.

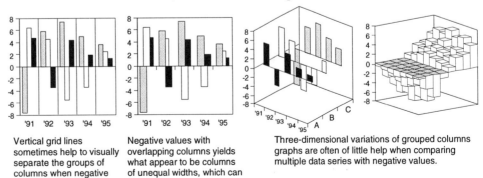

Vertical grid lines sometimes help to visually separate the groups of columns when negative values are present.

Negative values with overlapping columns yields what appear to be columns of unequal widths, which can be distracting.

Three-dimensional variations of grouped columns graphs are often of little help when comparing multiple data series with negative values.

Grouped column graphs displaying both positive and negative values

Overlapped and three-dimensional grouped column graph

Several techniques have been developed in an effort to make grouped column graphs easier to read, to enable them to convey additional information, and to make them more attractive. Shown below are several of these techniques.

Grouped column graph for reference

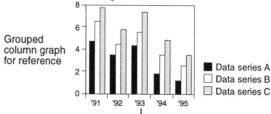

One technique that is widely used is to overlap the columns. In this method the columns representing complete data series are shifted sideways such that the columns of each successive data series are partially hidden by the columns of the data series in front of them. The columns can be shifted to the right or the left. They might be shifted by any amount ranging from barely to completely overlapped (100%). The two examples below show a 50% and a 100% overlap. The data series with the shortest columns is normally positioned in the front so as not to be hidden by the taller columns.

50% overlap   100% overlap

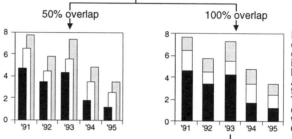

In this example, each column overlaps the column immediately behind it by 50%. As the amount of overlap increases, the width of the individual columns generally increases if the graph remains the same width.

Each column overlaps the column immediately behind it by 100%. All columns are the same width. This variation is easily confused with a stacked column graph.

When the columns are 100% overlapped, the graph resembles a stacked column graph. A special notation is generally included to alert the viewer, or additional graphical changes are made to assure no misunderstanding. Two examples are shown below.

Three-dimensional   Multiwidth columns

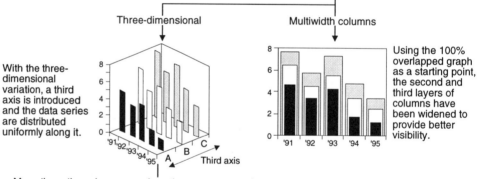

With the three-dimensional variation, a third axis is introduced and the data series are distributed uniformly along it.

Using the 100% overlapped graph as a starting point, the second and third layers of columns have been widened to provide better visibility.

Many times the columns are given the appearance of depth. Generally this is done for cosmetic purposes. In a few cases it is done because of the nature of the data. Frequently the appearance of depth is achieved at a sacrifice in accuracy of decoding.

Space between columns   No space between columns

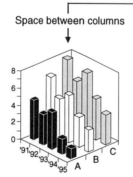

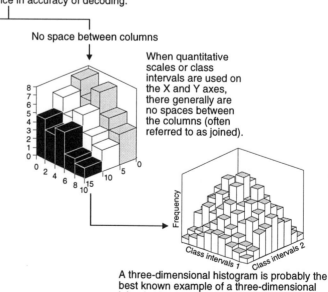

The depth of the columns can vary from almost none to the point where the columns touch one another. In this example the depth of the columns is about 30% of the distance between the column centerlines.

When quantitative scales or class intervals are used on the X and Y axes, there generally are no spaces between the columns (often referred to as joined).

A three-dimensional histogram is probably the best known example of a three-dimensional joined column graph using class intervals.

Data series A
Data series B
Data series C

Column Graph (continued)

Stacked column graph – (multiple data series positioned on top of one another)

Simple stacked column graph

Sometimes referred to as a segmented, extended, divided, composite, or subdivided column graph. Stacked column graphs are generally used to show such things as how a larger entity is divided into its various components, the relative effect that each component has on the total entity, and how the sizes of the components and the total change over time. A stacked column graph has multiple data series stacked on top of one another instead of being placed side-by-side as in a grouped column graph. This means that the tops of the columns represent the totals of all the components (data series) for each interval along the horizontal axis. For instance, if each of the components in the example above represents the sales for one of the three product lines a company sells, then the tops of the columns represent the total sales of the company. Each data series is identified by a different shade, color, or pattern, explained in a legend. Depending on the nature of the data and individual preferences, the largest, most important, or least variable component might be placed on the bottom of the stack. The scale on the vertical axis is always quantitative, linear, starts at zero, and has no breaks. Category or sequence scales are used on the horizontal axis.

Stacked column graph

Linked/connected column graph

In an attempt to make trends easier to see, lines are sometimes drawn connecting the boundaries between the data series, as shown at the right. When this is done the graph is generally referred to as a linked or connected stacked column graph. To further emphasize the trends or changes, the areas between the columns and the connecting lines can be filled, generally with the same shade, color, or pattern used in the columns. This linking technique can be used with the 100% stacked variation also.

Linked stacked column graph

Filled linked stacked column graph

100% stacked column graph

A variation of a percent-of-the-whole graph. With this type of graph, instead of plotting the actual values for each data series, the percents that the values represent of the total of all of the data series are plotted. For example, if the total sales for a company are $16 million and the sales for a product line in a given year are $4 million, 25% would be plotted for that product line instead of $4 million. In this type of graph, the sum of all the components, which is represented by the tops of the columns, is always 100%.

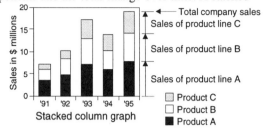

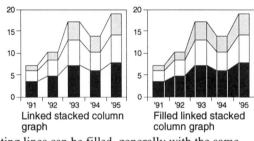

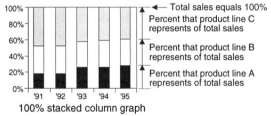

100% stacked column graph

Joined stacked column graph

Another technique sometimes used in an attempt to make the stacked graphs easier to read is elimination of the space between the columns. When the changes from column to column are small, this technique can be effective. If the changes are significant it is less effective.

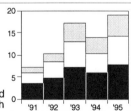

Joined stacked column graph

Three-dimensional stacked column

There are both two- and three-axes stacked column graphs. The two-axis variation is used primarily for its aesthetic value. With the three-axis variation, data graphics in the front may hide segments of the columns in the back.

Three-dimensional stacked column graphs

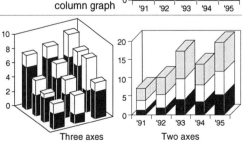

Floating column graph

Sometimes referred to as a two-way column graph. A floating column graph is a variation of column graph in which positive values are measured both up and down from a zero on the vertical axis. The major purpose is to compare multiple data series, with particular attention to correlations and relationships. The units of measure and scale intervals for the up and down scales may or may not be the same. For example, the number of males and females in five different professions might be compared, in which case the units of measure are the same on both scales (e.g., number of people). In another case, the number of people in five different departments might be compared with expenses incurred in each department, in which case one scale would be in number of people, and the other in dollars of expenses. Linear quantitative scales starting at zero are used on the vertical axis; generally only positive values are plotted. A category or sequence scale is used on the horizontal axis. Grid lines may or may not be used. The columns representing the different data series are generally filled with different colors, shading, or pattern to clearly differentiate them. Examples of the four major types of floating column graphs are shown below.

Simple floating column graph

There are two major applications for a simple floating column graph. In one, two things are compared in different situations at a given point in time (e.g., the number of males and females in several different professions in a particular year). In the second case, the same situation is recorded over time (e.g., the number of males and females in one profession over a five-year period). The example at the right illustrates this second situation.

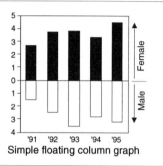
Simple floating column graph

Stacked floating column graph

This variation functions very much like the simple floating column graph, except that one or both of the data series are broken into subcomponents. Expanding on the example above, this type of graph might break the male and female values into those with and without college degrees. There is no limit on the number of subcomponents that each data series can be broken into. A different color, shade, or pattern is used for each subcomponent.

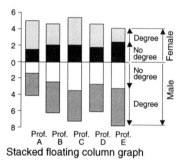
Stacked floating column graph

Grouped floating column graph

This format allows the incorporation of an additional variable into a single graph. For example, such a graph enables one to show the number of males and females employed in various professions at multiple periods of time. This example shows two different time periods; however, additional time periods could be shown by adding additional columns to each group of columns.

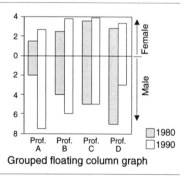
Grouped floating column graph

100% floating column graph

This type of graph compares two data series whose total at each interval equals a fixed value. That value can be any number, though it is typically 100%. One data series is plotted above the horizontal zero axis and the other below. Postive values are plotted in both directions. In the example at the right, the percents of applicants with and without high school diplomas is plotted over time. From the graph, one can see the actual percentage values for each year as well as the overall trends.

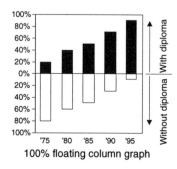
100% floating column graph

Column Graph (continued)

Range column graph

Range column graphs designate upper and lower boundaries of things by means of short columns (sometimes called bars). The boundaries of the ranges might be measured or calculated values, represent single or average values, designate confidence limits, indicate maximums and minimums, etc. Additional values, referred to as inner values, are sometimes also designated. When inner values are included the graph is sometimes called a high-low graph. Inner values many times designate such things as average, median, closing values, etc. They are the same values that the tops of the rectangles on a standard column graph typically designate. Inner values always lie at or between the upper and lower values and are typically indicated by a horizontal line across the column. Sample graphs are shown at the right. Examples of data that might have ranges designated are:

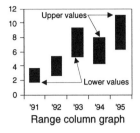

Range column graph

- Maximum, minimum, and average salaries by year
- Average test results and the 10th and 90th percentile values
- Median, high, and low temperatures by month

When a range graph is used to record stock prices it is sometimes called a high-low-close, open-high-low-close, or bar chart. In this specific application, the upper value represents the highest price paid for the stock during a given period. The lower value represents the lowest price and the inner value the price for the last sale of the period called closing value. See Bar Chart 2.

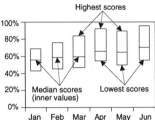

When inner values are added (median scores in this example) the graph is sometimes called a high-low graph.

Distribution information included

In some situations it is desirable to convey information about the distribution of the data as well as the upper, lower, and inner values. The amount of additional information and how it is communicated varies significantly. One example, sometimes called a box graph or box plot, is shown at the right. See Box Graph.

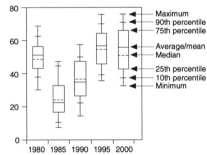

Range column graph with additional data added to give the viewer information about the distribution of the data as well as the upper, lower and inner values

Grouped range column graph

A grouped range column graph format can be used to compare the ranges of multiple entities. For example, the range of scores of other classes might be compared with the range of scores for the class under study. When the ranges of the multiple entities overlap they are plotted side-by-side. When the ranges do not overlap, the range bars can be plotted in line with one another. Examples of both variations are shown at the right.

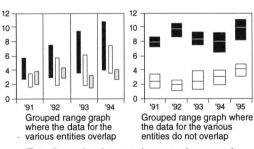

Grouped range graph where the data for the various entities overlap

Grouped range graph where the data for the various entities do not overlap

Two-dimensional grouped range column graphs

Three-dimensional range column graph

In some cases range column graphs comparing multiple data series are plotted on a three-axis graph as shown at the right. Such graphs are sometimes referred to as floating block or flying box graphs.

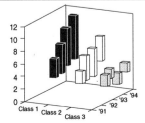

Three-dimensional range column graph

Graphs with ranges designated on columns

───────**Solid range symbol**───────

The tops of the columns on column graphs many times correspond to values such as average, median, etc. Range bars are sometimes added to the tops of the columns to indicate values that bracket the average or median values such as high and low values, plus and minus 1, 2, or 3 standard deviations, plus and minus 90% confidence limits, etc. When this is done the graph is sometimes referred to as a high-low graph. There are a number of different symbols used for this purpose. Examples of three of them are shown below. See the heading Range Symbol and Graph for other examples. High-low symbols are generally used with simple and grouped column graphs and almost never with stacked or three-dimensional column graphs.

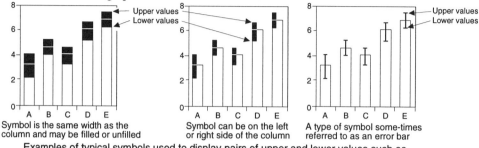

Symbol is the same width as the column and may be filled or unfilled

Symbol can be on the left or right side of the column

A type of symbol some-times referred to as an error bar

Examples of typical symbols used to display pairs of upper and lower values such as maximum and minimum, plus and minus 1, 2, or 3 standard deviations, confidence limits, etc. Such graphs are sometimes referred to as high-low graphs.

───────**Gradient range symbol**───────

When crisp horizontal lines are used to indicate confidence intervals or ranges in which a value probably lies, the viewer sometimes assumes there is a higher degree of certainty associated with the values than actually exists. This is true of the tops of the columns as well as the range symbols. To counter this misinterpretation, the crisp lines are sometimes replaced by graphics that are blurred. Without defining a specific number that the viewer can focus on, the blurred graphic encourages the viewer to think in terms of a range of values within which the actual value exists. In addition to designating a range instead of an exact value, some of the blurred graphics also indicate where the actual value most probably lies within the range. Several examples are shown below. One would interpret the data graphic associated with the column for week one in the graph on the left as indicating that, with a certain degree of confidence such as 90%, the actual value for A lies somewhere between two and four (with a greater probability that it is closer to four because the shading is darker there). The degree of confidence that applies is generally noted.

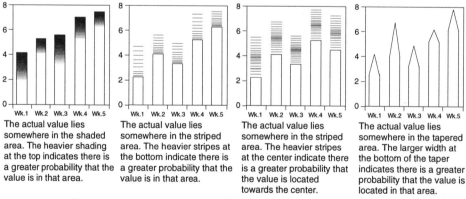

The actual value lies somewhere in the shaded area. The heavier shading at the top indicates there is a greater probability that the value is in that area.

The actual value lies somewhere in the striped area. The heavier stripes at the bottom indicate there is a greater probability that the value is in that area.

The actual value lies somewhere in the striped area. The heavier stripes at the center indicate there is a greater probability that the value is located towards the center.

The actual value lies somewhere in the tapered area. The larger width at the bottom of the taper indicates there is a greater probability that the value is located in that area.

Examples of techniques for indicating the potential range of values with some indication of the more probable location of the value within the range.

When the columns on the graph represent values such as averages or frequencies, as in a histogram, a similar technique is sometimes used. In these situations the blurred area might indicate the relative number of inputs used to arrive at the value and/or the degree of confidence in the value. For example, if the heights of 1,000 people in a congregation were averaged, that value would be closer to representing the average height of the entire congregation than if only 10 heights were averaged. The value arrived at by averaging 10 heights would have a broad blurred area, indicating that the actual average height of those in the congregation might vary significantly from the value shown, while the blurred area associated with the average of 1,000 would be narrow, indicating a higher degree of confidence in the value.

Column Graph (continued)

Difference column graph

There are three major types of difference column graphs: simple difference, change, and deviation. Each is reviewed separately below. In all cases a quantitative scale is shown on the vertical axis and a category or sequence scale on the horizontal axis.

——————**Simple difference column graph**——————

A simple difference graph compares two data series by plotting the actual values of the two data series and connecting the two values with a column. In some cases no effort is made to indicate which series has the largest value, and the columns are generally all the same color, shade, or pattern. When it is desirable to indicate which of the data series is the largest, the data series associated with the higher and/or lower values might be noted at the ends of the columns, or the information might be encoded into the data graphics. The actual numeric differences may or may not be noted on the columns. Examples of three different variations of simple difference column graphs are shown below. Typical information that might be plotted on a simple difference graph are:

- Average scores at the beginning and end of a semester
- Median incomes with and without a degree for various occupations
- Frequency of various diseases for two different countries
- Yearly rain fall in two countries for the past ten years

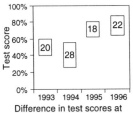

Difference in test scores at beginning and end of semester

In some cases, as shown here, there is no indication as to which data series value is the largest for each time interval.

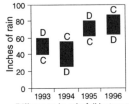

Difference in rain fall between countries C and D

Sometimes the data series associated with the largest and smallest values are identified by placing an abbreviation of the name of the data series adjacent to the values.

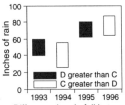

Difference in rain fall between countries C and D

Sometimes the data graphics are coded to indicate which data series value is the largest for each interval.

Simple difference column graphs

——————**Change column graph**——————

A change graph compares multiple factors at two points in time or under two different sets of conditions. Actual values are plotted. The values for the first time/condition and the second time/condition for each interval are connected with a column. With this type of graph, the direction of change is generally considered important and therefore a coding system is used to indicate direction. Color/shading and arrows are among the more widely used methods used to encode the direction of change. Since neither up or down is always favorable (desirable) or unfavorable (undesirable), an additional coding system is sometimes used to designate whether the change is favorable or not. The numeric value of the change may or may not be shown on the data graphics. Examples of several variations are shown below. Typical information that might be plotted on a change graph are:

- Changes in sales forecast numbers from one month's forecast to the next
- Changes in unemployment from one period to the next
- Changes in key business indicators from one period to another

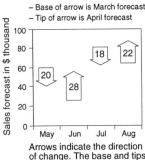

Arrows indicate the direction of change. The base and tips of the arrows indicate prior and subsequent values respectively.

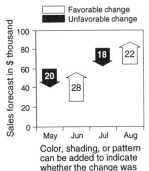

Color, shading, or pattern can be added to indicate whether the change was favorable (desirable) or unfavorable (undesirable).

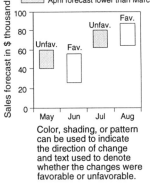

Color, shading, or pattern can be used to indicate the direction of change and text used to denote whether the changes were favorable or unfavorable.

Variations of change column graphs

Difference column graph (continued)

Deviation column graph

One type of deviation graph focuses on the differences between a data series and a reference. Frequently the actual values of the data series are not plotted. Instead, the difference or deviation values are plotted against a zero axis with columns connecting the deviation values and the zero axis. For example, if profit is being compared to budget, only the differences between the actual profit and the budgeted profit are shown. If actual profit exceeds budget, it is shown as a positive number. If it is below budget, it is shown as a negative number, if actual profit is the same as budget, it is shown as zero on the graph since there is no deviation from budget. When deviations are plotted along an axis with a sequence scale such as a time series, a cumulative deviation graph is sometimes also plotted. The cumulative data might be superimposed over the period data or plotted on a separate graph. Shown below is a graph of actual profit data with a reference line (budget) superimposed over it, a graph of the deviation of profit from the reference, and a graph of the cumulative deviation of profit from budget (sometimes referred to as year-to-date deviation graph).

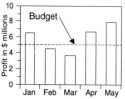

Graph of actual profit and a reference line designating budget. A reference can be constant from period to period or different for each time period.

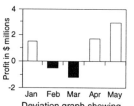

Deviation graph showing how actual profit differs or deviates from budget for each month.

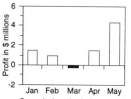

Cumulative deviation graph showing how year-to-date profits differ from budget.

Deviation column graphs in which a data series is compared against a reference other than another data series. The same data was used to generate all three graphs.

Gross and net deviation graphs

Sometimes the actual data consists of two data series, such as cash received versus cash dispersed. When both data series are plotted, the resulting graph is sometimes referred to as a gross deviation or over-under graph.

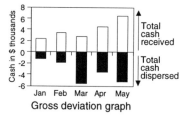

Gross deviation graph

When the differences between the two data series on a gross deviation graph are plotted, we have what is called a net deviation graph.

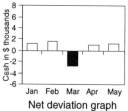

Net deviation graph

The gross and net deviation information can be combined onto one graph as shown at right.

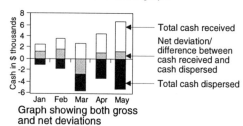

Graph showing both gross and net deviations

Cumulative net deviation graph

In many situations the deviation over a longer period of time is more important than the deviation in any particular period. For instance, if the goal is to have cash received exceed cash dispersed for the year, one might not be too concerned that in a particular month dispersements exceed receipts, if the cumulative year-to-data figures showed that cash received exceeded cash dispersed. A cumulative graph of the cash flow example above is shown at the right.

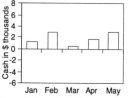

Cumulative deviation graph showing year-to-date whether more cash has been received or dispersed and how much.

Column Graph (continued)

Area column graph

Area column graphs are the only variation of column graphs in which the width of the columns has significance. In the area column graphs, the widths of the columns are proportional to some measure or characteristic of the data element represented by the column. For example, if the columns are comparing the average salary of different groups of people, the width of the columns might indicate how many people are in each group. If the columns are displaying the profitability of various product lines, the width of the columns might indicate what percentage of the total sales the various products represent.

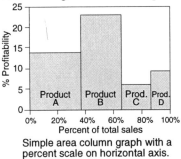

Simple area column graph with a percent scale on horizontal axis.

Column widths can be displayed along the horizontal axis in terms of percents (left) or units (right). There might be a scale on the horizontal axis or a scale may be provided in a legend.

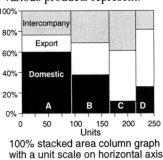

100% stacked area column graph with a unit scale on horizontal axis

In still other cases the values might be noted directly on the graph. Examples of all three variations are shown below.

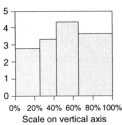 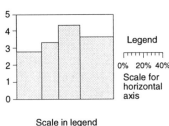 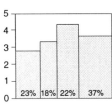

Scale on vertical axis   Scale in legend   Values noted on graph

Examples of how width of column information along the horizontal axis can be communicated

If the values along the horizontal axis are cumulative, the columns are generally joined so there are no unaccounted-for spaces. For example, if all of the values add up to a certain value, such as 100%, or if the sum of all of the values along the horizontal axis is important, no spaces are left between columns. Spaces can be used between the columns without affecting the accuracy of the graph, if the values are not cumulative, the total is unimportant, and the scale is part of the legend, An example is shown at the left. • Sometimes area column graphs are categorized based on the following criteria:

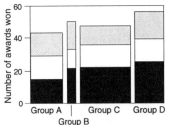

Area column graph with spaces between columns. In this type of graph the values along the horizontal axis are not cumulative, the total is unimportant, and the scale is shown in a legend.

1. The widths of the columns are basically descriptive with little or no relationship to the height of the columns. For example, width might denote the population of cities and height might designate the number of violent crimes. In this case the actual areas of the columns have little or no significance.

2. The widths and heights of the columns are related such that as one varies, the other varies also. For example, in a histogram, if the width of a column is increased to encompass a broader class interval, the height of the column is adjusted accordingly. In these cases the area of the column is a truer measure of the thing being represented than either the height or width.

3. In some cases the width of the column is only meant to convey approximate or relative values and no scale is shown.

• The concept of using the width of columns to convey additional information is generally applied only to simple, stacked, and range column graphs, including 100% stacked graphs. Perhaps the most widely used application of area columns is in histograms (left).

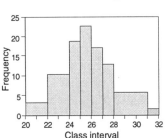

The histogram is probably the most widely used application of an area column graph. In a histogram, as the width of a column is expanded to cover a broader class interval, the height is changed accordingly.

Pictorial column graph

A pictorial column graph is a graph in which pictures, sketches, or icons are used to fill or replace the vertical rectangles. They are sometimes referred to as pictographs. There are many variations of this type of graph. Three widely used variations are shown below. The two major reasons for using pictorial graphs are to improve the appearance of the graph and to facilitate better communications. Three-dimensional column graphs are typically not done in pictorial format. See Pictorial Charts and Graphs.

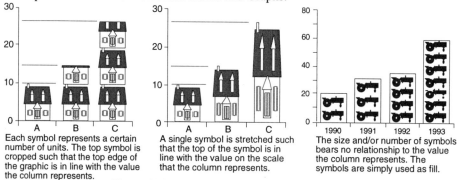

Each symbol represents a certain number of units. The top symbol is cropped such that the top edge of the graphic is in line with the value the column represents.

A single symbol is stretched such that the top of the symbol is in line with the value on the scale that the column represents.

The size and/or number of symbols bears no relationship to the value the column represents. The symbols are simply used as fill.

Three widely used variations of pictorial column graphs

Mosaic graph

A mosaic graph is a combination of 100% stacked column graphs and 100% stacked bar graphs. The major purpose of a mosaic graph is to display a system of interrelated values in such a way that groupings and relative sizes of the various elements can be seen graphically. For instance, in the example below one can see that 50% of the sales dollars go into producing the product (cost to manufacture). Within the cost to manufacture, about 55% goes into materials and about 36% of the material dollars goes into raw materials. See Mosaic Graph.

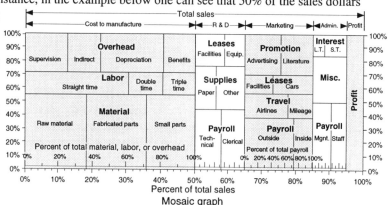

Mosaic graph

Column graph matrix

A column graph matrix consists of column graphs arranged in rows and columns in some organized fashion. The major purpose of a matrix is to simplify the analysis of large quantities of data by enabling the viewer to study multiple graphs at one time. For instance, in the example at the right one can note the trends of seven different measures in four divisions on a single page. See Matrix Display.

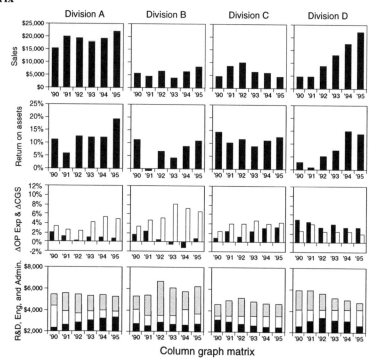

Column graph matrix

Column Graph (continued)

Column graph icons

Miniature (1" square or less) column graphs, frequently without titles, labels, tick marks, or grid lines, are sometimes called icons or symbols. The purpose of icons is not to convey specific quantitative information, but to show relative sizes, values, ratings, etc.; overall comparisons of multiple entities; relative trends for multiple entities; unusual patterns of information etc. Icons are used in a number of different applications. Three of these are shown below.

As symbols on graphs

Icon symbols can be used to communicate information single data points can not. For instance, in the example at the right, the segmented lines connect data points that represent five-year sales averages. The icons show the actual data from which the five-year averages were calculated. The segmented line shows October to have a higher five-year average than May. The icons provide the additional insight that for the last four years, sales in October have gone down while in May they have gone up.

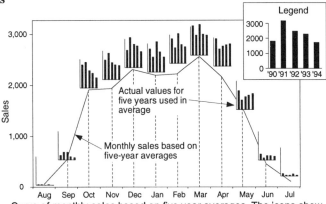

Curve of monthly sales based on five-year averages. The icons show the actual sales for each of the five years so trends can be noted.

As symbols on maps

Column icons are used on maps to convey two major types of information. One is to show how things change over time. In this type of application a time scale is used on the horizontal axis of the template or legend. The other major use is to compare multiple variables/characteristics between areas on a map – for example, the values of exports for each state to ten different countries, the production of six different crops in each state, population by ethnic background, etc. When used in this way the legend has a category scale on the horizontal axis. A legend is always required when icons are used.

Icons used as symbols to convey statistical information on a map

To compare multiple entities with three or more variables/characteristics

To compare a sizable number of entities simultaneously with regards to three or more variables (called multivariate), a display of column icons is sometimes used.

The example at the right compares ten desktop computers with regards to eight different characteristics. A template, as shown in the legend, is used when generating the icons to assure that the same characteristic is always in the same location on each icon and that the same value scale is used. Since values and units of measure can vary significantly, and some characteristics might be qualitative, a common scale of 0 to 1.0 is generally used for all characteristics. The highest value or rating in each characteristic is assigned the value of 1.0. The lowest value or rating is assigned some value less than 1.0, usually zero. Other than legibility, there is no limit on the number of variables that can be included in each icon. There

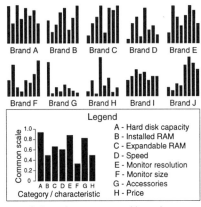

Column icons arranged in an icon comparison display

also is no limit as to how many icons can be included in the same display. The columns in the icons might be filled or unfilled, joined or not joined. Icons used in this way are sometimes referred to as icon comparison displays.

Column Graph (continued)

Circular column graph

Sometimes referred to as a star, radial line, or radial column graph. A circular column graph is generally used for aesthetic purposes. It has a quantitative scale on the radii which is generally linear and almost always progresses from the center out. It has a category or sequence scale on the circumference (circular axis) which can start at any point and progress in either direction. Radial and/or circular grid lines may or may not be used. Columns are typically equally spaced around the circumference. Although grouped, stacked, and range type graphs can be generated using the circular format, the circular technique is normally limited to simple column types. Circular column graphs are sometimes used to display repetitive information.

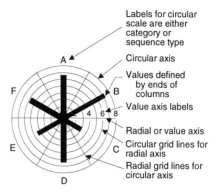

Circular column graph. Each column represents a separate data element. Values are measured from the center towards the circumference.

Column and symbol graph

Sometimes a series of symbols are shown on column graphs to designate reference points or points of comparison for each of the columns. The reference points might designate such things as budget, plan, last year's value, specification value, etc. Using the symbols, the viewer can easily make comparisons between the actual value (as shown by the top of the columns) and some other recognized measure. Any symbol that crisply identifies the reference value can be used. The symbol can be alongside or across the column. Multiple sets of symbols can be used on the same graph. Column and symbol graphs typically have only one data series represented by vertical rectangles. The technique is seldom applied to grouped or stacked type graphs. Two examples are shown at the right.

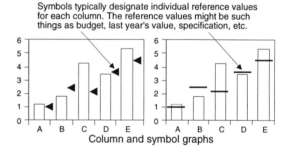

Symbols typically designate individual reference values for each column. The reference values might be such things as budget, last year's value, specification, etc.

Column and symbol graphs

Progressive column graph

Sometimes referred to as a stepped column, step-by-step, or staggered column graph. A progressive column graph is the equivalent of a stacked column graph with only one column and with the individual segments of that column each displaced sideways. Starting at the bottom, each successive segment is generally displaced to the right. The segments can be displaced by any amount. They are generally shifted slightly more than the width of the column. Displacing the segments adds visibility and emphasis to individual segments, while still maintaining the concept that the segments all add up to the whole. In the example at the upper right, the major overhead expenses for a hypothetical company are shown. The left-hand scale enables the viewer to determine overall expense dollars as well as individual expense dollars. The percent that each expense represents of the total is shown on each data graphic, and the cumulative percent is shown on the scale on the right axis. In the lower example, an additional column is included to represent the total of all of the individual segments.

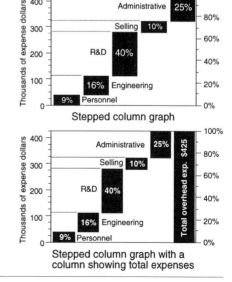

Stepped column graph

Stepped column graph with a column showing total expenses

Column Graph (continued)

Stacked column graph with a single column (one-axis graph)

A typical stacked column graph shows how the relative sizes of components change, particularly over time. Each column of such a graph shows the make up of the components at a given point in time or under one set of conditions. Sometimes the interest is in just one point in time or one condition in which case a single column is used. When used in this way the graph functions very much like a circle graph. When a single column is used, the quantitative scale can be placed directly on the side of the column. Examples are shown below. When this is done it is sometimes called a one-axis column graph. Actual value scales, percent-of-the-whole scales, or both might be used as shown below. Sometimes the scales are eliminated and the values placed on the data graphic, as shown at the right. Without the scale, the graphic is called a column chart or proportional chart instead of a column graph.

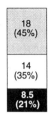

Column chart

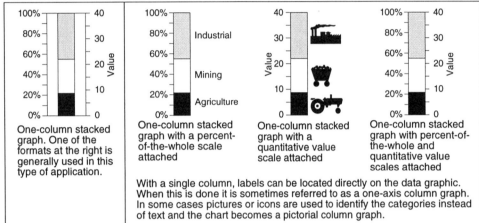

One-column stacked graph. One of the formats at the right is generally used in this type of application.

One-column stacked graph with a percent-of-the-whole scale attached

One-column stacked graph with a quantitative value scale attached

One-column stacked graph with percent-of-the-whole and quantitative value scales attached

With a single column, labels can be located directly on the data graphic. When this is done it is sometimes referred to as a one-axis column graph. In some cases pictures or icons are used to identify the categories instead of text and the chart becomes a pictorial column graph.

Curve fitting on column graphs

Curves can be fitted to column graphs, although the technique is used much less frequently with column graphs than with scatter and line graphs. During the fitting process the assumption is made that the data points are located in the center of the top of each column. See Curve Fitting.

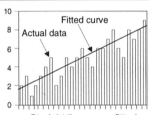

Straight line curve fitted to a column graph

Patch graph

For increased visibility of the hidden data, the sides of the columns are sometimes eliminated, as shown at the far right. Such a graph is sometimes called a patch graph. There are many variations of patch graphs using such things as color, shading, drop lines, and projections. See Patch Graph.

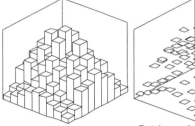

Standard column graph for reference

Patch graph resulting from the elimination of the sides of the columns in the graph at the left

Projection of columns onto ceiling

One hazard of three-dimensional column graphs is the potential for hidden columns. One technique for addressing this problem is to project the tops of the columns onto the ceiling. Such a process lets the viewer know whether or not there are hidden columns but does not indicate their values. Color coding is sometimes used for this purpose. In other cases the actual values are noted on the projection.

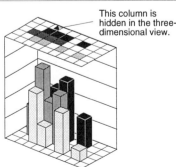

This column is hidden in the three-dimensional view.

Tops of columns projected onto the ceiling to determine if columns are hidden

Location of axis from which values are referenced

The most common location for the axis from which columns are drawn is at zero on the vertical scale (left below). In some cases the axis from which the vertical columns are drawn is located at a value other than zero. When this is done, the ends of the columns still correspond to the actual values as measured from the zero base line axis; though the value may be represented by either the top or the bottom of the column. For example, if one were plotting profit, the reference axis might correspond to budget. In this way, the columns for months in which profit exceeded budget would go up from the reference axis and the actual values would be designated by the tops of the columns. The columns for months that profit was below budget would go down from the reference axis and the actual values would be designated by the bottoms of the columns. Either the up or down columns might be filled to place special emphasis on them. The budget line might be straight (center below) if the budget is the same from month to month; or stepped (right below) if the budget varies from month to month. By using a reference other than zero, such as budget, the viewer can read budget and actual values directly from the graph. Differences from budget are visually highlighted as in a deviation graph.

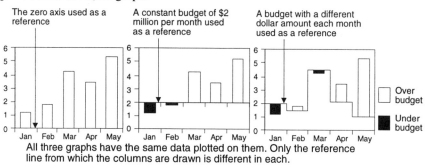

All three graphs have the same data plotted on them. Only the reference line from which the columns are drawn is different in each.

Column fill

– With a simple column graph and a category scale on the horizontal axis, all the columns might have the same fill. When the categories are distinctly different, each column might have a different fill to emphasize the differences.

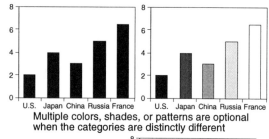

Multiple colors, shades, or patterns are optional when the categories are distinctly different

– When a single entity or characteristic is being plotted on a simple column graph with a sequence scale on the horizontal axis, the columns typically all have the same fill to convey the idea that variations of the same thing are being plotted, not a series of different things.

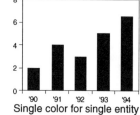

Single color for single entity

– An exception is when positive and negative values are plotted. In these cases the negative or unfavorable values may be filled with a different color, shade, or pattern to call attention to them.

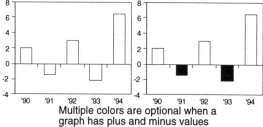

Multiple colors are optional when a graph has plus and minus values

– In grouped and stacked column graphs, all columns or segments of columns of a given data series typically are the same color, shade, or pattern.

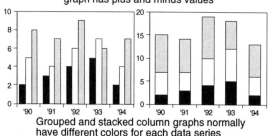

Grouped and stacked column graphs normally have different colors for each data series

Column Graph (continued)

Scales on column graphs

Although the actual value designated by a column is determined by the location of the top of the column, many people at first glance associate the length or area of the column with its value. As long as the scale is linear, starts at zero, is continuous, and the columns are the same width, this presents no problem. When any of these conditions are changed, the potential exists that the graph may be misinterpreted. For this reason the following general guidelines are offered regarding quantitative scales on the vertical axes of column graphs.

- Have all scales include zero unless it is clearly identified otherwise.
- Locate zero where the horizontal axis crosses the vertical axis.
- Have positive values increase upward and negative values downward.
- Make scales continuous (no breaks).
- Upper scale values should be larger than any value plotted.
- Lower scale values should be smaller than any value plotted.
- Use only linear scales (avoid scales such as logarithmic).
- Except for area column graphs, make all columns the same width.

There are exceptions to these guidelines.

- With range and difference graphs, the scale sometimes does not start at zero since the columns do not extend to the zero axis.
- With simple and grouped column graphs, the scales are sometimes broken, expanded, and/or cropped in order to make changes and differences from column to column more visible. Several methods for accomplishing this are shown below.

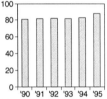 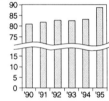 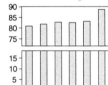 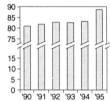

For reference, the scale starts at zero in this example. Variations from column to column are difficult to estimate.

These three examples expand the scale and then take a section out of the middle (sometimes referred to as a scale break) to maintain the same size graph. All three examples accomplish the same goal. The center example makes the break most obvious. The lower portion of the scale is retained to help the viewer more easily determine the size of the intervals.

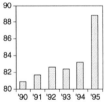 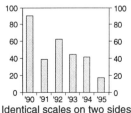 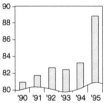 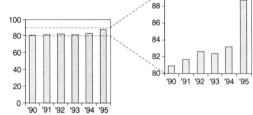

An alternative to a scale break is to eliminate the lower portion of the scale and start the scale at some value just under the lowest value being potted. This is one of the most dangerous variations since it is easy for the viewer to overlook the fact that the lower portion of the scale is missing, and to reach a conclusion based on the height or area of the columns instead of the value at the top of the column.

Making the lower portion of the graph uneven helps to call the viewer's attention to the fact that the lower portion of the scale is missing.

This alternative lets the viewer see each of the values in relation to one another using a zero axis. The enlargement on the right also lets the viewer more accurately see the detailed differences.

• There typically is only one scale on the horizontal axis. On the vertical axis the norm is to have just one scale; however, it is not uncommon to repeat the quantitative scale on the right side, particularly if the graph is wide and grid lines are not used. Two different quantitative scales can be used if necessary. If there are two different quantitative scales and both do not apply to all columns, care must be taken to clearly identify which scale applies to which columns. Examples of multiple quantitative scales are shown below.

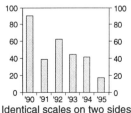 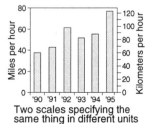 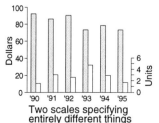

Identical scales on two sides

Two scales specifying the same thing in different units

Two scales specifying entirely different things

General

Grid lines

On two-dimensional graphs, grid lines, when shown, are typically displayed behind the columns. Two commonly used variations are shown below, as well as an example of the use of horizontal drop lines. Because it is sometimes difficult to estimate values on three-dimensional column graphs, grid lines are occasionally placed on the surfaces of the columns as well as on the back planes. See Grid Lines.

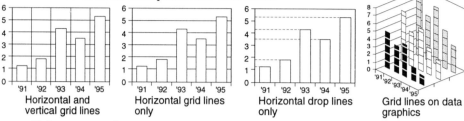

Horizontal and vertical grid lines | Horizontal grid lines only | Horizontal drop lines only | Grid lines on data graphics

Values on graphs

When it is desired to convey values more precisely, they are sometimes depicted directly on the data graphics as shown below. Generally only the value represented by the column is shown; however, it is possible to include additional values – for example, the percent-of-the-whole that each value represents, the percent or ratio the value is to some reference value, the amount the value deviates from a standard, etc. An example of multiple values is shown at the right.

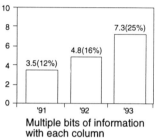

Multiple bits of information with each column

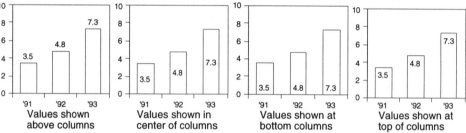

Values shown above columns | Values shown in center of columns | Values shown at bottom columns | Values shown at top of columns

Other shapes used on three dimensional column graphs

For presentation purposes, shapes other than rectangular columns are sometimes used in three-dimensional column graphs (examples at right). Generally speaking, the alternate shapes tend to reduce accuracy in reading the graph. In addition, the viewer is sometimes confused as to which point on the data graphic actually specifies the value.

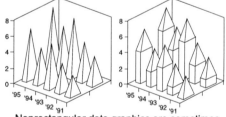

Nonrectangular data graphics are sometimes used for presentation purposes.

Frames

Because of the boldness of the columns, the frame around a column graph can sometimes be reduced significantly without detracting from the value of the graph. Four variations are shown below, ranging from a full frame on the left to no frame on the right.

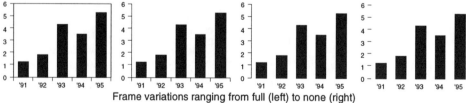

Frame variations ranging from full (left) to none (right)

Column chart

For presentation or publication purposes, the vertical scale is sometimes eliminated as well as the frame. When this is done the values are generally noted on or adjacent to the data graphics. Without a quantitative scale, this is no longer a graph. Instead it becomes one of a large family of proportional area charts in which the sizes/areas of the data graphics are proportional to the data element they represent. See Proportional Chart.

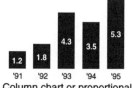

Column chart or proportional chart (a column graph without a quantitative scale)

Combination Graph

Sometimes referred to as a mixed, composite, or overlay graph. A combination graph displays multiple data series using two or more different types of data graphics. The different data graphics are said to be superimposed or overlaid on one another. The major reason for using combination graphs is to improve clarity and highlight relationships between the various data series. An example incorporating three types of data graphics is shown on the left below. Alongside it are two conventional graphs displaying the same data, each using a single type of data graphic for comparison.

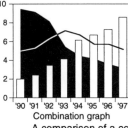
Combination graph

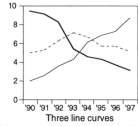
Three line curves

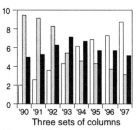
Three sets of columns

A comparison of a combination graph using three different types of data graphics and the same data plotted on graphs using only one type of data graphic.

Combination graphs generally use two-dimensional rectangular formats. For aesthetic purposes they occasionally are generated as two-axis, three-dimensional graphs, as shown at the right. Combination graphs are seldom used with three-axis, three-

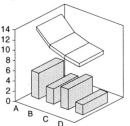
Three-dimensional, two-axis combination graph

dimensional graphs except for curve fitting. • The bottom data graphic on a two-dimensional combination graph is called the base or main graph, and the data graphics that are placed on top of it are called overlays. The sequence in which the data graphics are overlaid can effect the legibility of the combination graph as illustrated by the examples at the left.

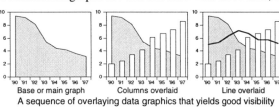
Base or main graph   Columns overlaid   Line overlaid
A sequence of overlaying data graphics that yields good visibility

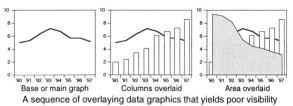
Base or main graph   Columns overlaid   Area overlaid
A sequence of overlaying data graphics that yields poor visibility

An illustration of the importance of the sequence in which the data graphics are overlaid. The same data series and data graphics are used in both examples.

Types of data graphics used

Other than legibility, there are no restrictions as to number and types of data graphics that can be used on the same graph. To illustrate, one of the graphs at the right displays six data series and the other uses box, stepped line, and stepped area data graphics.

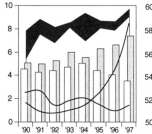

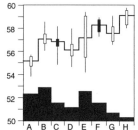

Scales

Other than legibility, there is no limitation on the number of vertical scales that can be used on a single combination graph. The example shown below has three scales. Values, units of measure, and length can be different for each scale. Data graphics are sometimes coordinated with the different scaled. For example, all data series that use the same scale might use the same type of data graphic. • Generally there is only one horizontal scale. Category and sequential types are most widely used. If there is more than one scale on the horizontal axis, different types of scales are typically not mixed. Different units of measure might be used on multiple horizontal scales, but are generally proportional to one another or use alternate terms for the same category.

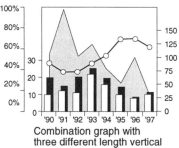
Combination graph with three different length vertical quantitative scales

In one sense, many graphs with two or more data series plotted on them might be considered comparative. The group of graphs typically called comparative graphs simply focuses more on the comparative aspects and less on the actual values. Comparisons might be made on either a quantitative or qualitative basis. Comparative graphs might be used to compare such things as performance, ranking, changes, or characteristics. They frequently compare things at two different times, under two different conditions, by two different organizations, etc. Typically the entries on the horizontal axis are dates, conditions, locations, or organizations. The scales on the vertical axes are frequently percents, index numbers, or ranking numbers. Several examples are shown below.

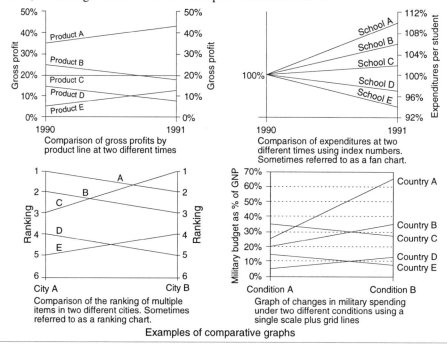

Comparison of gross profits by product line at two different times

Comparison of expenditures at two different times using index numbers. Sometimes referred to as a fan chart.

Comparison of the ranking of multiple items in two different cities. Sometimes referred to as a ranking chart.

Graph of changes in military spending under two different conditions using a single scale plus grid lines

Examples of comparative graphs

Comparative graph with many items

Although generally used with just a few entities, the comparative graph can sometimes be used effectively when comparing many entities, especially when the emphasis is on overall patterns, trends, and relationships. For instance, in the example below, it might be noted that between 1979 and 1989 the average incomes went from a single, relatively continuous cluster to several smaller clusters. At the same time, the range from the lowest to the highest expanded significantly. In addition to reading the actual values on the scales, the viewer gets a graphical feel for the relative change of the various entities by noting the slopes of the lines (i.e., the steeper the slope the greater the change). One additional observation is the individual lines that cross many other lines. Such lines mean that the relative ranking of the entities the lines represent have made a marked change in their relative ranking within the total group of entities being plotted.

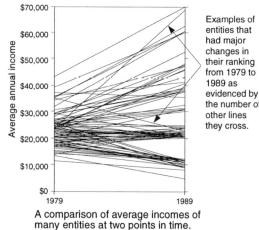

Examples of entities that had major changes in their ranking from 1979 to 1989 as evidenced by the number of other lines they cross.

A comparison of average incomes of many entities at two points in time.

Three or more labels on the horizontal axis

Although comparative graphs generally have only two points on the horizontal axis, they sometimes are extended to include three or more. In the example at the right, as the number of labels increases along the horizontal axis, the graph looks more and more like a standard segmented line graph.

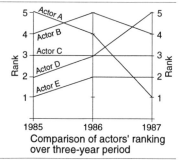

Comparison of actors' ranking over three-year period

Comparative Graph (continued)	**Parallel coordinate or profile graph**

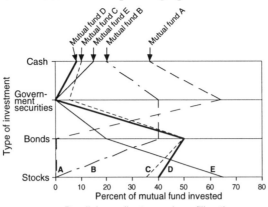

Parallel coordinate graph profiling the
investment patterns of five mutual funds

Instead of comparing a single characteristic over time, several characteristics are sometimes compared at a given point in time. The example at the left compares the investments of five mutual funds to see if their investment strategies are the same. From the example, it is clear that funds C and D invest very much the same way. Such graphs, which are sometimes referred to as a parallel coordinate graphs or profile graphs, are many times constructed horizontally. When a large number of entities are plotted on the same graph, all of the crossing lines may become confusing. Coloring selected lines generally helps. An alternate method is to place each entity being studied on a separate graph as shown below. Since the key purpose of this type of chart is to note similarities in profiles, the size of the graphs can be greatly reduced and scales might be deleted. In this format, the series of small graphs resemble an icon comparison display.

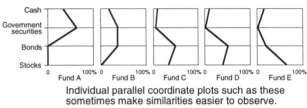

Individual parallel coordinate plots such as these
sometimes make similarities easier to observe.

Composite Bar Graph	Sometimes referred to as a divided, extended, segmented, stacked, or subdivided bar graph. A composite bar graph has multiple data series stacked end-to-end. This results in the far right end of each bar representing the total of all the components contained in that bar. Composite bar graphs are used to show how a larger entity is divided into its various components and the relative effect that each component has on the whole. See Bar Graph.	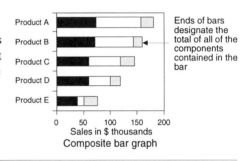 Composite bar graph

Composite Column Graph	Sometimes referred to as a divided, extended, segmented, stacked, or subdivided column graph. A composite column graph has multiple data series stacked on top of one another,

so that the top of each column represents the total of all the components shown in that column. Such graphs are generally used to show how a larger entity is divided into its various components, the relative effect that each component has on the whole, and how the sizes of the components and the total change over time. See Column Graph.

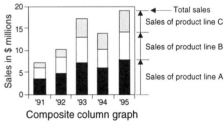

Composite column graph

Composite Graph	Sometimes referred to as a combination, mixed, or overlay graph. A composite graph displays multiple data series and uses two or more types of data graphics to represent them. See Combination Graph.

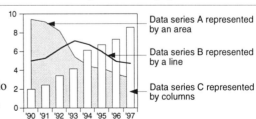

Composite graph using area, line, and column data graphics

Compound Line Graph	Sometimes referred to as a dual, grouped, multiple, or overlapped line graph. A line graph with multiple data series, all of which are measured from the same zero base line axis. See Line Graph.

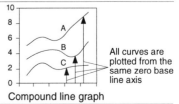

Compound line graph

99

Conceptual Diagram

Sometimes referred to as an explanatory or summary diagram. Conceptual diagrams provide a graphical overview of how people, things, ideas, influences, actions, etc., interrelate. They often condense large amounts of information into a single chart and can be effective in presenting material that is still in the idea stage. They work equally well in communicating simple and complex concepts. They typically are not quantitative and generally do not get into much detail. Conceptual diagrams are applicable to any subject, whether concrete or abstract, theoretical or actual, current, historical, or future. The distinction between conceptual diagrams and certain other types of diagrams such as flow charts and organizational charts is not always clear-cut. If the diagram becomes fairly detailed, it generally is not considered a conceptual diagram. Some block diagrams and Venn diagrams are classified as conceptual diagrams. Conceptual diagrams are sometimes used in presentations because they can be rapidly understood, tend to have a significant impact on the audience, and may be remembered longer than the same subject presented using words and numbers alone. Conceptual diagrams consist of geometric shapes, lines, arrows, symbols, pictures or

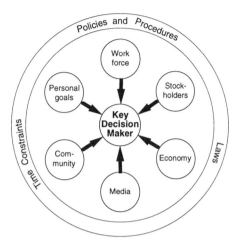

A conceptual diagram indicating the relationship between a key decision maker, influencing factors, and some of the constraints that the entire system operates within.

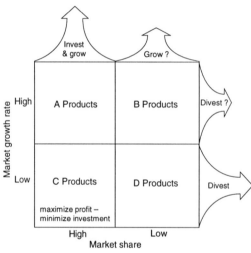

A conceptual diagram indicating how the future of product lines might relate to current market evaluations

sketches, and text. When used in conceptual diagrams, standard symbols sometimes take on different meanings than when used in other diagrams. For example, in most diagrams, arrows are used to call attention, indicate a direction, or depict flow from one thing or place to another. In conceptual diagrams; however, an arrow might mean any of those things and in addition, might mean that one thing has an influence or impact on another. Arrow heads pointing in both directions might mean mutual, interdependent influences. Other arrows might mean an action is to be taken, as opposed to no action, or they might mean that something shifts or changes (such as an asset shifting to a liability) even though nothing has physically moved or flowed. Despite the varied potential meanings of symbols, legends are seldom used to explain them. Well designed conceptual diagrams are many times self-explanatory; however, text, explanations, and/or discussions normally accompany them. The four figures on this page are examples of conceptual diagrams.

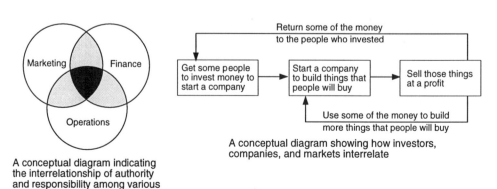

A conceptual diagram indicating the interrelationship of authority and responsibility among various departments within an organization

A conceptual diagram showing how investors, companies, and markets interrelate

**Confidence Band
or
Confidence Interval
and
Confidence Limit**

Sometimes referred to as possible margin of error. In terms of charts and graphs, a confidence interval graphically indicates the range of values within which a given value probably exists. For example, if an organization wants an estimate of the percent of members that will participate in a given activity, they might send a questionnaire to 10% of the members asking if they plan to participate. If 15.2% of the people receiving the questionnaire indicate they will participate, it may be assumed that about 15.2% of all members will participate. Even though it is highly unlikely the overall participation will be exactly 15.2%, one can calculate with a given degree of confidence, such as 90%, the range of values within which overall participation will probably lie. This range of values is called a confidence band or confidence interval. If the range of values extended from minus 3.4% to plus 3.4% of the sample value, the confidence interval extends from 11.8% (15.2% - 3.4%) to 18.6% (15.2% + 3.4%). • A confidence interval can be shown graphically several different ways. One widely used method employs an error bar as shown at the right. The top and bottom of confidence intervals are generally called the upper and lower confidence limits, respectively. When confidence intervals are shown, two additional bits of information are sometimes included:

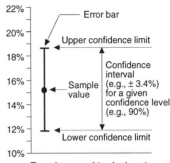

Error bar used to designate sample value, confidence interval, and upper and lower confidence limits

- The level of confidence associated with the interval. For example, if the confidence level is 50%, there are 50 out of 100 chances that the actual value will lie within the confidence interval. Frequently used confidence percentages are 50, 68, 90, 95, and 99.
- The reasons for the potential spread in values, such as sample size, sample-to-sample variations, biases of the people being interviewed, or methods of processing the data.

Confidence intervals can be displayed for a single set of data, as in the example above, or for multiple sets of data, as shown below. When comparing multiple data sets, confidence intervals are often used to decide whether the averages of the two sets of data are significantly different. If the confidence intervals of two sets of data overlap, the averages of the two sets of data may not be significantly different. • Examples of how confidence intervals might be graphically displayed are shown below.

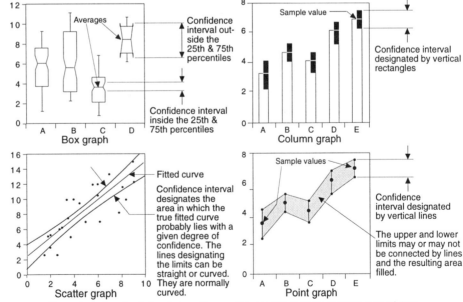

Examples of methods for designating confidence intervals for multiple sets of data

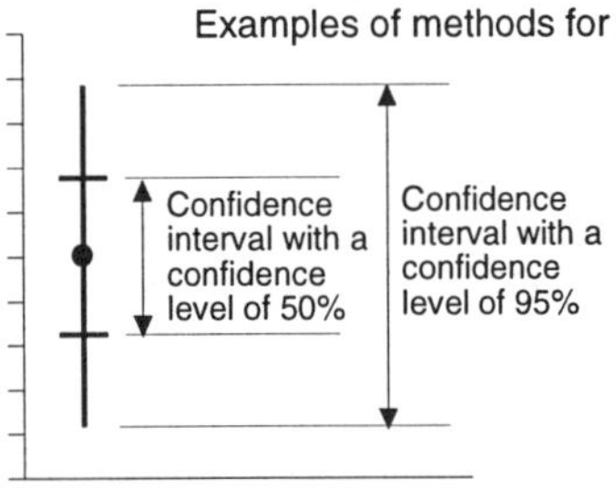

In some cases two confidence intervals are shown for the same set of data. These are sometimes referred to as two-tiered confidence intervals. An example is shown at the left. • Confidence intervals can be shown in units or percent. The upper and lower limits may or may not be equidistant from the reference data point. For additional examples as to how confidence intervals might be graphically displayed, see Range Symbols and Graphs.

Two-tiered confidence interval

101

Connected Bar Graph

When the bars on a bar graph become so wide there are no spaces between them, the graph is sometimes called a joined bar graph, connected bar graph, stepped bar graph, or histogram. See Bar Graph.

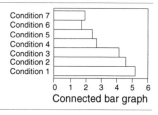

Connected bar graph

Connected Column Graph

When the columns on a column graph become so wide there are no spaces between them, the graph is sometimes called a joined column graph, connected column graph, stepped column graph, or histogram. See Column Graph.

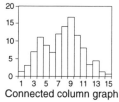

Connected column graph

Connected Stacked Graph

In an attempt to make relationships easier to see on stacked bar and column graphs, lines are sometimes drawn connecting the boundaries between the data series. When this is done the graph is referred to as a linked or connected stacked graph. Connecting lines can be used with simple or 100% stacked graphs. The area between the links may or may not be filled. See Bar Graph and Column Graph.

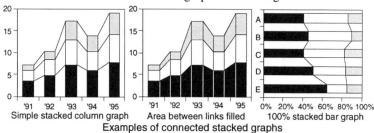

Simple stacked column graph   Area between links filled   100% stacked bar graph

Examples of connected stacked graphs

Continuous Data

Sometimes referred to as indiscrete data. Continuous data is data that transitions smoothly without interruption from one value to another. A commonplace example of continuous data is the temperature in a room. For the temperature to go from 65 to 70 degrees, it must pass through every degree and fraction of a degree between the two temperatures. Measurement of any value in-between is possible. The opposite of continuous data is discrete data, which has no transitions or intermediate values. For example, families have whole numbers of children (e.g., 1, 2, 3); there are no possible values in between (e.g., no fractions such as one-and-a-half or two-and-three-quarters children).

Contour Graph
or
Contour Line Graph

A contour graph is one form of isoline graph and functions like a contour map. There are two- and three-dimensional contour graphs. The three-dimensional version is a variation of a surface graph (See Surface Graph). In both cases, three variables are plotted. A series of lines are drawn to connect points of equal value on the Z or vertical axis. These lines are called contour lines or isolines. In two-dimensional contour graphs, the Z-axis is perpendicular to the plane of the paper and therefore all the contour lines appear in the same plane. A two-dimensional contour graph is the equivalent of looking directly down on a three-dimensional contour graph. The example at the right illustrates this by projecting the contour lines of a three-dimensional graph onto its ceiling, thus forming a two-dimensional contour graph of the same data.

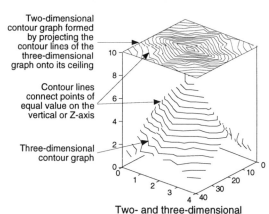

Two-dimensional contour graph formed by projecting the contour lines of the three-dimensional graph onto its ceiling

Contour lines connect points of equal value on the vertical or Z-axis

Three-dimensional contour graph

Two- and three-dimensional contour graphs of the same data

Designating values

When the only purpose of the contour graph is to display the general patterns of the data, Z-axis values are sometimes not shown. When the purpose of the graph is to analyze specific values, a Z-axis scale might be included on a three-dimensional graph, or values of the contour lines might be noted directly on the data graphic in either the two- or three-dimensional case. Values for Z-axis contour lines are generally selected with equal intervals between them such as 2, 5, 10, etc. Two examples are shown on the following page.

**Contour Graph
or
Contour Line Graph**
(continued)

Designating values (continued)

When there are many contour lines, every fifth line or so is sometime highlighted to visually assist the viewer. Scales on the Z-axis are always quantitative. Scales on the X and Y axes might be quantitative, categorical, or sequential.

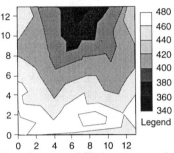

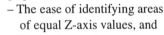

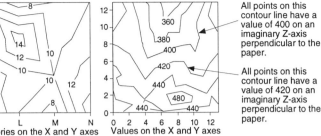

All points on this contour line have a value of 400 on an imaginary Z-axis perpendicular to the paper.

All points on this contour line have a value of 420 on an imaginary Z-axis perpendicular to the paper.

Categories on the X and Y axes / Values on the X and Y axes
Two-dimensional contour graphs

Filled contour graph

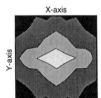

Two-dimensional filled contour graph with a legend identifying the values of the contour lines

An alternative solution for identifying values on the Z-axis is to fill the areas between the contour lines, thus forming what is called a filled contour graph (left). A legend is provided to relate the fill colors, shades, or patterns to Z-axis values. Major advantages of a filled contour graph include:

– The ease of identifying areas of equal Z-axis values, and
– The uncluttered appearance resulting from the fact that the Z-axis values do not have to be shown in the plot area.

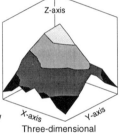

X-axis

Y-axis

Two-dimensional filled contour graph

The boundaries between the different fills represent the lines of equal values. • There are two- and three-dimensional variations of filled contour graphs. Sometimes they are used in conjunction with one another. An example is shown at the right. Included with the two contour graphs is a side view of the three-dimensional graph to illustrate how the boundaries between the fills (lines of equal value on the Z-axis) form straight lines.

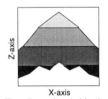

Z-axis

X-axis
Two-dimensional side view of the three-dimensional graph at the right

Z-axis

X-axis / Y-axis
Three-dimensional filled contour graph

Two- and three-dimensional filled contour graphs plus a side view illustrating the lines of equal value on the Z-axis

---

Contour Line

Contour lines are a form of isoline. On maps they connect points of equal elevation. On graphs they connect points of equal values on the vertical or Z-axis. They can be used with two- or three-dimensional charts and maps. See Isoline.

Contour Map

Sometimes referred to as a relief map. A variation of a topographic map, a contour map shows the surface features of terrain such as locations, configurations, and elevations. This is accomplished largely through the use of contour lines which connect points of equal elevation. The areas between contour lines are sometimes filled with uniform or varying shades or tints of color (sometimes called hypsometric map). Unlike statistical and descriptive maps where the emphasis is on the information superimposed on a deemphasized base map, the emphasis of contour maps is on the base map itself. Contour maps may be two- or three-dimensional, though two-dimensional are the most common. See Topographic Map.

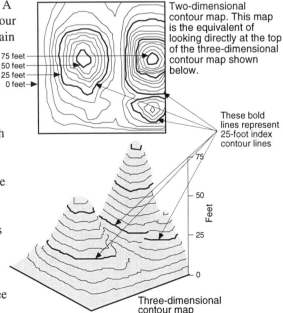

75 feet
50 feet
25 feet
0 feet

Two-dimensional contour map. This map is the equivalent of looking directly at the top of the three-dimensional contour map shown below.

These bold lines represent 25-foot index contour lines

Feet

Three-dimensional contour map

Control Charts

Sometimes referred to as Skewhart charts. Shown below are examples of the nine most widely used types of control charts. Each type has its specific applications. For example, the type of graph differs depending on whether the data is continuous or discrete, individual or grouped, or whether the graph is to record ranges of values or deviations. There are a number of variations possible within each type of chart. (Basic elements of control charts are defined on the following pages.)

				Symbols sometimes used to identify chart	Description	Example of chart
For use with variable data — Variable data is data that can be measured and expressed numerically (sometimes referred to as continuous or indiscrete data). Examples are length, diameter, weight, temperature, strength, etc.	**Individual data** — Used when data for individual units are plotted as opposed to average data for a group of units.		Value	**X, M**	• The measured value of each individual unit is plotted. • Used when subgroups are not practical or possible.	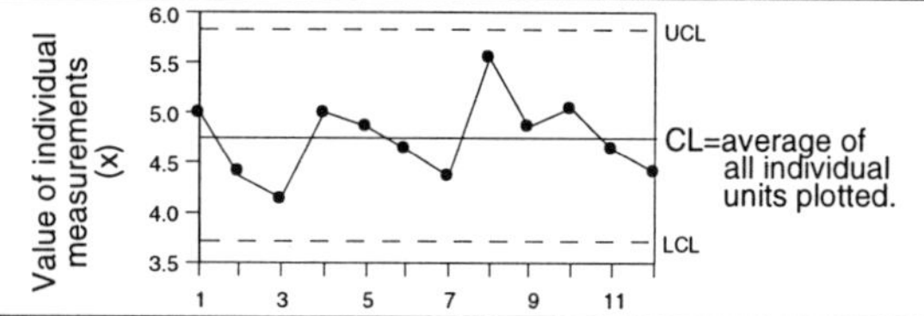
			Moving Range	**MR**	• The difference between each value and the preceding value(s) (range) are plotted.	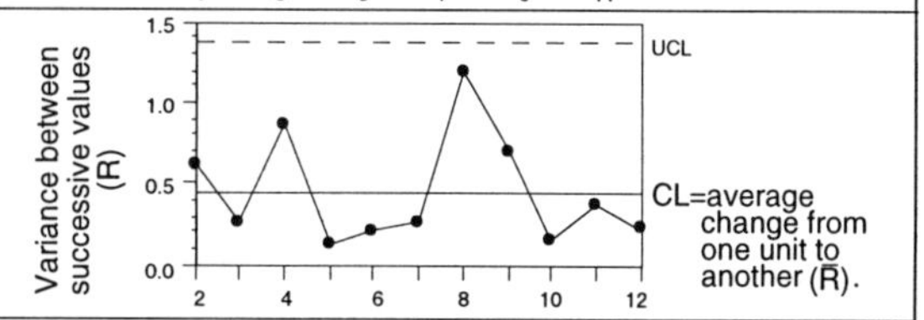
	Grouped data — Used when data is grouped, as in sample lots or periods of time	**Spread** — Measures of dispersion or spread of data	Averages	**X-Bar**	• The average of each subgroup is plotted. • Frequently used in conjunction with an R or S type chart.	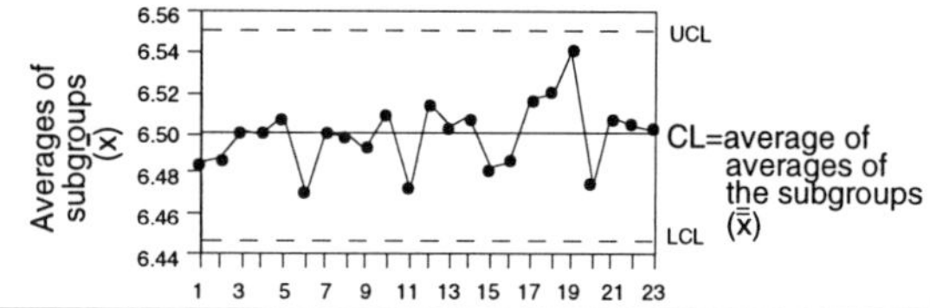
			Range	**R**	• Range values are obtained by subtracting the smallest value in a subgroup from the largest value in the same subgroup. • Generally used with smaller subgroup sizes. • Frequently used in conjunction with an X-Bar chart.	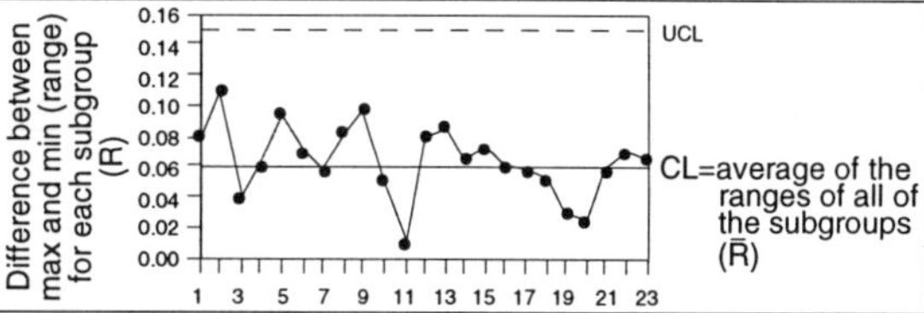
			Standard deviation	**S, s, σ Sigma**	• Standard deviations are calculated and plotted for each subgroup instead of ranges. • Generally used with larger sample sizes (more than 10). • Considered slightly more accurate than the R type chart. • Frequently used in conjunction with an X-Bar chart.	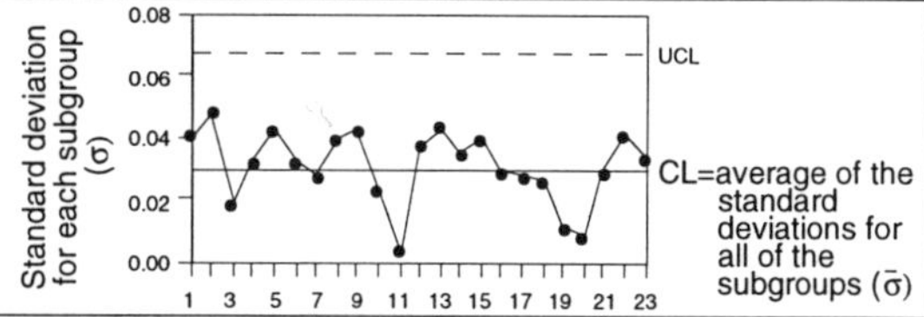
For use with attribute data — Attribute data is data that can be counted but not measured (sometimes referred to as discrete data). Examples are quantity of acceptable parts, percent that pass a no-go gage, number of scratches, defects per unit, etc.	**Grouped unit data** — Applicable only when data is grouped, as in sample lots or periods of time		Quantity of units with defects that exceed specification	**NP, Np, np**	• Plots the number of units in each subgroup that have numbers of defects/imperfections that exceed specification. • Generally used when the number of units in each subgroup is the same.	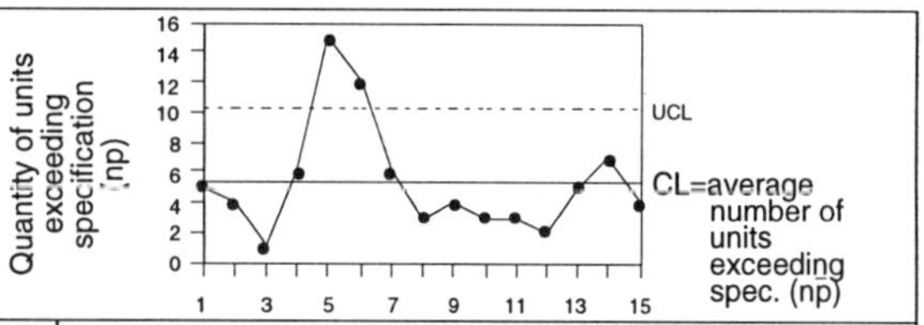
			Fraction or percent of units with defects that exceed specification	**P, p**	• Plots the quantity of nonconforming units in each subgroup as a fraction or percent of the total number of units in the subgroup. • Can be used with subgroups of equal or unequal size. • Example is based on subgroups of unequal size.	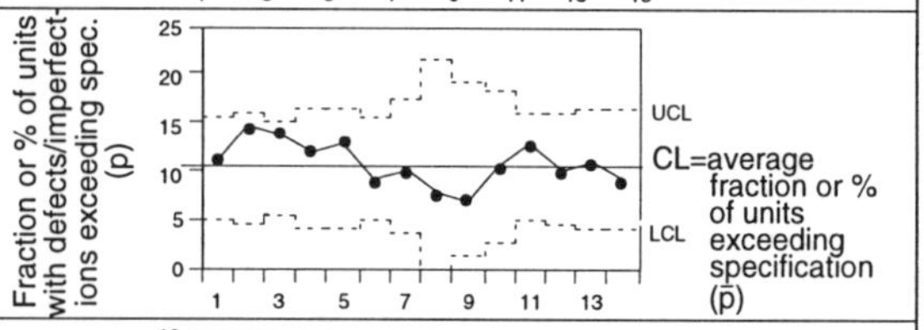
	Defect / imperfection data — Used when multiple defects/imperfections are not uncommon		Quantity of defects/imperfections	**C, c**	• The value obtained by counting the total number of defects/imperfections in the subgroup is plotted. • Generally used when sizes of subgroups are the same.	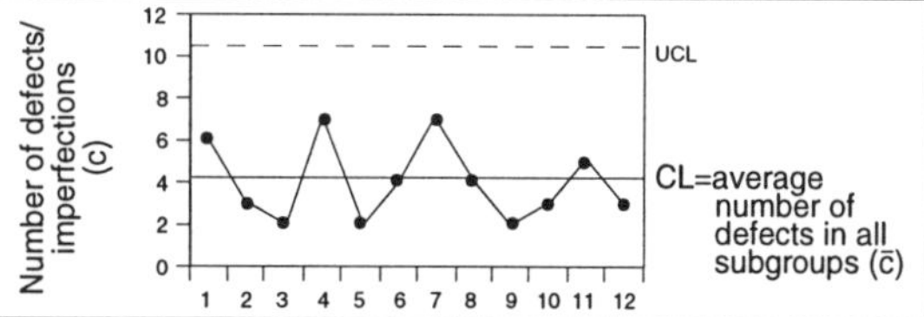
			Quantity of defects/imperfections per unit measure	**U, u**	• The number of defects/imperfections per unit of measure such as square foot, meter, piano top, etc., is plotted. • Can be used with subgroups of equal or unequal size. • Example is based on subgroups of unequal size.	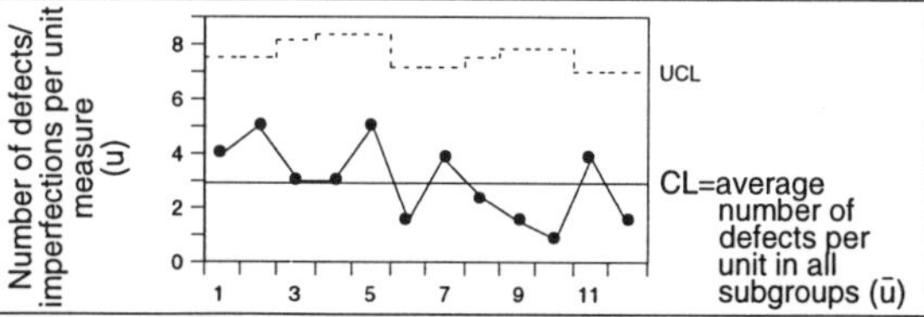

Control Charts (continued)

Overview

Most control charts operate on the principle of monitoring some quantifiable characteristic of a repetitive process or operation. Actual values, average values, spreads (ranges), etc., of those characteristics are recorded and plotted on either a point or line graph. All control charts have a sequential scale on the horizontal axis, a quantitative scale on the vertical axis and one or more horizontal lines indicating some sort of limits for control purposes. When the values and ranges stay within the prescribed limits and the data does not indicate the process to be drifting out of limits, the process or operation being monitored is generally said to be in control. Originally control charts were used almost exclusively in manufacturing. They are now used in many different fields including medical, financial, cost reduction, etc.

Basic elements of an X-Bar/R chart

Frequently pairs of control charts are used in conjunction with one another. One of the most widely used combinations is the chart displaying subgroup averages (denoted as X-bar) and the range of the values in the subgroups (denoted as R). The combination of the two is referred to as an X-Bar/R chart. The following series of graphs illustrate the elements that go into an X-Bar/R chart. These steps are shown here for illustrative purposes only. In practice, the procedure is to go directly from the raw data to the X bar/R chart shown at the bottom of the next page. Any one of the intermediate graphs can, and sometimes are, used by themselves if only that specific information is desired.

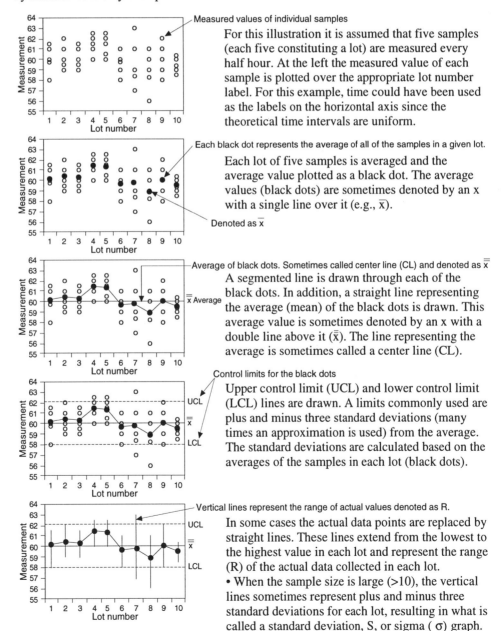

Measured values of individual samples

For this illustration it is assumed that five samples (each five constituting a lot) are measured every half hour. At the left the measured value of each sample is plotted over the appropriate lot number label. For this example, time could have been used as the labels on the horizontal axis since the theoretical time intervals are uniform.

Each black dot represents the average of all of the samples in a given lot.

Each lot of five samples is averaged and the average value plotted as a black dot. The average values (black dots) are sometimes denoted by an x with a single line over it (e.g., $\bar{x}$).

Denoted as $\bar{x}$

Average of black dots. Sometimes called center line (CL) and denoted as $\bar{\bar{x}}$

A segmented line is drawn through each of the black dots. In addition, a straight line representing the average (mean) of the black dots is drawn. This average value is sometimes denoted by an x with a double line above it ($\bar{\bar{x}}$). The line representing the average is sometimes called a center line (CL).

$\bar{\bar{x}}$ Average

Control limits for the black dots

Upper control limit (UCL) and lower control limit (LCL) lines are drawn. A limits commonly used are plus and minus three standard deviations (many times an approximation is used) from the average. The standard deviations are calculated based on the averages of the samples in each lot (black dots).

Vertical lines represent the range of actual values denoted as R.

In some cases the actual data points are replaced by straight lines. These lines extend from the lowest to the highest value in each lot and represent the range (R) of the actual data collected in each lot.
• When the sample size is large (>10), the vertical lines sometimes represent plus and minus three standard deviations for each lot, resulting in what is called a standard deviation, S, or sigma (σ) graph.

(This sequence is continued on the next page.)

105

(This sequence is continued from the previous page.)

Control Charts (continued)   At this point in the sequence, the range values (left) and the average values (right) are sometimes placed on separate graphs.

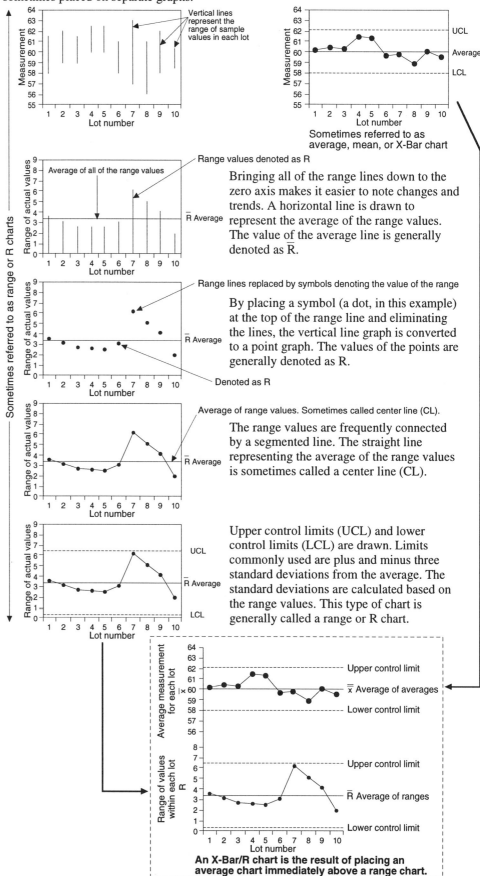

Sometimes referred to as average, mean, or X-Bar chart

Range values denoted as R

Bringing all of the range lines down to the zero axis makes it easier to note changes and trends. A horizontal line is drawn to represent the average of the range values. The value of the average line is generally denoted as $\bar{R}$.

Range lines replaced by symbols denoting the value of the range

By placing a symbol (a dot, in this example) at the top of the range line and eliminating the lines, the vertical line graph is converted to a point graph. The values of the points are generally denoted as R.

Average of range values. Sometimes called center line (CL).

The range values are frequently connected by a segmented line. The straight line representing the average of the range values is sometimes called a center line (CL).

Upper control limits (UCL) and lower control limits (LCL) are drawn. Limits commonly used are plus and minus three standard deviations from the average. The standard deviations are calculated based on the range values. This type of chart is generally called a range or R chart.

An X-Bar/R chart is the result of placing an average chart immediately above a range chart.

The combined pair of graphs at the end of the series is called an X Bar/R chart, X-bar and range chart, or X-bar/range chart. If deviation values had been used instead of range values it would be called an X-bar sigma chart, X-bar/deviation chart, or X-bar/S chart.

Control Charts (continued)

Relationship of control limits, specifications, zones or regions, etc.

Upper and lower control limits are typically more restrictive than specification values for a given operation. Control limits are sometimes looked upon as early warning systems so problems can be resolved before values outside the acceptable specifications occur. In some cases, multiple control limits are indicated on the chart to further assist in early detection of process shifts. This second set of control limits is sometimes referred to as the upper and lower warning limits. For instance, if the upper control limit is three standard deviations over the average, the upper warning limit (UWL) might be one-and-one-half or two standard deviations over the average. The areas that these multiple limits generate on the graph are sometimes called zones or regions. The example below depicts the relationships of the various guideline values and zones or regions.

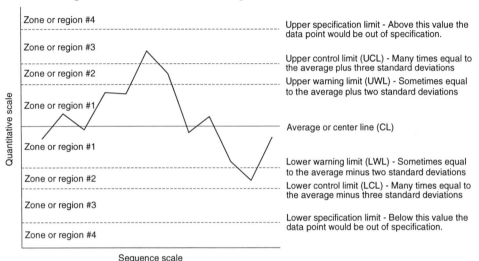

Illustration showing the relationships between the various control lines and zones or regions

Just because a process stays within its control limits does not necessarily mean it will produce acceptable results. Staying within the control limits generally only means the process is continuing to perform as it was when the averages and control limits were established. If, for example, the upper specification limit is lower than the upper control limit, parts exceeding the specification can be produced, yet the process might still be considered in control. This condition frequently exists when a specification is more restrictive than a piece of equipment is capable of producing.

Lots with different numbers of samples

Since the size of the standard deviation for a lot is a function of the number of samples in the lot, when lots containing different numbers of samples are plotted on the same control graph, the different control limit values for the various lot sizes are sometimes shown. Three methods are shown below.

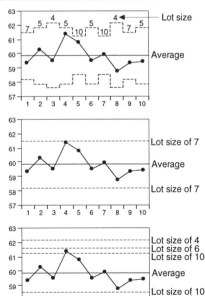

In one method, a separate control limit is shown for each lot. The limit shown for the lot is proportional to the number of samples in that particular lot. The numeric value for the number of samples in each lot may or may not be shown.

In another method, the single control limit shown is a blend of all the different lot sizes, – for example, average lot size, weighted average lot size, median lot size, etc.

In a third method, control limits for representative lot sizes are shown.

Control Charts (continued)

Construction details

Grid lines

Grid lines may or may not be used in control charts. Vertical grid lines are not uncommon, but horizontal grid lines are much less frequently used.

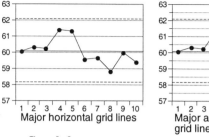
Major horizontal grid lines

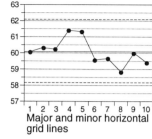
Major and minor horizontal grid lines

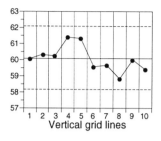
Vertical grid lines

Graph frame

Frames may or may not be used with control charts. They are sometimes eliminated so as to place emphasis on the average and control limits.

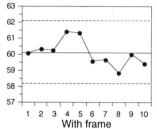
With frame

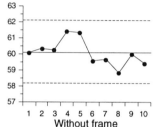
Without frame

Expanded scale

Expanded scales are frequently used on the vertical axis of control charts. Showing zero is only a secondary consideration and in many cases may detract from the major function of the chart. For instance, it would be much more difficult to monitor and control a process using the chart at the right, with the zero shown, than with the charts immediately above, which show the same data using an expanded scale.

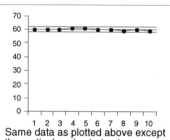
Same data as plotted above except the vertical scale starts at zero

Vertical orientation

Traditionally, control charts are oriented horizontally, primarily because sequence scales, particularly time series, almost always progress from left to right. Occasionally, however, there is some reason or advantage to orienting the chart vertically. Scales on the vertical axis of such graphs might progress from top to bottom or from bottom to top.

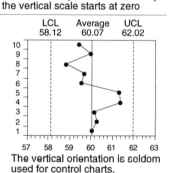
The vertical orientation is seldom used for control charts.

Cumulative sum chart

Sometimes referred to as a cusum chart. A cumulative sum chart is a variation of a control chart that is used when a high degree of sensitivity to small changes in a process is desired. Unlike most control charts, which plot actual values or averages of actual values, a cusum graph plots the accumulation of the differences or deviations of individual values from a

reference value. For example, if the historical average of a process is 10 and the first sample measured 10.5, 0.5 would be plotted on the graph. If the second sample tested 10.2, the difference from specification of 0.2 would be added to the 0.5 for the first sample and 0.7 would be plotted. If the third sample measured 10.7, the deviation of 0.7 from the center line value would be

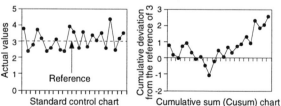
Comparison of a standard control chart plotting actual values and a cumulative sum graph plotting the sum of the differences between the actual values and the reference (3 in this example). The same data is plotted on both graphs.

added to the previous total of 0.7, and 1.4 would be plotted. Establishing control limits for cumulative sum charts is more difficult than with the other types of control charts since the angle of the line must be taken into consideration as well as the actual values. V-shaped templates are sometimes used for this purpose.

Coordinate	A coordinate is a number or value used to locate a point with respect to a reference point, line, or plane. Generally the reference is zero. If a data point has coordinates of X=3 and Y=4, the point would be located three units in the positive direction from zero on the X-axis and four units in the positive direction from zero on the Y-axis. Its coordinates would generally be written as 3,4. Coordinates on graphs can have positive or negative values. On maps, the directional coordinates are generally all positive. Elevations on maps are generally positive, but in some situations can be negative (e.g., points below sea level). The major function of coordinates is to provide a method for encoding information on charts, graphs, and maps in such a way that viewers can accurately decode the information after the graph or map has been generated. Examples of coordinates are shown below.

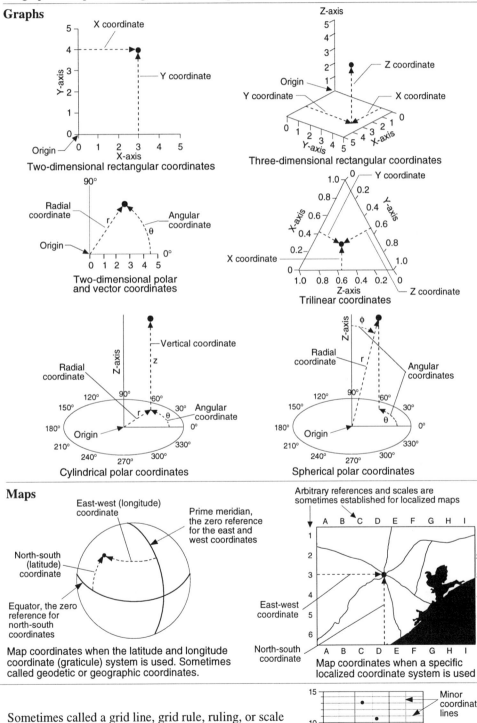

Coordinate Line	Sometimes called a grid line, grid rule, ruling, or scale line. Coordinate lines are thin lines used on graphs as visual aids for the purpose of encoding and decoding information. See Grid Line.

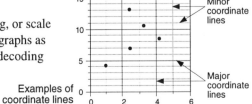

Correlation Graph

A graph whose primary purpose is to determine whether two or more sets of quantitative information are correlated (i.e., when one set of information changes, the other one simultaneously changes in some orderly fashion). Correlation graphs are used extensively to explore areas where meaningful associations might exist and as tools for looking into possible cause and effect relationships. Correlation graphs fall into two major categories: (1) where two or more individual data sets are analyzed, and (2) where two or more data series (each series consisting of two or more data sets) are analyzed.

Analysis of data sets for correlations

———Using scatter graphs to check for correlations———

In this type of correlation graph, specific data sets are plotted, such as pollen count and number of sinus pills sold, or selling price of a product and the volume of the product sold.

Scatter graphs are most widely used in this application. One data set (generally the independent variable) is plotted on the horizontal axis and the other data set on the vertical axis. For example, pollen count could be plotted on the horizontal axis, and sinus pills sold on the vertical. A line is frequently fitted to the resulting group of data points to assist in the analysis. An example is shown at the right. The pattern formed by the data points determines whether a correlation exists, what type of correlation it is, and how strong the correlation is. For example, the pattern of the data points in the example at the right indicates a direct, linear, and strong correlation. It is called a direct correlation

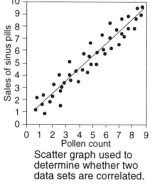

Scatter graph used to determine whether two data sets are correlated.

because as the pollen count increases, the sales of sinus pills also increase. It is described as linear since a straight line best fits the data points. It is called a strong correlation because

the data points are tightly clustered around the inclined fitted line. • There are several different patterns the data points can form; in many cases it is a judgment call as to whether correlation is weak or non-existent and whether the relationship is linear or nonlinear. Examples illustrating some of the key types of patterns are shown here.

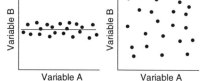

Two examples of no correlation. On the left, as variable A changes there is little on no change in Variable B. On the right, as A changes, B changes, but in no meaningful or orderly pattern.

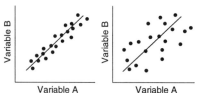

Two examples of direct correlations. The example on the left shows a strong correlation because the points are tightly clustered about the fitted line. The example on the right shows a weak correlation.

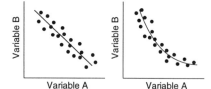

Two examples of inverse or negative correlation (e.g., as variable A increases, variable B decreases). The example on the left is considered linear because the fitted curve is a straight line. The example on the right is nonlinear or curvilinear, as indicated by the curved line.

———Using scatter graph matrixes to check for correlations———

When there are more than two variables and one wants to see if there is a correlation between any given pair, it is necessary to generate a separate graph for each combination of variables. Technically, the number of graphs required can be reduced by plotting multiple sets of data points on a single graph. However; this technique generally produces a very cluttered graph that is hard to interpret. An alternate method is to arrange graphs of two data sets each into a matrix, as shown at the right. There is no limit as to how many individual graphs can be incorporated into such a matrix. Since the viewer is generally looking only for patterns of data points, the small size of the graphs is seldom a problem. The major advantages of matrixes are their compact size and ease of analysis. See Matrix Display.

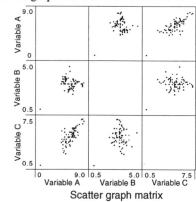

Scatter graph matrix

Correlation Graph
(continued)

Analysis of data sets for correlations (continued)

The use of graphs other than scatter graphs to check for data set correlations

Graphs other than scatter graphs can also be used to check for correlations between sets of data. The three graphs below display the same data as shown in the scatter graph at the top of the previous page. In these three graphs, the data points are grouped into intervals along the horizontal axis, and the data elements in those intervals are averaged and plotted along the vertical axis. In each case the same correlation can be seen. The two examples on the left give no indication of how tightly the data points are clustered around the averages. The example on the right overcomes this shortcoming by indicating the range of values, using the short vertical lines to designate the upper and lower values in each interval. A similar technique might be used with the column graph.

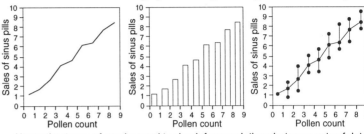

Alternative types of graphs used to check for correlations between sets of data

Analysis of data series for correlations

Time series graphs are frequently used to determine whether correlations exist between multiple data series. In addition to exploring areas where meaningful associations might exist and looking for clues to possible cause and effect relationships, time series graphs are also used for establishing leading indicators in order to make projections. Although the sequence in which data is collected for the scatter graph applications described above is generally not important, with time series it is very important. Three of the major types of correlations that are often looked for in this type of graph are shown below.

Direct and inverse correlations

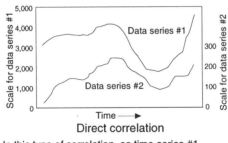

Direct correlation

In this type of correlation, as time series #1 increases, series #2 also tends to increase; when #1 decreases, #2 tends to do the same. The amount or rate of change in the two series may or may not be proportional.

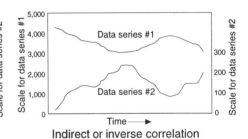

Indirect or inverse correlation

In this type of correlation, as time series #1 increases, #2 decreases and when #1 decreases, #2 increases. The amount or rate of change in the two series may or may not be proportional.

Correlations that are shifted in time

Sometimes the curves of two sets of data that are correlated are shifted in time. For example, the curve of retail prices on a given product often correlates with the curve of wholesale prices on the same product, except that changes in the retail curve lag behind changes in the wholesale curve by some period of time. Data series with this relationship are frequently referred to as leading or lagging indicators. Sometimes it is necessary to shift one of the curves on the graph to discover that a correlation actually exists. In the example at the right, data series #1 is a leading indicator of data series #2. What that means in this example is that two months after series #1 changes, series #2 will change in the same way. • Direct or inverse correlations can be shifted in time.

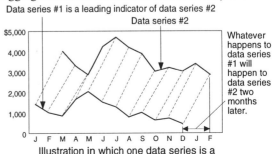

Illustration in which one data series is a leading indicator of another data series

111

Correlation Graph
(continued)

Analysis of data series for correlations (continued)
———Correlations with a particular curve pattern ———

In some cases a correlation does not exist between two entire data series (curves) but instead between one curve and a specific portion or pattern of another curve. One type of pattern that is often watched for is a reversal in the slope of a particular curve, whether it be positive to negative or negative to positive. This is illustrated in the example at the right. In this example, each time there is a reversal in the slope of data series A, there is a short term increase in data series B. For example, each time there is a change in interest rates, whether it be up or down, the sales of certain stock brokers experience a short term increase.

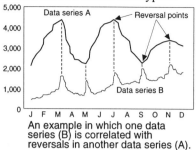

An example in which one data series (B) is correlated with reversals in another data series (A).

Analysis of data sets and data series for correlations

Frequently one is interested in potential correlations between data series as well as individual data sets – for example, the relationship between selling price and total sales in terms of dollars and units, as shown at the right. This graph illustrates a direct linear correlation between selling price and total sales dollars (data sets), an inverse linear correlation between selling price and units shipped (data sets), and an inverse correlation between units shipped and total sales dollars (data series). • A correlation does not necessarily imply a cause-and-effect relationship. For instance, in this example, the increased selling prices did cause the total sales dollars to go up and the units to go down. In this case there is both a correlation and a cause-and-effect relationship. In the same graph, there is an inverse correlation between total dollars of sales and units shipped; however, there is no cause-and-effect relationship. The lower number of units shipped did not cause the total sales to go up. In this case they both happened to be affected by the same third variable, which was selling price.

Paired bar graph

A paired bar graph is occasionally used to check for correlations. The two horizontal scales may or may not be the same. For example, one might be in terms of number of arrests and the other in number of convictions. Or one might be in terms of number of arrests and the other in dollars spent on crime prevention.

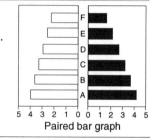

Paired bar graph

Corridor Buffer

A variation of a buffer used with maps. See Buffer Map.

Cosmograph

A chart showing how equal inputs and outputs are distributed – for example, how much money a government agency or company receives from various sources and what portions of those total monies are spent where. Input is normally on the left and output on the right. Values might be in any units of measure such as dollars, percent, or units. Cosmographs were originally generated by arranging mechanical models and then photographing them. There are many potential variations. Two are shown at the right.

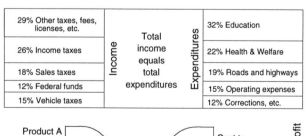

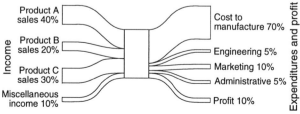

Cosmographs where total input on left equals total output on right.

Count

Sometimes referred to as frequency or frequency of occurrence. See Histogram.

Critical Path Method (CPM) **or** **Critical Path Analysis (CPA)**	Sometimes referred to as an arrow diagram. Critical Path Method (CPM) charts are time and activity graphic networks that represent the major activities and events of a large program and show their interrelationships. They are used to plan, analyze, and monitor programs. Among other things, they are helpful in determining how programs can be shortened and identifying which of the many subprograms are the most critical in assuring that an overall program is completed on schedule. CPM charts accomplish basically the same functions as PERT charts. The major difference is that PERT charts focus on events (e.g., start pouring foundation or foundation completed), while CPM charts focus on activities (e.g., pour the foundation). A CPM chart diagrams a program or project using a separate arrow (sometimes referred to as link, branch, or arc) to represent each of the major subprograms or activities. All arrows start and end at a node (exceptions are sometimes the first and last arrows). Nodes represent events. An event may be a review meeting, a decision point, or simply the time at which one activity is completed and another begins. Nodes are represented by geometric symbols such as squares, circles, ovals, etc., and are numbered sequentially. The main body of a CPM chart generally starts and ends with a single activity (arrow) or node.

Arrangement of arrows and nodes

The arrows and nodes are arranged in the same order they actually occur. The first activity is represented by the first arrow on the left, the last activity by the last arrow on the right, and all others are placed in their proper sequence in between. If an activity (B) can not be started until another activity (A) is completed, the arrows representing these two activities would be drawn such that the tip of the arrow for activity A joins with the base of the arrow for activity B with a node symbol in between. If neither activity is dependent on the completion of the other, the tip of one arrow and the base of the other arrow do not meet. • The bases of multiple arrows often meet at the right side of a common node, indicating that the activity(s) on the left side of the node must be completed before any of the activities on the right side can be started. When the heads of multiple arrows meet at the left side of a common node, it means that all these activities must be completed before the activity(s) on the right side of the node can proceed. Examples of all three variations are shown at the right, and an abbreviated critical path method diagram is shown below.

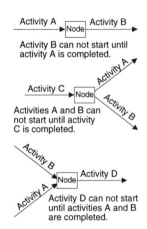

Examples of how relationships between activities are represented graphically

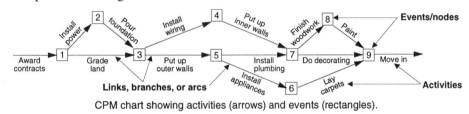

CPM chart showing activities (arrows) and events (rectangles).

Methods for identifying activities

In actual CPM charts it is not uncommon to have dozens or even hundreds of activities. A word description of the activity represented by the arrow can be placed alongside of the arrow (as shown above) or code numbers or letters can be used (below). Events (nodes) can be coded the same way. When a coding system is used, a cross-reference table is included.

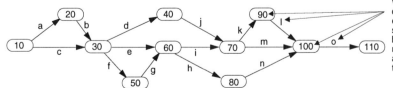

Word descriptions of activities and events are sometimes replaced by reference numbers or letters and explained in a table.

CPM chart with events and activities coded. In an actual chart, a table or legend would match a description with each code number or letter.

An alternative way of identifying activities is by the numbers at the ends of the arrows. For example, activity g above might be referred to as activity 50-60 and activity m as activity 70-100.

**Critical Path Method (CPM)
or
Critical Path Analysis (CPA)**
(continued)

Incorporating time into a critical path method chart
Times noted on arrows

Small numbers alongside the arrows sometimes denote information regarding the timing of the activity. Four methods currently used are illustrated in the examples below. In one case a single number is used to represent the estimated time required to accomplish an activity.

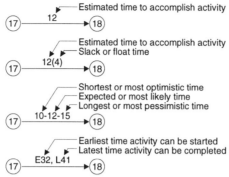

Examples of how times (days, weeks, months, etc.) to complete an activity might be noted

Another method is to show two numbers, the first being the estimated time required to complete the task and the second (many times in parentheses) the amount of additional time (called float or slack) available before that activity or the series of activities in that chain delays the overall program. A third method is to show three times, the shortest (most optimistic), the expected, and the longest (most pessimistic). The fourth alternative is to indicate the earliest time an activity can be started (E), and latest time an activity can be completed (L), without delaying the overall completion of the program. The times noted in the first three methods are incremental and apply to the length of time the activity will take. In the fourth alternative, the times are referenced from the time the overall program was started.

Times noted on nodes

In addition to incremental times for each activity, cumulative times for events are sometimes also shown. Two numbers that are sometimes used are the earliest and latest time each event can occur. For instance, in the example below, event #5 cannot occur before the eighteenth week after the start of the overall program, and if it occurs after the twenty-fifth week, the critical events behind it will be delayed.

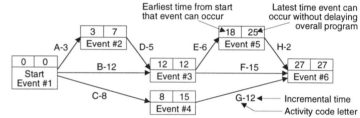

Chart showing cumulative times for events and incremental times for activities

Time scale for reference

The length of arrows are normally not proportional to the length of time the activity is estimated to take. Because of this, the viewer normally cannot estimate visually how long an activity will take or when an event will occur in real time versus elapsed time. In an attempt to give some visual indication of relative times, CPM charts are sometimes arranged along a time scale, as shown below.

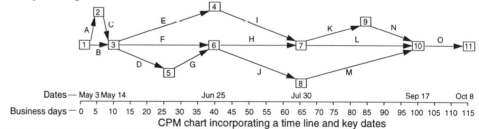

CPM chart incorporating a time line and key dates

Critical path

Once times are entered, the longest time path can be determined. This is called the critical path and is generally designated on the chart with bold or colored arrows. This is the path along which a delay in any event will cause a delay in the overall program. An example with times designated and the critical path noted is shown at the right.

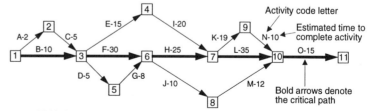

CPM chart showing code letters, activity times, and critical path.

In some cases the second most critical path is indicated by arrows drawn with a width of line somewhere between that of a standard path and a critical path.

114

**Critical Path Method (CPM)
or
Critical Path Analysis (CPA)**
(continued)

Dummy arrows

Dummy arrows (indicated with dashes) are sometimes used to overcome some layout and identification problems. For example, if two activities occur between events 7 and 8 as shown at the left below, there is no way to differentiate the two activities if one uses the convention of referring to activities by their start and finish event numbers (i.e., both would be referred to as 7-8). A dummy arrow, as shown at the right, that has no time associated with it solves the problem. After the insertion of a dummy arrow, one activity is still referred to as 7-8 while the other gets the new designation of 7-9. The dummy activity of 8-9 is ignored.

Without the use of a dummy arrow, the two activities have the same designation, in this example 7-8.

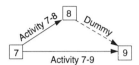

With the use of a dummy arrow, the two activities each have a unique designation and the dummy path/activity is ignored.

General

– Arrows generally proceed from left to right, run parallel when possible, and seldom cross.
– Where possible, the number at the base of an arrow is smaller than the number at the head.
– In some cases the charts are updated periodically to reflect new information and/or the status of the program.
– More and more CPM and CPA charts are being generated and maintained on computers. When done on computers, in addition to the graphical diagram, the information is frequently also available in tabular form.
– Several optional construction details are shown below.

Additional information is sometimes placed beside the activity arrows. Examples of such information are costs, departments responsible for coordinating the activity, people and disciplines involved, documentation needed, etc.

Straight or curved arrows can be used although straight arrows are generally preferred.

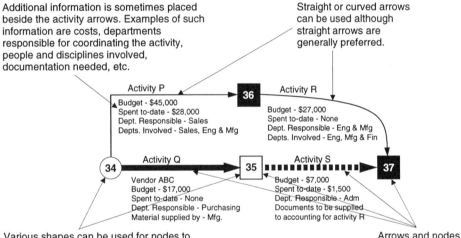

Various shapes can be used for nodes to designate similar events (e.g., review meetings, decisions, etc.) or responsibilities (e.g., person, department, etc.).

Arrows and nodes can be shaded, colored, or patterned to encode additional information.

Optional construction details

Crosshatch

The term hatch is used to refer to a pattern of fill that consists of evenly spaced parallel lines. If there are two or more sets of parallel lines intersecting one another, the pattern is referred to as crosshatch.

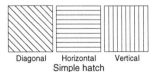

Diagonal　Horizontal　Vertical　　　Diagonal　Horizontal
　　Simple hatch　　　　　　　　　　Crosshatch
Patterns referred to as hatched and crosshatched

Crosshatched Map

Sometimes referred to as a choropleth, shaded, or textured map. A crosshatched map is a variation of a statistical map that displays area data by means of shading, color, or patterns. The areas (sometimes called areal units) might be countries, states, territories, counties, zip codes areas, trading areas, etc. The data are generally in terms of ratios, percents, rates, etc., as opposed to absolute units. For example, incomes will typically be given in terms of dollars per capita, dollars per household, etc., as opposed to total dollars for the area. Data is often organized into class intervals, as in the example at the right. See Statistical Map.

350 to 400
300 to 349
250 to 299
200 to 249
150 to 199
100 to 149
50 to 99
0 to 49

Crosshatched map

Cumulative Frequency Graph

Sometimes referred to as an ogive (pronounced oh-jiv) or summation graph. A cumulative frequency graph is the equivalent of a cumulative histogram with the columns replaced by a smooth curve (See Histogram). This type of graph has four major purposes:

– To provide a graphical means for determining the percent and/or numeric probability of a given value or set of values occurring
– To graphically categorize the data elements in meaningful compartments such as quartiles, deciles, percentiles, etc.
– To determine whether a data set has a normal distribution
– To provide a convenient method for comparing the distributions and/or probabilities of multiple data sets

Individual values or class intervals may be used when constructing a cumulative frequency graph. In one of the more widely used variations of a cumulative frequency graph the individual values or class intervals are plotted on the horizontal axis and the cumulative frequencies on the vertical axis. The frequency scale might be in terms of cumulative frequency, cumulative percent (sometimes called cumulative relative frequency), or both. When constructing a cumulative frequency graph, one prepares the data just as thought a standard histogram or frequency polygon was being generated. The key difference is that instead of plotting the individual frequency for each class interval, the cumulative frequency is plotted. The cumulative frequency is equal to the individual interval's frequency plus the frequencies for all preceding intervals. An example of a cumulative frequency graph is shown at the right. A histogram and cumulative histogram for the same data are included for reference. • When individual values are plotted on the horizontal axis, data points are located directly above the values. With class intervals the points are located over the boundaries.

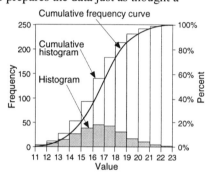

Graph showing cumulative frequency curve, cumulative histogram, and a standard histogram for the same data

"More than" and "less than" graphs

When class intervals are used, the cumulative data points are generally plotted in line with either the upper or lower boundaries of each class interval. A frequency scale, percent frequency scale, or both can be used on the vertical axis. The cumulative histograms in the examples below are for reference only and are not used in actual applications.

"Less than" cumulative frequency graph

When the cumulation process proceeds from the smallest value to the largest, the data points are plotted in line with the upper boundary of the class interval. This variation is often called a "less than" graph since these are the values that can be read directly from the scale on the graph. For instance, in the example at the right, 63% of all the data elements have values equal to or less than 2.1.

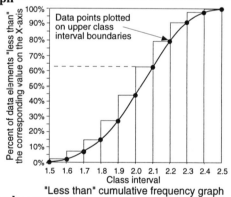

"Less than" cumulative frequency graph

"More than" cumulative frequency graph

When the cumulation process proceeds from the largest value to the smallest, the data points are generally plotted in line with the lower boundary of the class interval. This variation is frequently called a "more than" graph since these are the values that can be read directly from the scale on the graph. For instance, in the example at the right, 37% of all the data elements have values equal to or more than 2.1.

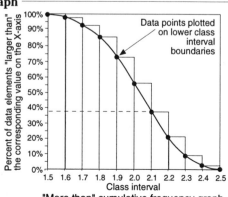

"More than" cumulative frequency graph

"More than" and "less than" graphs (continued)

────── **Combining the "more than" and "less than" curves on a single graph** ──────

In the graph at the right the two types of curves are combined onto a single graph by adding a second scale. The "less than" values are read directly from the scale on the left and the "more than" values directly from the scale

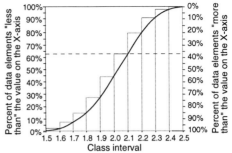

"More than" and "less than" graph with one curve and two scales

on the right. The graph at the left combines the two curves by superimposing both on the same graph. In this case only a single scale is required.

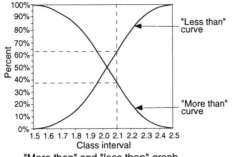

"More than" and "less than" graph with two curves and one scale

---

Cumulative frequency curves used for percentile and probability graphs

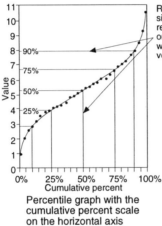

Reference lines simplify the task of relating percentiles on the horizontal axis with values on the vertical axis.

Percentile graph with the cumulative percent scale on the horizontal axis

Although cumulative percents are frequently plotted on the vertical axis, the axes can be reversed and percents plotted on the horizontal axis, as shown at the left. When this is done the graph is sometimes referred to as a percentile or probability graph. If one of the major purposes of the graph is to categorize the data into meaningful compartments such as quartiles and deciles, reference lines are sometimes added to more easily relate percentiles on the horizontal axis with values on the vertical axis. A further potential step is to align the data points vertically and convert the information to a one-axis graph, as shown at the right. Values should not be interpolated using this type of one-axis graph.

One-axis percentile graph using the same data as the graph at the left. In this graph, all of the data points have been vertically aligned and the cumulative percent values transferred from the horizontal to the vertical axis.

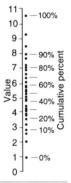

Probability graph using normal probability grid

Knowing whether a data set has a normal distribution contributes to the reliability of assumptions, predictions, mathematical analyses, etc., that are associated with the data. One use of a cumulative relative frequency graph is to determine whether a data set has a normal distribution. This is done by plotting the cumulative percents on a special grid referred to as a normal probability grid. If the data set has a normal distribution, the data points cluster around a straight line as shown at the right. The straight line characteristic of a normal probability grid enables one to more easily make projections and interpolations. Because of the nonlinear scale on the cumulative percent axis of a normal probability graph, this type of curve is commonly generated with a computer or by using special preprinted graph paper. Cumulative percent scales can be shown on either axis. They are most often shown on the horizontal axis. Due to a quirk of mathematics, there is no zero or 100% on the cumulative percent scale of a normal probability grid. The scale on the vertical axis is generally linear.

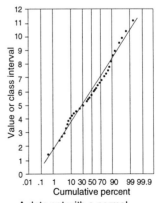

A data set with a normal distribution approximates a straight line when plotted on a normal probability grid

Comparisons using cumulative frequency curves

Cumulative frequency graphs enable the viewer to make comparisons of multiple data sets that would not be possible with simple frequency distribution graphs. For example, the graph below displays the cumulative percent of failures for two different product lines against years of service. Typical observations based on this graph might be:

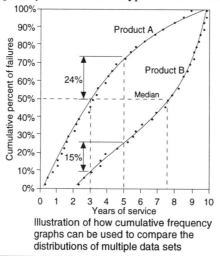

Illustration of how cumulative frequency graphs can be used to compare the distributions of multiple data sets

– Although both product lines had a few units that survived almost 10 years, the median life (50% cumulative failures) of product A was only three years, versus about 7.5 years for product B.

– Between years three and five, 24% of product A failed, versus approximately 15% for product B.

– With product A there is only a 26% probability a given unit will live beyond five years, versus a 75% probability for B type products.

Other than readability, there is no limitation on the number of curves that can be compared on a single graph.

Graphs with cumulative frequency scales on both axes

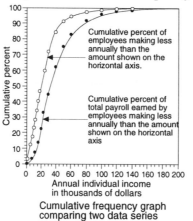

Cumulative frequency graph comparing two data series

In most cases, values or class intervals are shown on one axis and cumulative frequency or cumulative percent frequency on the other. For special purposes, cumulative percent frequency scales are occasionally displayed on both axes. Such a graph is sometimes referred to as a Lorenz graph. The traditional graph at the left displays the cumulative percent frequency curves for total income and total numbers of employees for a hypothetical company. It may be noted that the cumulative number of employees rises faster than the cumulative income; however, it is difficult to get a good appreciation for the relationship between the two sets of data. The combined graph in the upper right-hand corner of the box below has the same two data series plotted on it; however, a cumulative percent scale is used on both axes. The graphs alongside and below it are included for reference only. A graph with two cumulative scales sometimes illustrates differences between two sets of data more clearly. For example, it can be noted in the combined graph that the lower 50% of the employees earned about 24% of the total income, while the upper 10% of the employees earned about 30% of the total income. If the two distributions had been the same, the data points would have clustered around the diagonal line, sometimes referred to as the line of equal distribution.

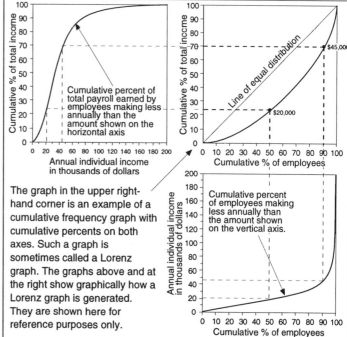

The graph in the upper right-hand corner is an example of a cumulative frequency graph with cumulative percents on both axes. Such a graph is sometimes called a Lorenz graph. The graphs above and at the right show graphically how a Lorenz graph is generated. They are shown here for reference purposes only.

Cumulative Graph

All values on a cumulative graph reflect totals with respect to some common starting point. For example, the value plotted for any given interval or period is equal to the incremental value for that period plus the sum of the incremental values for all of the prior periods back to the common starting point. Month-to-date (MTD) and year-to-date (YTD) graphs are among the better known types of cumulative graphs. In a YTD graph the common starting point is frequently the first of the year. The value for the first month, say January, is plotted by itself. In February the value for February plus January is plotted. In March it is March plus February plus January. This continues throughout the year until, for the month of December, the sum of all 12 months is plotted. In January the whole process starts again. This procedure applies whether actual or deviation values are being graphed. Examples of both are shown at the right, with the actual values shown for reference.

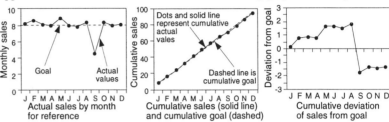

Comparison of a graph of actual data (left) versus two forms of cumulative data graphs. Cumulative deviation or difference is generally more sensitive for monitoring purposes.

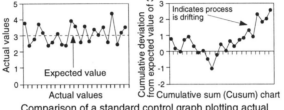

Comparison of a standard control graph plotting actual values and a cumulative sum chart plotting cumulative deviations from a reference (a value of 3 in this example)

Cumulative graphs can be generated with almost any type of sequential scale on the horizontal axis. The examples above use time scales. The two examples at the left use the order in which samples were collected in a quality control process as the sequence. For reference, one graph displays actual values. The other, called a cumulative sum (cusum) chart, displays cumulative differences between the actual value and an expected value. Cusum graphs are more sensitive to small, gradual variations than a graph of actual values. As a general rule, cumulative deviation or difference graphs tend to be more sensitive, for monitoring purposes, than overall cumulative graphs. • If one wants to display individual increments for each period as well as cumulative values, a column graph is sometimes used, as shown at the right. In this type of graph the cumulative values are designated by the tops of the columns, while the incremental values are indicated by a shaded or colored component at the bottom of each column. • An algebraic summing process, taking plus and minus signs into consideration, is used when determining the values for cumulative graphs; therefore, both positive and negative values can be used in the calculation as well as on the graph. • Multiple data series can be plotted on the same cumulative graph. When multiple sources contribute to the values being cumulated, stacked column or stacked area graphs are sometimes used to indicate how much each source contributes to overall values.

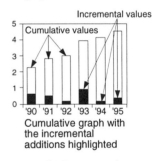

Cumulative graph with the incremental additions highlighted

Cumulative Histogram

A variation of a histogram in which cumulative values are plotted instead of incremental values. The frequency or percent to be plotted for a given class interval is determined by adding the frequency value or percent for that class interval to the frequencies or percents for all class intervals preceding it. The example below shows only cumulative percents. A graph of cumulative frequency is identical except for the scale. In some graphs, frequency and percent frequency are both shown. In this example, the values are cumulated from the smallest value to the largest. An alternative method is to cumulate from the largest value to the smallest. Sometimes cumulative histograms are used to determine what percentage of elements in a data set are more than or less than a particular value; however, a cumulative frequency graph is generally recommended for this purpose. See Histogram and Frequency Polygon as well as Cumulative Frequency Graph.

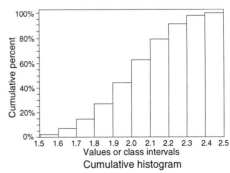

Cumulative histogram

Cumulative Sum Chart

Sometimes referred to as a cusum chart. A cumulative sum chart is a variation of a control chart that is used when a high degree of sensitivity to small changes in a process is desired. Such charts offer three major advantages over standard control charts.

    – Small changes can be detected more easily.

    – Changes that occur over long periods of time are easier to spot.

    – It is easier to determine when a change in the process began.

Unlike most control charts that plot actual values or averages of actual values, a cusum chart plots the accumulation of the differences or deviations of actual values from a reference value. For example, if the historical average of a process is 10 and the first sample measured 10.5, 0.5 is plotted on the graph. If the second sample tested 10.2, the difference from specification of 0.2 is added to the 0.5 for the first sample, and 0.7 is plotted. If the third sample measured 10.7, the deviation of 0.7 from the center line value is added to the previous total of 0.7 and 1.4 is plotted, and so on. • Establishing control limits for cumulative sum charts is more difficult than with the other types of control chart since the angle of the line must be taken into consideration as well as the actual values. V-shaped templates are sometimes used for this purpose.

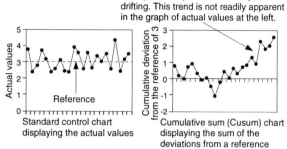

Comparison of a standard control chart plotting actual values (left) and a cumulative sum graph (right) plotting the sum of the differences between the actual values and a reference (3 in this example). The same data is plotted on both.

Curve

In the context of graphs, the term curve refers to any line on a graph used to represent a set or series of data. The table below shows the major types of lines that are most frequently used to generate curves.

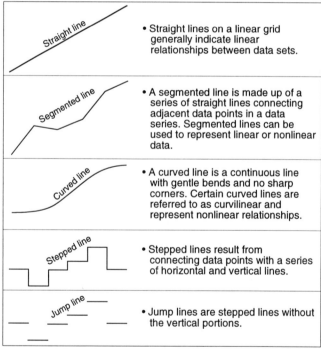

Straight line	• Straight lines on a linear grid generally indicate linear relationships between data sets.
Segmented line	• A segmented line is made up of a series of straight lines connecting adjacent data points in a data series. Segmented lines can be used to represent linear or nonlinear data.
Curved line	• A curved line is a continuous line with gentle bends and no sharp corners. Certain curved lines are referred to as curvilinear and represent nonlinear relationships.
Stepped line	• Stepped lines result from connecting data points with a series of horizontal and vertical lines.
Jump line	• Jump lines are stepped lines without the vertical portions.

Variations of curves used on graphs to represent data

Curve Difference Graph

Sometimes referred to as an intersecting silhouette graph, intersecting band graph, or difference line graph. A curve difference graph is a graph on which the curves representing two data series intersect one another and the areas between the two curves are highlighted by means of shading or coloring. For example, one curve might represent the dollar value of exports each year and the other the dollar value of imports, as shown in the example at right. When exports have a higher value than imports, one shade or color is used. When exports are lower than imports, another shade or color is used.

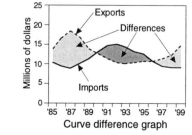

Curve difference graph

Curve Fitting

Sometimes referred to as smoothing or line-of-best fit. When used with a sequence scale on the horizontal axis fitted curves are sometimes called trend lines. Curve fitting is a process in which a curve or surface that most closely approximates a data series is superimposed over a plot of the data points of that data series. Fitted curves may be drawn free hand based on visual estimates or plotted using well-established mathematical procedures. There are a number of reasons fitted curves are used including:

– Establishing the general trend of data;
– Determining the type of relationship between two variables such as linear or exponential;
– Comparing data series that are intermingled;
– Making forecasts or projections;
– Determining the degree of variation of individual data points from a theoretical or expected curve; and
– Determining whether data points vary randomly, symmetrically, uniformly, etc., from a theoretical curve.

Multiple fitted curves for same data

Computers can generate many different curves for the same data. It is up to an individual to decide which of the curves either most closely approximates the actual data or is best for a given purpose. To illustrate the point, four different computer-generated curves based on the same data are shown at the right. The person using the data would have to decide which of these four, plus many other possible curves, is appropriate for the application. In some cases it is not known what the relationship is or should be between the two variables; therefore, many types of curves might be reviewed to learn which best fits the data. In other cases, it is known that if a process is working properly, the relationship

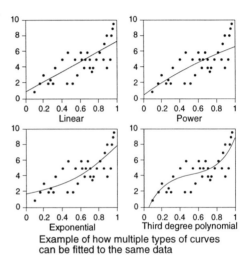

Example of how multiple types of curves can be fitted to the same data

of the two variables should cluster around the theoretical curve that defines the process or relationship. In these cases, perhaps only the one theoretical curve is fitted. If the data points approximate that curve, the process might be consider to be functioning properly.

Spline type fitted curve

It is rare for a fitted curve to intersect every data point. Among the few exceptions are certain variations of spline curves. With this type of fitted curve, the fitted line passes through every point as shown at the right. The data used in the example at the right are the same as used in the four examples above.

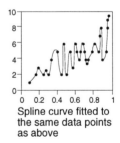

Spline curve fitted to the same data points as above

Linear regression line

The straight line variation of a fitted curve is sometimes referred to as linear regression line. When the location of the line is developed mathematically, a technique called least-squares is commonly used and the curve is occasionally called a least squares line. Linear

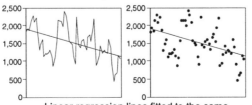

Linear regression lines fitted to the same data on two different types of graphs

regression lines can be used with most types of graphs. They are often used with scatter and line graphs. Two examples are shown at the left. The same data is plotted on both. See Linear Regression Line.

121

Curve Fitting (continued)

Residuals

After curves are fitted to a data series, certain techniques may be used to determine such things as which type of curve best fits the data, how well the curve fits, and whether the data points differ from the fitted curve in some regular pattern. Residuals, defined as the distances between the data points and the fitted curve, are often used for this purpose. The residuals may be observed on the original graph or, for a more detailed analysis, may be plotted on a separate graph by themselves. See Residuals.

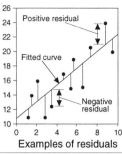

Examples of residuals

Independent versus dependent variable

When there are two quantitative variables, the decision as to which variable is a function of the other (i.e., which is the independent and dependent variable) can have an effect on the location of the fitted curve as shown at the right. In this illustration, the same data is plotted in both graphs. In one graph, the Y variable is dependent. In the other, the X variable is dependent.

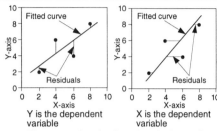

Y is the dependent variable

X is the dependent variable

The same data is plotted on both graphs

Projections

Fitted curves are often used for making projections. The examples shown here illustrate

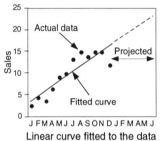

Linear curve fitted to the data

how important it is that the proper fitted curve be selected when making projections. In the example on the left, the linear fitted curve would indicate continued growth in sales in the immediate future. But the polynomial fitted to the

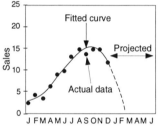

Polynomial curve fitted to the data

same data on the right indicates a potential serious downturn in sales.

Fitted surfaces and three-dimensional graphs

Surfaces can be fitted to three-dimensional point data similar to the way lines are fitted to two-dimensional data. An example is shown at the right. Three-dimensional fitted surfaces are much more difficult to interpret that two-dimensional fitted curves.

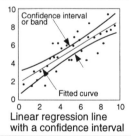

Surface fitted to three-dimensional scatter data

Dots designate actual data points

Confidence interval or band

Fitted curves are best estimates based on the data available. Many times the available data is a sampling of a larger family of data. A confidence interval indicates the region in which the fitted curve would probably lie, with a given degree of confidence such as 90% or 95%, if information for the entire family of data was available. See Confidence Interval.

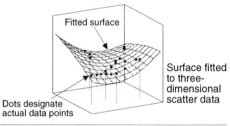

Linear regression line with a confidence interval

Curves fitted to multiple data series

Curves can be fitted to multiple data series on the same graph. This is particularly advantageous when data series are interspersed as in the examples at the right. In the example with only the data points, the difference in the general slopes of the two data series is difficult to discern. With the fitted curves, the difference stands out more clearly.

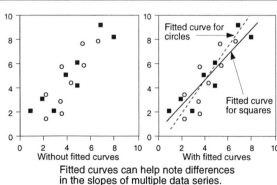

Without fitted curves

With fitted curves

Fitted curves can help note differences in the slopes of multiple data series.

Curve Fitting (continued)

Curve fitting applied to various types of graphs

Although most frequently applied to scatter graphs, curve fitting can also be used with other types of graphs such as column, bar, line, and area.

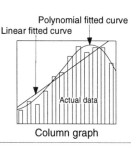
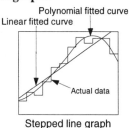

Effect of unusual data points on fitted curves

Unusual data points (sometimes called outliers) can sometimes have a significant effect on the curve fitted to the data. For example, a single unusual data point can cause a dramatic shift in the angle of the regression line, as shown at the right. Only those familiar with the data can determine whether the line with or without the unusual data point is more representative.

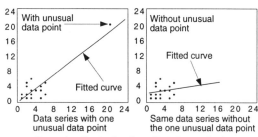

Examples illustrating the effect that a single unusual data point can have on a fitted curve

Equations for fitted curves

When curve fitting is done with a computer, the equation for the fitted curve is sometimes accessible to the operator. With this equation, the fitted curve can be reproduced on other graphs; mathematical calculations can be performed; and/or specific observations about the curve can be noted. Two examples of fitted curves and their equations are shown at the left.

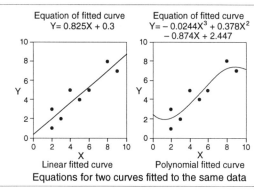

Equations for two curves fitted to the same data

Curved Line
or
Curvilinear Line

A curved line is one of five major types of lines used to represent data on graphs. The curved line is generally continuous with gentle bends and no sharp corners. Such lines frequently represent nonlinear data. Examples are shown at the right.

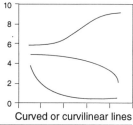

Curved or curvilinear lines

Curve Graph

Sometimes referred to as a line graph. Curve graphs are a family of graphs that display quantitative information using lines. They are used extensively for plotting quantitative information against a time or other sequential scale. See Line Graph.

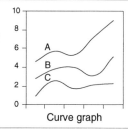

Curve graph

Cusum Chart

An abbreviation for cumulative sum chart. A cusum chart is a variation of a control chart used when a high degree of sensitivity to small changes in a process is desired. The examples at the right illustrate the increased sensitivity of a cusum chart by plotting the same data on a standard control chart and a cusum chart. See Cumulative Sum Chart.

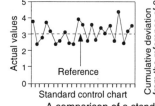
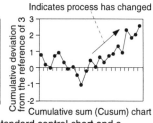

A comparison of a standard control chart and a cusum chart with the same data plotted on both

Cycle

Sometimes called bank, deck, phase, or tier. The major interval on a logarithmic scale.

Data Distribution Graphs

Many times an average or median value is inadequate to describe a set of data, and a graph showing the distribution of the data is necessary. For example, in addition to an average value, one might want to know whether the data elements are distributed evenly or clustered. If clustered, how many clusters and where? What are the maximum and minimum values and are there any unusual data elements? What percent of the data elements are over a certain value

and which value occurs the most frequently? Shown on this page and the next are typical examples of data distribution graphs used to answer these types of questions. The information here is meant only as an overview; detailed information on each graph is provided under its individual heading.

			Distribution of an individual data set	Distributions of multiple data sets
One-axis data distribution graphs	**With every data element represented**	**Point graph** Sometimes referred to as a one-axis point graph. In this type of graph every element of a data set is plotted as a separate point. Occasionally used in conjunction with other types of graphs such as in the borders of scatter graphs.		
		Stripe graph Similar to a one-axis point graph except the points are replaced by short lines (stripes). Frequently used in conjunction with other types of graphs such as in the borders of scatter graphs.		
	Data summarized	**Box graph** Summarizes key features of the distribution of a data set. The example at the right displays one combination of key elements that might be shown. There are many other possible variations. Certain variations are sometimes referred to as percentile plots or graphs.	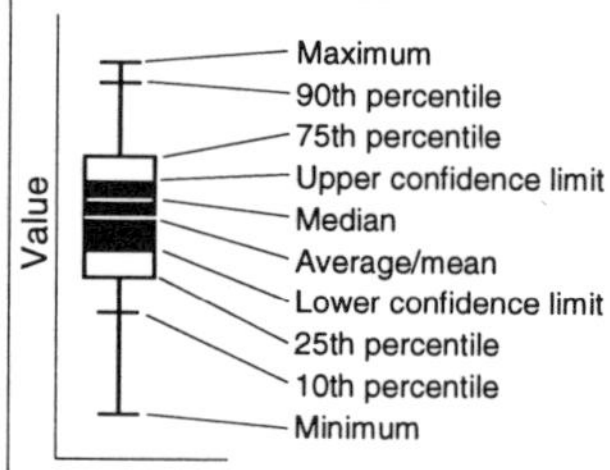	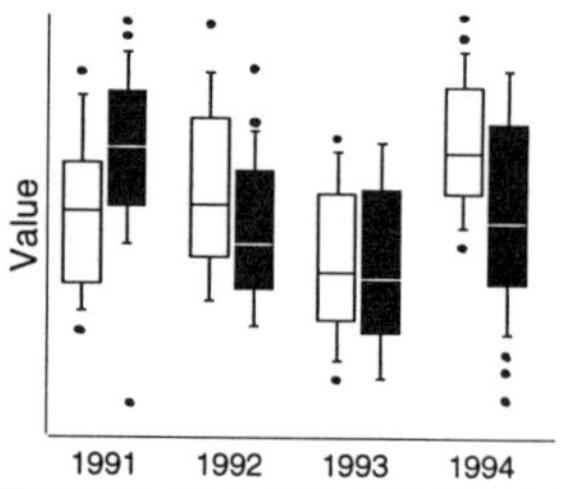
Frequency distribution graphs with every data element represented. The major emphasis is on concentrations of data elements and skewness.		**Stem and leaf chart** The first digit(s) of each data element in a data set is shown to the left of a vertical line. The last significant digit(s) are shown to the right of the vertical line. The digits to the right of the vertical line form a sort of histogram. This type of chart enables the viewer to identify every data element in the data set.		
		Tally chart Values or class intervals are shown to the left of a vertical line. A hatch mark is added to the right of the vertical line each time the value occurs in the data set. The hatch marks form a sort of histogram.	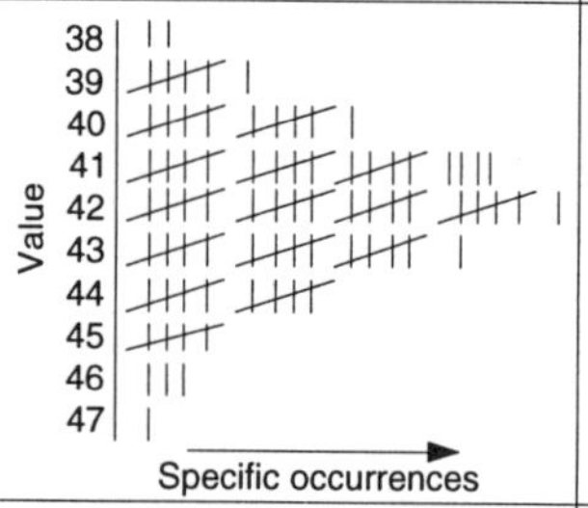	Multiple charts can be displayed side-by-side or back-to-back; however, tally charts are not normally used for comparing multiple data sets.
		Dot array chart Values or class intervals are shown to the left of the vertical line. A dot is added to the right of the vertical line each time the value occurs in the data set. The dots form a sort of histogram. Dots may be stacked on top of one another or entered on a grid as shown here.	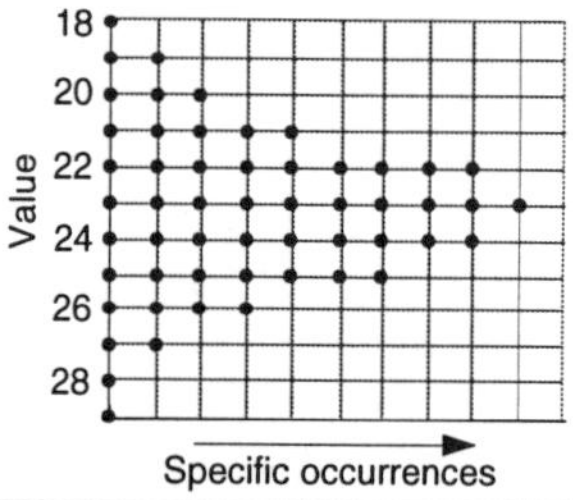	Multiple charts can be displayed side-by-side or back-to-back; however, dot array charts are not normally used for comparing multiple data sets.

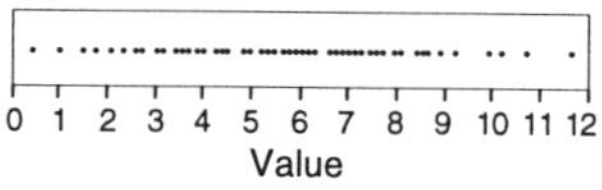

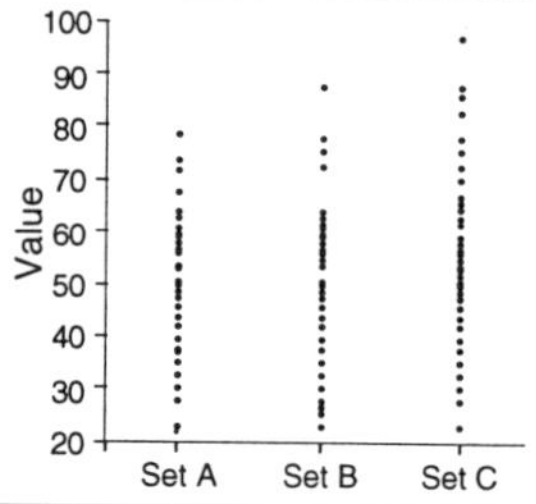

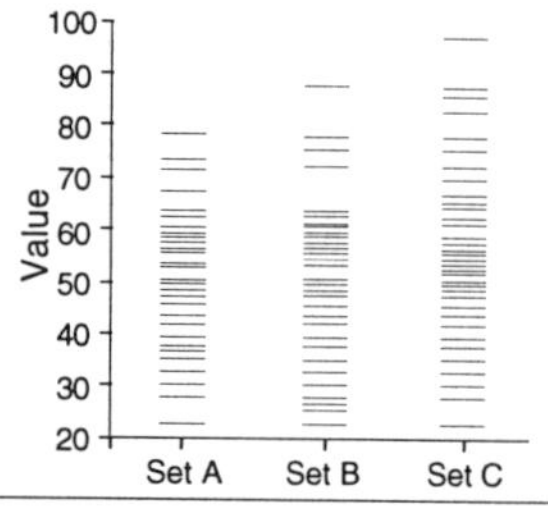

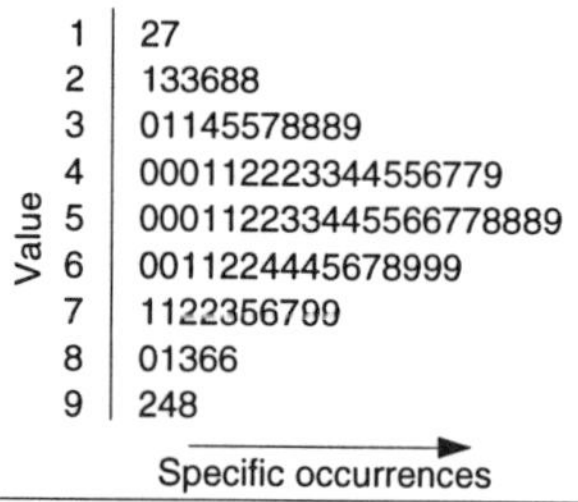

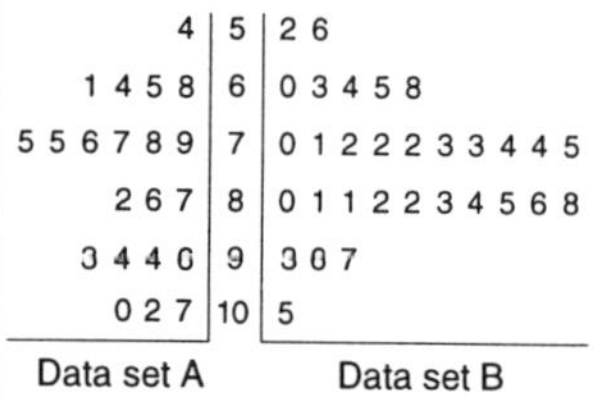

Data Distribution Graphs (continued)

		Distribution of an individual data set	Distribution of multiple data sets
Frequency distribution graphs with data grouped. Major emphasis is on concentrations of data elements and skewness.	**Histogram** Probably the most widely used method for displaying data distributions. Can be used for discrete or continuous data. Sets of data can be of any size. Percent frequency can be shown on the vertical axis in addition to or in lieu of frequency. Simple column or bar graphs (with space between the columns or bars) are sometimes used as histograms.		
	Frequency polygon The data points on a frequency polygon represent the same values as the tops of the columns in a histogram. The data points are plotted at the midpoint of the class intervals. Polygons are often better than histograms for comparing multiple distributions. Percent (relative frequency) can be shown on the vertical axis in addition to or in lieu of frequency.		
Cumulative frequency distribution graphs. The major emphasis is on groupings such as deciles and quartiles and how many data elements are above or below given values.	**Cumulative histogram** In a cumulative histogram the value plotted for a given class interval is equal to the sum of the incremental value for that class interval plus the values for all previous class intervals. Cumulative frequency and/or cumulative percent frequency can be plotted on the vertical axis.		
	Cumulative frequency graph Sometimes referred to as Ogive curve or cumulative relative frequency (percent) graph. The values on both axes are the same as those in a comparable cumulative histogram. A key difference is that the data points are plotted at the class interval boundaries. Cumulative frequency and/or cumulative percent frequency can be plotted on the vertical axis.		
	Quantile graph Quantile graphs function similar to cumulative frequency graphs but are constructed very differently. With this type of graph, specific data points are plotted as opposed to the class interval summaries used by cumulative frequency graphs. Quantile numbers (0 to 1) or their percent equivalent (0 to 100) are frequently plotted on the horizontal axis.		
Normal probability graphs. The major emphasis is on determining whether a data set has a normal distribution.	**Cumulative frequency graph** To determine whether a data series has a normal distribution, cumulative relative frequency data are plotted on a special normal probability grid. The grid has a non-linear percent axis and therefore a computer or special graph paper is normally used. If the distribution is normal, the data points cluster around a straight line.		
	Quantile graph When using a quantile graph to determine whether a data series has a normal distribution, the data series under study is compared against a theoretical data series with a known normal distribution, a median value of zero, and a standard deviation of one. If the distribution of the data series under study is normal, the data points will cluster around a straight line.		

Data distribution charts and graphs

	Data Element	Data elements are bits or groups of data. As used in this book, data elements exist in three forms:

Data Element

Data elements are bits or groups of data. As used in this book, data elements exist in three forms:

1. A single isolated piece of information such as 37, James, 3PM, Ohio, or 220°
2. A part of a data set (generally a single bit of information called a data set element)
3. A part of a data series (generally two or more bits of information called a data series element). A data series consists of two or more data sets.

The illustration shown here provides examples as to how data elements relate to four major types of charts.

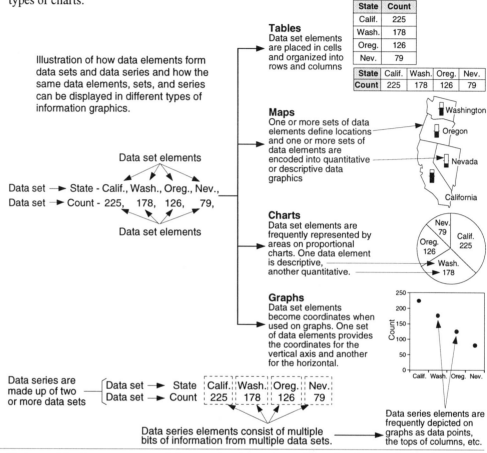

Illustration of how data elements form data sets and data series and how the same data elements, sets, and series can be displayed in different types of information graphics.

Data set elements

Data set → State - Calif., Wash., Oreg., Nev.,
Data set → Count - 225, 178, 126, 79,

Data set elements

Tables
Data set elements are placed in cells and organized into rows and columns

State	Count
Calif.	225
Wash.	178
Oreg.	126
Nev.	79

State	Calif.	Wash.	Oreg.	Nev.
Count	225	178	126	79

Maps
One or more sets of data elements define locations and one or more sets of data elements are encoded into quantitative or descriptive data graphics

Charts
Data set elements are frequently represented by areas on proportional charts. One data element is descriptive, another quantitative.

Graphs
Data set elements become coordinates when used on graphs. One set of data elements provides the coordinates for the vertical axis and another for the horizontal.

Data series are made up of two or more data sets

Data set →	State	Calif.	Wash.	Oreg.	Nev.
Data set →	Count	225	178	126	79

Data series elements consist of multiple bits of information from multiple data sets.

Data series elements are frequently depicted on graphs as data points, the tops of columns, etc.

Data Graphic
or
Data Marker
or
Data Measure

Data graphic, data measure, and data marker (sometimes abbreviated to marker) refer to the dots, lines, bars, columns, symbols, or other graphic configurations used to represent information, frequently quantitative, on charts, graphs, and maps. For example, in a bar graph the bars are referred to as data graphics. In a pie chart it is the wedges. In a point graph it is the dots or symbols. In a map the data graphic might be a symbol whose size is proportional to the value it represents.

Examples of data graphics, data markers, or data measures

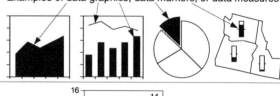

Data Graphic Label
or
Data Point Label

Labels placed within the plot area, on, or immediately adjacent to the data points they apply to are called interior labels, data graphic labels, or data point labels.

Interior labels, data graphic labels, or data point labels

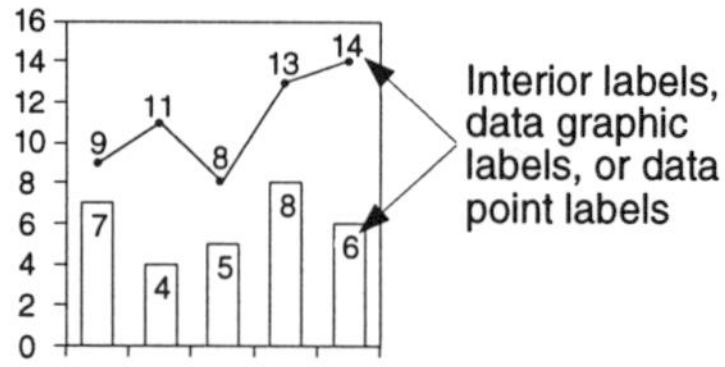

Data-Ink Ratio

An expression coined in conjunction with the concept that, for clarity, it is often advantageous to keep the amount of unnecessary graphics to a minimum when generating a chart or graph.
The expression can be shown as:

$$\text{Data-Ink ratio} = \frac{\text{Amount of ink essential to communicate the information}}{\text{Amount of ink actually used in the chart}}$$

The proposed objective is to keep the ratio as large as possible by reducing what is sometimes called chartjunk. See Chartjunk.

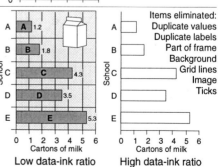

Items eliminated:
Duplicate values
Duplicate labels
Part of frame
Background
Grid lines
Image
Ticks

Low data-ink ratio   High data-ink ratio

| Data Map | Sometimes referred to as a statistical or quantitative map. Data maps are used to convey statistical information with respect to areas and locations. Two examples are shown at the right. See Statistical Map. |

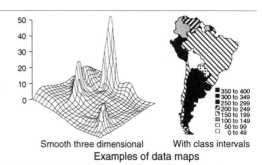

Smooth three dimensional   With class intervals
Examples of data maps

Data Point

Technically, a data point is the point on a graph that corresponds to a data element in a data set, data series, equation, electronic input device, etc. On certain graphs the definition and location of data points is clear. For example, if a data element has the values of 3,4, the data point on a scatter graph for that element is located three units along the horizontal axis and four units up the vertical axis. Since data points are technically dimensionless, the data point 3,4 is designated by a plotting symbol such as a dot, square, triangle, etc. Generally the size of the plotting symbol has no relationship to the value or location of the data point.
• On many other types of graphs the definition and location of data points is more difficult. For example, on a two-dimensional column graph, any or all points along the top of the column might qualify as data points. On a three-dimensional column graph, any or all points on the top surface of the column might be considered data points. On curves that are plotted from equations, technically there are an infinite number of data points and therefore, every point along the curve might be considered as a data point. Until some industry standards are established, the following guide lines are generally applicable.

– With point and scatter graphs, the values of the data elements become coordinates which define the location of the data point.
– On most other graphs, if the exact location of a data point is important, it is advisable to use plotting symbols to identify the points.
– On column and bar graphs, unless otherwise specified, the data points are generally assumed to be in the center of the tops of columns and ends of bars.
– On segmented line and area graphs the data points are generally assumed to be located where the line segments meet.
– On most smooth curved lines and surfaces without plotting symbols, unless otherwise specified or made obvious by the scale and the nature of the data, it might be assumed there are an infinite number of data points and any point on the line or surface can be considered to be a data point.
– On surfaces made up of multiple flat planes, the data points are frequently located somewhere along the intersections of the planes.

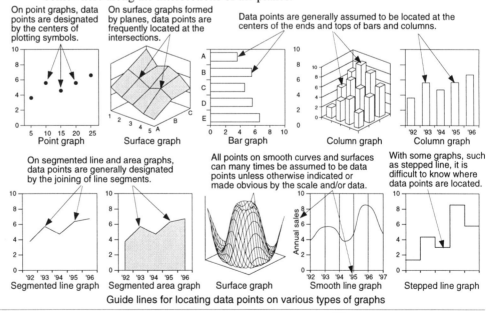

Guide lines for locating data points on various types of graphs

Data Region

Also called plot area. The data region of a graph is the area in which the data is plotted. It is frequently bounded by a border or frame. Grid lines normally end at the edges of the data region. The data region is shaded gray on the example at the right.

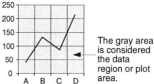

The gray area is considered the data region or plot area.

Data Series and Data Set

Data set

In this book, a data set is a group of individual bits of information (called data set elements) that are related in some way or have some common characteristic or attribute. The information in a data set can be quantitative or nonquantitative. For example, the heights of all students in a school comprise a set of quantitative data. The names of the fifty states in the United States are a set of nonquantitative data. Each individual height and the name of each state is considered a data set element. The information in a row or column of a table or spread sheet is generally considered a data set. The information in each cell of the table is considered a data set element. Data series are made up of multiple data sets.

Data series

In this book, data series consist of two or more data sets with two being the most common. When data series are plotted on a two-axis graph, one data set supplies the coordinate information for the horizontal axis and the other data set the coordinate information for the vertical axis. If three data sets are used, the information from the third data set is encoded by a third means such as a third axis or the size of the symbol. When a data series is plotted on a quantitative map, one or more data sets designate the location in terms of city, state, geographic coordinates, etc., and the other data set supplies the information such as population, birth rate, or type of crop. Examples are shown below.

These two data sets constitute a data series.

Data set 1	Temperature	10	18	30	55
Data set 2	Pressure	0.8	2.7	4.5	6.4

Example of a data series with two sets of data (two variables)

These three data sets constitute a data series.

Data set 1	Selling price	$18.00	$19.50	$20.25	$22.20
Data set 2	Quantity	2	10	4	8
Data set 3	Size	A	B	C	D

Example of a data series with three sets of data (three variables)

These two data sets constitute a data series.

Data set	State	Calif.	Wash.	Oreg.	Nev.
Data set	Count	225	178	126	79

Example of a data series with two sets of data (two variables)

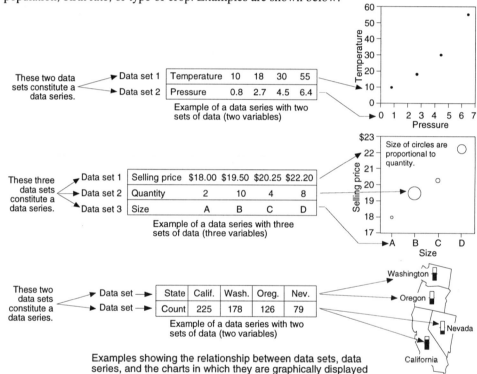

Examples showing the relationship between data sets, data series, and the charts in which they are graphically displayed

The fact that multiple data sets are combined to form a data series does not necessarily imply a relationship between the data sets. In fact, two data sets are frequently combined into a data series and plotted on a scatter graph for the express purpose of determining whether a relationship does exist between the two data sets.

Data elements

Data sets and data series are made up of individual data elements. In data sets, each individual bit of information in the set is a data element. In data series the data elements consist of as many bits of information as there are data sets in the series. In the example at the right, each data series element consists of three bits of information, since the data series is made up of three data sets. Each data graphic on a two- or three-dimensional graph or map has a corresponding data series element. See Data Element.

Size	A	B	C	D
Selling price	$18.00	$19.50	$20.25	$22.20
Volume	2	10	4	8

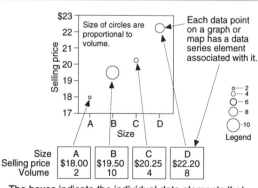

The boxes indicate the individual data elements that make up the data series. The arrows designate which data graphic on the graph represents each of the data elements.

The relationship between tables, data series, and data sets

In most cases the information for data series is contained in tables or spreadsheets. The example below shows what combinations of data sets in the table on the left were used to draw the curves on graph on the right. In this example, the first column is used as one data set for all data series displayed on the graph. This is frequently the case but does not have to be.

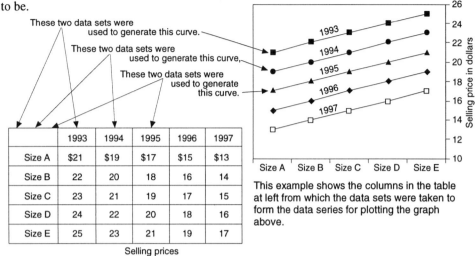

These two data sets were used to generate this curve.

These two data sets were used to generate this curve.

These two data sets were used to generate this curve.

	1993	1994	1995	1996	1997
Size A	$21	$19	$17	$15	$13
Size B	22	20	18	16	14
Size C	23	21	19	17	15
Size D	24	22	20	18	16
Size E	25	23	21	19	17

Selling prices

This example shows the columns in the table at left from which the data sets were taken to form the data series for plotting the graph above.

The data sets for data series can be taken from either the rows or columns of a table. In the example above, the data in the columns were used to form the data sets. In the example below, the data from the table's rows are used for generating a graph. In this case, the top row is used as one data set for all the data series displayed on the graph.

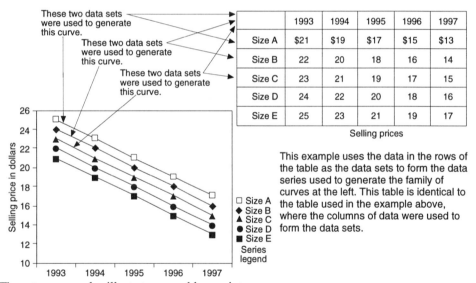

These two data sets were used to generate this curve.

These two data sets were used to generate this curve.

These two data sets were used to generate this curve.

	1993	1994	1995	1996	1997
Size A	$21	$19	$17	$15	$13
Size B	22	20	18	16	14
Size C	23	21	19	17	15
Size D	24	22	20	18	16
Size E	25	23	21	19	17

Selling prices

This example uses the data in the rows of the table as the data sets to form the data series used to generate the family of curves at the left. This table is identical to the table used in the example above, where the columns of data were used to form the data sets.

These two examples illustrate several key points.

– Data sets can be taken from columns or rows of a table or spread sheet.

– The exact same bits of data can be used in more that one data set or data series.

– Whether one uses the rows or the columns to form the data series can significantly affect the primary visual message the resulting graph conveys. For example, in the upper graph the eye focuses on the fact that prices increase uniformly with size and have done so for five years. Using the exact same data, the primary visual message from the lower graph is that prices on all sizes have decreased uniformly for the five years.

Summary illustration

The example at the right illustrates the key terms associated with date sets and data series as discussed in this section.

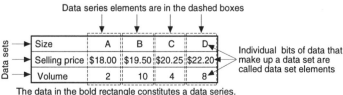

Data series elements are in the dashed boxes

Individual bits of data that make up a data set are called data set elements

The data in the bold rectangle constitutes a data series.

Illustration of the relationship between data elements, data sets, and data series

Datum	Sometimes referred to as a reference or base line. A line, plane, or point on a chart, graph, or map from which other values are compared, referenced, or measured. In many cases the datum is zero. Other values are also used such as budget, industry average, the edge of a property, sea level, an arbitrary elevation, etc.

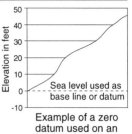

Example of a nonzero datum used on a graph

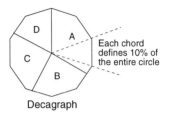

Example of a zero datum used on an elevation map

Decagraph	A seldom-used variation of a pie chart in which the circle around the pie is replaced by a ten-sided polygon. With the circumference made up of ten chords of equal lengths, the points where the chords meet automatically designate every 10% of the complete circle. The objective of the decagraph is to enable the viewer to more accurately estimate the sizes of the individual segments with the assistance of the chords that indicate 10% increments.

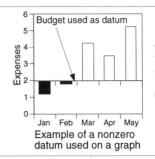

Each chord defines 10% of the entire circle

Decagraph

Decision Diagram or **Decision Chart** or **Decision Tree** or **Decision Flow Chart**	A decision diagram is a graphic representation of alternative decisions or actions that might be taken, plus potential outcomes resulting from those decisions and actions. The ability to see options and estimated outcomes before decisions are made is one of the main advantages of decision diagrams. Such diagrams can be used by almost any type of organization to make simple or complex decisions on a qualitative or quantitative basis. These diagrams might be used to establish a pricing policy, decide the size of a new plant, determine which research projects to pursue, decide which proposal to accept, decide whether to change jobs, etc. The examples that follow illustrate some of the many variations of decision diagrams in use today.

Binary Decision Diagram

If each decision point on a diagram allows only one of two decisions (e.g., yes or no), the chart is called a binary decision diagram.

In addition to the decision that is to be made, additional information is sometimes included, such as who will make the decision and when it will be made.

If space allows, the basic question about which the decision is to be made can be described inside the symbol. Alternately, a coding system can be used and explained in a legend.

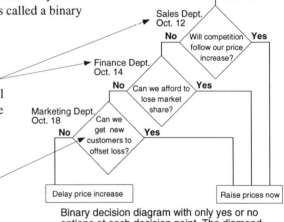

Binary decision diagram with only yes or no options at each decision point. The diamond shaped symbols represent decision points.

Multiple-Choice Decision Diagram

If more than two options are possible at decision points the chart is sometimes called a multiple-choice decision diagram. Decision diagrams might run horizontal or vertical. If horizontal, they progress from the left to right. If vertical, they proceed from top to bottom.

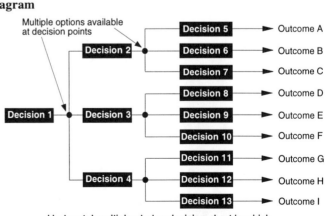

Horizontal multiple-choice decision chart in which three or more options are available at decision points

**Decision Diagram
or
Decision Chart
or
Decision Tree
or
Decision Flow Chart**
(continued)

Combination decision and flow chart

In some cases decision diagrams include only symbols that represent decisions. In other cases symbols representing other events and activities are also included. This latter variation is frequently called a flow chart.

Arrow heads are sometimes added to the lines connecting the symbols to indicate direction of flow. This is particularly important when feedback loops are included.

The significance or symbol shapes, if any, is generally explained in a legend. Diamond shapes are frequently used to indicate decision points. Occasionally flow chart symbols are used in decision diagrams.

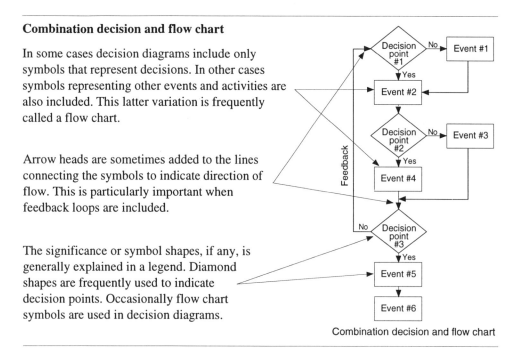

Combination decision and flow chart

Passive decision diagram

In addition to aiding individuals or organizations in making their own decisions (an active decision diagram), decision charts are sometimes used to estimate outcomes based on decisions made by others (passive decision diagram). For example, to a company that supplies raw material such as cement to subcontractors, the specific subcontractor awarded a contract is very important. The cement company, however, has no influence on the decisions as to which subcontractor will get the contract. Since subcontractors are many times closely affiliated with prime contractors, the probability of the company selling cement is further dependent on the decision as to who the prime contractor will be. If the cement company managers want to estimate the probabilities of the various subcontractors getting the order, they might construct a passive decision diagram, as shown below. Based on this example, the cement company might focus attention on subcontractors C and E (25% and 31% probability they will receive a subcontract order, respectively) and expend very little effort on subcontractor D (6% probability of receiving a subcontract).

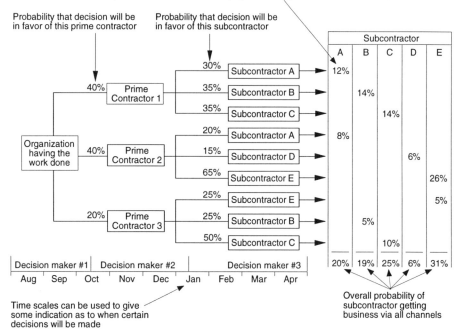

Passive decision chart indicating who is making the decisions,
approximately when and the probability associated with each option

131

**Decision Diagram
or
Decision Chart
or
Decision Tree
or
Decision Flow Chart**
(continued)

Passive decision diagram (continued)

In addition to illustrating a passive decision diagram, this example also introduces two other concepts:

– The idea of quantifying the chances of certain things happening so that the probability of a given result occurring can be estimated; and

– The use of a time scale to communicate when certain decisions might be made.

Both of these techniques can be used with most types of decision diagrams.

Tree diagram

One variation of decision diagram that has been highly developed is sometimes referred to as a tree diagram or decision tree. In addition to utilizing probabilities, the decision tree also incorporates monetary values and combinations of the two. This type of diagram is sometimes used when making risk decisions in which probabilities are assigned to each uncontrollable event. Such diagrams are typically horizontal, constructed from left to right, and are analyzed from right to left. For example, in analyzing a diagram, the most desirable outcome is selected from all possible outcomes (as listed in the far right column). Next, the path is followed from right to left to see what decisions have to be made to achieve the desired outcome. Some of the key elements and terminology used with decision trees are shown below.

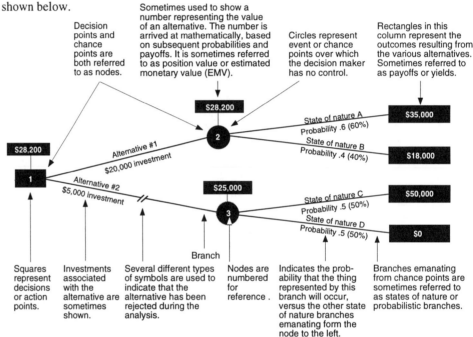

Terminology used with decision trees

This example illustrates one series of a decision and subsequent chance events. In an actual situation one series of decision and chance events may be all that is involved. In some cases, however, two or more such sequences follow one another. These are called sequential decisions or multiperiod decision processes.

Decision Matrix

A decision matrix is a method for compiling quantitative information on alternatives or options to aid in a decision-making process. There are many variations, however, most use a matrix similar to the one shown below for compiling the quantitative information.

			– Potential conditions options might encounter, or – Events that might impact outcome of options			
			Number 1	Number 2	Number 3	
– Probability the conditions, events, etc., will happen (total = 100%) or – Weighting factor applied to criteria						Total weighted value
Action, alternative, proposal, options, etc.	Option A	Unweighted value				
		Weighted value				
	Option B	Unweighted value				
		Weighted value				
	Option C	Unweighted value				
		Weighted value				

Basic format used with most decision matrixes

Deck Sometimes called bank, cycle, phase, or tier. The major interval on a logarithmic scale.

Decoding	The process of converting information into charts and graphs is sometimes described as encoding. For example, the number 37 might be encoded into a graph by drawing a certain size column on a set of grid lines (column graph), or the state of Missouri might be encoded by drawing a certain shape of polygon on a map. As the viewer looks at the charts and graphs and mentally converts the graphical representations back into meaningful information (e.g., the column back to the number 37 and the polygon back to Missouri), the process is sometimes described as decoding. The processes of encoding and decoding are equally important in making a chart or graph successful.

Decomposition Graph

Sometimes the data from which a graph is plotted is made up of several components. For example, a five-year sales curve might include the continuing growth of the company, the cyclical fluctuations of the economy, seasonal fluctuations, and random fluctuations due to such things as special promotions. When the data for those components are separated out and plotted individually, the resulting set of graphs is sometimes called a decomposition graph. The examples at the right show a composite graph on the left and graphs of the four components that make it up on the right.

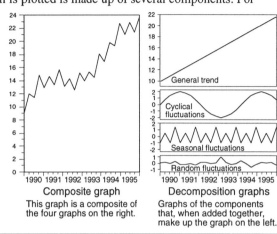

Composite graph
This graph is a composite of the four graphs on the right.

Decomposition graphs
Graphs of the components that, when added together, make up the graph on the left.

Demographic Map

A demographic map is a specific application of a statistical or thematic map on which population information such as age, sex, race, nationality, etc., are displayed by area.

Dendrite Diagram

Sometimes referred to as a tree diagram or tree chart. See Tree Diagram

Dendrogram

Sometimes referred to as a tree diagram, linkage tree, or cluster map. A dendrogram is a graphical means of organizing information to establish groupings and/or categorize individual elements (many times referred to as objects or single member clusters). For example, one might study different groups of consumers to analyze how they perceive a product, or one might use a dendrogram to look for meaningful groupings in a collection of archeological specimens. In constructing a dendrogram, the data regarding each individual object is first organized into a matrix. From this matrix and a series of calculations, numeric values are established that are proportional to the similarities between the objects and/or groups of objects. These numeric values are used to determine the level at which the various objects and subclusters will be connected in the dendrogram. After the numbers are calculated, the stems (the lines leading from the objects) of the two or more objects with the greatest similarity are connected (fused or linked) forming a subcluster. Next the objects and/or subclusters with the second greatest similarity are connected. This process is continued until all objects are connected into a single cluster similar to the example shown below.

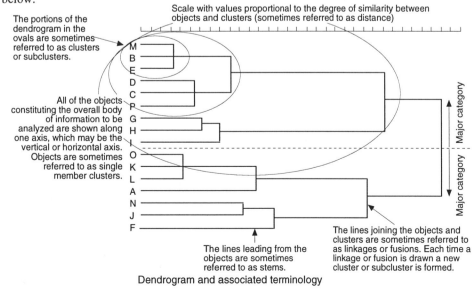

Dendrogram and associated terminology

Dendrogram (continued)	The completed diagram is sometimes referred to as a hierarchy of clusters. • The value at which a linkage crosses from one element or subgroup to another is generally proportional to the degree of similarity between the elements or subgroups. From the example shown, one might conclude that objects M, B, and E have the greatest similarity. Objects O,K, and L have the next largest similarity, etc. • After the dendrogram is complete, overall groupings or families of elements sometimes stand out. In the example, there are two major clusters or categories that the objects fall into – one above the horizontal dashed line and one below. • There are several methods used to quantify the degree of similarity between the various elements as well as how the linkages are drawn between the elements and clusters. • Dendrograms are one facet of the broader subject of cluster analysis.
Density Graph	The term density graph is sometimes used when referring to the family of graphs whose primary purpose is to display the density or distribution of a set or sets of data. These graphs are also sometimes called data distribution graphs. See Data Distribution Graphs.
Density Stripe Graph	Sometimes referred to as a stripe graph. A density stripe graph is a variation of a one-axis data distribution graph. The major purpose of this type of graph is to show graphically how the data elements of a data set are distributed. See Stripe Graph.

Density Trace

A density trace smooths the jaggedness that occurs in may histograms and minimizes the impression that data elements are evenly distributed across the width of class intervals. There are several methods for generating density trace graphs, all of which work on the principle of overlapping class intervals. In standard histograms, class intervals do not overlap. In density trace graphs, the number and width of class intervals are purposefully chosen to assure that every interval overlaps with two or more other class intervals. To illustrate, the density trace shown below and the conventional histogram shown at the right were both generated using the same data.

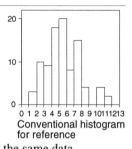

Conventional histogram for reference

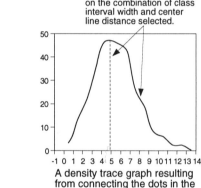

The dots designate the centers of the class intervals. They are the points connected by the line that forms the density trace.

In this particular example, class intervals are three units wide and on center distances of one unit which results in the overlap of four other class intervals.

The exact location of the mode as well as the shape of the curve vary depending on the combination of class interval width and center line distance selected.

Values are larger than those in the standard histogram shown above because of the duplicate counts resulting from the overlapping.

An intermediate step in the process of generating a density trace. It is included here for illustrative purposes only.

A density trace graph resulting from connecting the dots in the graph at the left.

Listed below are some of the variations used in the various methods for generating density trace graphs.
- The number of class intervals is optional and might range as high as 50 to 100.
- The width of the class intervals is optional and might be specified in units as used on the horizontal axis or percent of the range from minimum to maximum of the data set.
- A vertical scale may or may not be displayed.
- If displayed, the vertical scale might be in terms of frequency, density (frequency divided by the total number of elements in the data set), or density per unit of measure on the horizontal axis.
- All elements in a class interval might be considered equal or a weighting method employed so those elements further from the center of the interval have less impact.

Dependent Variable

Two related variables are many times designated as dependent and independent with respect to one another. There is no single or simple definition as to which variable should receive which designation or whether the terms are even applicable to a given situation. In some cases, the decision as to which is the dependent variable is easy. For example, category and sequence variables are generally considered independent variables and the values or measurements associated with them as the dependent variables. In other cases it is more difficult. For example, if one is plotting the performance of one group of students against another, there may or may not be any interdependency and therefore it would be difficult to decide which, if either, is the dependent variable. Dependent variables are generally plotted on the vertical axis of graphs, except for bar graphs. See Variable.

Depth

In order to improve the appearance of charts, what many times is referred to as depth is added to all or portions of the objects on the chart. When this is done it gives the objects the appearance of being three-dimensional. When done for cosmetic purposes, the amount and direction of the depth typically has no significance. When depth is the result of data actually being plotted on three axes, the amount of depth can have significance.

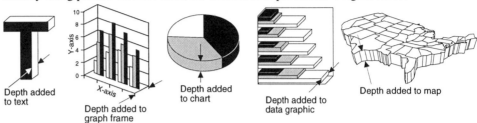

Depth added to text

Depth added to graph frame

Depth added to chart

Depth added to data graphic

Depth added to map

Examples of depth being added to give objects a three-dimensional appearance

Derived Value

A derived value is one that is obtained by performing one or more mathematical operations on an original set of data. Percents, ratios, differences, indexes, etc., are sometimes referred to as derived values since a mathematical operation has to be performed on the original data to arrive at these values. Derived values can seldom be measured directly and in most cases do not physically exist. They can be used in tables, charts, graphs, or maps. Reasons for using derived values include:

- To convert data to a common basis (e.g., values by unit time, quantity, or area) for more meaningful comparisons.
- To avoid viewers having to perform mathematical calculations in their head (e.g., dividing assets by profit to determine return on assets).
- To reduce the number of bits of data the viewer must deal with (e.g., bushels per acre versus bushels and acres).
- To make trends more obvious by segregating out known variables such as seasonal and cyclical fluctuations.
- To make deviations from some reference such as budget or plan easier to spot (e.g., plot only plus and minus deviations from budget).
- To make the charts easier to interpret.

Several examples comparing original and derived data are shown at the right. Other than the source of the data, the graphs, maps, and charts of derived data are constructed the same as graphs using original data.

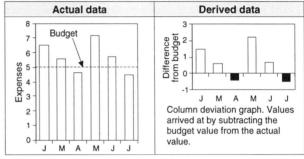

Column deviation graph. Values arrived at by subtracting the budget value from the actual value.

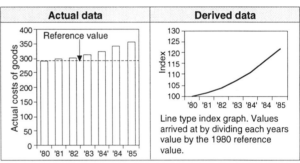

Line type index graph. Values arrived at by dividing each years value by the 1980 reference value.

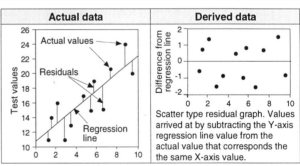

Scatter type residual graph. Values arrived at by subtracting the Y-axis regression line value from the actual value that corresponds the the same X-axis value.

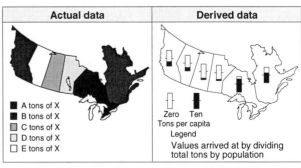

Values arrived at by dividing total tons by population

Examples of charts using original and derived data

135

Descriptive Map

Sometimes referred to as an explanatory, nonquantitative, or qualitative map. This type of map is generally used to show such information as where things are located, how areas are organized or subdivided, and where the routes are located that get people and things from one place to another. Descriptive maps may include grain fields, buildings, rivers, cities, roads, land masses, power lines, sewer lines, sales areas, public facilities, voting precincts, schools, political divisions, resorts, pet stores, sales areas, etc. There are no guidelines as to what should or should not be included on descriptive maps. Generally only those things that are germane to the theme or purpose of the map and the viewers' understanding of the map are included. Information might be encoded using standard geometric symbols, industry established symbols, pictures, icons, lines, shaded areas, or text. The map may be a small one showing how to get to the local shoe store, or it might be a world map covering an entire wall. Although descriptive maps sometimes have numeric data included, it is generally incidental to the major purpose of the map as opposed to statistical maps where the numeric data is the primary focus of the map. Examples of several different types and functions of descriptive maps are shown below.

Define and identify territories, regions, plots, areas, etc.

Descriptive maps are used extensively to identify the boundaries of specific areas. Examples are countries on continents, states in countries, ZIP code areas, school districts, sales areas, plots of land, etc. Many have descriptive names, such as world map or sales territory map. Others have specialized names and functions, such as a plat map which very accurately identifies the boundaries of specific units of land. The amount of detail varies from slight to elaborate. Selected areas are frequently color-coded, and occasionally portions of the map will be exploded or separated to accentuate particular portions.

Western Region   Central Region   Eastern Region

Descriptive map identifying sales regions

Approximate the size, shape, and location of surface and subsurface features

Sometimes referred to as blot or patch maps. If the map focuses primarily on things above the surface such as farmland, waste sites, forest preserves, wetlands, or industrial parks it is sometimes referred to as a land use and land cover map. If the map's focus is primarily on things below the surface such as layers of soil and clay, bedrock formations, major sediment deposits, minerals, etc., it is frequently called a geologic map. The size and shape of the areas on the map representing specific features generally approximate the size and shape of the features they represent. The meaning of the fills and colors used with the various entities are normally explained in a legend.

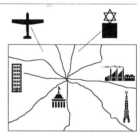

Resource A
Resource B
Resource C

Descriptive map identifying surface and subsurface features

Locate structures, facilities, activities, etc.

Icons, pictures, or symbols, are widely used on descriptive maps to show where things are located. When the icons or pictures that are used closely resemble the things they represent, the map is sometimes called a pictorial map. The pictures or icons might be located on the map or in the border around the map. If unique symbols are used, they are normally explained in a legend or note. The pictures, icons, and symbols are generally not to scale and the distances between the various symbols typically are only close approximations.

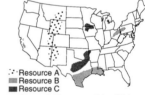

Pictorial map where the symbols resemble the things they represent

Relationships of things to one another

In addition to showing the relationships of things to features on the base map, descriptive maps are occasionally used to show the relationship of various entities to one another. For example, a map may show the relationship of a sales office to the territory it manages, a factory to its warehouses, a church to its members, or a bus terminal to the cities it serves (sometimes called a ray map). The focus is usually not be on actual distances or locations, but on patterns and distributions of the entities, whether or not they overlap, how many there are, their relationship to customers and resources, etc.

Terminal

Cities served

Descriptive map showing the relationship of a bus terminal to the cities it serves. Sometimes called a ray map.

Descriptive Map (continued)	**The paths that people and things use to get from one place to another** Some descriptive maps display the paths people and things use to travel from one point to another. By contrast, flow maps show what and sometimes how much moves from one place to another. The paths or routes might be walkways, corridors, roads, railroad tracks, streets, shipping lanes, overhead cables, buried pipes, etc. The most common such map, of course, is the road map, which shows paths (roads, routes, or streets) for vehicles. Extensive amounts of information are often encoded into the lines representing the paths, such as type pavement, number of lanes, diameter of pipe, voltage of cable, etc. When the paths connect multiple locations they are sometimes called networks.  Example of a road map
Deviation Graph	A variation of a difference graph. See Difference Graph.
Diagram	Diagrams are charts made up primarily of geometric shapes such as circles, rectangles, or triangles, connected by lines or arrows. One of the major purposes of diagrams is to show how people, things, ideas, functions, etc., interrelate. Text is frequently used both inside and outside the geometric shapes. Numerical values are used to a much smaller extent, since diagrams are generally nonquantitative. Flow charts, PERT and CPM charts, organization charts, network charts, decision charts, and conceptual charts are examples of what are frequently referred to as diagrams. Occasionally graphs, maps, and illustration charts are referred to as diagrams. Schematics are a subcategory of diagrams. Several examples of diagrams are shown here.

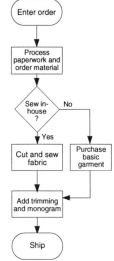

Flow diagram or chart

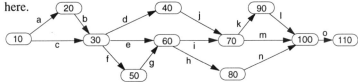

PERT chart, sometimes referred to as an arrow diagram

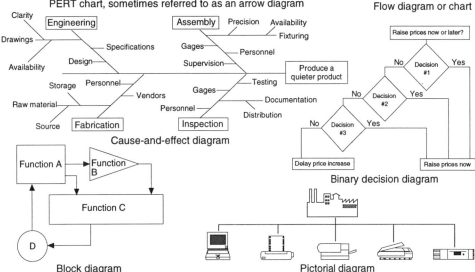

Cause-and-effect diagram

Binary decision diagram

Block diagram

Pictorial diagram

Examples of charts that are typically referred to as diagrams

Diagrammatic Map	A diagrammatic map resembles a diagram. The purpose of such maps is generally to convey only a minimum amount of information on a very specific subject. In most cases any information or graphic convention not germane to the narrow focus of the map is either not shown, deemphasized, or drawn in such a way as to minimize distraction from the main objective of the map. In some cases, detail is reduced to such an extent that the base map is eliminated completely and little or no attempt is made to locate points according to their actual physical locations. A public transportation map is an example, some of which consists of nothing more than the stops on the route equally spaced along a straight line, as shown below. Diagrammatic maps are sometimes classified as cartograms or abstract maps.

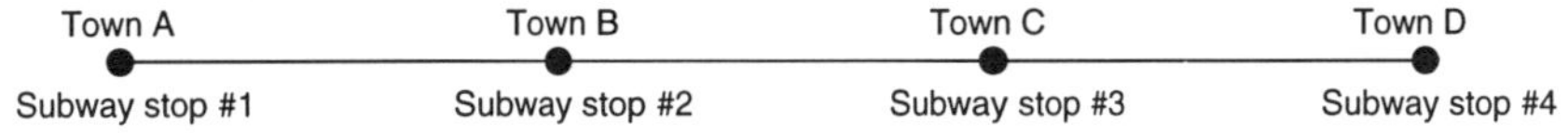

Town A	Town B	Town C	Town D
●	●	●	●
Subway stop #1	Subway stop #2	Subway stop #3	Subway stop #4

Diagrammatic map similar to those used in many public transportation vehicles

Difference Graph

The terms difference and deviation are sometimes used to describe the same graph. Generally a deviation graph is considered a variation of a difference graph that displays the differences between a data series and some known reference such as budget, industry standard, etc. The graphs that are most frequently referred to as difference and deviation are described in this section. In each case the difference or deviation is between the elements of two data series or a data series and a known reference such as budget, standard, or last year.

Major variations of difference graphs

Difference between two data series with positive values

With this type of data the actual values might be plotted and the differences highlighted or just the numeric differences might be plotted from a zero axis. One of the data series is generally used as a reference. For instance, in the examples below, 1994 is used as the reference. This means that in the two graphs on the right, when the rain in 1995 is less than 1994, the value is either highlighted (filled with black in the center graph), or shown as negative (graph on right). If 1995 were considered the reference, the absolute difference values (ignoring plus and minus signs) would be the same but the highlighting and plus and minus values would be reversed

Actual data for reference

Difference graph
Actual values plotted and the differences highlighted

Difference graph
Only the differences have been plotted in this graph

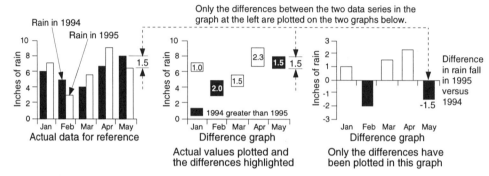

Differences between two data series with positive and negative values (gross deviation and net deviation graphs)

In this variation, both positive and negative values are compared: cash in and cash out, expenses that increased and those that decreased, value of stocks that went up and value of stocks that went down. In these applications, the graph displaying the actual data is often referred to as a gross deviation graph. The graph showing the algebraic sum of the positive and negative values that appear on the gross deviation graph is called a net deviation graph. The column graph variation of a gross deviation graph is sometimes called an over-under graph. The bar graph variation of a gross deviation graph is sometimes called a split bar graph.

Gross deviation graph displaying actual values for two data series

Net difference or deviation graph displaying the algebraic sums of the two series displayed in the example at the left

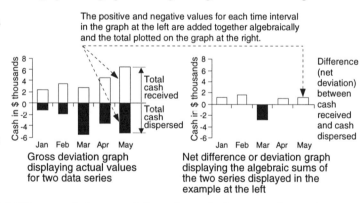

Deviation graph (differences between a data series and a known reference)

A typical deviation graph displays the differences between a data series and a fixed or known reference such as budget, average, industry standard, etc. In this example budget is used as the reference. The reference might be the same or different for each interval. When the actual profit value is above budget, the difference is denoted as positive; below budget as negative; and when profit is equal to budget, the deviation is zero.

Actual data

Deviation graph

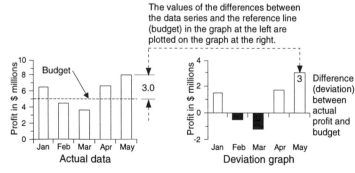

Variations of bar and column difference/deviation graphs

Gross and net deviation information combined

The information for gross and net deviation graphs can be plotted on the same bar or column graph so that the viewer sees both types of information at the same time.

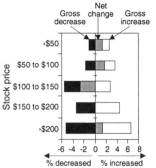

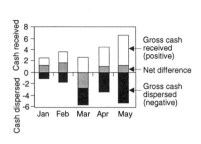

Graphs combining gross and net differences/deviations

Labels between data graphics

Labels can be located between the data graphics for the two data series. When the graph is particularly tall or wide, this placement can make it easier for the viewer to relate the labels with the proper data graphic.

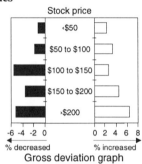

Gross deviation graph

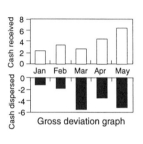

Gross deviation graph

Bar and column graphs with the labels internal

Comparison of the most frequently used variations

Difference graphs are typically generated using column, bar, or line type graphs. Examples of each are shown below with the same data plotted on all of the graphs. Also included are examples of common cumulative difference graphs. The cumulative variation is many times referred to as month-to-date or year-to-date difference or deviation graphs.

Type of information plotted	Column graph	Bar graph	Line graph
Actual values for reference This row of graphs is shown for reference purposes only. These graphs display the actual values of two data series as they would appear in a standard grouped graph.	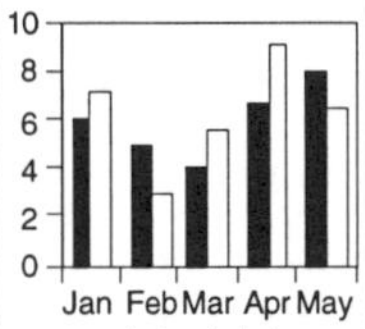 Actual data	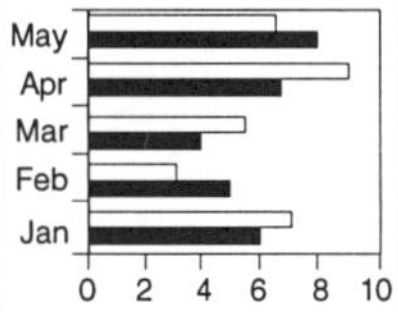 Actual data	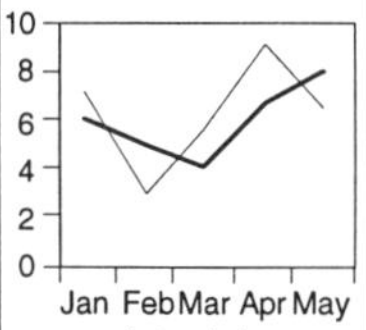 Actual data
Differences showing actual values In this variation the actual values of the two data series are shown and the differences between them highlighted.	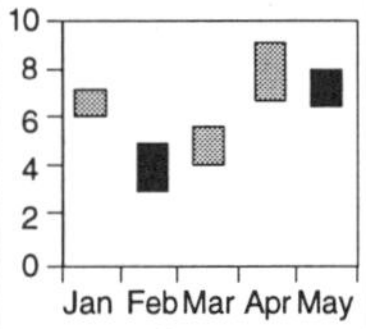 Differences	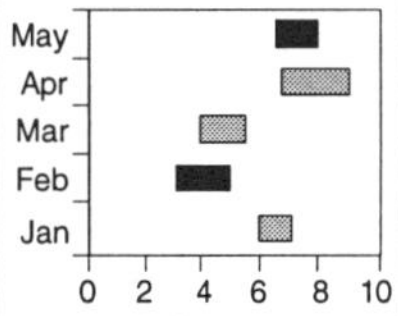 Differences	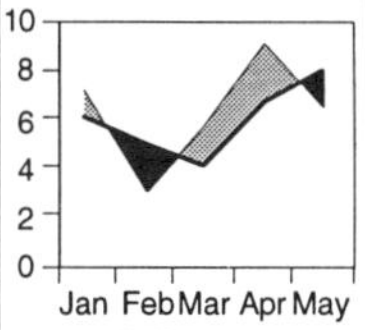 Differences
Differences showing only the deviations In this variation only the differences between two data series are plotted against a zero axis.	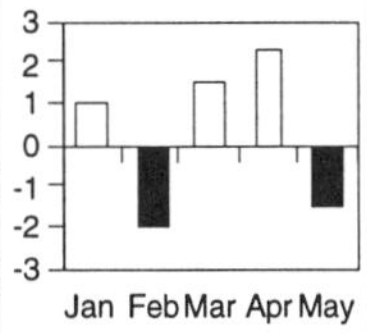 Deviations	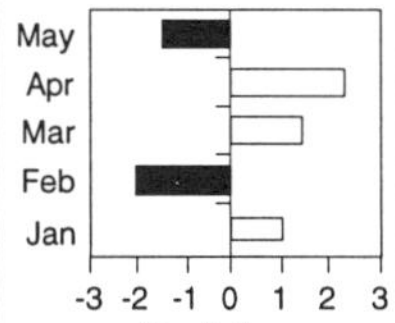 Deviations	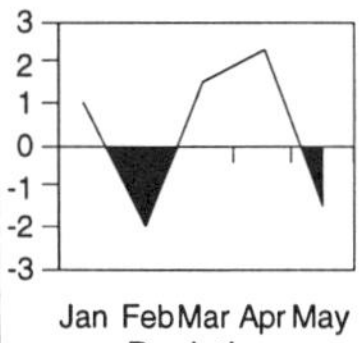 Deviations
Cumulative differences or deviations shown In this type of graph the value plotted for a specific time period is equal to the difference or deviation for that period plus the algebraic sum of all the similar values of the preceding time intervals.	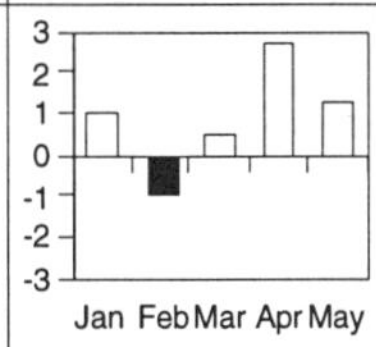 Cumulative differences	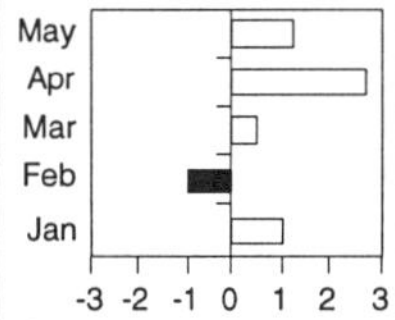 Cumulative differences	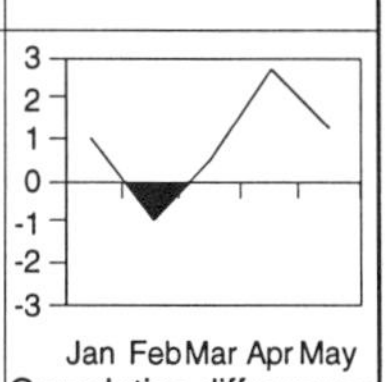 Cumulative differences

Variations of difference graphs for the three major graph styles used for this purpose

Difference Graph (continued)

Change Graph

A change graph is a variation of a difference graph that compares multiple factors at two points in time or under two different sets of conditions. Actual values are plotted. With this type of graph the direction of change or difference is generally considered important; therefore, a coding system is used to indicate direction of change. Three ways of accomplishing this are shown below. • Since no particular direction is always favorable (desirable) or unfavorable (undesirable), an additional coding system is sometimes used to designate whether the change is favorable or not (e.g., profit over budget is favorable. Expenses over budget are unfavorable). The numeric value of the change may or may not be shown on the data graphics. Examples of data that might be plotted on a change graph are changes in sales forecasts from one month to another, changes in employment from one period to another, and changes in economic indicators from one period to another. Examples of both bar and column change graphs are shown below.

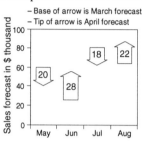
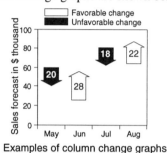
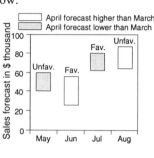

Examples of column change graphs

Difference Line Graph

Sometimes referred to as an intersecting silhouette graph, intersecting band graph, or curve difference graph. A difference line graph is a graph on which the curves representing two data series intersect one another and the areas between the two curves are highlighted by means of shading or color. For example, one curve might represent the dollar value of exports each year and the other the dollar value of imports, as shown in the example at the right. When exports have a higher value than imports, one shade or color is used. When exports are lower than imports, another shade or color is used.

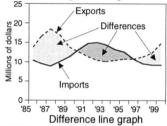

Difference line graph

Digraph

When the graphs used in graph theory have arrows on the lines, the diagrams are called digraphs. When the term graph is used in graph theory it has no relationship to any of the graphs described in this book. See Graph Theory.

Dimension

A term sometimes used when referring to variables. See Variable.

Dingbat

Dingbats are typographical ornaments or symbols. They are frequently used for such things as differentiating one data series from another on graphs, emphasizing key points on text charts, identifying specific items, making borders, or simply improving the appearance of charts. Dingbats are readily available in most computer applications and are generally applied by means of keystrokes. Previously dingbats were relatively standard shapes as shown at the left. They now include a limitless variety of shapes such as those shown at the right.

Traditional dingbats

Non-traditional dingbats

Discrete Data

Data in which only specific values can exist without the possibility of intermediate steps or values between the permissible values. For example, families have whole numbers of children (e.g., 1, 2, 3, etc.). There are no possible values in between (e.g., no fractions such as one-and-a-half or two-and-three-quarters children). • The opposite of discrete data is continuous (sometimes called indiscrete) data, which transitions smoothly from one bit of data to another and any value in between is possible. • There is disagreement whether discrete data should be connected by lines to form line or area graphs – for example, connecting data points for the populations of Paris, London, Rome, and Moscow. Some feel that since the four data points do not represent things that transition smoothly from one to the other, they should not be represented by a line or area graph. Instead, they recommend a point, column, or bar graph. Others feel it is OK to connect the data points to form a line or area graph reasoning that it helps the viewer interpret the information and the viewer will know that the line is simply a visual aid. When the points are connected, stepped line or stepped area graphs are sometimes used to emphasize the fact that the graph is showing values for comparison purposes and not to establish a trend. The horizontal portion of the line or area can also be darkened to emphasize the values instead of the transitions.

Distorted Map	Sometimes called a proportional or value-by-area map. A distorted map is a variation of a statistical map in which the sizes of the entities on the map are proportional to values other than the land areas of the entities. For example, if a particular resource is being studied by country, the size of each country might be proportional to the availability of the resource in that country. The country with the largest supply of that resource would have the largest area on the map even if it had the smallest physical area in terms of actual land. Distorted maps are sometimes classified as cartograms, abstract, or thematic maps. See Statistical Map. Distorted map

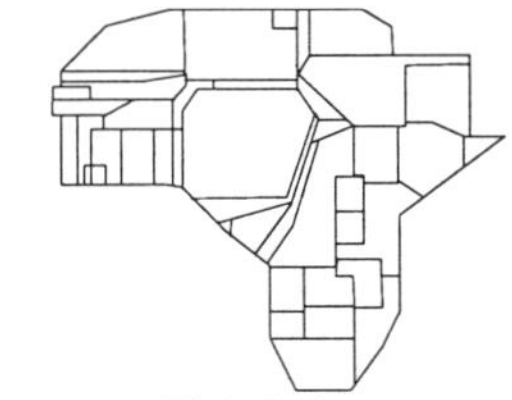

Distribution Channels Chart A chart to graphically display the channels used to get something from its source to an end recipient or user. This type of chart is most widely used to study the distribution of product; however, it can be used in other situations. The chart can be applied to a total organization or a subsection of a larger organization. Distribution channels charts can run vertically or horizontally, and flow is normally from top to bottom or left to right. When constructed vertically, the source of the product is shown in a rectangle at the top. The ultimate recipient or user of the product is shown in a rectangle at the bottom. If there are multiple categories of ultimate users, such as domestic and foreign customers, multiple rectangles might be shown or multiple charts might be generated. Between the upper and lower rectangles are other rectangles representing individuals or organizations such as distributors and retailers through which the product might pass. Arrows are used to depict the flow of the product through the various entities. An example is shown below.

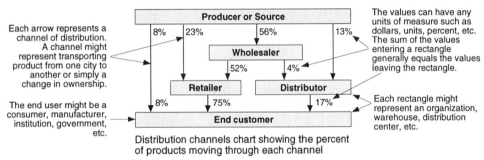

Distribution channels chart showing the percent of products moving through each channel

Percentages of product or dollars that pass through each channel may or may not be indicated beside each arrow that represents a channel. For planning purposes, multiple charts are sometimes made to indicate what type of distribution a company has today and what it plans to have at some point in the future, or to compare one organization's distribution with that of another.

Divided Area Graph	Sometimes referred to as a stacked, strata, multiple-strata, stratum, layer, subdivided area, or subdivided surface graph. A divided area graph is an area graph in which the data graphics for multiple data series are stacked on top of one another such that the top curve on the graph represents the sum or total of all the data series shown on the graph. Such graphs are many times used to show the relationship between the parts of the whole and how those relationships change over time, – for example, to show how much each product line contributes to the overall sales of the company from year to year. See Area Graph. 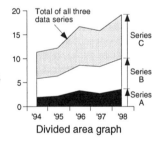 Divided area graph

Divided Bar Graph	Sometimes referred to as a composite, extended, segmented, stacked, or subdivided bar graph. A divided bar graph has multiple data series stacked end-to-end. This results in the far right end of each bar representing the total of all the components contained in that bar. Divided bar graphs are used to show how a larger entity is divided into its various components and the relative effect that each component has on the whole. See Bar Graph. 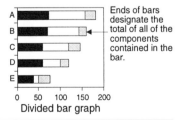 Divided bar graph

Divided Circle Chart	Sometimes referred to as a pie chart, cake chart, circular percentage chart, sector chart, circle diagram, sectogram, circle graph, or segmented chart. Divided circle charts are a category of proportional area charts consisting of a circle divided into wedge-shaped segments. Each segment is the same percent of the total circle as the data element it represents is of the sum of all the data elements in its data set. See Pie Chart.

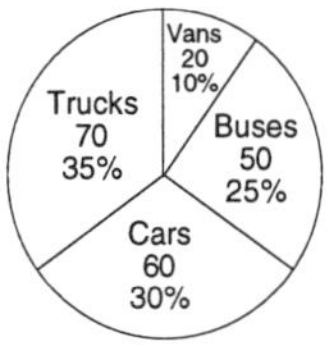

Divided circle chart

Divided Column Graph	Sometimes referred to as a composite, extended, segmented, stacked, or sub-divided column graph. A divided column graph has multiple data series stacked on top of one another. This results in the top of each column representing the sum or total of all of the components shown in that column. Such graphs are generally used to show how a larger entity is divided into its various components, the relative effect that each component has on the whole, and how the sizes of the components and the total change over time. See Column Graph.

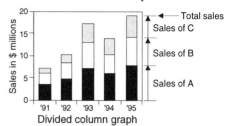

Divided column graph

Documentation Chart	A term sometimes used when referring to charts that record complex projects, sequences, relationships, etc. Examples are PERT charts, critical path method charts; complex flow or network charts; organization charts; etc.

Donut Chart	A donut chart is a pie chart with an area blanked out in the center so information such as the overall value of all the pieces of the pie can be shown. One of the criticisms sometimes expressed about pie charts is that they focus on the relative sizes of the components to one another and to the whole, but give no indication of changes in the whole when two or more pie charts are shown. The donut chart partially addresses this issue by somewhat more forcefully bringing changes in overall values to the viewer's attention. The size, shape, nor color of the blanked area typically have any significance.

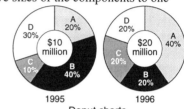

Donut charts

Dot Array **or** **Dot Diagram**	A type of frequency distribution chart in which dots are used to show the distribution of data elements in a data set. In constructing a dot array, values or class intervals are shown on either the horizontal or vertical axis. Each time a value occurs in the data set, a dot is added to the chart in line with the appropriate value. When all the data elements are accounted for, the resulting series of dots forms the equivalent of a simple histogram. In some cases the symbols are plotted on a numbered grid, as shown at the left. In other cases, the symbols are stacked on top of or beside one another and no grid is used, as shown at the right. When the main purpose of the array is simply to locate the value or class interval with the highest frequency of occurrence (mode), scales on the frequency axis may not be necessary. Almost any type of symbol, filled or unfilled, can be used.

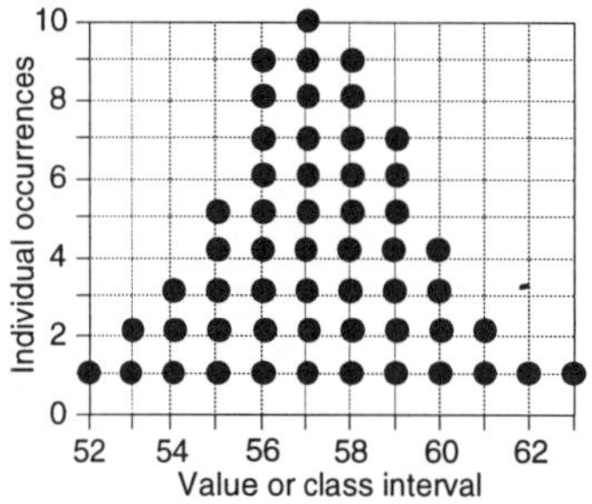

Dot array using grid lines and a scale for determining frequency of occurrence

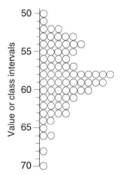

Dot array with no grid lines or frequency scale, unfilled circles, and the circles stacked beside each other

Dot Density Map	In a dot density map, each dot stands for a certain number of the things it represents. For example, if each dot represents 100 cats, a neighborhood with 10 dots would indicate a cat population of 1,000. One of the advantages of displaying statistical information in this way is that in addition to graphically showing the total number, the location of the dots also shows how the things are distributed. For instance, in the map at the right, the things represented tend to concentrate at highways and intersections.

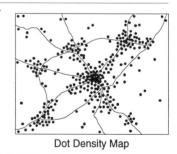

Dot Density Map

Dot Graph	Sometimes referred to as a point, scatter, or symbol graph. Dot graphs are a family of graphs, including scatter graphs, that display quantitative information by means of data points represented by dots or other symbols. Dot graphs, particularly scatter graphs, are frequently used for analyzing the relationships between two or more variables. They are considered one of the best types of graphs for investigating potential correlations between two data sets. See Point Graph.

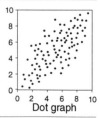

Dot graph

Draftsman's Display

When a draftsperson makes an engineering drawing of an object, a standard procedure is to draw views of two sides that are at right angles to one another, plus a view of the top. With these three views a person looking at the drawing can generally get a reasonable understanding of the appearance and features of the object. When that same procedure is applied to a three-dimensional graph, the result is three separate two-dimensional graphs called a draftsman's display. The example below uses a three-dimensional contour graph to illustrate the technique. The process works equally well with almost any type of graph. The same basic procedure can be expanded to include views of the other two sides, the bottom, and sections through the three-

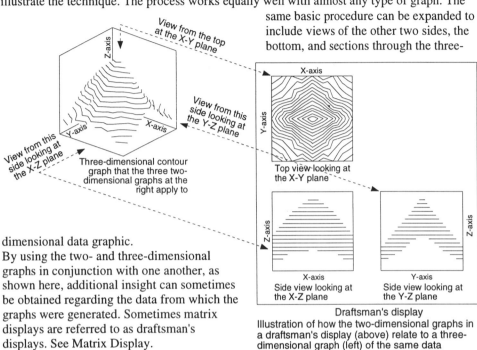

Three-dimensional contour graph that the three two-dimensional graphs at the right apply to

Top view looking at the X-Y plane

Side view looking at the X-Z plane

Side view looking at the Y-Z plane

Draftsman's display
Illustration of how the two-dimensional graphs in a draftsman's display (above) relate to a three-dimensional graph (left) of the same data

dimensional data graphic.

By using the two- and three-dimensional graphs in conjunction with one another, as shown here, additional insight can sometimes be obtained regarding the data from which the graphs were generated. Sometimes matrix displays are referred to as draftsman's displays. See Matrix Display.

**Drop Line
or
Drop Grid**

Sometimes referred to as a tether. Drop lines are thin lines drawn from data points to a reference point, line, or plane to assist the viewer in determining the value or location of the data point. They are sometimes described as connecting or anchoring data points to some reference. The value highlighted by the drop line might be the actual value of the data point or the difference between the data point and a reference or some other data point. The examples below illustrate some of the various ways drop lines are used.

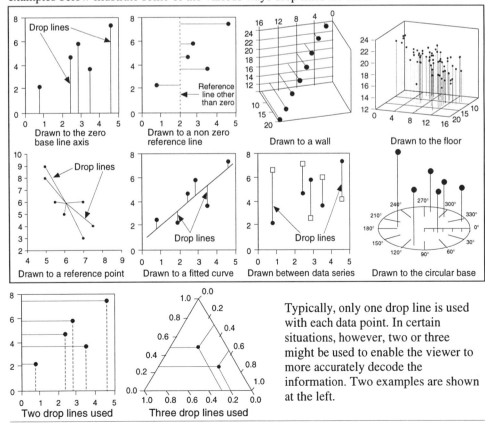

Typically, only one drop line is used with each data point. In certain situations, however, two or three might be used to enable the viewer to more accurately decode the information. Two examples are shown at the left.

143

Drop Line or Drop Grid (continued)	Grid lines may or may not be used in conjunction with drop lines. Drop lines can be solid or dashed. If multiple data series are plotted on the same graph, the drop lines might be coded to differentiate the data series. Drop lines are frequently used with point graphs but can be used with almost any type of graph as shown below.

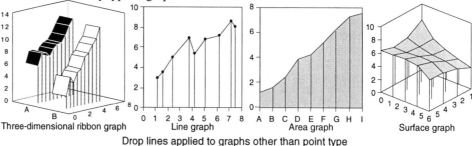

Drop lines applied to graphs other than point type

Drop Line Graph	A variation of a point graph in which two or more data series are plotted on the same graph, generally with a time series scale on the horizontal axis. After plotting, the data points for each time interval are connected with vertical lines. The major purpose of drop line graphs is to highlight the differences between the individual data points of two or more data series.

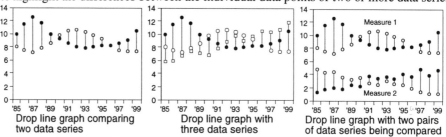

Drop Shadow	A variation of shadow used extensively in information graphics. See Shadow.
Dual Axis or Dual Scale **Includes:** • Dual Category • Dual Vertical • Dual X-axis • Dual Y-axis	Each of the terms at the left means there are two scales of the type described. For example, dual Y-axis means there are, or can be, two vertical (Y-axis) scales. Dual X-axis means there are, or can be, two horizontal (X-axis) scales. These general phrases do not restrict the type or location of the scales; however, specific applications often do. For example, in the broadest sense, the phrase "dual vertical scale" means there can be two vertical scales of any type, anywhere. In more restrictive cases, it often means one scale on either side, but not two scales on the same side. Each individual case must be studied to determine how the term is being used in that particular situation.

Dual Line Graph	Sometimes referred to as a compound, grouped, multiple, or overlapped line graph. A dual line graph is a line graph with multiple data series, all of which are measured from the same zero base line axis. See Line Graph.

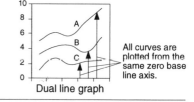

All curves are plotted from the same zero base line axis.

Dual line graph

Dummy Variable	Sometimes a variable is designated to get more insight into a data series, as opposed to seeing how it varies with respect to some other variable. For example, if a study is done to see how reaction time varies with age, a group of people of all ages might be tested and the data plotted on a scatter graph, as shown at the right. In this example, age and reaction times are the independent and dependent variables respectively. As an extension of the

study one might identify those data points that represent individuals with some particular characteristic, such as those who are active in sports. This would be called a dummy variable. If, in the graph above, circles were used to identify those individuals active in sports, the graph might look like the one at the left, which indicates that those individuals active in sports tend to score higher than those who are not. • There is no limit as to how many dummy variables can be designated for a given data series; however, generally only one is shown per graph.

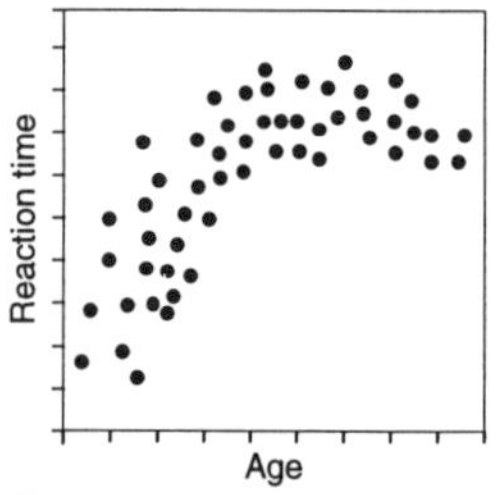

Results of tests on a cross-section of the population

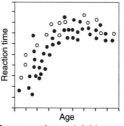

Same graph as at right, except individuals active in sports are identified with circles

| **Economists' Use of Graphs** | Graphs are used extensively in the field of economics. Although the applications are many times unique, most of the graphs used are traditional types such as line, column, and bar. In addition to plotting empirical data, graphs are also widely used to analyze, understand, and interpret information, to assist in formulating concepts and overall relationships, and as an aid in communicating ideas, theories, and findings. When the major purpose of a graph is to convey a concept, scales, grid lines, and tick marks are often omitted, and straight or smooth lines are frequently used to represent variables even though with actual data the line might be irregularly shaped. Such graphs are sometimes referred to as abstract graphs or conceptual diagrams. Examples of graphs developed by economists are shown below. |

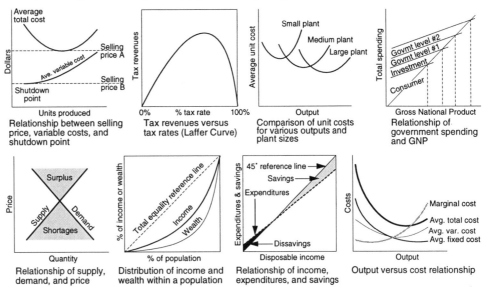

Abstract graphs developed and used in the field of economics.

| **Elevation Map** | A term used with maps when referring to the vertical distance a point or place is above a reference line or plane. For example, a point on a hill might be described as having an elevation of 2,543 feet which means it is 2,543 feet above a reference plane. The reference plane is generally sea level; however, any reference point can be used as long as it is stated on the map. Elevations are generally positive numbers. Points below the reference point (depressions) sometimes have negative values. |

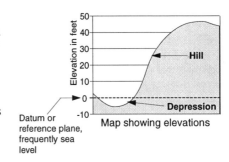

Map showing elevations

| **Encoding** | The process of converting information into charts and graphs is sometimes described as encoding. For example, the number 37 might be encoded into a chart by drawing a certain size rectangle on a set of grid lines (column graph); or the state of Missouri might be encoded by drawing a certain shape of polygon (map). As the viewer looks at the charts and graphs and mentally converts the graphical representations back into meaningful information (e.g., the rectangle back to the number 37 and the polygon back to Missouri), the process is sometimes described as decoding. The processes of encoding and decoding are equally important in making a chart or graph successful. |

| **Envelope** | An envelope is generated when a pair of lines are drawn at the top and bottom or on both sides of a series of data points. The lines may be drawn free hand or plotted based on some mathematical procedure. They may or may not encompass all of the data points. The purpose of the lines is to approximate the boundaries of the data to assist the viewer in estimating the general shape and trend of a group of data points. A regression line may or may not be used in conjunction with the envelope. Envelopes are generally used only with point and line graphs. |

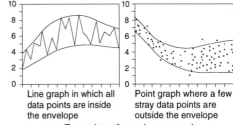

Examples of envelopes used with two types of graphs

Equation/Function Graph

Most equations (many of which are called functions) can be displayed on a graph by plotting a number of data points and connecting them with lines. Frequently, the equation is entered into a computer and the curve is automatically drawn. Graphs for equations generally have quantitative scales on all axes. Graphs of several equations of varying complexity are shown here.

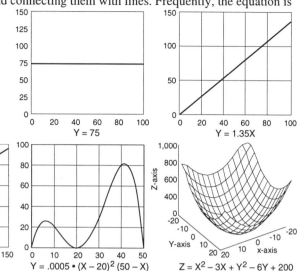

$Y = 75$

$Y = 1.35X$

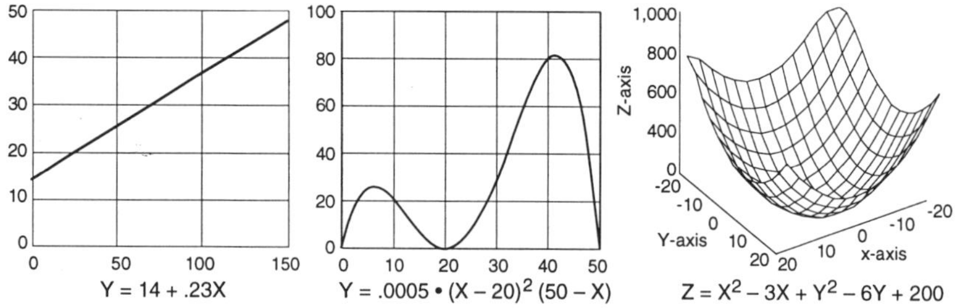

$Y = 14 + .23X$

$Y = .0005 \cdot (X - 20)^2 (50 - X)$

$Z = X^2 - 3X + Y^2 - 6Y + 200$

Examples of graphs that have equations displayed on them

Equations that describe three-dimensional surfaces

When graphs of equations are generated by computer, technically, every point that satisfies the equation is plotted. In two-dimensional graphs, this results in continuous lines rather than occasional data symbols connected by lines. If an equation describes a surface on a three-dimensional graph and all points that satisfy the equation are plotted, a solid data graphic results, as shown at the left. This many times makes it difficult to interpret the graph. To overcome this problem, only a limited number of values along the X and Y axes are typically plotted, which results in a mesh or fishnet appearance, as shown at the right. This type of graph is generally referred to as a surface or lined surface graph. The spacing of the lines (sometimes called isolines) is optional and can be increased or decreased to enhance the appearance or ability to see fine detail.

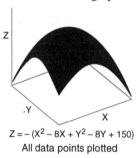

$Z = - (X^2 - 8X + Y^2 - 8Y + 150)$
All data points plotted

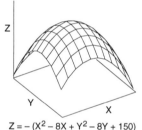

$Z = - (X^2 - 8X + Y^2 - 8Y + 150)$
Same graph as at the left except only selected lines are plotted to identify the surface

Specifying limits

When graphing equations, limits must be established. For example, in the simple equation of $X = Y$, the values of the two variables can technically range from minus more than a billion to plus more than a billion. In most cases this would be far outside the values of interest. Therefore, one selects limits that accomplish the purpose of the graph. If the range of values involved is known, the limits might be selected slightly above and below that range of values. If the equation is repetitive, limits might be chosen to restrict the graph to only a few cycles. In other cases, exploring must be done to select the proper limits. The four graphs below illustrate the effect that using different limits can have on the appearance of a curve and are representative of graphs that might be generated when exploring an unknown equation. Each of the four graphs is a plot of the same equation, except the limits on the two axes are different. The person doing the analysis generally decides which graph is most meaningful or if several are required to fully understand the nature of the equation and the phenomenon it describes.

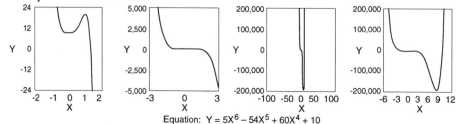

Equation: $Y = 5X^6 - 54X^5 + 60X^4 + 10$
All four graphs are plots of the same equation. Only the upper and lower limits are different.

Equation/Function Graph
(continued)

Example relating total sales and units sold

Equations can be written to describe many day-to-day situations. A simple example is the equation describing the total sales of a product based on the units sold and the selling price of the product. The equation reads:

Total sales dollars (Y) = Units sold (X) times the unit selling price ($7.85)

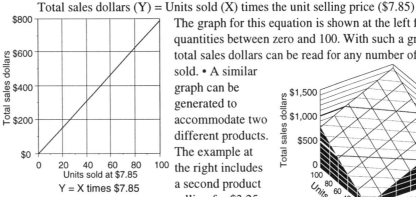

Y = X times $7.85

The graph for this equation is shown at the left for quantities between zero and 100. With such a graph, the total sales dollars can be read for any number of units sold. • A similar graph can be generated to accommodate two different products. The example at the right includes a second product selling for $3.25

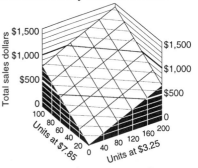

Z = X times $7.85 + Y times $3.25

with units up to 200. With this graph, the total sales for any combination of unit sales can be estimated.

Families of equations

Graphs bases on equations are many times used for planning purposes since they present options for the viewer to chose from. For example, if a company has an earnings per share (EPS) value of $2.00 and wants to increase it to $2.50, the company might increase the earnings, reduce the number of shares outstanding, or choose a combination of the two.

EPS (Y) = $\dfrac{\text{Annual earnings (\$80 million)}}{\text{Shares (X)}}$

Assuming an earnings of $80 million and 40 million shares outstanding, the graph on the left shows the effect of reducing the number of shares. The one on the right shows the effect of increasing earnings. Based on these two graphs, the earnings per

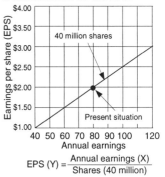

EPS (Y) = $\dfrac{\text{Annual earnings (X)}}{\text{Shares (40 million)}}$

share might be increased to $2.50 by either reducing the shares outstanding from 40 to 32 million, or by increasing earnings from $80 to $100 million. • In order to consider combinations of changes in the two variables, multiple equations (sometimes called a family of equations) are plotted on the same graph. The two graphs below are similar to the graphs above except families of curves are shown. Both show which combinations of improved earnings and reduced number of shares might achieve the goal of $2.50 EPS. In an actual situation, only one of the graphs would be used. Based on these graphs, the viewer can select from a series of alternatives, such as reducing shares to 34 million and increasing earnings to $85 million, reducing shares to 36 million and increasing earnings to $90 million, etc.

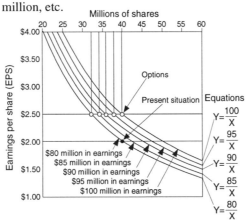

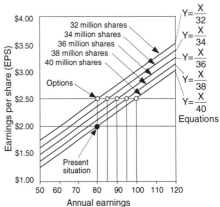

These two graphs are the same as the two graphs shown above, except a family of curves representing multiple equations is added so that many potential solutions can be seen.

147

Error Bar

Error bars are short lines used to convey statistical information about the data displayed on graphs. For example, if a data point represents the average of a set of numbers, the error bar might denote additional information about the same set of numbers, such as confidence interval, plus and minus 1, 2, or 3 standard deviations, standard errors, or the 10th and 90th percentiles. Error bars run parallel to quantitative scales and therefore can run either vertical or horizontal. When there are two quantitative scales, two sets of error bars are sometimes used. Symbols or short perpendicular lines are located at the ends of the error bars. Error bars are members of a family of range symbols. See Range Symbols and Graphs.

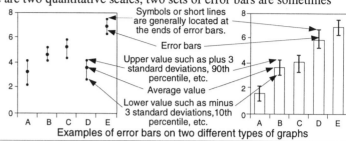
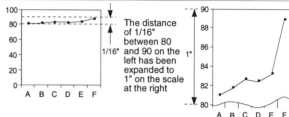

Examples of error bars on two different types of graphs

Expanded Scale

When a scale is expanded, the physical distance between the labels is increased. For example, if a scale has one-sixteenth of an inch between two values such as 80 and 90 before expanding, after expanding the two values might be separated by one full inch, as shown at the right. See Scales.

The vertical scale of the graph on the right is an expanded portion of the vertical scale of the graph on the left.

Explanatory Diagram

Sometimes referred to as a conceptual or summary diagram. Explanatory diagrams are made up of geometric shapes, lines, arrows, sketches, and/or text to aid in the visualization and understanding of relationships. The key purpose of explanatory diagrams is to provide a graphical overview of how people, things, ideas, influences, actions, etc., interrelate. They can effectively condense large amounts of information into a single chart. See Conceptual Diagram.

Example of a simple explanatory diagram

Explanatory Maps

Explanatory maps are sometimes referred to as descriptive, nonquantitative, or qualitative maps. This type of map is generally used to show nonquantitative information such as where things are located, how areas are organized and/or subdivided, and where the routes are located that get people and things from one place to another. The types of things included on explanatory maps might be grain fields, buildings, rivers, cities, roads, land masses, power lines, sewer lines, sales areas, public facilities, voting precincts, schools, political divisions, resorts, pet stores, sales areas, etc. See Descriptive Map.

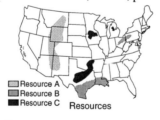

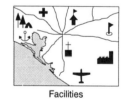

Examples of explanatory maps

Exploded Diagram/Chart

A picture, sketch, or drawing, of a device, assembly, mechanism, etc., showing all or some of the pieces/parts as though they were moved radially or axially away from an imaginary point somewhere within the device. Such a diagram not only lets the observer see the specifics of all the parts that normally would be hidden from view, but also shows how all the parts fit together in relationship to one another. Such diagrams are very useful for training manuals, repair instructions, spare parts lists, etc.

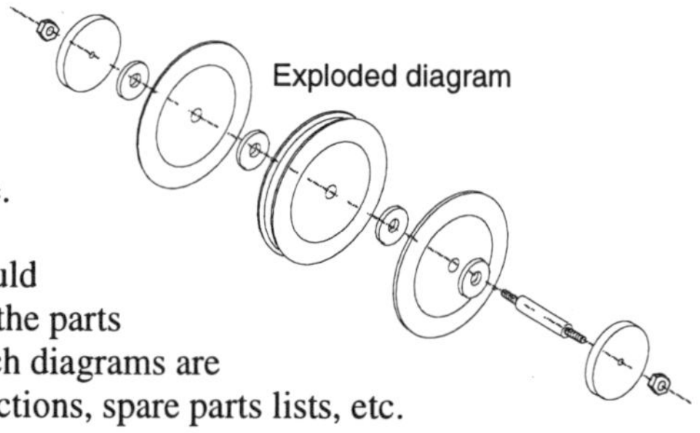

Exploded diagram

Exploded Map

A map in which portions of the map have been graphically separated from one another. The separations are typically made along man-made boundaries such as countries, states, territories, or counties. Maps are exploded to add emphasis to certain areas or groups of areas or to help focus the viewer's attention as to where certain boundaries are located.

Exploded map

Exploded Pie Chart	Sometimes referred to as a separated pie chart. An exploded pie chart is a variation of a pie chart in which one or more wedges of the pie are moved radially away from the center of the circle for visual emphasis. Examples are shown at the right.	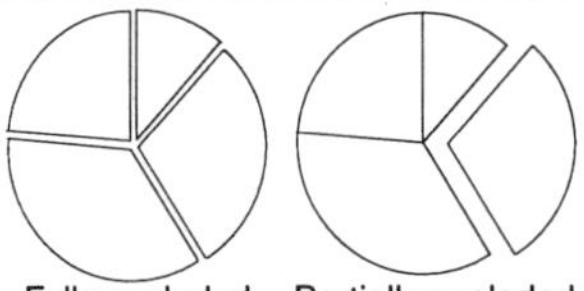 Fully exploded Partially exploded
Extended Bar Graph	Sometimes referred to as a divided, composite, segmented, stacked, or subdivided bar graph. An extended bar graph has multiple data series stacked end-to-end. This results in the far right end of each bar representing the total of all of the components contained in that bar. Extended bar graphs are used to show how a larger entity is divided into its various components and the relative effect that each component has on the whole. See Bar Graph.	 Extended bar graph
Extended Column Graph	Sometimes referred to as divided, composite, segmented, stacked, or subdivided column graph. An extended column graph has multiple data series stacked on top of one another. The top of each column represents the total of all components in that column. Such graphs are used to show how a larger entity is divided into its various components, the relative effect that each component has on the whole, and how the sizes of the components and the total change over time. See Column Graph.	 Extended column graph
Extent	The point on a graph diagonally opposite the origin. In a two dimensional graph it is normally the upper right hand corner of the plot area.	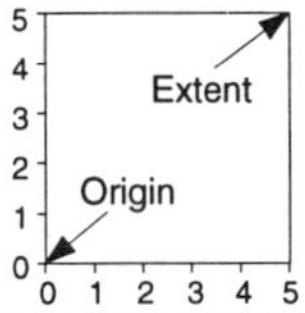
Faces	Sometimes called cartoon faces or Chernoff faces. Faces are icons that are occasionally used as symbols to encode three or more variables into charts, graphs, maps, and icon comparison displays. The variables are encoded by assigning values, or characteristics to variations in facial features. See Chernoff Faces.	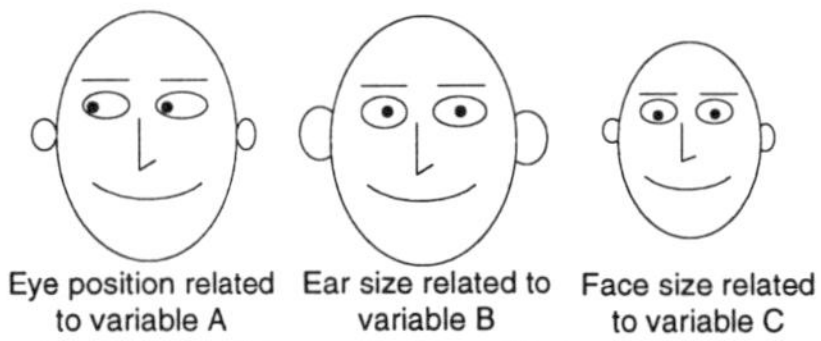 Eye position related Ear size related to Face size related to variable A variable B to variable C Faces used to encode multiple variables
Fan Chart/Graph	A fan chart is a variation of a comparative graph. It is basically a line graph with only two entries on the horizontal axis. Typically the entries are two times, two conditions, two locations, etc. The vertical scales are generally percents, index numbers, or ranking numbers. Fan charts are used to compare the performance, ranking, characteristics, etc., of several entities at different times or under different conditions. See Comparative Graph.	 Fan chart/graph
Feedback Path or **Feedback Loop**	The return or counter flow of product, materials, information, etc., to a previous part of a flow pattern as shown on a flow chart or diagram. Typical applications are the path used for the return of defective parts found during inspection in a manufacturing process, or the feedback of information from buyers to sellers. Feedback paths generally go in the opposite direction of the basic flow. For instance, in the flow chart below, the normal flow of the process is from left to right. The feedback path for defective product is from right to left. If the basic flow chart proceeds from top to bottom, the feedback path typically proceeds from bottom to top. There are no limits on the number or length of feedback paths that can be used. Feedback path (loop) as used in a flow chart	
Fever Graph	Sometimes referred to as a segmented line, broken line, thermometer or zigzag graph. A fever graph is a variation of a line graph in which the data points are connected by straight lines as shown at the right.	 Fever graph

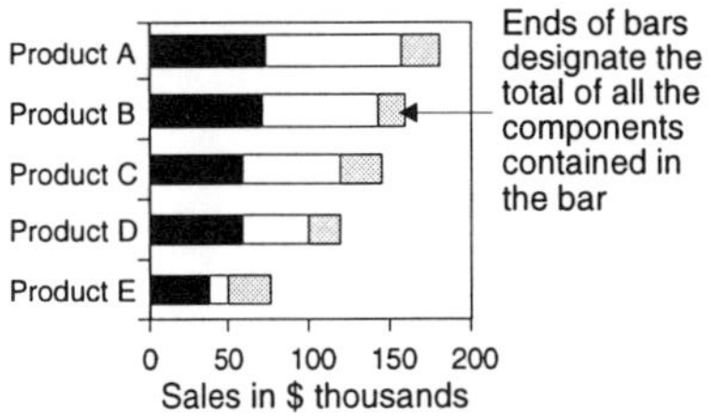

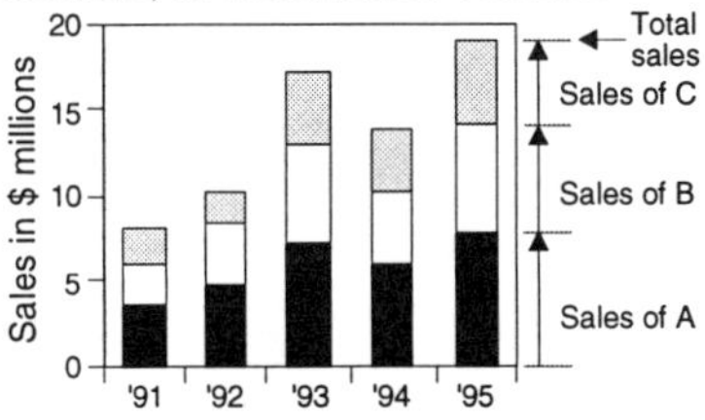

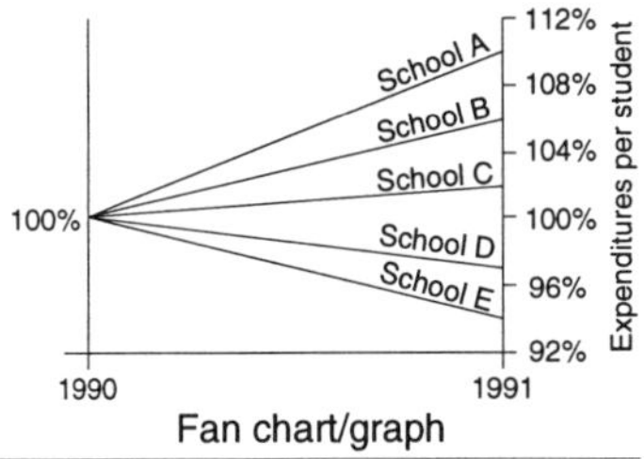

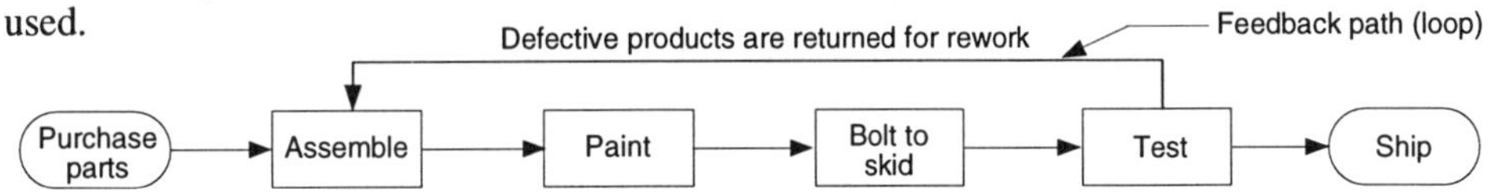

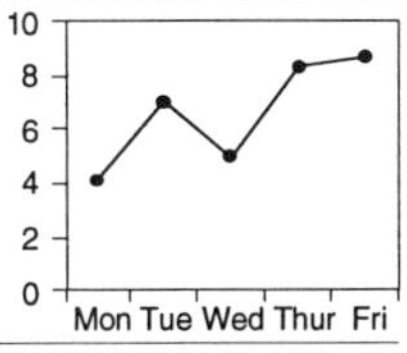

Fill

The opaque material applied to areas and lines for identification, differentiation, encoding values, emphasis, or improving appearances. Fills have four distinguishing characteristics: color, tone, pattern, and graduation. Within each characteristic there are many variations. Examples of some of those variations are shown below. • Patterns formed by parallel lines are many times referred to as hatched. Patterns with two or more sets of intersecting parallel lines are often referred to as crosshatched. (Example at right)

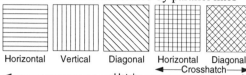

Horizontal Vertical Diagonal Horizontal Diagonal
◄———————— Hatch ————————► ◄— Crosshatch —►

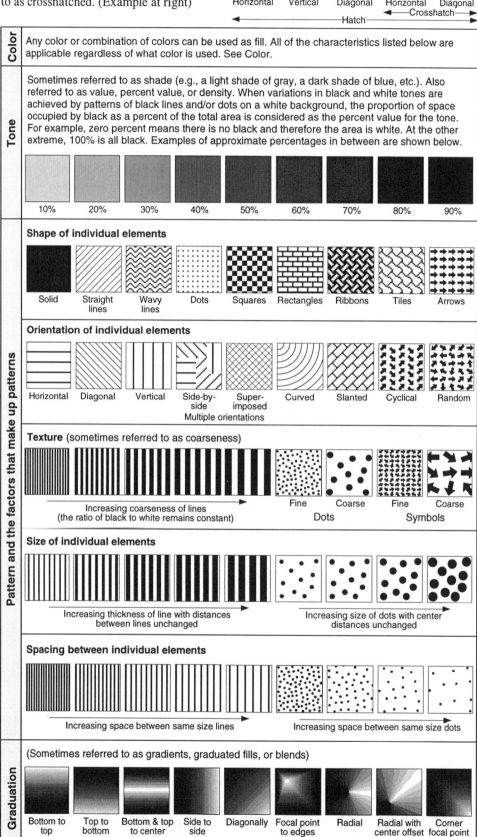

Color

Any color or combination of colors can be used as fill. All of the characteristics listed below are applicable regardless of what color is used. See Color.

Tone

Sometimes referred to as shade (e.g., a light shade of gray, a dark shade of blue, etc.). Also referred to as value, percent value, or density. When variations in black and white tones are achieved by patterns of black lines and/or dots on a white background, the proportion of space occupied by black as a percent of the total area is considered as the percent value for the tone. For example, zero percent means there is no black and therefore the area is white. At the other extreme, 100% is all black. Examples of approximate percentages in between are shown below.

10% 20% 30% 40% 50% 60% 70% 80% 90%

Pattern and the factors that make up patterns

Shape of individual elements

Solid Straight lines Wavy lines Dots Squares Rectangles Ribbons Tiles Arrows

Orientation of individual elements

Horizontal Diagonal Vertical Side-by-side Super-imposed Curved Slanted Cyclical Random
Multiple orientations

Texture (sometimes referred to as coarseness)

Increasing coarseness of lines
(the ratio of black to white remains constant)

Fine Coarse Fine Coarse
Dots Symbols

Size of individual elements

Increasing thickness of line with distances between lines unchanged

Increasing size of dots with center distances unchanged

Spacing between individual elements

Increasing space between same size lines

Increasing space between same size dots

Graduation

(Sometimes referred to as gradients, graduated fills, or blends)

Bottom to top Top to bottom Bottom & top to center Side to side Diagonally Focal point to edges Radial Radial with center offset Corner focal point

Fill (continued)

Fill used in lines

In addition to their use with areas, fills may also be used with lines, borders, and frames. When used with lines, the exact patterns are sometimes not distinguishable because only a portion of the design appears, depending on the width of the line. In many cases this presents no problem, as long as the fill successfully differentiates one line from another. Several examples of filled lines are shown below.

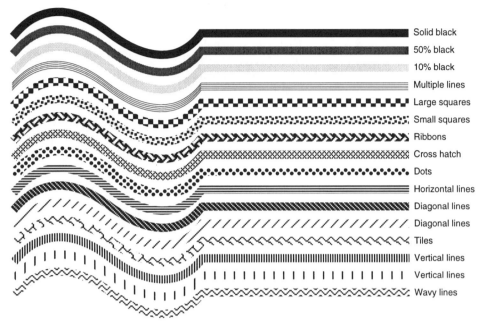

Solid black
50% black
10% black
Multiple lines
Large squares
Small squares
Ribbons
Cross hatch
Dots
Horizontal lines
Diagonal lines
Diagonal lines
Tiles
Vertical lines
Vertical lines
Wavy lines

Examples of lines incorporating several different types of fill

General guidelines

The following is a list of some general guidelines regarding the use of fills:
– Solids such as black and white are many times reserved for small areas.
– Lighter colors and finer patterns are generally used in large areas.
– Coarse patterns and bold vertical or horizontal lines sometimes give a harsh appearance.
– Many copiers cause faint patterns to disappear and dense patterns to turn to solid black.
– Small differences in tones are generally not used to convey quantitative information because of the difficulty of discriminating one from the other.

Filled Contour Graph

A filled contour graph can be a two- or three-dimensional contour graph with the areas between the contour lines filled with shading, color, or patterns. Three variables are plotted on such graphs and fill is used to identify points of equal value on the Z-axis. The boundaries between the filled areas define the points of equal value. The use of fill avoids the clutter of having the numeric values shown on the graph. See Contour Graph.

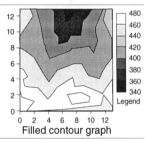

Filled contour graph

Filled Line Graph

Sometimes referred to as a silhouette or simple area graph. A filled line graph is a simple line graph (one data series) with the area between the line and the axis filled with a color, pattern, or shading. When filled with vertical lines, it is occasionally referred to as a histogram (unrelated to the traditional histogram). See Area Graph.

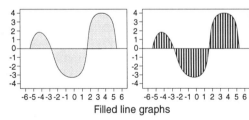

Filled line graphs

Filled Surface Graph

A variation of a three-dimensional surface graph in which the areas between the lines are filled with color or shading. Patterns are seldom used. There may be one or more types of fill and the fill may or may not depict quantitative information. See Surface Graph.

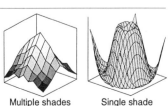

Multiple shades   Single shade
Filled surface graphs

151

Fishbone Diagram

Sometimes referred to as a cause-and-effect, CE, Ishikawa, or characteristic diagram. The purpose of a fishbone diagram is to provide a method for systematically reviewing all factors that might have an effect on a given quality objective or problem. This is accomplished by assigning each factor to a line or arrow and arranging the lines and arrows into meaningful clusters in a hierarchal fashion. The lines and clusters are arranged so that they all feed into a central line that, in turn, feeds into a symbol representing the overall object or problem. See Cause-and-Effect Diagram.

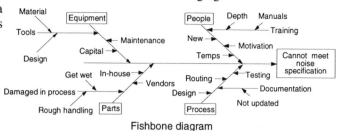

Fishbone diagram

Fishnet Graph

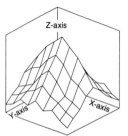

Three-dimensional surface graph using lines of equal values along the X and Y axes. Sometimes called a fishnet graph.

A fishnet graph is a three-dimensional wireframe or surface graph that is represented by lines of equal value along the X and Y axes. An example is shown at the left. The lines are equally spaced along both axes such that a view directly from the top reveals a fishnet pattern, as shown at the right. See Surface Graph.

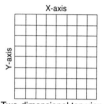

Two-dimensional top view

Example of the lines of equal values along the X and Y axes as seen when looking at the top of the three-dimensional fishnet graph shown at the left

Fitted Curve

A curve generated by a process called curve fitting. See Curve Fitting.

Floating Block Graph

A three-dimensional, three-axis range column graph is sometimes referred to as a floating block or flying box graph. On this type of graph the ranges (from lowest value to highest value) of each data set of one or more data series are compared.

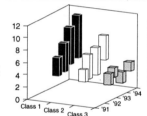

Floating block or flying box graph that is a three-axis, three-dimensional range column graph

Floating Column Graph

Sometimes referred to as a two-way column graph. A floating column graph is a variation of column graph in which positive values are measured both up and down from a zero on the vertical axis. Their major purpose is to compare multiple data series with particular attention to meaningful relationships such as correlations. Linear quantitative scales starting at zero are used on the vertical axis; generally only positive values are plotted. A category or sequence scale is used on the horizontal axis. Examples of the four major types of floating column graphs are shown at the right. See Column Graph.

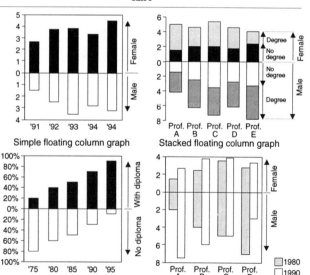

The four major variations of floating column graphs

Floor

Sometimes referred to as a bottom or platform. The term floor is used to identify the lower plane of a three-dimensional graph.

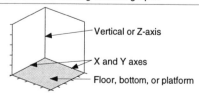

Example of floor on a three-dimensional graph

**Flow Chart
or
Flow Diagram**[1]

A flow chart is a diagram that visually displays interrelated information such as events, steps in a process, functions, etc., in an organized fashion, such as sequentially or chronologically. The things being represented can be tangible or intangible. A flow chart can be constructed for products being manufactured, currency moving from country to country and bank to bank, the development of an idea, how paperwork progresses through an organization, the procedure for getting a request approved for capital expenditures, etc. Flow charts can be very general, for overview purposes, or very detailed with lots of supporting and auxiliary information for use in day-to-day activities. They can describe an entire sequence of events from start to finish or address just a portion of the overall sequence. An example of one of the more widely used types of flow chart is shown at the right.

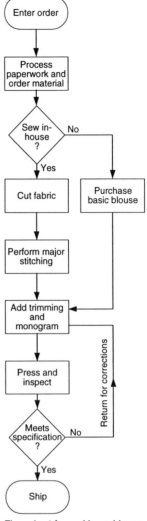

Flow chart for making a blouse

Key features of flow charts

- Word descriptions of events, activities, steps, or functions are typically enclosed by symbols and connected by lines or arrows.
- Generally two-dimensional. Those that are three-dimensional are generally pictorial.
- Typically not hierarchical or quantitative.
- Typically plotted sequentially.
- Typically not plotted against a time scale.
- Can run vertically or horizontally. Large flow charts generally run horizontally because of space considerations (some charts are several feet long).
- Normally proceed from top to bottom or left to right.
- On very large programs, individual charts are made for each subprogram and all of them cross-referenced.
- The major information is conveyed by text; however, significant additional information can be encoded by symbols, lines, colors, and images.
- Flow charts are applicable to large and small activities.

Variations of flow charts

There are a number of charts that are normally referred to by other names but may also be considered variations of flow charts. Several of these are listed below.

Name of chart	What the chart displays
• Organization chart	- Flow of authority and responsibility
• Decision Tree	- Flow or sequence of decisions
• Time and activity charts	- Sequence or flow of events
• Some "how-to" charts	- Sequence of tasks to achieve an objective
• Conceptual charts	- Flow of intellectual information and/or ideas
• Process charts	- Step-by-step description of a process
• Procedural chart	- Procedures to accomplish a particular goal
• Flow diagram	- Internal logic of a software system

Reasons for using flow charts

- Describe processes, ideas, networks, etc., particularly complex and abstract ones
- Define, analyze and better explicate processes, procedures, sequences, etc.
- Improve communications
- Help to clarify ideas
- Aid in trouble shooting
- Serve as a tool in planning and forecasting
- Reduce misunderstandings and conserve time
- Simplify training
- Document procedures
- Illustrate cross-functional relationships and responsibilities

Terminology and construction options

• **Symbols, enclosures, boxes, or blocks**

– The shape of the symbol sometimes has significance. The four symbols shown below, for example, generally have the meanings as noted. Other symbols are shown at the end of this section.

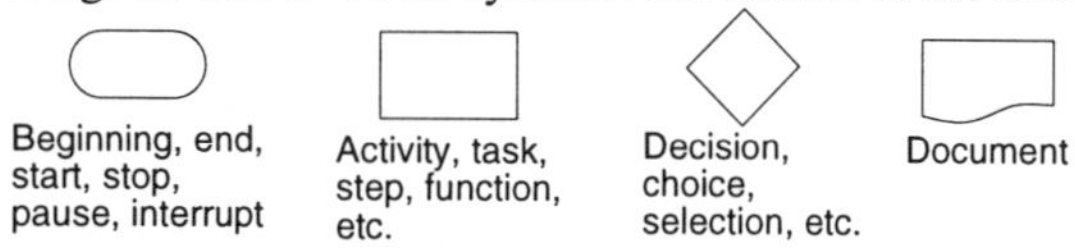

Beginning, end, start, stop, pause, interrupt

Activity, task, step, function, etc.

Decision, choice, selection, etc.

Document

There are dozens such symbols used in different applications. Some organizations have established standards for their specific needs.

– In addition to shape, other variations in symbols can also be used to encode information such as the department responsible, critical operation, pending, proposed, to be eliminated, new, etc. The symbol variations include:

-- Color, shade, or pattern of fill
-- Whether the outline of the symbol is solid, dashed, bold, thin, etc.
-- Size of symbol
-- Whether or not it has a shadow

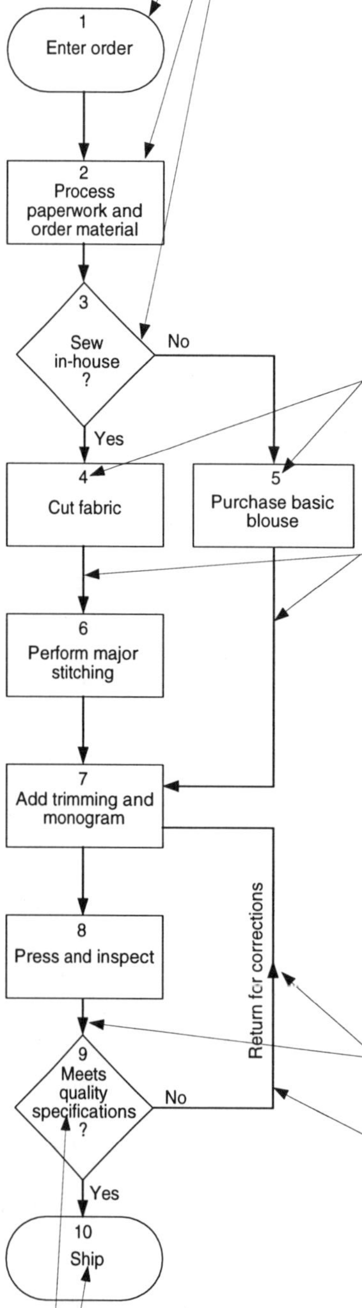

• **Reference numbers**

Sometimes included with each symbol, the numbers are normally consecutive. They are generally used for ease of communications and computerization of the data.

• **Lines, line paths, links, or branches**

– Variations in the type of line are sometimes used to encode additional information such as:

-- What the line represents (material, information, paperwork, responsibility, recommendations, communication, etc.)
-- Method of flow (electronic, telephone, paper forms, verbal, etc.)
-- Status (existing, proposed, eliminated, etc.)

– Variations sometimes used include:

-- Width of line. When line widths are proportional to the thing they represent, the chart is sometimes called a proportional flow chart.
-- Type of line such as solid, dashes, dotted, etc.
-- Color, shading, or pattern

– Arrow heads are generally used to indicate the direction of flow. Major flow is normally from top to bottom or left to right. Secondary flow paths such as feedback loops, two-way flow, and return paths go in the opposite direction. An example of a feedback path or loop is shown at the left. Arrow heads are generally at the ends of the lines but can be at any point along the line.

– Horizontal, vertical, diagonal, and curved lines are all used with flow charts. Horizontal and vertical are the most widely used.

– Lines can connect with the symbols at any point on the periphery of the symbol. The most common technique is to connect the lines at the vertical or horizontal midpoints.

1 Enter order
2 Process paperwork and order material
3 Sew in-house ?
No
Yes
4 Cut fabric
5 Purchase basic blouse
6 Perform major stitching
7 Add trimming and monogram
Return for corrections
8 Press and inspect
9 Meets quality specifications ?
No
Yes
10 Ship

Flow chart for making a blouse.

• **Text**

The text in the symbols normally describes the event, step, operation, action, position, or question the symbol represents.

Flow Chart
or
Flow Diagram[1] (continued)

Examples

Because of the versatility of flow charts and the lack of universally accepted standards for constructing them, their appearance varies significantly. Shown below are a few popular variations.

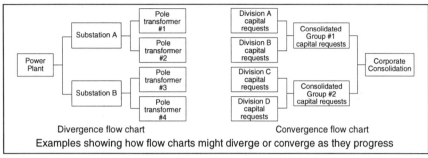

Divergence flow chart   Convergence flow chart

Examples showing how flow charts might diverge or converge as they progress

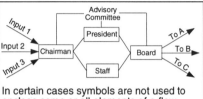

In certain cases symbols are not used to enclose some or all elements of a flow chart. Descriptive information is sometimes simply placed alongside of the arrows.

Example where no lines are used

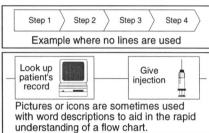

Pictures or icons are sometimes used with word descriptions to aid in the rapid understanding of a flow chart.

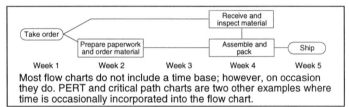

Week 1   Week 2   Week 3   Week 4   Week 5

Most flow charts do not include a time base; however, on occasion they do. PERT and critical path charts are two other examples where time is occasionally incorporated into the flow chart.

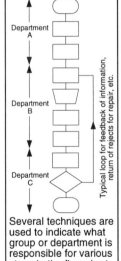

Several techniques are used to indicate what group or department is responsible for various steps in the flow chart. This example uses arrows on the left-hand side to denote which department has responsibility for specific steps.

Pictorial flow charts are sometimes used, particularly where the subject matter is complex or the viewer might be more familiar with the visual image than with the name or word description of the thing being represented.

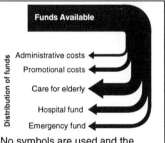

No symbols are used and the thickness of the line is proportional to the value it represents.

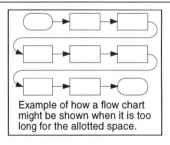

Example of how a flow chart might be shown when it is too long for the allotted space.

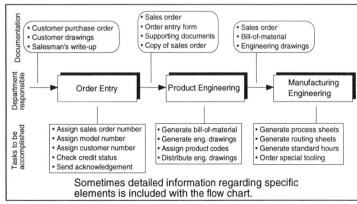

Sometimes detailed information regarding specific elements is included with the flow chart.

**Flow Chart
or
Flow Diagram[1]** (continued)

Symbols

Some symbol standards have been established by professional organizations such as the International Organization for Standardization (ISO) and the American National Standards Institute (ANSI). Some organizations have established their own standards. However, because the same symbol sometimes means different things in different context (in a process flow chart, a computer program flow chart, a machine operation flow chart, etc.), legends help minimize misunderstandings. The table below displays some of the most widely used symbols and their most common meanings. The fact that some symbols have multiple definitions or there are two or more symbols that mean the same thing is a reflection of the multiple standards. Additional information can be encoded into symbols by means of color, fill, dashes, etc.

	Specific step, activity, procedure, etc., in the process being diagramed. Examples: painting a fender or typing a letter.		Auxiliary operation such as inspection, examination, coding, calibration, etc.		Input or output to a computer or a system, such as receipt of orders or sending out of invoices
	Manual operation such as operating a punch press or signing of letters. In some cases it means information to be obtained by computer or individual		Correspondence between noncomputer entities: mailing of sales invoice to customers, financial reports from stores to headquarters, etc.		An idle period, waiting period, or delay: parts waiting to be processed on a machine, patients waiting to be seen by a doctor.
	Specific operation, step, activity, etc. Other times used to indicate start and stop, origin and termination, beginning and end, etc.		Transport, move, go from one place to another. Might be by truck, conveyor, bus, walking, etc.		An operation, process, activity performed while something is in motion, being transferred, or transported.
	Information is or can be displayed on a computer or television screen.		Manual input such as entering data into the computer using a keyboard.		Information transferred to storage within the computer.
	Designates the ends of a process such as the beginning and end, start and finish, stop, interrupt, etc. In other cases it may mean an exception, clerical, or terminal.		Document such as a purchase order, sales order, etc. Sometimes it refers to the actual document. Other times it refers to the need or activity of generating the document.		Filing or storage of information or material except in a computer. When used in a flow diagram of a computer system it sometimes means to merge multiple files.
	Combined activity such as an inspection conducted concurrent with another operation or activity.	White / Yellow / Blue copy	Multiple documents such as three copies of a purchase order. The individual copies may or may not be identified on the flow chart.		Designates an operating procedure; an input or output to a magnetic tape; or a key punch operation.
Collate Sort Communication link			Prepare or modify a report, product, piece of equipment, etc.		A point in the process where a decision is to be made or where a selection between alternatives is possible.

Symbols sometimes used in flow charts

Flow Diagram[2]

A graphic representation of the physical route or flow of people, materials, paperwork, vehicles, or communications associated with a process, procedure, plan, or investigation. A flow diagram is often the counterpart of a flow or process chart. The flow or process chart indicates the sequential order of activities. The flow diagram indicates the location of these activities and how the physical flow of people, material, etc. occurs between them. Flow diagrams are frequently used to study the movement patterns of people, animals, automobiles, etc. For example, a flow diagram of the way customers wander through a store or department might indicate to merchandisers how and where to display certain products or how to rearrange counters to get the customer to spend more time in the store or department. A flow diagram may or may not be coded to correspond to a flow chart, process chart, plan, or research study. Flow diagrams are frequently superimposed on floor plans or maps of the area where the activities take place.

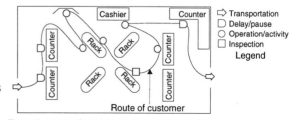

Flow diagram of a customer in a clothing store

Flow Map

Route maps show the paths from one point to another but generally do not indicate what or how much moves or flows in what direction along the paths. Flow maps, on the other hand, say little or nothing about the exact path but do indicate such things as:

– What it is that flows, moves, migrates, etc.
– What direction the flow is moving and/or what the source and destination are.
– How much is flowing, being transferred, transported, etc.
– General information about what is flowing and how it is flowing.

Each is discussed below.

Methods for designating what is flowing, moving, migrating, etc.

Flow maps can be used to show movement of almost anything, including tangible things such as people, products, produce, natural resources, weather, etc., as well as intangible things such as know-how, talent, credit, or goodwill. Sometimes the subject of a flow map is indicated in the title of the map. In other cases it is noted directly on the map, as shown

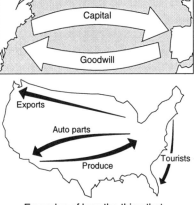

at the right. In still other cases, it is denoted by symbols, as shown at the left. The symbols may be used as the data graphic or they

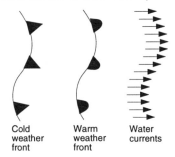

Cold weather front   Warm weather front   Water currents

Symbols are sometimes used to identify what is flowing.

Examples of how the thing that is flowing might be identified

may complement another data graphic, such as a line. When many different things are flowing, a legend is sometimes required.

Methods for designating the direction and/or destination of the thing that flows

How precisely the direction and location of flow is indicated may range from very general to very detailed. The example at the right is very general, indicating simply that something is flowing from one country to three other parts of the world in varying amounts. The map below gets more detailed by indicating movement

Flow map showing approximate amounts and destinations

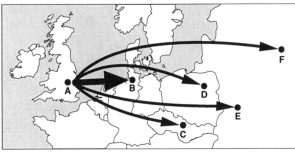

Flow map in which specific cities are designated

from one specific city to five other defined cities. The flow map below renders the detail in yards, as in a tactical combat map.

• In some situations, such as a railroad, flow is concentrated with limited options as to the number of directions it can move. In other cases the movement is distributed over a large area, with movement sometimes occurring in many directions

Example of movement over a long distance in many different directions

at once. In these cases, special data graphics are sometimes developed, such as the weather front at the left. Arrows are almost always used to indicate direction of flow.

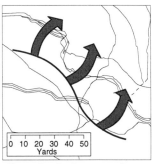

0 10 20 30 40 50
Yards

Example of a flow map where directions and distances are very detailed

Flow Map (continued)

Methods for designating the quantity, weight, or value of the thing that flows

In many cases the amount of the thing that is being moved is simply noted on the arrow, as shown at right, or placed in a legend or note on the map. In these cases the width of arrows is usually the same.

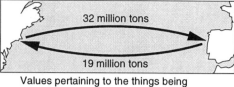

Values pertaining to the things being moved are sometimes placed adjacent to the appropriate flow lines.

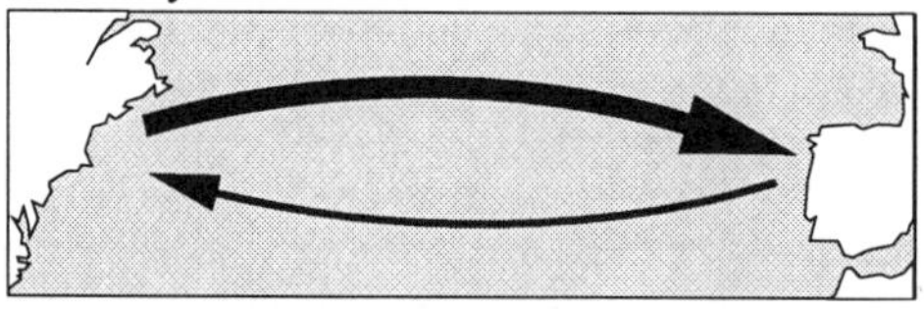

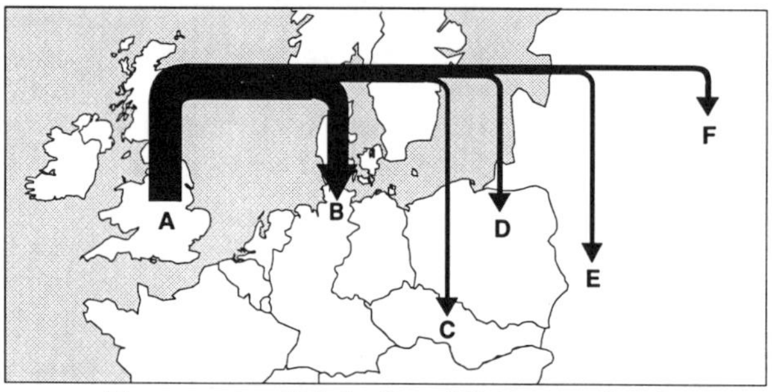

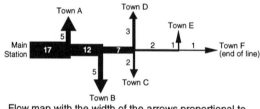

Flow maps in which the line width is proportional to the amount of material flowing in that branch

In other cases the width of the arrows is proportional to the quantity, value, etc., of the thing being moved, as shown in the examples at left. When the size of the arrows is meant only to be approximate, a scale is generally not used. If the viewer is expected to estimate values based on the width of the arrows, a scale is generally included. Examples of two types of scales are shown below.

Sometimes the actual values are placed adjacent to the flow lines, even though their widths are drawn proportional to their values. This is particularly useful when there are many values, as in the example below. This example illustrates a special type of flow map where values change as things progress, move, or flow, from point to

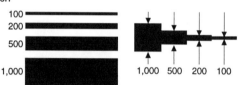

Scales that are sometimes used when flow line widths are drawn proportional to the values of the thing being transferred

Flow map with the width of the arrows proportional to the values they represent, with actual values also noted.

point. For example, subway passenger density may vary in relation to the distance from a main station. A delivery truck's weight varies as it progresses through its route, and the volume of natural gas varies with the distance from the source of supply.

Methods for encoding general information

Frequently the map can be made more informative by coding the flow lines to indicate specific characteristics, either about the entity that moves or the surface, or medium by which it moves. For example, a transcontinental phone call might travel part of the way by local phone lines, part by long distance cable, part by microwave and part by satellite. A flow map can display all of these to give the viewer more of an overview than simply a straight line from one phone to another.

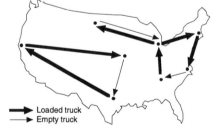

Example of a flow map including information about the conveyance

Example of a flow map including information about responsibility for different parts of flow

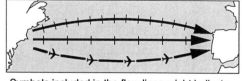

Symbols included in the flow lines might indicate how the entity moved (e.g., by air, sea, microwave, pipe, etc.), how fast it moved, who was responsible for managing the movement, etc.

Flying Box Graph

Sometimes referred to as a three-dimensional range column graph or floating box graph. See Column Graph.

Font	Sometimes referred to as typeface. When a set of printed letters, numbers, and characters have the same design features, the style or design of the type is identified as the font or typeface. There are hundreds of styles to choose from. They have such names as Helvetica, Times, Geneva, etc. Three major categories of typefaces are serif, sans serif, and decorative. Examples of each are shown below. The size of the type in each example is nine-point.

Serif typeface	This is an example of a serif typeface called Times. Serif letters have small lines projecting from the ends of each of their main lines or strokes.	With serifs T
Sans serif typeface	This is an example of a sans serif typeface called Helvetica. Sans (without) serif letters do not have the small lines projecting as the serif typefaces do.	Without serifs T
Decorative typeface	*This is an example of a decorative typeface named Zapf Chancery. Decorative typefaces are many times used for headings, special effects, and/or to improve the appearance of charts.*	T

Four-Fold Chart/Graph

Sometimes referred to as a two-by-two chart. A variation of a sector graph when scales are used on the radii, or a variation of a proportional chart when scales are not used. Four-fold charts use four 90 degree segments to represent four different categories. The radius or area of each segment is proportional to the value of the category it represents. Unless the chart is only meant to approximate values, the numeric values are frequently shown on or adjacent to the data graphics. Four-fold charts are often used when four conditions or situations are generated by two variables. For example, if a weight loss study is done where there were two distinct variations of diet and exercise, the following four categories of participants would be generated: those participants on a strenuous exercise program with and without a special diet and those on a mild exercise program with and without a special diet. Plotting the results of this type of study on a four-fold chart can sometimes make the results stand out clearer, as seen at the right.

Four-fold graph

Frame

The terms frame and border are often used interchangeably in charts and graphs. The illustration below depicts the major chart elements the terms apply to.

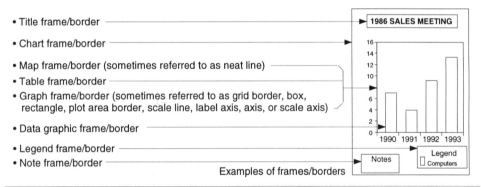

- Title frame/border
- Chart frame/border
- Map frame/border (sometimes referred to as neat line)
- Table frame/border
- Graph frame/border (sometimes referred to as grid border, box, rectangle, plot area border, scale line, label axis, axis, or scale axis)
- Data graphic frame/border
- Legend frame/border
- Note frame/border

Examples of frames/borders

Location of frame with respect to data graphics

Although there are many ways to position frames with respect to data graphics, in most cases the frame has little or no effect on the chart as a whole. In a few cases frames can add to the effectiveness of the chart. Four examples are shown below.

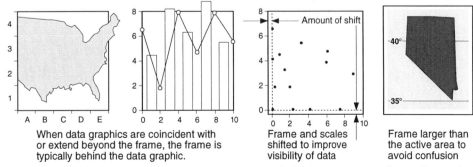

When data graphics are coincident with or extend beyond the frame, the frame is typically behind the data graphic.

Frame and scales shifted to improve visibility of data

Frame larger than the active area to avoid confusion

The key considerations in applying a frame are ease of viewing, accuracy of encoding and decoding, and attractiveness.

Frame (continued)

Complete, partial, or no frame

Sometimes a complete frame (two sides, a top and a bottom) is used with a graph, map, or table. Other times a partial frame or no frame is used. Factors affecting the decision as to the type of frame to use include:

- How many scales are used and where they are located
- Whether or not grid lines are used
- How cluttered the chart is
- The size of the chart
- The desired appearance of the chart

The examples below illustrate the use of varying amounts of a frame ranging from a complete frame on the left to no frame on the right.

	1993	1994	1995
Japan	18	21	23
Korea	12	17	19
Vietnam	15	25	21
Taiwan	14	17	18
China	18	20	22

	1993	1994	1995
Japan	18	21	23
Korea	12	17	19
Vietnam	15	25	21
Taiwan	14	17	18
China	18	20	22

	1993	1994	1995
Japan	18	21	23
Korea	12	17	19
Vietnam	15	25	21
Taiwan	14	17	18
China	18	20	22

	1993	1994	1995
Japan	18	21	23
Korea	12	17	19
Vietnam	15	25	21
Taiwan	14	17	18
China	18	20	22

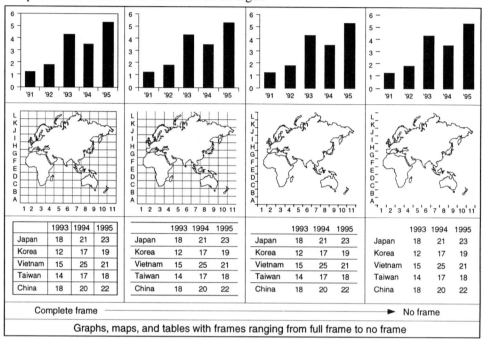

Complete frame ——————► No frame

Graphs, maps, and tables with frames ranging from full frame to no frame

Framed Rectangle

A symbol used primarily to designate values of a single variable. It is particularly useful for encoding quantitative information onto maps, as shown at the right. The symbol consists of a rectangle that is taller than it is wide. There are tick marks on both sides that designate a distance half way up the rectangle. The symbol can be any size. Typically all rectangles on the same chart or map are the same size, except when the width of the rectangle is used to encode an additional variable. When the rectangle is empty (all white), a lower value is indicated. When the rectangle is full (all black), an upper value is indicated. When it is half full (the lower half black), a value midway between the lower and upper values is indicated. The same reasoning applies to any fractional fill of the rectangle. The upper and lower values are indicated in a legend as shown below. Black and white are used in these examples. Any combination of colors and/or fills works equally well.

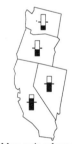

Map using framed rectangles to encode statistical information

Examples of specific values

• Althought the standard framed rectangle has only one set of tick marks and an upper and lower value, other numbers of tick marks and values can be used if it makes it easier for the viewer to decode the data. The examples at the right show variations of tick marks and values.

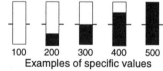

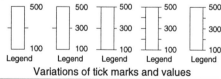

Variations of tick marks and values

Frequency or Frequency of Occurrence

Sometimes referred to as count. When histograms are generated, the first step is to set up class intervals that are ranges of values. The second step is to assign each data element to the range (class interval) in which its value lies. The next step is to count the number of data elements in each range or class interval. The number of data elements in a given class interval is called frequency, count, or frequency of occurrence. For example, if the values of 14 data elements are between 27 and 27.9, the frequency for the class interval of 27 to 27.9 is 14. See Histogram.

Frequency Bar Graph

Generally column graphs are used to generate histograms. Occasionally bar graphs are used. When they are, they sometimes are called frequency bar graphs. Almost all of the information that is applicable to column type histograms applies to bar type histograms (frequency bar graphs). See Histogram and Frequency Polygon.

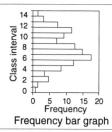

Frequency bar graph

Frequency Diagram

Sometimes referred to as a histogram. See Histogram and Freqency Polygon.

Frequency Distribution Graphs

Most frequency distribution graphs are two-dimensional graphs used to show how data elements are distributed in data sets. They are a variation of data distribution graphs. See Data Distribution Graphs. Within the frequency distribution graph family, there are two major categories: those that indicate incremental frequencies for individual values or class intervals and those that plot cumulative values. Examples of each of the major types of frequency distribution graphs are shown below. Each is discussed in detail under its individual heading.

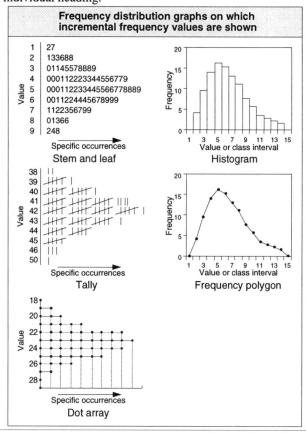

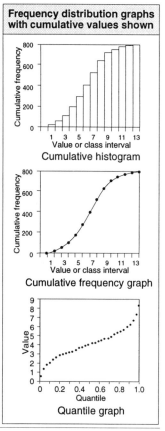

Frequency Polygon

A variation of a histogram where the columns are replaced by a curve that connects the centers of the tops of the columns of a histogram. In the example at the right, the columns of the histogram are included for reference only. In an actual situation the columns are generally not shown. See Histogram and Frequency Polygon.

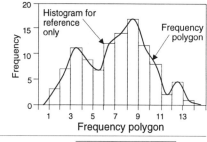

Frequency polygon

Frequency Table

When a table displays how frequently various things occur or how many there are of different things, the table is sometimes referred to as a frequency table. For example, in the body of the table at right, the entries show the frequency that each type of reject occured for each type of product. In addition, the column on the far right indicates the total frequency of rejects for each product and the bottom row shows the total frequency of rejects by type.

	Reject type A	Reject type B	Reject type C	Total
Product A	2	4	6	12
Product B	10	12	17	39
Product C	8	7	6	21
Product D	22	29	20	71
Total	42	52	49	143

A frequency table noting the number of times various things occurred or how many there are of different things

161

Function Graph

In many graphs, specific data points are plotted such as performance measurements, financial results, budgets, forecasts, test results, etc. These data points are frequently joined by lines. Another way to generate graphs is to plot curves based on equations. These are sometimes referred to as function or equation graphs. Such graphs are frequently used by engineers and scientists. See Equation/Function Graph.

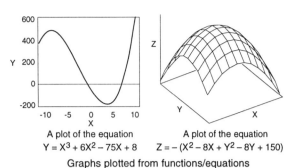

A plot of the equation
$Y = X^3 + 6X^2 - 75X + 8$

A plot of the equation
$Z = -(X^2 - 8X + Y^2 - 8Y + 150)$

Graphs plotted from functions/equations

Fuzzygram

Fuzzygrams result from a technique occasionally used with histograms in which the tops of the columns are replaced by blurred or fuzzy areas to indicate the probability of the values plotted. The degree of fuzziness is inversely proportional to the sample size. For example, the smaller the sample size, the larger the fuzzy area. There are mathematical methods for determining the number and spacing of the lines used to generate the fuzz. A vertical line at the top and in the horizontal center of the columns indicates where the top of the column would be on a conventional histogram for the same data without fuzz. In the figure below, two fuzzygrams are shown with a standard histogram for reference. The data for the two fuzzygrams were chosen so that percentage wise, both would have the same standard histogram, even though one represents a data set size of 50 and the other 200.

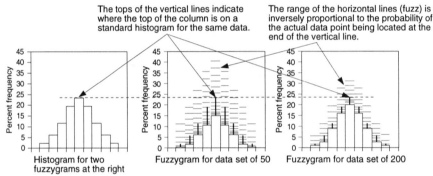

The tops of the vertical lines indicate where the top of the column is on a standard histogram for the same data.

The range of the horizontal lines (fuzz) is inversely proportional to the probability of the actual data point being located at the end of the vertical line.

Histogram for two fuzzygrams at the right

Fuzzygram for data set of 50

Fuzzygram for data set of 200

A comparison of two fuzzygrams and a standard histogram for the same data

Gantt Chart

A time and activity bar chart that is used for planning, managing, and controlling major programs that have a distinct beginning and end. In this type of chart, each major subprogram or activity involved in the completion of an overall program is represented by a horizontal bar. The two ends of the bar represent the start and finish of the activity. If the chart is for planning purposes only, the start and finish times are estimates or projections. If the chart is for both planning and tracking purposes, the start and finish times for future activities are estimated or projected and the historical portions are actual times. See Time and Activity Bar Chart.

	January				February				March				April			
Week	1	2	3	4	5	6	7	8	9	10	11	12	13	14	15	16
Subprogram #1																
Subprogram #2																
Subprogram #3																
Subprogram #4																
Subprogram #5																
Subprogram #6																
Subprogram #7																

Gantt chart used for planning and/or monitoring major programs

Gap, Data

Generally refers to the void left in a data series due to missing data. For example, if sales are noted for each of ten years except the third year, that missing third year would be said to generate a data gap. The term gap is used regardless of whether the data was left out intentionally or accidentally. In most cases, when a gap exists in the data, it will be reflected in a graph of the data. Ways to handle data gaps in graphs is discussed under the individual graph headings.

Gap, Graph

The term gap is sometimes applied to the spaces between bars and columns or groups of bars and columns as indicated in the illustrations shown here. The term is applicable to both two- and three-dimensional graphs.

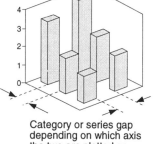

Spaces between the columns within a group are sometimes called series gaps or inter-column spaces. When columns overlap, they are said to have negative gaps or negative inter-column spaces.

Spaces between groups/clusters/categories are sometimes called cluster gaps, category gaps, or inter-group spaces.

Spaces between the first and last columns and the frame are sometimes called margin gaps or column margins.

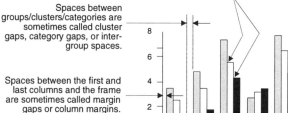

Category or series gap depending on which axis the two are plotted on

Examples of gaps on column graphs. The same terminology applies to bar graphs, except the term "column" is replaced by the term "bar".

Sizes of the gaps are generally specified in percents. When there is a single data series, the percents might be in terms of column width or the combination of the space and column width. The examples below for one data series are based on percent of column width.

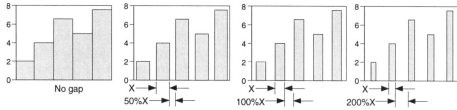

The same simple column graph with varying gap widths

When there are multiple data series, the space between the clusters might be based on a single column width, the cluster width, or a combination of the cluster width and the space between the clusters. Examples of all three are shown below.

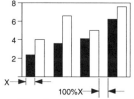 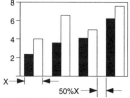 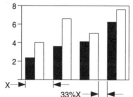

When the gap is based on a percent of a column width, a gap equal to the width of a column is described as 100%.

When the gap is based on a percent of the group or cluster width, a gap equal to the width of one column of a two column cluster is described as 50%.

When the gap is based on a percent of the distance from one side of a cluster to the same side of the next cluster, a gap equal to the width of one column of a two column cluster is described as 33%.

Three different ways the same gap between clusters can be specified

Overlapping of bars or columns is specified in terms of percent of column or bar width. The overlapping is sometimes referred to as negative gap but still in terms of column width.

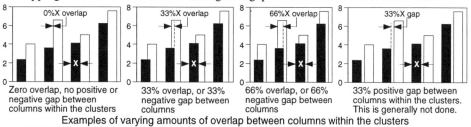

Zero overlap, no positive or negative gap between columns within the clusters

33% overlap, or 33% negative gap between columns

66% overlap, or 66% negative gap between columns

33% positive gap between columns within the clusters. This is generally not done.

Examples of varying amounts of overlap between columns within the clusters

Geodetic Coordinates and Geographic Coordinates	The values of latitude and longitude used to locate places and things on maps. See Map.
Geographic Grid	Sometimes called graticule. The network of lines of latitude and longitude used to locate places and things on maps and globes. See Map.
Geologic Map	A map that shows the composition and structure of earth materials and their distribution across and beneath the earth's surface. Geologic maps may show such things as soil and clay layers, bedrock formations, sediment deposits, minerals, and major folds and faults.

Graph

Sometimes referred to as a plot. A graph is a chart that graphically displays quantitative relationships between two or more groups of information – for example, the relationship between cities and their populations, a car's speed and its efficiency, or the buying power of a dollar over time. • Graphs have combinations of one, two, or three straight or circular axes utilizing one or more quantitative scales. This definition clearly differentiates graphs from other charts such as diagrams, tables, text charts, proportional charts, illustration charts, and most maps. In a few cases there is a slight overlap between graphs and certain three-dimensional statistical maps, which may be called either a graphs or maps. Since graphs constitute one of the major categories of charts, graphs are frequently referred to as charts. Graphs offer many important features, a few of which are:

– Large amounts of information can be conveniently and effectively reviewed
– Overall patterns of data stand out more clearly than in tabulated form
– Deviations, trends, and relationships many times are more noticeable
– Comparisons and projections can many times be made more easily and accurately
– Anomalies in data frequently become obvious
– Viewers can more rapidly determine and absorb the essence of the information
– When used in presentations, graphs shorten meetings and expedite group decisions

The following pages discuss the overall aspects of graphs. Detailed information is presented under individual graph headings and under headings that encompass families of graphs.

Methods for categorizing graphs

Graphs are categorized several different ways, each focusing on one of their many facets. Shown on this and the following pages are five methods used to categorize the major graphs in use today. The graphs included in each category are quite subjective. • Detailed information on each category as well as on individual graphs are included under their separate headings.

Categorized by major family or configuration

Most graphs can be categorized into one or more of the families or configurations shown below. There is considerable overlapping. For example, a polar vector graph could be included in the vector, circular, and/or line graph family. • Within each graph family there are generally several major subcategories. For instance, in the column graph family, there are subcategories of simple, grouped, overlapped, stacked, floating, range, difference, deviation, pictorial, and progressive column graphs. • In some cases, specialized graphs are constructed by combining features from two or more families of graphs.

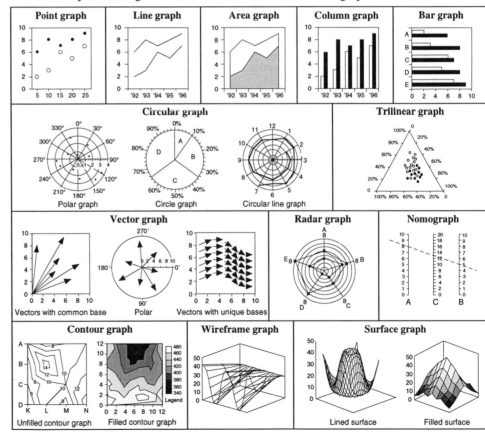

Categorized by area or field in which the graph is used

Three primary usage categories sometimes referred to in literature and software applications are technical/scientific, data analysis (statistical), and business, though they are seldom clearly defined. The examples shown below are representative of the types of graphs included in these categories. Standard graphs such as line, column, and bar are not included, since they are applicable to all categories. The categories into which graphs are placed is subjective, with considerable overlap. For example, histograms are used by data analysts, technical, and business personnel and therefore histograms could have been placed in any, or all, of the three categories.

———— Technical/scientific graphs ————

These graphs are many times very specialized and often have a higher percentage of three-dimensional graphs than either of the other two categories. Large amounts of data are common and the data is frequently entered via equations or directly from electronic data collection devices (sometimes referred to as real time data).

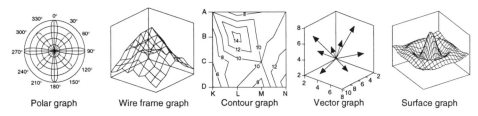

Polar graph　　Wire frame graph　　Contour graph　　Vector graph　　Surface graph

———— Data analysis graphs (sometimes referred to as statistical) ————

These graphs are used for such things as determining whether meaningful relationships exist between various data, how the data is distributed, whether or not differences are significant, etc.

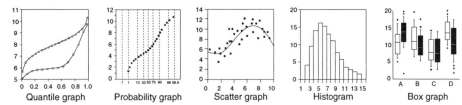

Quantile graph　　Probability graph　　Scatter graph　　Histogram　　Box graph

———— Business graphs ————

Business graphs sometimes fall into the following three major categories.

• Analyzing and planning

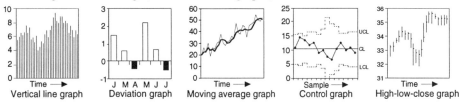

Index graph　　Bubble graph　　Supply and demand graph　　Spider graph　　Pareto graph

• Monitoring and controlling (often time series graphs)

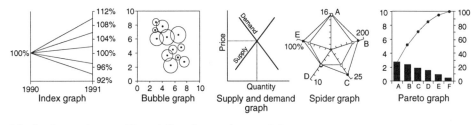

Vertical line graph　　Deviation graph　　Moving average graph　　Control graph　　High-low-close graph

• Presentation - Although used by all disciplines, presentation graphs are frequently classified as business graphs. There are no special types of graphs in this category, only variations of graphs in other categories with special emphasis on aesthetics, ease of understanding, and a focused message.

Categorized by type of data plotted

Original data

These graphs are constructed with data in its unmodified form. Many times the data is empirical, such as number of people, temperature, pressure, sales, income, defects, or salaries: things that can be observed or measured. The majority of graphs are based on this type of data.

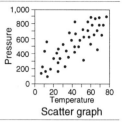

Scatter graph

Derived data

Examples of graphs plotted from derived data are index, ratio, deviation, residual, difference, change, and relative value graphs. Derived data is generated by mathematically manipulating original data. For example, index values may be derived by dividing all values by a common reference value. Deviation values are derived by subtracting some reference value, such as budget, from the actual values. Derived

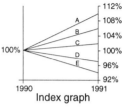

Index graph

data generally can not be observed or measured directly, although sometimes it is more meaningful than the original data. For example, when comparing crime in various cities, total crimes are generally not as meaningful as crimes per thousand inhabitants.

Abstract data

Abstract graphs are generally used to illustrate such things as concepts, principles, theories, or generalities. For example, an abstract graph such as a break-even graph is many times used to show how sales, volume of product, and costs interrelate to generate a profit or loss. The axes generally have titles but minimal or no labels. Since these graphs often do not have a true quantitative scale, some people prefer to call them charts or diagrams.

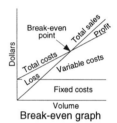

Break-even graph

Categorized by the number of axes

Most graphs are either rectangular, circular, or triangular. With few exceptions, graphs have one, two, or three primary axes depending on their geometric shape, or in a few cases, the type of scales. The table below indicates the primary axes for each of the major geometric configurations. See Axis, Graph.

	One Axis Sometimes referred to as one-dimensional	**Two Axes** Sometimes referred to as two-dimensional	**Three Axes** Sometimes referred to as three-dimensional
Rectangular graphs	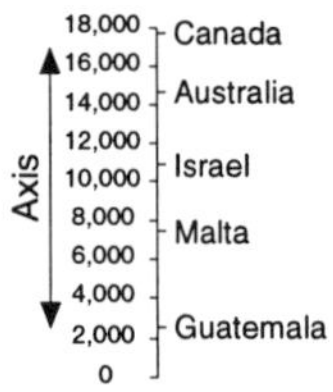 Circle graph data (Canada, Australia, Israel, Malta, Guatemala)	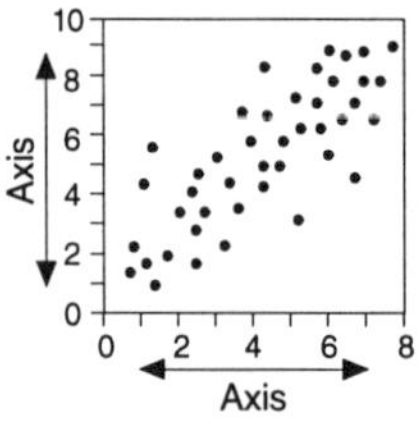 Scatter graph	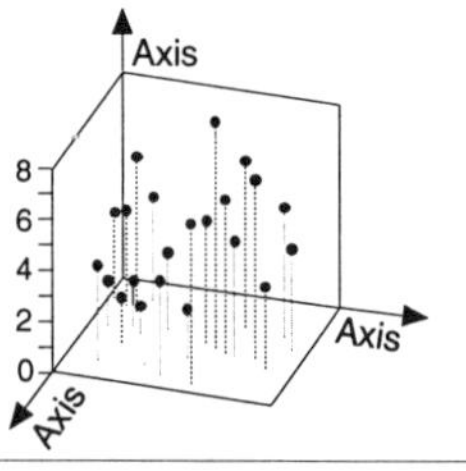 3-D scatter graph
Circular graphs	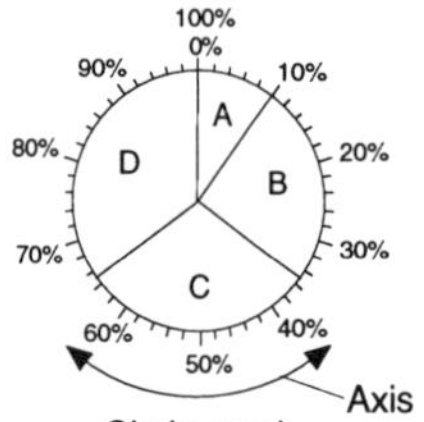 Circle graph	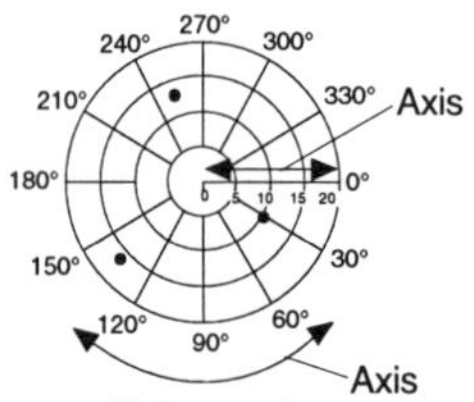 Polar graph	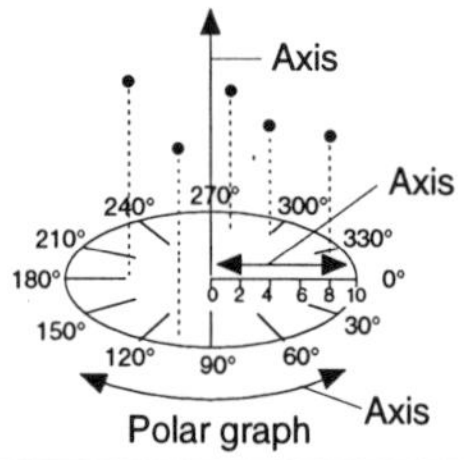 Polar graph
Triangular graphs	Not applicable	Not applicable	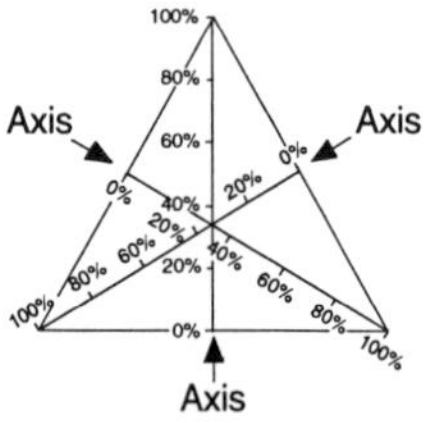 Triangular graph

Categorized by purpose

Analyzing, exploring, planning, etc.

– Major focus is on extracting the information contained in the data so that meaningful observations can be made, conclusions reached, and/or plans made.
– Almost every form of graph is used in these types of applications except those specifically designed for presentation or publication purposes, such as pictorial graphs.
– Graphs that compare multiple sets of data such a scatter, grouped, and combination graphs are widely used to look for correlations and meaningful relationships.
– Techniques such as curve fitting, tolerance intervals, and three-dimensional rotation are used more frequently here than in other applications.
– Graphs such as surface, contour, probability, and abstract graphs are used more frequently in these types of applications.
– There is no limit as to how much data can be included on graph as long as it is legible.
– Greater emphasis is placed on content than appearance.
– Color may or may not be used.
– Final graphs are generally in the form of hard copy (a paper printout) or on a computer screen.

Monitoring and controlling

– Major focus is on observing one or more measurements/values over time (monitoring) to determine whether actions should be taken (controlling).
– Sequence graphs are generally used, particularly time series.
– Line and column are the most widely used, point types to a lesser extent, and area types occasionally.
– When a reference such as budget, plan, or goal is known it is often included.
– When a reference exists, deviation graphs are sometimes used in which only the difference between the actual and reference are shown. Both period and cumulative data are sometimes used.
– Greater emphasis is placed on content than appearance.
– Color may or may not be used.
– Final graphs are generally either on hard copy (a paper printout) or computer screen.

Presentation

– Major focus is on communicating specific information from one person or group of people to another person or group of people.
– Graphs can contain large amounts of data if they are being presented to a small group using overhead transparencies or handouts. When presenting to a large audience using 35mm slides, the material on each slide is greatly reduced.
– Any type of graph can be used as long as it is appropriate for the audience and legible to everyone.
– Techniques such as the inclusion of pictures, the use of perspective, graduated fills, etc., are used more frequently in these applications than in the two applications above.
– About equal emphasis is placed on content and appearance.
– Color is used extensively.
– Final graphs are generally in the form of 35mm slides, overhead transparencies, or video projection.

Publications, reports, reference documents, etc.

– Applications range from simple bar graphs for newspapers to complex graphs for technical publications.
– The major focus varies with the application.
– The types of graphs, enhancements, amount of detail, size, etc., vary depending on the subject, the audience, or the publication.
– About equal emphasis is placed on content and appearance.
– Color is used extensively.
– Final graphs are generally in the form of hard copy.

Methods for improving the value and ease of understanding of graphs

Many methods have been developed to increase the amount of information encoded into graphs and to make them easier to understand. Some of the techniques apply only to specific types of graphs; others are applicable to almost all types. Summarized below, alphabetically, are representative examples of many of the major methods used today. For detailed information see the entries referenced with the examples.

Additional variable encoded with symbols

An additional variable can be encoded in many ways. In this example it is done by varying the sizes of the plotting symbols (circles). Varying color, shading, patterns, shapes, or line thickness, are other methods. See Symbol.

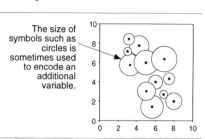

The size of symbols such as circles is sometimes used to encode an additional variable.

Confidence intervals to show probabilities

Methods have been developed for indicating confidence limits and intervals for many types of graphs. See Confidence Intervals and Range Symbols.

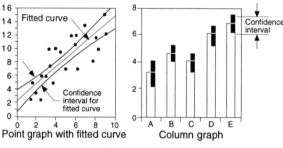

Point graph with fitted curve Column graph Box graph

Direction of changes

By means of arrow heads and color coding, the direction of change and whether or not it is favorable can be encoded. See Change Graph.

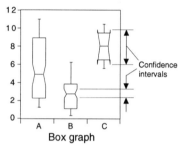

Change graph

Drop lines used to help locate data points

Thin lines help the viewer's eye relate a data point and its value or location. See Drop Line.

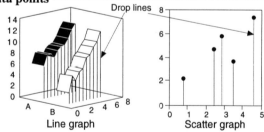

Line graph Scatter graph

Fills are used to highlight particular data

A wide variety of fills are available to differentiate data series, highlight differences, or call attention to particular data. See Fill, Silhouette Graph and Difference Graph. Examples of fill are also presented in the reviews of most of the major graphs.

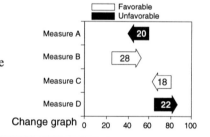

Focus placed on ranges

When ranges or spreads are important, the traditional bars or columns are sometimes eliminated and only the ranges plus some inner value such as average or median are shown. See Bar Graph and Column Graph.

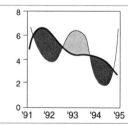

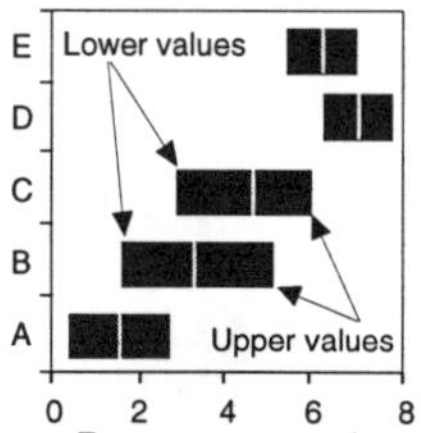

Bar range graph 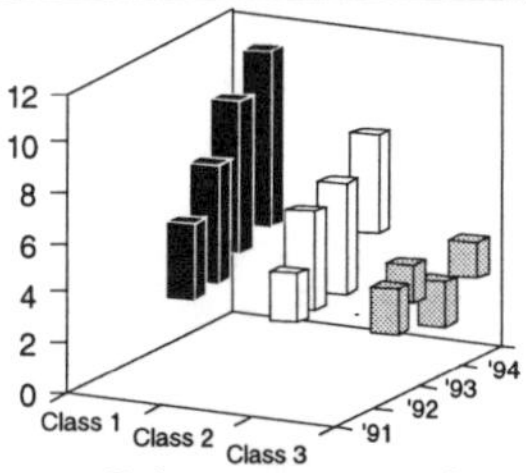 Column range graph

Methods for improving the value and ease of understanding of graphs (continued)

Frame shifted for clarity

To avoid having data points appearing on the axes or scale lines, the frame and scale lines are sometimes shifted slightly horizontally and/or vertically. See Shifted Frame and Scales.

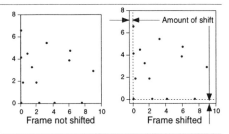

Multiple variables encoded into symbols

When several variables are to be encoded, the symbols become more complex. Symbols that are capable of encoding large numbers of variables are sometimes called icons. See Icon.

Complex symbols, such as faces, can be used to encode multiple variables. See Chernoff Faces.

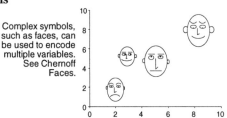

Pictures or images used to expedite orientation and improve appearance

Pictures, images, or icons are sometimes used to orient the viewer and improve the appearance of the graph for presentation purposes. See Pictorial Graph.

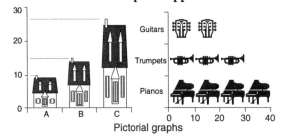

Pictorial graphs

Projections of data

Graphs are often used to project or estimate the future values of a data series. The projected portions of the graph are generally clearly differentiated. See Projection.

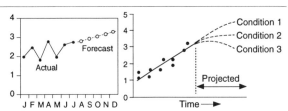

Reference lines to make deviations stand out

Reference lines can be used to compare actual data against or simply to point out a value of interest. Drop lines, columns, or bars, might extend all the way to the zero axis or stop at the reference line to highlight the deviations. See Reference Line.

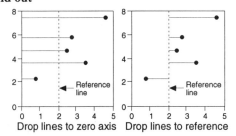

Scale bars used to indicate equal increments on graphs with different scales

When scales are expanded, comparisons between graphs can be misleading. Scale bars help reduce this problem. See Scale Bars.

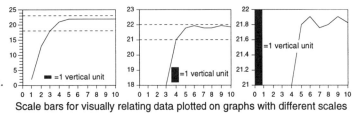

Scale bars for visually relating data plotted on graphs with different scales

Scales expanded for clarity

In order to magnify that portion of the graph that contains the data under study, the scales are sometimes expanded. See Scale.

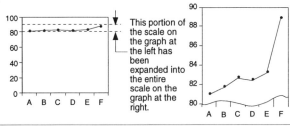

This portion of the scale on the graph at the left has been expanded into the entire scale on the graph at the right.

Methods for improving the value and ease of understanding of graphs (continued)

Shadows projected onto flat surfaces

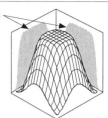

Three-dimensional graphs are many times difficult to visualize. One method for assisting in the visualization process is to project a shadow or outline of the data onto the walls of the graph. See Three-Dimensional Graph.

Showing only slices or profiles of three-dimensional graphs

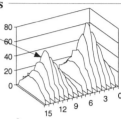

An additional method for gaining a better understanding of three-dimensional data graphics is to display only selected planes, slices, or profiles. See Slice Graph and Profile Graph.

Sketching boundaries to show patterns of upper and lower values

Lines are sometimes drawn that approximate the major upper and lower values of a data series to emphasize how the outer values change and how the spread changes. See Envelope.

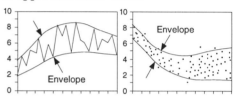

Smoothing to make data easier to analyze

When data points are spread out or if they have large fluctuations from time period to time period, it is sometimes helpful to superimpose a line or surface over the data to approximate the general shape, or slope of the data. Shown below are three examples of such techniques. See Curve Fitting and Moving Average.

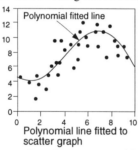

Polynomial line fitted to scatter graph

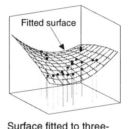

Surface fitted to three-dimensional scatter data

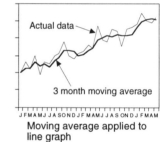

Moving average applied to line graph

Supplementary grid and scale added

To condense two or more graphs into one and make certain relationships easier to spot, a supplementary grid and scale are sometimes added. See Supplementary Scale/Grid.

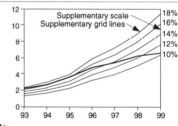

Symbol type and placement to improve readability

New symbols have been developed as well as ways of displaying them. The three examples shown here address the problem of overlapping plotting symbols. See Symbol.

Sunflower symbol

Data points stacked

Data points jittered

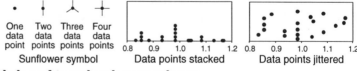

Symbols used to replace bars or columns

Instead of displaying two data series as two sets of bars or columns, symbols are occasionally used to represent one of the data series. This technique is most widely used when one of the data series is to be used as a reference. See Bar Graph and Column Graph.

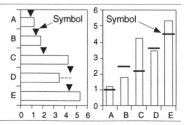

Graph (continued)

The use of multiple graphs in conjunction with one another

Using graphs in conjunction with one another sometimes has a synergistic effect, in that the value of the multiple graphs is greater than the sum of the value of the individual graphs. The examples below demonstrate some of the many ways multiple graphs are used together. They also illustrate some of the advantages of such a technique.

Multiple graphs superimposed on one another

One of the most widely used methods for combining multiple graphs is to superimpose two or more graphs on top of one another. The resulting graph is sometimes referred to as a combination graph. See Combination Graph.

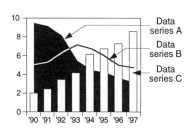

Multiple graphs juxtaposed to one another (with complementary data)

When two or more graphs are placed immediately adjacent to one another they are often referred to as juxtaposed. They may be above and below or side-by-side. With complementary data, the graphs are assisting one another to present broader, more meaningful information to the viewer. Three well-known examples illustrate the idea.

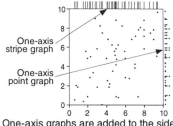

One-axis graphs are added to the side and/or top of scatter graphs to show how data is distributed along an individual axis. See Marginal Frequency Distribution Graph.

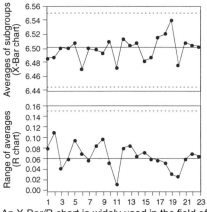

An X-Bar/R chart is widely used in the field of quality control. The upper graph displays sample averages and the bottom graph the spread of sample values. See Control Charts.

A price-volume chart is used extensively in the field of investment. The upper graph plots the key stock prices for a given period (normally a day). The lower graph plots the number of shares of the stock sold during the same period. See Bar Chart 2.

Multiple graphs juxtaposed to one another (with comparative data)

With a juxtaposed group of graphs, the primary purpose is to compare the data on each graph with the data on each of the other graphs. The alternative to juxtaposing the graphs is to superimpose them on top of each other. An example of four juxtaposed graphs with a different data series on each is shown at the left. For ease of comparison, the same scales are used on all four graphs. The same four graphs (data series) are superimposed in the example at the right. Each method has its advantages. In these examples, the juxtaposed graphs are less cluttered, but on the superimposed graph it is easier to note that all four data series had approximately the same value sometime between 1994 and 1995.

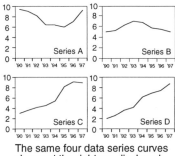

The same four data series curves shown at the right are displayed here on juxtaposed graphs.

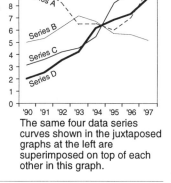

The same four data series curves shown in the juxtaposed graphs at the left are superimposed on top of each other in this graph.

171

The use of multiple graphs in conjunction with one another (continued)

──────**Multiple graphs in matrix form**──────

In situations where multiple variables are to be monitored, compared, analyzed, etc., matrixes are often advantageous. The examples shown here represent three distinctly different ways that matrixes are used.

At the right is a typical matrix of graphs. Column graphs are used in this example; however, almost any type of graph will work. Common scales may or may not be used. See Matrix Display.

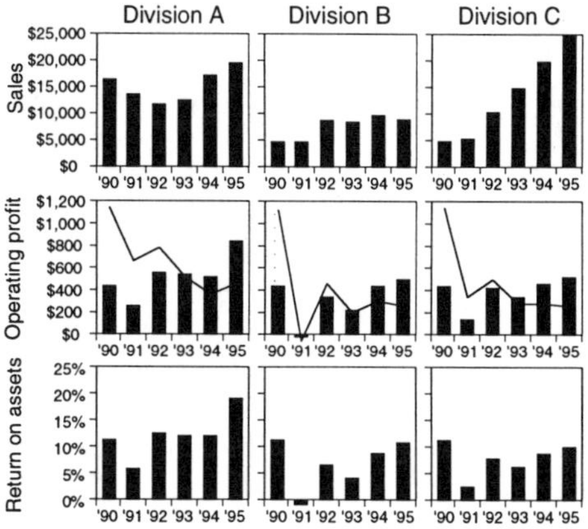

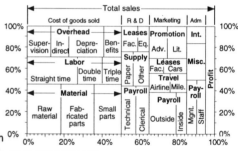

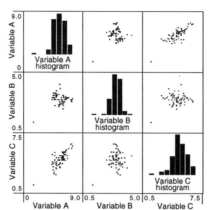

Scatter graph matrices are used to investigate correlations between multiple variables. They frequently utilize a bidirectional tabular format. The histograms in the diagonal squares are optional. They illustrate another case of multiple graphs that complement one another.

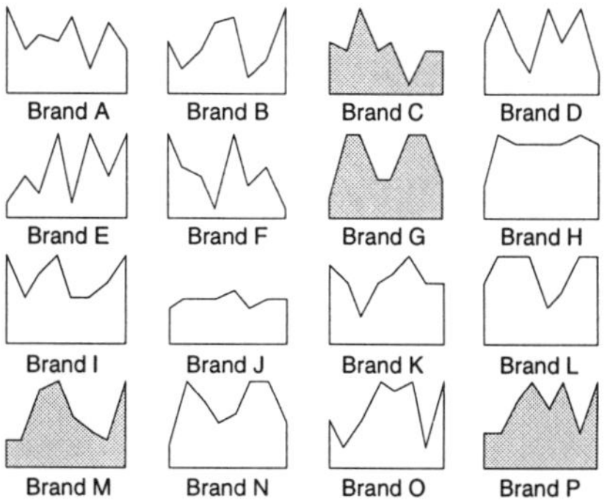

Above, each icon represents a graph of eight characteristics for 16 different brands of desktop computers. An icon comparison display, as this is called, is used to compare multiple variables for multiple entities. A legend is always required for this type of matrix. See Icon Comparison Display.

──────**Multiple graphs intermingled**──────

A mosaic graph, as shown at the right, is an example of how two different types of graphs can be intermingled to form a single meaningful graph. In this case it is a mixture of 100% bar and 100% column graphs. See Mosaic Graph.

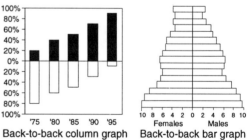

Mosaic graph

──────**Multiple graphs plotted back to back**──────

In order to make comparisons clearer, two bar graphs or two column graphs may be constructed using the same zero axis. See Bar Graph and Column Graph.

Back-to-back column graph   Back-to-back bar graph

──────**Graphs used as insets for other graphs**──────

Insets are used to provide additional insight into the data or to help orient the viewer. Sometimes an inset magnifies particular data in a graph. Other times it lets the viewer see the data in the basic graph in terms of a larger family of data. See Inset.

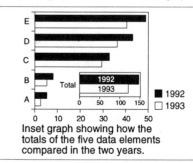

Inset graph showing how the totals of the five data elements compared in the two years.

The use of multiple graphs in conjunction with one another (continued)

──── **Multiple graphs representing the elements of a composite graph** ────

Sometimes the data series from which a graph is plotted is the sum of several component data series. When those component data series are separated out and plotted individually, the result is sometimes called a decomposition graph. The examples at the right show the composite curve as well as graphs of the four component data series that comprise it. See Time Series.

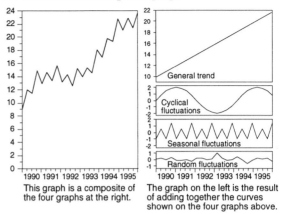

This graph is a composite of the four graphs at the right.

The graph on the left is the result of adding together the curves shown on the four graphs above.

Decomposition graph

──── **Multiple cross sections of a three dimensional graph** ────

The data points in three-dimensional scatter graphs are sometimes grouped into a series of planes or cross-sections so the viewer can get a better idea as to how the data is distributed and the patterns it forms. These two-dimensional cross-sections are then sometimes rotated into the plane of the paper for easier viewing, as shown on the right below. A group of such graphs is sometimes called a casement plot. See Slice Graph and Casement Display.

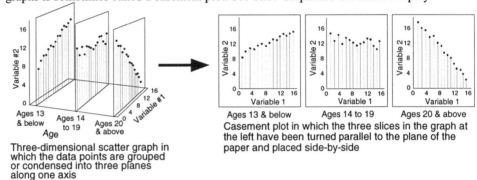

Three-dimensional scatter graph in which the data points are grouped or condensed into three planes along one axis

Casement plot in which the three slices in the graph at the left have been turned parallel to the plane of the paper and placed side-by-side

──── **Multiple graphs formed by projections of a three-dimensional graph** ────

With three-dimensional graphs, certain features of the graph can be projected onto the floor, ceiling, and/or walls, forming two-dimensional views of the data. Such projections can help in the understanding and interpretation of the data. See Three-Dimensional Graphs.

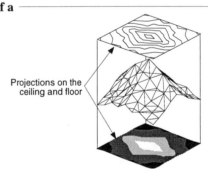

Projections on the ceiling and floor

──── **Multiple graphs representing different views of a three-dimensional graph** ────

An additional method to help viewers understand the data in a three-dimensional graph is to show multiple views of the data graphic. With a computer, this can be done by rotating the three-dimensional graph. When hard copy is used, this is sometimes done by showing the three-dimensional view, plus selected side and top views. The example shown here includes views of the top, side, and front. These three views are sometimes called a draftsman's display. See Draftsman's Display.

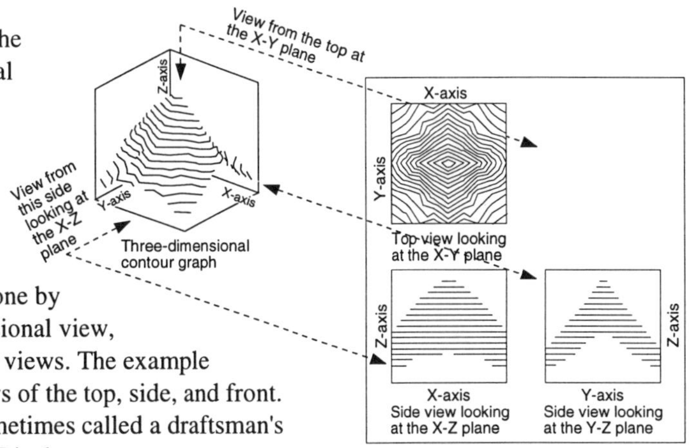

View from the top at the X-Y plane

View from this side looking at the X-Z plane

Three-dimensional contour graph

X-axis

Y-axis

Top-view looking at the X-Y plane

X-axis

Side view looking at the X-Z plane

Y-axis

Side view looking at the Y-Z plane

Draftsman's display

The use of multiple graphs for analyzing and presenting the same data

When analyzing data it is many times advantageous to generate a variety of graphs using the same data. This is true whether there is little or lots of data. Reasons for this are:

- Frequently, all aspects of a group of data can not be displayed on a single graph
- Multiple graphs generally result in a more in-depth understanding of the information.
- Different aspects of the same data often become apparent.
- Some types of graphs cause certain features of the data to stand out better
- Some people relate better to one type of graph than another.

To illustrate, the graphs on this and the following page were all generated using the data shown in the table at the right. Some of the graphs bring out information about the data not shown in the other graphs. Some simply show the same data in a different format.

Elected Officials in Some County				
	Male		Female	
Year	Under 40 years old	Over 40 years old	Under 40 years old	Over 40 years old
1980	49	69	27	21
1990	66	41	52	43

This **line graph** indicates that three of the four subgroups increased roughly the same amount from 1980 to 1990 while the group of males over 40 went down significantly. In addition, the approximate sizes of the subgroups can be estimated.

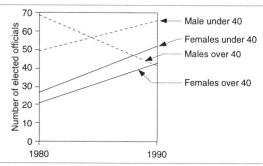

In this **grouped column graph** the changes in age groups are not as apparent, but the comparisons between the males and females in each age bracket tend to stand out clearer than in the line graph above. Since this graph includes totals, we can see that the total number of female officials went up significantly, while the total for the males went down slightly from 1980 to 1990.

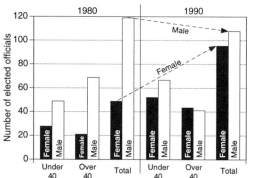

This **grouped bar graph** contains exactly the same information as the column graph immediately above. Some people prefer to have similar data lined up as it is here instead of having to search out the columns that represent comparable data. For example, the bar showing the number of females under 40 in 1990 is directly across from females under 40 in 1980. In addition, some people prefer to have the labels running horizontal.

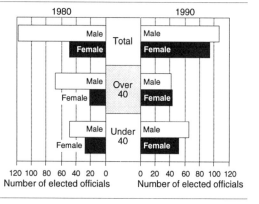

When the number of data points is small, **three-dimensional graphs** can sometimes make relative relationships easier to see although exact data can seldom be read accurately.

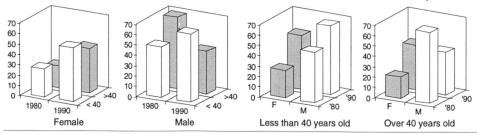

The use of multiple graphs for analyzing and presenting the same data (continued)

This **index comparison graph** makes it easy to compare the percent changes in the overall number of officials as well as all of the subgroups. In addition to presenting the exact percents for each entity directly on the graph, the slopes of the lines give a good visual overview of the changes relative to one another.

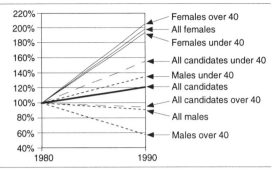

If one is primarily interested in the differences between males and females, a **difference graph** can be generated as shown at the right. From these two graphs we can see that a major change occurred in the over-40 age group, resulting in more female than male officials in this age group in 1990.

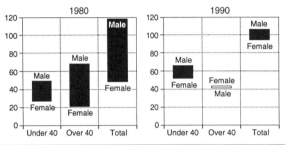

The same information shown in the difference graph can be plotted on a **deviation graph** where the differences are plotted along a zero axis. In this graph the differences from group to group and period to period can more easily be quantified and compared.

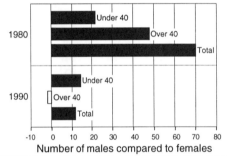

By plotting **multiple graphs,** the changes within subgroups such as male and female can be seen in more detail. For example, the number of females increased from about 50 to about 95. Since the increase in both age brackets was about the same percent, the mix of the two age groups remained about the same. With males the overall number dropped from about 120 to about 110. Because one age group of males increased while the other decreased, the mix almost reversed itself from about 40% under 40 and 60% over 40 in 1980 to 60% under 40 and 40% over 40 in 1990.

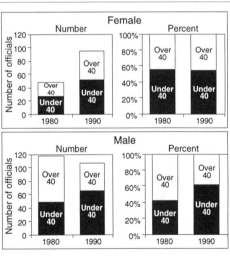

In these **stacked area graphs** each of the subgroups is plotted against the whole. The upper graph is in terms of number of officials, while the lower one, a **100% stacked area graph**, is in terms of percent of the whole, using all officials as the whole. In both graphs the sexes are combined in order to focus on the overall changes by age group. From these graphs we can see that the younger group increased in numbers while the older group, as a whole, remained about the same resulting in an increase in officials younger than 40 from about 45% in 1980 to just under 60% in 1990. It can also be seen that the major cause for the large shift was the significant decrease in the male officials over 40.

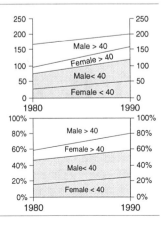

175

Terminology

With more and more graphs being generated on computers, the name a graph is referred to by is becoming increasingly important. To match the type of data to an appropriate graph, and to obtain the maximum benefit from graphical software, one has to understand the terminology in the users manual.

• The name of a graph might be affected by such things as its configuration (e.g., line, column, area, etc.); the number of data series (e.g., singled or grouped); the type of data (e.g., quantitative, category, time series, etc.); or the arrangement of the data series (e.g., stacked, overlapped). Major variations

are addressed under individual graph headings. The table below summarizes variations of graphs that have a quantitative scale on the vertical axis (the norm for these types of graphs) and on which all data points are referenced from the horizontal axis (e.g., excludes variations such as stacked and range). The bold type above each graph or group of graphs indicates a name frequently used for that type of graph. Most graphs are referred to by more than one name. Only representative names are shown in this table. Alternate names are listed under the individual graph headings.

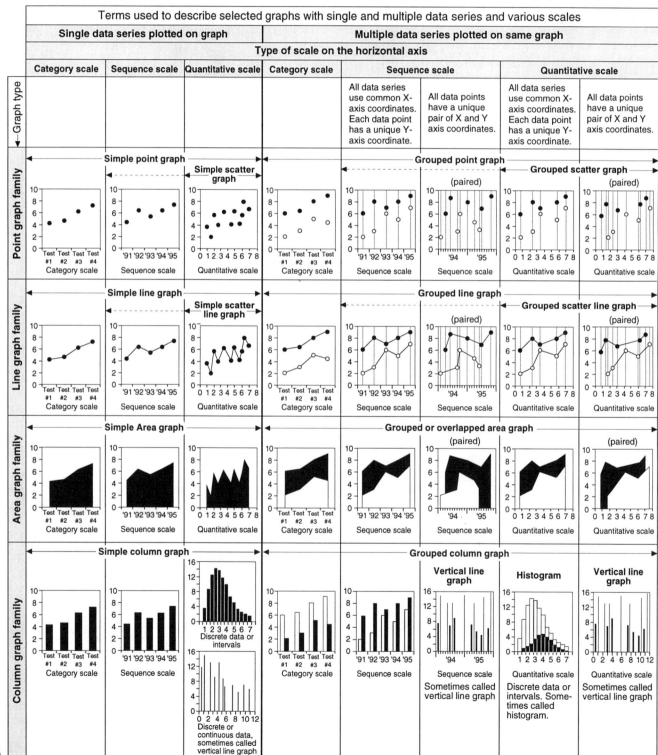

Graph (continued)

Terminology and location of key graph elements

The illustrations below outline the major terms used with graphs. When multiple terms are widely used to refer to the same thing, each of the terms is included. All of the parts of a graph such as grid lines, axes, labels, tick marks, legends, data graphics, notes, etc., are referred to as graph elements. For more detail see Grid Lines; Tick Marks; Scales; Axis, Graph; Legend; Symbols; Data Graphic; Frame; Data Point; Data Set; Data Series; Data Element; Background; and Three-Dimensional Graph.

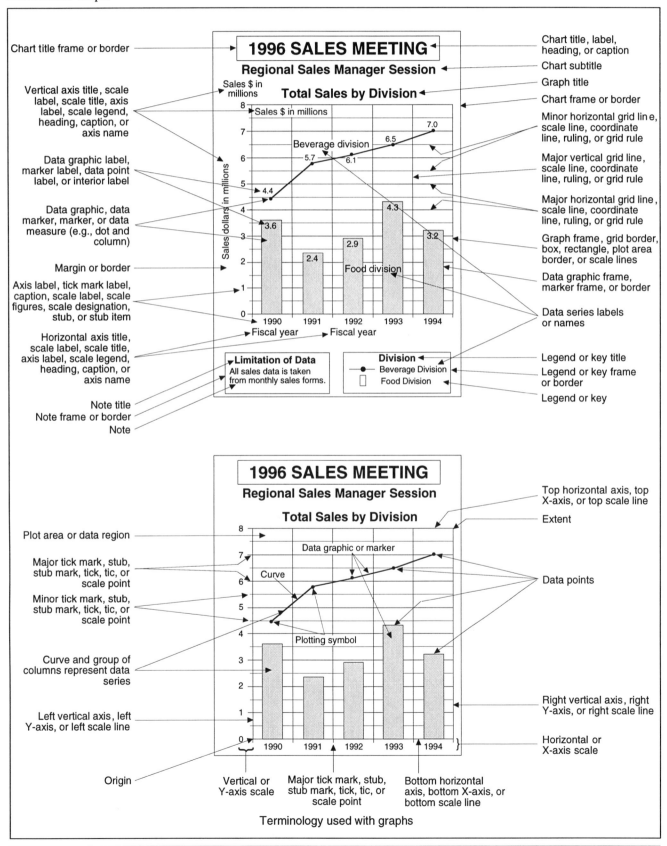

Terminology used with graphs

Graphical Linear Programming

Graphical linear programming is a graphical method for solving problems using the linear programming technique. It is basically restricted to problems that have only two variables, no negative numbers, and all linear functions. Problems that lend themselves to this technique typically have two things in common. First, they have a single objective of maximizing or minimizing something (referred to as objective function) and second, they have limitations (called constraints) that many times are stated in terms of equal to or greater than or equal to or less than. Linear programming is used for such applications as resource allocation, distribution problems, mixture and blend decisions. Some of the terminology used with linear programming graphs is shown at the right.

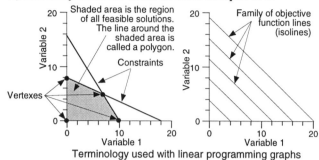

Terminology used with linear programming graphs

• There are generally three major steps to solving a problem using graphical linear programming.

1. Determine the mathematical expressions for the constraints and plot these equations to establish the region of feasible solutions.
2. Determine the expressions for the objective funtion and plot a series of parallel lines with the slope of this equation. (In practice, a single line can be plotted to establish the slope and then a number of techniques used to develop a series of parallel lines to determine the optimum data point in the feasible solution area.)
3. Overlay the two plots and determine the coordinates for the data point where the objective function line and the feasible solutions region just meet. The X and Y values for that point yield the optimum solution. When maximizing something, one looks for the objective function line farthest from the origin that just touches the feasible solution area. That point at which the line just touches will be a vertex and is the optimum solution. When minimizing something, one looks for the objective function line closest to the origin that just touches the feasible solution area. That too will be a vertex and will yield the optimum minimization solution.

Examples of the graphical steps involved in solving a graphical linear programming problem are shown below.

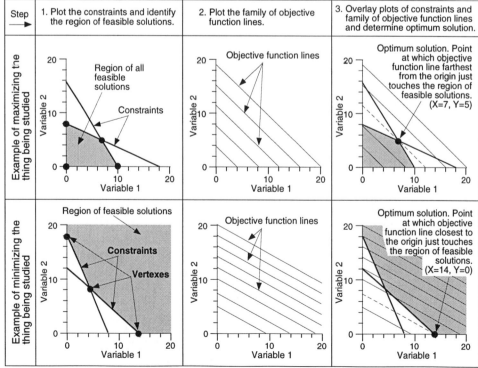

Steps involved in solving a problem using the graphical linear programming technique

| **Graphic Table** | When graphics are used in a table instead of words or numbers, the result is sometimes referred to as a graphic table. Examples are shown below. See Table. |

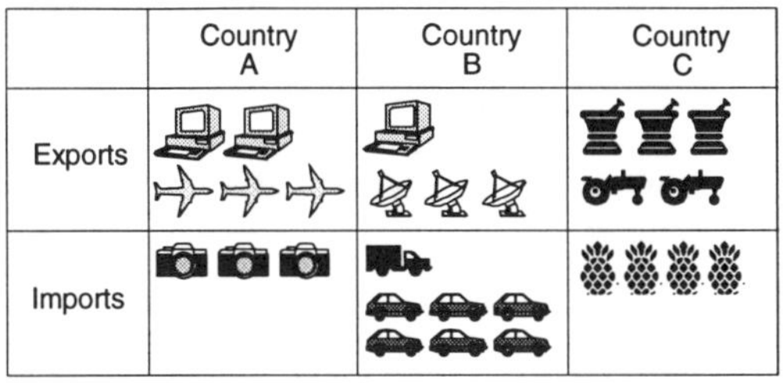

	Country A	Country B	Country C
Exports	💻💻 ✈✈✈	💻 📡📡📡	⚖⚖⚖ 🚜🚜
Imports	📷📷📷	🚚 🚗🚗🚗	🍍🍍🍍🍍

	Brand A	Brand B	Brand C	Brand D
Product 1	○□△	○△	○	□△
Product 2	○	○□	○□△	

	1990	1991	1992	1993
Company 1	↗	↓	→	→
Company 2	↘	→	↙	↗

| **Graph Matrix** | A variation of a matrix display. A graph matrix consists of multiple graphs, frequently of the same basic type, arranged in rows and columns in some organized fashion. Their major purpose is to simplify the analysis of large quantities of data by enabling the viewer to study multiple graphs at one time. See Matrix Display. |

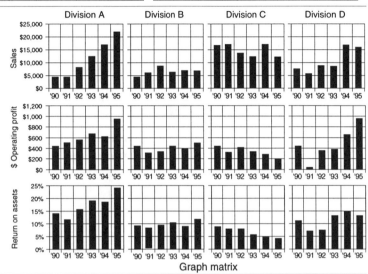

Graph matrix

| **Grapho** | A slang word for graphical error. It is the equivalent of typo, which means typographical error in word processing. |

| **Graph Paper** | Graph paper consists of commercially available, preprinted graph forms with grid lines and sometimes scales. The paper is used for manually generating charts and graphs. The grid line patterns available range from very simple to very complex. The list below gives some idea of the more popular grid line patterns available. |

- Circle graphs   – Probability
- Isometric orthographic   – Semi Log
- Log Log   – Simple square sections
- Perspective   – Time series
- Polar coordinate   – Triangular

Within each pattern there are generally many variations available. For example, in the simple square pattern, one can get coarse grids with as few as one or two squares per inch up to fine grids of 20 or more squares per inch. Patterns are available in metric as well as inch dimensions. Many specialized technical and statistical patterns are also available as well as preprinted forms for maps, charts, planners, etc.

| **Graph Theory** | Graph theory is considered a branch of mathematics. It has practical applications in many different areas, ranging from determining the most efficient routes for letter carriers to the design of complex computer circuits. The term graph, as used in the context of graph theory, has an entirely different meaning than the word graph as used anywhere else in this book. In graph theory, graphs are sometimes described as diagrams, networks, or models and consist almost entirely of dots and lines similar to the examples shown below. The dots are normally referred to as vertices, points, or nodes. The lines, which might be straight or curved, are referred to as edges, arcs, links, or curves. The dots or vertices may represent almost anything including people, places, things, events, etc. The lines may represent specific entities or simply indicate a relationship between the two points they connect. When the lines or edges have arrows indicating direction, the diagrams are frequently referred to as digraphs. For example a diagram used in the critical path method (CPM) is sometimes considered a digraph if it has arrows. |

Examples of graphs as referred to in graph theory

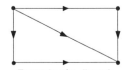

Example of a digraph

179

Graticule	Sometimes called a geographic grid. A network of lines on maps and globes used to locate places and things, a graticule consists of two major sets of lines, one set running north and south (called meridians or longitudinal lines) and the other running east and west (called parallels or latitudinal lines). See Map.

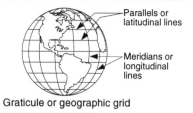

Graticule or geographic grid

Grid
and
Grid Line
or
Grid Rule
	Thin lines forming grids that are used as visual aids in many types of charts. They have many functions including:
	  – Serving as a tool for the encoding and decoding of information
	  – Helping locate items
	  – Orienting the viewer and giving a visual indication as to how the chart is laid out
	  – Providing guide lines for the viewer to use in visually following data
	  – Compartmentalizing information such as forming quadrants, columns, cells, etc.
	  – Providing a reference so the viewer can make visual comparisons
	Although grid lines can be used on all types of charts, they are most frequently used on graphs, maps, and tables.

Grid lines on graphs

Sometimes called rulings, grid rules, coordinate lines, or scale lines.

Types of grid lines

The three major types of grid lines used with graphs are major, minor, and intermediate. The major grid lines align with the major labels on the scale and are the boldest in appearance. Intermediate and minor grid lines align with the lesser tick marks and/or labels. If lesser tick marks or labels are not shown, the grid lines denote where they would be if they were shown. Intermediate grid lines are only occasionally used. Minor grid lines are the least bold of the three types. All three types of grid lines can be used on either the vertical or horizontal axis.

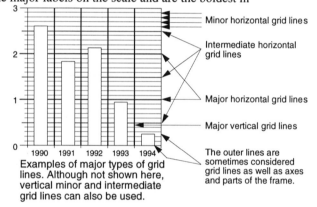

Examples of major types of grid lines. Although not shown here, vertical minor and intermediate grid lines can also be used.

Alignment of grid lines

The examples shown here indicate how grid lines are most frequently aligned with regards to labels, tick marks, and data points. Vertical grid lines are used in each of the examples. The same principles apply to horizontal grid lines.

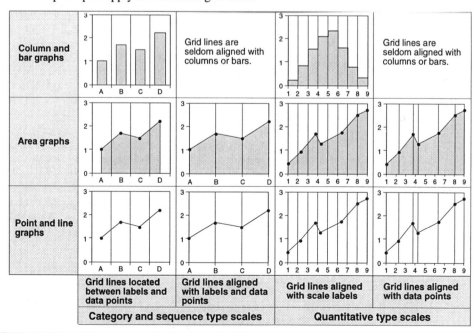

180

Grid lines on graphs (continued)

Spacing

There are four major variations of spacings used with grid lines on graphs.

- On most graphs the grid lines are equally spaced. Major grid lines are normally aligned with tick marks and/or scale labels and since most scales are linear, the spacings between most labels, tick marks, and grid lines are equal.
- When nonlinear scales are used, the spacings between grid lines are uniform but generally not equal (e.g., logarithmic).
- In some cases grid lines are shown only where data points exist; therefore, the spacing of the grid lines is dependent on the spacing of the data points.
- In the majority of cases, the space between any two parallel grid lines is the same for the full length of the grid lines. Exceptions are grid lines on graphs drawn in perspective, grid lines for certain supplementary scales, and the radial grid lines of circular graphs.

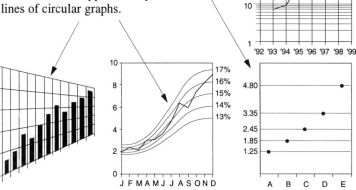

Type of grid line

Major, intermediate, and minor grid lines are differentiated from one another by means of line width, dashes, dots, color, or shading. Major grid lines are always the boldest and minor grid lines the faintest, whichever method is used.

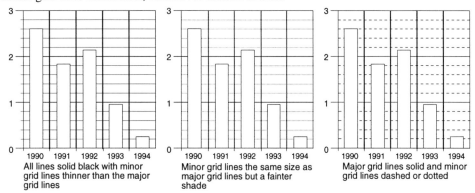

All lines solid black with minor grid lines thinner than the major grid lines

Minor grid lines the same size as major grid lines but a fainter shade

Major grid lines solid and minor grid lines dashed or dotted

Variations sometimes used to differentiate various types of grid lines

Length of grid lines (partial grid lines, drop grid lines, and drop lines)

When a grid line does not extend completely across the graph it is sometimes referred to as a partial grid line, a drop grid, or a drop line. Partial grid lines can be used with any type of graph, including three-dimensional. They can be used with single or multiple data series, on the horizontal or vertical axis, and as major, minor, or intermediate type grid lines. The use of partial grid lines can sometimes enhance the readability of the graph by reducing the number of unnecessary lines.

Examples of graphs using partial grid lines, sometimes called drop grids or drop lines

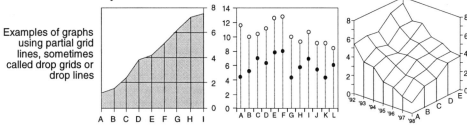

Grid lines on graphs (continued)

────Number of grid lines────

The two factors having the greatest influence on the number of grid lines used on graphs are the number of labels and the degree of accuracy expected in the decoding process. Major grid lines are generally associated with scale labels; therefore, the more labels, the more grid lines. Two key ways of increasing the accuracy of decoding are (1) to increase the number of scale labels and their associated major grid lines and/or (2) to increase the number of minor and intermediate grid lines between the labels. Both actions increase the total number of grid lines. Generally an effort is made to strike a balance between degree of accuracy and number of grid lines. Too few lines reduces the accuracy, and too many causes clutter and makes the graph harder to read. Fewer grid lines tend to be used with three-dimensional graphs than with two-dimensional graphs, because added grid lines generally do little to improve the readability of three-dimensional graphs.

• At the right are six examples of the same data with the number of horizontal grid lines ranging from zero to 29, not counting the frame lines.

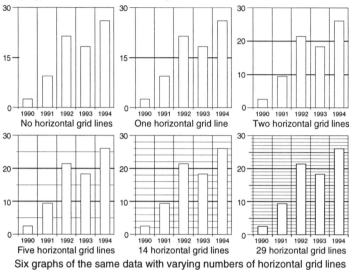

Six graphs of the same data with varying numbers of horizontal grid lines

────Location of grid lines────

– Typically grid lines are located behind the data graphics in both two- and three-dimensional graphs, as shown in the four examples below.

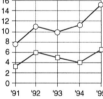
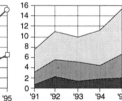
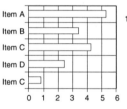
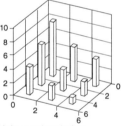

Examples of grid lines located behind the data graphics, the most widely used variation

– Occasionally the grid lines are located in front of or on the data graphics, as shown at the right. This is done for two main reasons: to aid the viewer in estimating data point values, particularly with data graphics that have large surface areas such as area graphs; and to improve the appearance of the graph.

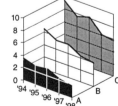
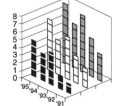

Grid lines located in front of or on data graphics

– In still other cases, the grid lines are shown both on and behind the data graphics. This technique is used more frequently with three-dimensional graphs, since they are more difficult to read. When multiple types and locations of grid lines are used, it is a matter of judgment as to which grid lines should go where.

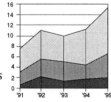

Graphs on which gird lines are located behind and on the data graphics

────Split grid────

Sometimes called scale break. In order to expand a scale without increasing the size of the graph, discontinuities are sometimes intentionally introduced in the scale and grid. When discontinuities are introduced the major considerations involve the scale; therefore, this subject is discussed in detail under the heading Scale as well as the major graph headings.

Grid lines on graphs (continued)

Circular graphs

Most circular graphs have two sets of grid lines. One set, called radial grid lines, marks off distances (generally of equal length) around the circumference. The other set, called circular grid lines, marks off distances along the radius. Examples of both are shown below. Major, minor, and intermediated grid lines may be used as with rectangular graphs. If a circular graph has only one axis, as in the case of circle graphs, there are no circular grid lines and only the radial grid lines are used. See specific circular graph headings for more information.

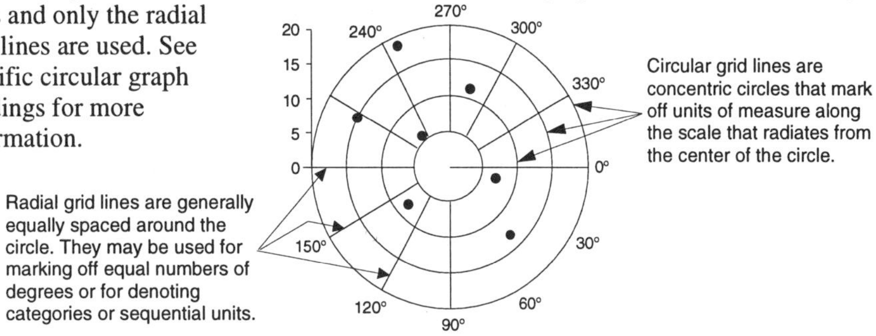

Circular grid lines are concentric circles that mark off units of measure along the scale that radiates from the center of the circle.

Radial grid lines are generally equally spaced around the circle. They may be used for marking off equal numbers of degrees or for denoting categories or sequential units.

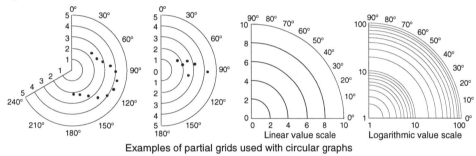

Examples of grid lines typically used on circular graphs

In some cases only partial grids are used with circular graphs to enlarge the area of interest, improve accuracy in decoding, make scales more visible, and focus the viewer's attention on areas where the data is plotted. Three examples are shown below. • Linear or logarithmic scales can be used with either a full or partial grid. An example of a partial grid with a logarithmic scale is also shown below.

Linear value scale

Logarithmic value scale

Examples of partial grids used with circular graphs

Trilinear graphs

There are three sets of grid lines on trilinear graphs, one for each axis. The grid lines are drawn perpendicular to the axes, which are the altitudes of the triangle. Major grid lines (shown at right), minor grid lines (shown below) and intermediate grid lines (not shown) can be used with trilinear graphs. In some cases, no grid lines are used; in other cases partial grid lines (drop lines) are used. See Trilinear Graph.

Lines from the vertexes to the bases opposite (called altitudes) form the axes for trilinear graphs.

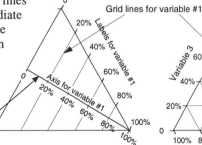

Grid lines are drawn perpendicular to the axis to which they apply. Axes are lines that run from a vertex to the base opposite it and are perpendicular to that base. Only one axis and set of grid lines is shown here.

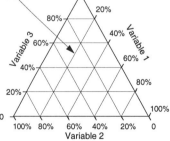

Many times the axes are not drawn, and only the three sets of grid lines and scales are shown.

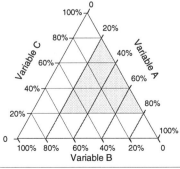

When a small portion of a trilinear grid (gray area on left) is expanded (example on right), minor grid lines can sometimes be included for greater accuracy in decoding.

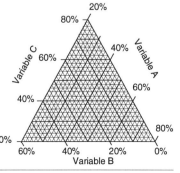

Grid
and
Grid Line
or
Grid Rule
(continued)

Grid lines on tables

Sometimes referred to as rules. Grid lines are used on tables to help organize and compartmentalize information, assist the viewer in visually tracking data, emphasize key information, indicate where totals occur, and improve the appearance of the table. The vast majority of grid lines are either vertical or horizontal. The number of lines used on a given table ranges from none to one or more between every row and column. Different weights and types of lines are sometimes used to differentiate types or groupings of data. Shown below are six of the many combinations of grid lines/rules used on tables. See Table.

	1993	1994	1995	1996	1997	Total
Product A	23.1	23.7	24.2	24.9	25.6	121.5
Product B	2.7	3.1	2.5	1.8	0.9	11.0
Product C	10.7	11.2	11.5	11.9	12.5	57.8
Product D	5.9	7.2	9.8	12.4	15.7	51.0
Total	42.4	45.2	48.0	51.0	54.7	241.3

No lines (for reference)

	1993	1994	1995	1996	1997	Total
Product A	23.1	23.7	24.2	24.9	25.6	121.5
Product B	2.7	3.1	2.5	1.8	0.9	11.0
Product C	10.7	11.2	11.5	11.9	12.5	57.8
Product D	5.9	7.2	9.8	12.4	15.7	51.0
Total	42.4	45.2	48.0	51.0	54.7	241.3

Horizontal lines with a double line indicating totals

	1993	1994	1995	1996	1997	Total
Product A	23.1	23.7	24.2	24.9	25.6	121.5
Product B	2.7	3.1	2.5	1.8	0.9	11.0
Product C	10.7	11.2	11.5	11.9	12.5	57.8
Product D	5.9	7.2	9.8	12.4	15.7	51.0
Total	42.4	45.2	48.0	51.0	54.7	241.3

Vertical lines only

	1993	1994	1995	1996	1997	Total
Product A	23.1	23.7	24.2	24.9	25.6	121.5
Product B	2.7	3.1	2.5	1.8	0.9	11.0
Product C	10.7	11.2	11.5	11.9	12.5	57.8
Product D	5.9	7.2	9.8	12.4	15.7	51.0
Total	42.4	45.2	48.0	51.0	54.7	241.3

Vertical and horizontal lines

	1993	1994	1995	1996	1997	Total
Product A	23.1	23.7	24.2	24.9	25.6	121.5
Product B	2.7	3.1	2.5	1.8	0.9	11.0
Product C	10.7	11.2	11.5	11.9	12.5	57.8
Product D	5.9	7.2	9.8	12.4	15.7	51.0
Total	42.4	45.2	48.0	51.0	54.7	241.3

Partial horizontal lines with a gap indicating totals

Line Item	Quantity Ordered	Description	Unit Price	Amount

Grid lines/rules drawn independent of text

Grid lines on maps

One of the primary functions of grid lines on maps is to locate entities. When precise locations are required, a coordinate system of latitudes and longitudes (sometimes referred to as a geographic grid or graticule) is frequently used, as shown below. The grid lines that run north and south are called meridians or longitudinal lines. They are used to measure distances east and west from a line called the prime meridian, which passes through Greenwich, England. The grid lines that run east and west are called parallels or latitudinal lines and are used to measure distances north and south of the equator. Major and minor grid lines can be used with either latitudes or longitudes. Generally only the major grid lines are used at uniform intervals. See Map.
• With many maps, latitude and longitude are not important. Of greater importance are things like the relative location of a place or thing with respect to other places or things, distances between various locations in units such as miles or kilometers, statistics with regards to various areas, etc. For these purposes grid lines may

Examples of key grid lines used with a latitude and longitude coordinate system.

not be necessary at all. In other cases, a map is divided into rows and columns by means of horizontal and vertical grid lines, as shown at the left, that are unrelated to lines of latitude and longitude. The grid lines seldom have a particular mileage distance between them, although the maps may have scales for measuring distances. Each vertical column is designated by a number or letter, and each horizontal row by a letter or number, respectively. A table accompanying the map cross references each major item with its appropriate number and letter combination so that the viewer can locate it on the map. • Supplementary grid lines, such as a series of circles, may be added to bracket a specific point and indicate distances from that point to various locations. See example at right.

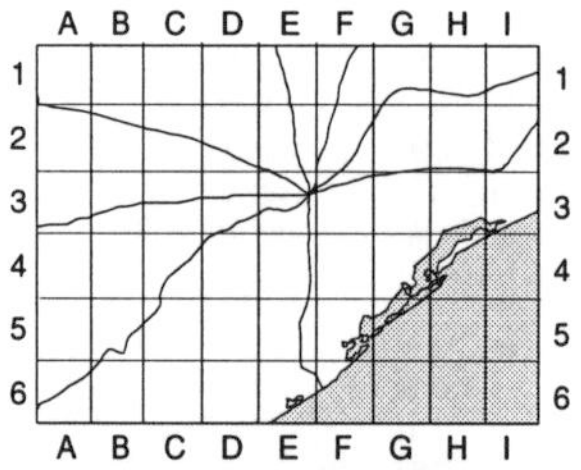

On localized maps, grid lines frequently have no relationship to lines of latitude and longitude.

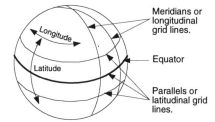

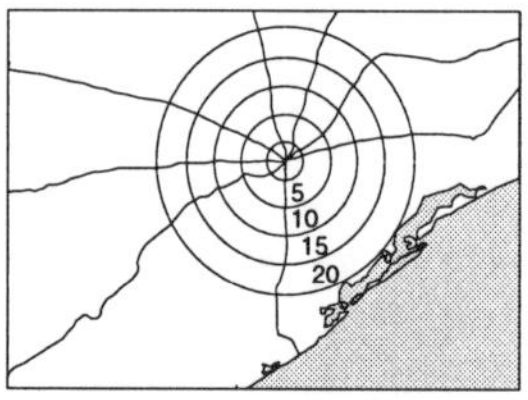

A supplementary set of grid lines is sometimes used for designating distances from specific locations.

Gross Deviation Graph	A variation of a difference graph. See Difference Graph.
Grouped Area Graph	Sometimes referred to as a multiple or overlapped area graph. A grouped area graph displays multiple data series, all of which are measured or referenced from the same zero base line, usually the horizontal axis. As a result of all of the data series extending down to the zero base line, the values for each data series can be read directly from the graph. This type of area graph should be used with caution because it can easily be confused with the more familiar stacked area graph. See Area Graph.

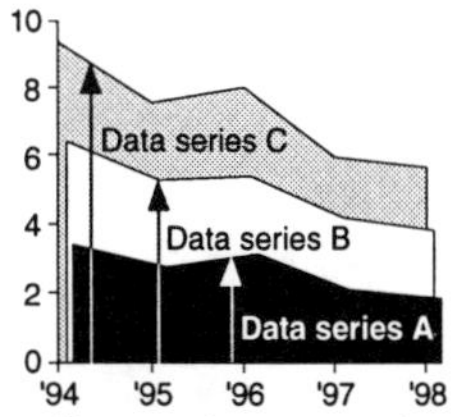

Grouped area graph

Grouped Bar Graph	Sometimes called a clustered bar, multiple bar, or side-by-side bar graph. A grouped bar graph is a bar graph on which two or more data series are plotted side-by-side. The bars for a given data series are always in the same position in each group throughout a given graph. Each data series is typically a different color, shade or pattern. See Bar Graph.

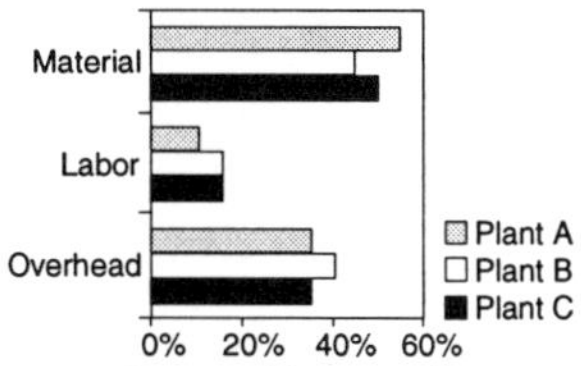

Grouped bar graph

Grouped Box Graph	Sometimes referred to as a clustered box graph. A grouped box graph is a box graph on which two or more data series are plotted side-by-side. Each data series is typically a different color, shade, or pattern. See Box Graph.

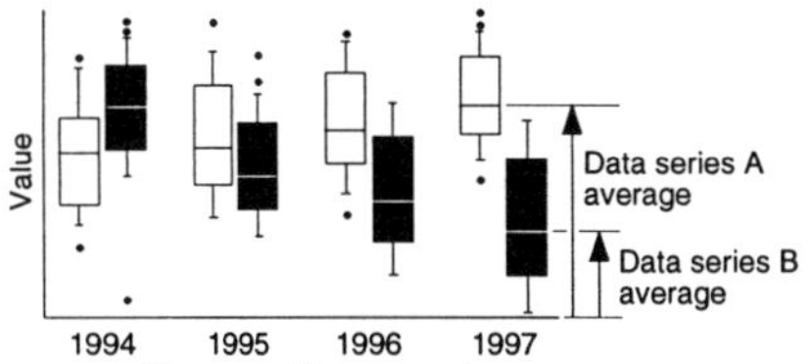

Grouped box graph

Grouped/Clustered Graphs	Using groups or clusters of two or more graphs on the same or related subject many times offers synergistic advantages over the use of a single graph. See Graph.

Grouped Column Graph	Sometimes called a clustered, multiple, or side-by-side column graph. A grouped column graph is a column graph on which two or more data series are plotted side-by-side. The columns for a given data series are always in the same position in each group throughout a given graph. Each data series is typically a different color, shade or pattern. See Column Graph.

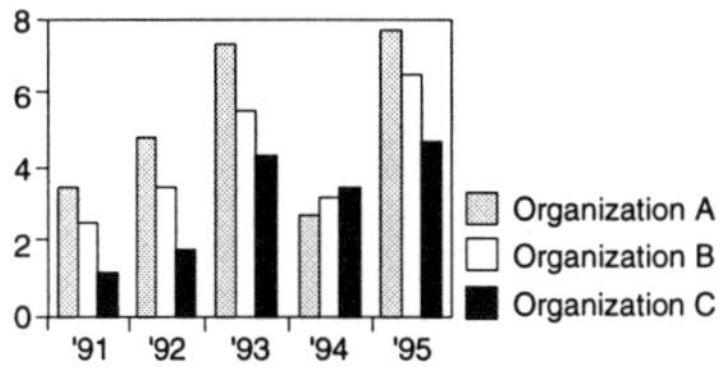

Grouped column graph

Grouped Line Graph	Sometimes referred to as a compound line, dual line, multiple line, or overlapped line graph. A grouped line graph is a line graph with multiple data series, all of which are measured from the same zero base line axis. This is the most widely used variation of a line graph. See Line Graph.

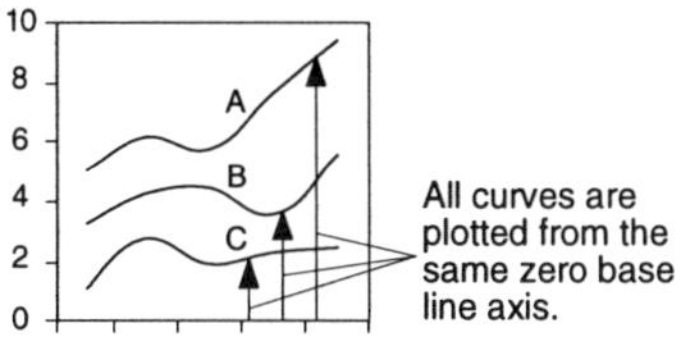

Grouped line graph

Grouped Point Graph or **Grouped Scatter Graph**	A point graph with multiple data series, all of which are measured from the same zero base line axis, is referred to as a grouped or multiple point graph if one of the axes has a category scale. If there are quantitative scales on both the vertical and horizontal axes, graph is generally called a grouped or multiple scatter graph. If there is a sequence scale on the horizontal axis, the chart may be called either a grouped point graph or grouped scatter graph. See Point Graph and Scatter Graph.

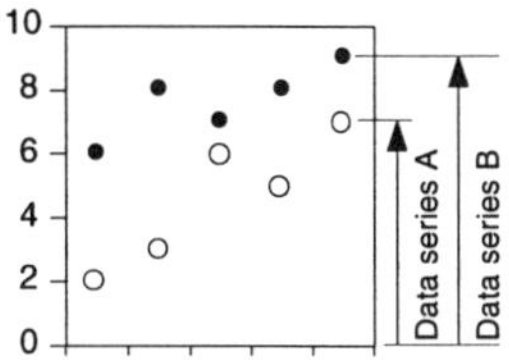

Grouped point graph or grouped scatter graph

Grouped Surface Graph	A surface graph with multiple data series, all of which are measured from the same zero base line axis (XY plane). See Surface Graph.

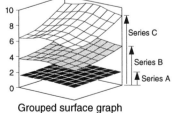

Grouped surface graph

Group Interval	Sometimes referred to as a class, class interval, bin, or cell. Group intervals are the groupings into which a set of data is divided for purposes of generating any of several data distribution graphs or maps. See Histogram and Statistical Map.

185

	Hachures	Short lines used to indicate the slope of the terrain on topographic maps. The length, width, color, etc., are varied to convey information such as steepness or direction of slope. An example is shown at the right. Hachures have largely been replaced by the use of contour lines and shading.

Hachures

Short lines used to indicate the slope of the terrain on topographic maps. The length, width, color, etc., are varied to convey information such as steepness or direction of slope. An example is shown at the right. Hachures have largely been replaced by the use of contour lines and shading.

Examples of hachures

Half-Bidirectional Table or Half-Matrix

In certain matrixes, particularly bidirectional tables, information is occasionally duplicated as a result of the type and arrangement of headings. In order to minimize the space required for the table or to make the table easier to read, the duplicated information is sometimes eliminated. When this is done the result is called a half-bidirectional table or half-matrix. Examples of a full bidirectional table and a half bidirectional table are shown below.

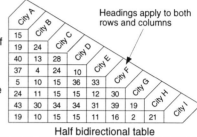

The numbers in the grayed area are duplicates of the numbers in the nongrayed areas. They are eliminated in the table on the right.

Headings apply to both rows and columns

Full bidirectional table

Half bidirectional table

Hanging Column

A column that extends below a horizontal axis or reference line.

Hard Copy

The term hard copy refers to information printed on paper as contrasted to information delivered by verbal, on-line, computer disk, film, tape, video, or other format.

Hatch

The term hatch is used to refer to a pattern of fill that consists of evenly spaced parallel lines. If there are two or more sets of parallel lines intersecting one another, the pattern is referred to as crosshatch.

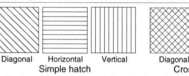
Diagonal   Horizontal   Vertical   Diagonal   Horizontal
Simple hatch   Crosshatch
Patterns referred to as hatched and crosshatched

Hierarchical Organization Chart

A widely used variation of organization chart in which people, positions, organizations, ideas, facilities, or equipment are categorized by their authority, responsibility, ability, status, power, etc. In a hierarchical chart the entities with the most of whatever is used as the criteria are located at one end of a diagram, and the entities with the least are located at the other end. All other entities are spaced proportionately in between. For example, if authority is the criteria and the organization chart is running horizontally, the person with the most authority is shown at the left and those with the least authority are shown at the right. Those with medium authority are shown towards the middle. Hierarchical organization charts can be any shape. The most typical one runs vertically with the entity highest in the hierarchy at the top, as shown at the left. Sometimes the shape of an organization chart is selected to deemphasize its hierarchical nature, as in the circular chart at the right. See Organization Chart.

Vertical hierarchical organization chart with the entity with the most authority at the top

Example of a circular hierarchical organization chart with the entity with the most authority at the center

High-Low Graph

Includes:
High-Low-Close (HLC) Graph and High-Low-Close-Open (HLCO) Graph

A variation of a range graph in which symbols represent sets of data, with one end indicating the value of the largest or highest data element in the set of data and the other end the smallest or lowest. With vertical symbols, the top represents the largest value. With horizontal symbols the right end designates the largest value. One or more midvalues are frequently included, such as average, median, first, etc. The symbols representing the ranges might be displayed independently (left) or at the ends of standard bars or columns (right). See Range Symbols and Graphs. A similar type of graph is frequently used for recording the price of stocks. When used for this purpose, although the symbols run vertically, it is generally referred to as a bar chart and sometimes as a high-low-close (HLC) graph or high-low-close-open (HLCO) graph. See Bar Chart .

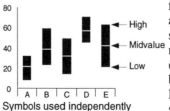
High
Midvalue
Low
Symbols used independently

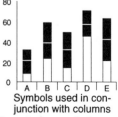
Symbols used in conjunction with columns

								4
1.9	2.56	2.84	2.985	3.08	3.24	3.43	3.6	4.02
2.2	2.6	2.86	3	3.1	3.26	3.44	3.65	4.06
2.25	2.7	2.88	3.01	3.12	3.28	3.45	3.7	4.1
2.3	2.72	2.9	3.02	3.14	3.3	3.46	3.75	4.2
2.4	2.74	2.91	3.03	3.16	3.35	3.48	3.8	4.4
2.43	2.76	2.93	3.04	3.18	3.4	3.5	3.85	4.43
2.46	2.8	2.95	3.06	3.2	3.41	3.53	3.9	4.55
2.53	2.82	2.97	3.07	3.22	3.42	3.58	3.95	4.7

Histogram and Frequency Polygon

Sometimes referred to as frequency diagram. The histogram is the best known member of the family of data distribution graphs. A frequency polygon is a segmented or smooth line version of a histogram. Histograms and frequency polygons show the frequency with which specific values (referred to as data elements) or values within ranges (referred to as class intervals) occur in a set of data.

Description of a histogram

The set of numbers at the right is typical of a group of values whose distribution requires evaluation. One method is to plot all the values along an axis

The data set of 73 numbers used to generate the graphs shown below

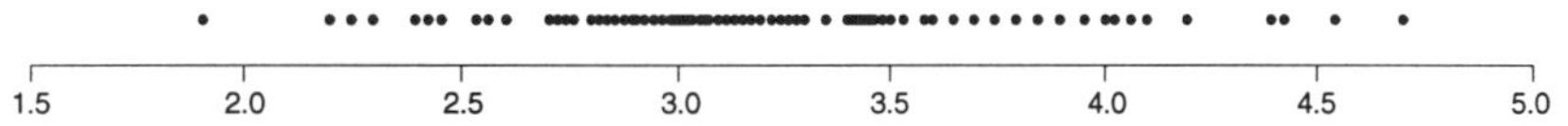

(sometimes called a one-axis data distribution graph), as shown below. Based on this graph, an estimate can be made as to where the values are concentrated, what the extremes are, and whether there are any gaps or unusual values. Overlapping data points can become a problem with such a graph, even with a relatively small number of data elements.

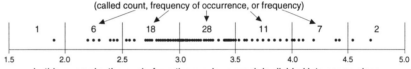

One-axis data distribution graph displaying the data shown above

To address the problem of overlapping and at the same time quantify the distribution, one may divide the axis into five to ten equal intervals (called class intervals, classes, group intervals, bins, or cells) and, using the actual data, count the number of data elements in each interval. These counts are noted in the individual class intervals below. In the set of data used in this example, the largest number of data points are located in the interval between 3.0 and 3.5, and there are slightly more points to the left of that interval than to the right.

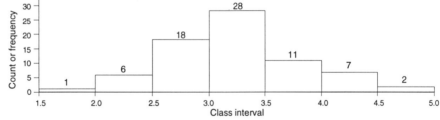

Number of data elements in each interval (called count, frequency of occurrence, or frequency)

In this example, the scale from the previous graph is divided into seven class intervals and the number of data elements in each class interval noted.

A further step is to make a joined column graph of the counts or frequency in each interval, as shown below. Such a graph is called a histogram.

Joined column graph of values shown in class intervals above. This type of graph is called a histogram.

Many times, in addition to knowing how many data elements fall into each interval, it is desirable to know what percentage those quantities represent of the total number of data elements in the data set. This information can be displayed on a histogram either by showing the percent in or adjacent to each data graphic and/or adding a percent scale to the graph. Both methods are shown in the example below. These percent figures are called percent frequency or relative frequency. When the percent is shown in its decimal equivalent format [i.e., 0(0%) to 1(100%)], it is sometimes referred to as proportion.

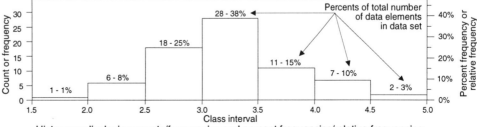

Histogram displaying counts/frequencies and percent frequencies/relative frequencies

Class intervals

Number of class intervals

The decision as to how many class intervals to sort the data into is an important consideration. There are no rigid guidelines to follow and each situation is generally handled on an individual basis. As the number of intervals increases, the more accurately the graph depicts the actual data; however, at some point, distinctive patterns blur, and gaps begin to appear because there are no data elements in some of the class intervals. (When there are no values in a given class interval, the interval is shown with a value of zero.) On the other hand, if there are too few intervals, accurate representation is impaired and the shape of the histogram can be distorted. The use of between 5 and 40 class intervals is most common. The examples below illustrate the effect of changing the number of class intervals. The same data is used in each example.

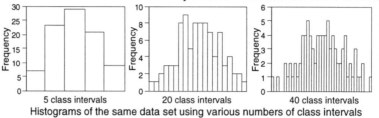

Histograms of the same data set using various numbers of class intervals

Width of class intervals

The widths of class intervals are determined in several different ways, including:

- Subtract a value slightly less than the lowest value in the data set from a value slightly larger than the highest value in the data set, and divide the difference by the number of intervals desired.
- Use whole numbers (integers) as the boundaries. For example, if the data series ranges from 17.3 to 21.8, the boundaries might be set at 17, 18, 19, 20, 21, and 22.
- Specify the interval width. For example, if the data series ranges from 0.50 to 0.59 and the interval width is specified as 0.02, the boundaries might be 0.50, 0.52, 0.54, 0.56, 0.58, and 0.60.
- Specify specific upper and lower boundaries for each interval. This method sometimes results in different widths for some or all of the columns. For example, if the data series ranges from 1 to 20, the boundaries might be 1,5,8,10,11,13,16, and 20.
- If every data element in the data series is allocated a column, the available space is divided by the number of different data elements.
- When nominal or ordinal information is plotted, the available space is divided by the number of categories, or alternately, the widths are made proportional to another variable.

Terminology

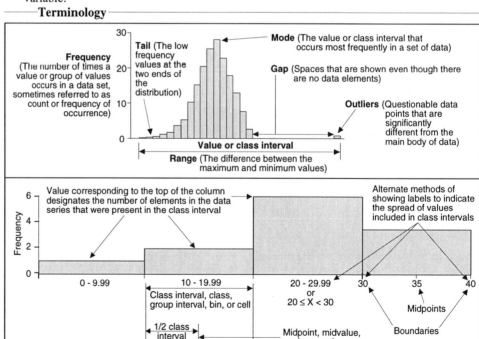

Illustration of terms frequently used with histograms

Frequency polygon

Sometimes the midpoints of the columns of a histogram are connected by straight lines and the columns eliminated. This type of graph is generally called a frequency polygon. A frequency polygon can make the visualization of data easier and is helpful when comparing multiple data series. In the upper example at the right, the columns of the histogram are included only to show the relationship between the histogram and the polygon. Typically, the columns are not shown with a frequency polygon. An additional interval is often added to both ends of a histogram when it is converted to a frequency polygon, so the curve starts and ends at zero. The straight lines connecting the data points are sometimes converted to smooth lines to improve the appearance of the curve (as shown in the lower example at the right). Polygons are not used with irregular width intervals.

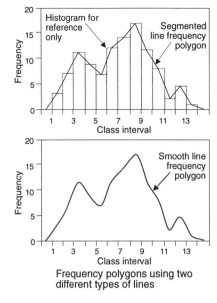

Frequency polygons using two different types of lines

Distinctive shapes of histograms and polygons

Of particular, and many times prime interest when using histograms and frequency polygons are such questions as how and where the data elements are concentrated, whether they are spread out or clustered, whether the data elements are distributed normally, whether the data is skewed and if so in which direction, etc. Such observations sometimes give insight into the nature of the data that the curve represents. Shown below are examples of typical frequency polygon shapes. Most of the examples include terms used to describe their unique shape, a potential interpretation of the shape of the curve, and/or some observations about what might cause such a shape.

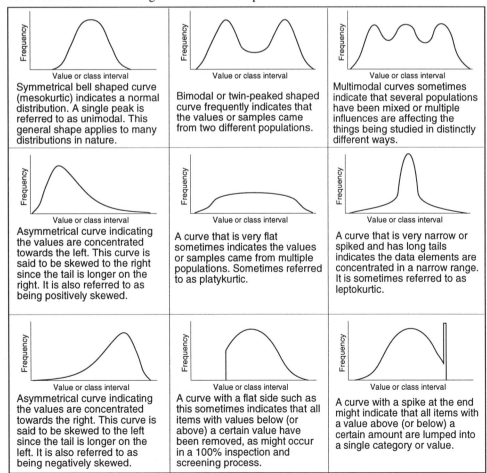

Symmetrical bell shaped curve (mesokurtic) indicates a normal distribution. A single peak is referred to as unimodal. This general shape applies to many distributions in nature.

Bimodal or twin-peaked shaped curve frequently indicates that the values or samples came from two different populations.

Multimodal curves sometimes indicate that several populations have been mixed or multiple influences are affecting the things being studied in distinctly different ways.

Asymmetrical curve indicating the values are concentrated towards the left. This curve is said to be skewed to the right since the tail is longer on the right. It is also referred to as being positively skewed.

A curve that is very flat sometimes indicates the values or samples came from multiple populations. Sometimes referred to as platykurtic.

A curve that is very narrow or spiked and has long tails indicates the data elements are concentrated in a narrow range. It is sometimes referred to as leptokurtic.

Asymmetrical curve indicating the values are concentrated towards the right. This curve is said to be skewed to the left since the tail is longer on the left. It is also referred to as being negatively skewed.

A curve with a flat side such as this sometimes indicates that all items with values below (or above) a certain value have been removed, as might occur in a 100% inspection and screening process.

A curve with a spike at the end might indicate that all items with a value above (or below) a certain amount are lumped into a single category or value.

Typical shapes associated with histograms/frequency polygons

Histogram values are proportional to the area of the column

Technically the area of the column of a histogram is proportional to the value it represents. When the columns are the same width, the heights of the columns automatically become proportional to the values they represent. In some cases the width of the columns are varied (irregular class intervals) either to convey additional information, to emphasize a particular aspect of the data, or because of the nature of the data (e.g., gaps in the data). When this is done, the height of the columns must be adjusted to keep the areas proportional. For example, if the widths of most of the columns in the histogram are one unit wide and it is decided to have one column represent two class intervals, the values for the two intervals are added and divided by two (averaged) to arrive at the height of the two-unit-wide column. If one column is to represent five class intervals, the values of all five intervals are added and divided by five (averaged) to arrive at the height of the five-unit-wide column. A comparison of a histogram with all equal column widths and one with irregular widths is shown below. Since it is the areas of the columns that are proportional to the values they represent, it is important that the vertical or frequency scale always start at zero and that no

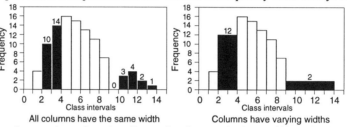

scale breaks be introduced. Starting the horizontal or class interval scale at something other than zero normally has no adverse effect as long as the lowest value on the scale is less than the smallest value in the data set.

All columns have the same width   Columns have varying widths

A comparison of graphs with equal and unequal column widths using the same data. The two sets of black columns in the graph at the left have been replaced by the two wide, black columns in the graph at the right.

Comparing multiple histograms or frequency polygons

Multiple data sets can be compared using histograms or polygons. The comparisons can be based on actual frequencies or on percents. When there are differences in the number of data elements in the data sets being compared, the percent comparison is generally recommended so the vertical sizes of the columns are more comparable. Examples of some of the most widely used methods for comparing frequency distributions are shown below.
• Pyramid graphs (not shown here), use histograms to compare two sets of data. They are sometimes referred to as two-way histograms. See Pyramid Graph.

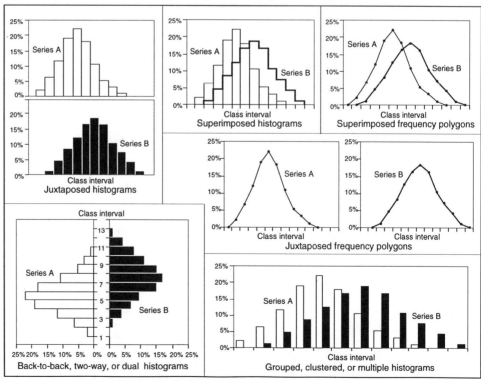

Examples of methods for comparing multiple frequency distributions

Histogram and Frequency Polygon
(continued)

Stacked histogram

Stacked histograms are sometimes used to show how subcategories of a data series affect the overall frequency distribution pattern. For example, a histogram similar to the one at the right might be generated to study the distribution of orders based on dollar value. The black portions might represent the number of export orders in the various sales dollar ranges and the white portions might represent the number of domestic orders. The tops of the columns indicate the overall number of orders. More than two data series can be plotted on this type of graph; however, interpretation sometimes becomes difficult.

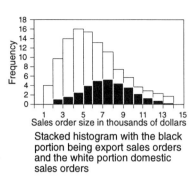

Stacked histogram with the black portion being export sales orders and the white portion domestic sales orders

Other types of graphs sometimes referred to as histograms

Simple column graph

The term histogram is sometimes loosely used to refer to any simple column graph that displays the distribution of something, for example, the distribution of population by country or the distribution of income by age. When used this way there may or may not be spaces between the columns and class intervals may or may not be used.

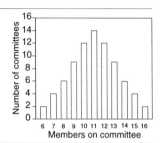

Bar graph

Sometimes called frequency bar graph. Joined and unjoined bar graphs are occasionally used as histograms. Almost all of the information on column type histograms also applies to bar graphs as histograms.

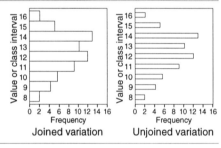

Joined variation   Unjoined variation

Charts that record specific occurrences

Stem and leaf, tally, and dot array charts are sometimes referred to as simple histograms.

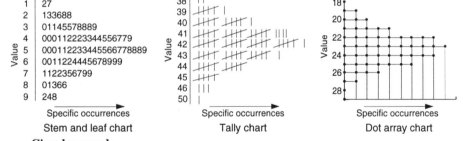

Stem and leaf chart   Tally chart   Dot array chart

Circular graph

On rare occasions, a circular histogram similar to the ones shown below is used. In both examples, class intervals are plotted on the circumference of the circle. In the example on the left, the frequencies are noted by the ends of the radial columns. In the other case, the size of the wedges are proportional to the frequencies. Other than a unique appearance, circular histograms offer no advantages over rectangular histograms.

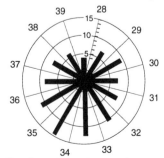

Class intervals are shown on the circumference. Frequencies are indicated by the ends of the radial columns.

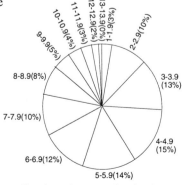

Class intervals are noted on the circumference. The sizes of the sectors or wedges are proportional to frequencies.

Examples of circular graphs used as histograms

Three-dimensional histogram

Histograms have traditionally been two-dimensional. With the developments in computers and software, three-dimensional histograms are becoming more common. An example is shown at the right. Three-dimensional histograms work on the same principles as two-dimensional histograms, except there are two sets of categories or class intervals, one on the X-axis and one on the Y-axis. Frequencies are plotted on the vertical or Z-axis. Unless one has the capability to rotate the graph, the hidden columns in a three-dimensional histogram make accurate analysis difficult.

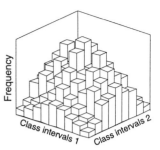

Three-dimensional histogram

Cumulative histogram

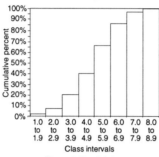

Cumulative histogram

In a cumulative histogram, the value to be plotted for a particular class interval is determined by adding the frequency value or percent for that class interval to the frequencies or percents for all class intervals preceding it. Frequencies can be cumulated from the smallest value to the largest, or vice-versa. The example at the left shows cumulative frequency percents with the percents being cumulated starting with the smallest value. If actual frequencies were used instead of percents, the graph would remain the same except for the scale. • A cumulative histogram allows one to determine what percentage of the data elements in the data set are above or below selected values. Those selected values are the upper and lower boundaries of the class intervals (upper boundaries when cumulated

from smallest to largest; lower boundaries when cumulated in the reverse). For example, using the class interval of 2.0 to 2.1 in the graph at the right, one may determine that 63% of the data points are equal to or less than 2.1, and 37% of the data elements are equal to or greater that 2.1. To make such observations possible at any point along the horizontal axis, the boundary values for all class intervals are sometimes connected with a curve (below). If the columns are eliminated, the resulting graph is called a

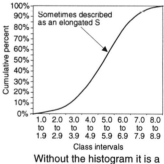

A cumulative histogram can be used to determine the percent of data elements in a data series that are greater than or less than selected values.

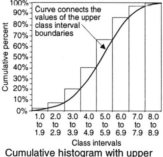

Cumulative histogram with upper boundaries connected with a curve

cumulative frequency graph (right). Although a cumulative histogram can be used for determining greater-than and less-than information at a few points, a cumulative frequency graph is normally used for this purpose. See Cumulative Frequency Graph.

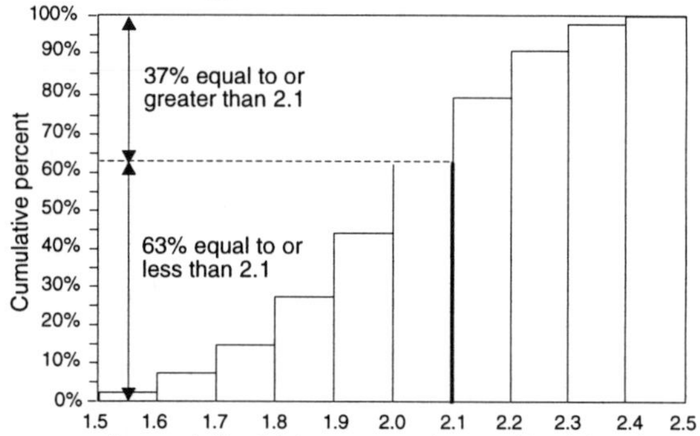

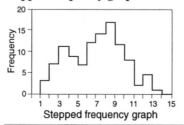

Without the histogram it is a cumulative frequency graph

Stepped frequency graph

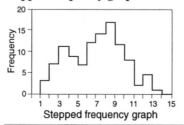

Stepped frequency graph

Typically a histogram is made up of a series columns. Occasionally, primarily for appearance, the lines between the columns are eliminated and the figure becomes a stepped frequency graph as shown at the left. Functionally a stepped frequency graph is the same as a histogram.

**Histogram and
Frequency Polygon**
(continued)

Methods for smoothing histograms

In addition to frequency polygons, there are several other techniques that eliminate the stepped appearance of histograms. Reasons for using such techniques include:

- Reducing the impact that the selection of class interval width has on the shape of the histogram;
- Eliminating the sharp steps or discontinuities that occur from one interval to another; and
- Overcoming the impression that the data are distributed uniformly over the widths of the intervals.

Three variations are noted below.

Density trace curves

In a standard histogram, class intervals do not overlap. In a density trace, the number and width of the intervals is purposefully chosen to assure that every interval overlaps two or more other intervals. For instance, in the example at the right, the class interval is two units and the center lines of the class intervals are two units apart. Thus, there is no overlapping. In a density trace, the class interval might also be two units, but the centerlines might be a half-unit apart, resulting in considerable overlapping. Instead of plotting the columns as in a standard histogram, only the values are plotted at the centerline location of the class intervals and connected by a smooth line. • The overlapping results in multiple counting of most values; consequently the values on the vertical scale are larger than on the standard histogram. Since the shape of the curve and the location of the modes is generally of more interest than the actual values, the difference in counts generally causes no difficulty. In fact, to avoid confusion, the vertical scale is sometimes omitted. • Varying the width and number of the intervals changes the shape of the curve and sometimes slightly shifts the apparent location of the mode. See Density Trace.

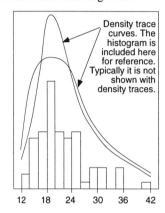

Density trace curves. The histogram is included here for reference. Typically it is not shown with density traces.

Hand fitted curve

The simplest, although least accurate, smoothing technique involves hand-fitting a curve to a histogram. When this is done, an effort is made to have the area under the fitted curve the same as the area of the histogram. When curves are hand-fitted, the histogram is frequently displayed along with the curve.

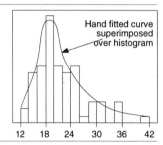

Hand fitted curve superimposed over histogram

Normal curve

In some situations a normal distribution curve is superimposed over a histogram. The normal curve is frequently based on the average and standard deviation of the data series plus the assumption that the distribution of the data set is normal. This can be misleading if the actual distribution is significantly different from normal.

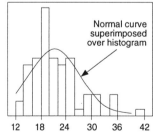

Normal curve superimposed over histogram

Scales (based on column type histogram)

Vertical axis

- Quantitative and linear with positive values only
- Starts at zero (which is coincident with the horizontal axis) without scale breaks
- Can be in terms of units, percents, or both

Horizontal axis

- Scales can be quantitative, category, or ordinal, it generally is quantitative
- If quantitative, the scale is linear and values can be positive or negative
- If quantitative, the scale can use individual values or class intervals
- Lower scale value is less than smallest value in data set; it frequently is not zero
- Largest scale value is larger than largest value in data set
- If not quantitative, categories or ordinal units are organized in some meaningful fashion

Histogram and Frequency Polygon (continued)	**Labels**

Labels

Labels on the vertical axes are straightforward and need no discussion. There are several variations possible on the horizontal axis. When each interval represents a single value or a category, the labels are centered below the columns. When class intervals are used, labels are many times placed directly below the boundaries between the class intervals. This can cause a slight confusion as to where data elements with the same values as the boundaries are counted. Sometimes a note states where such data elements are included. An alternative is to label the intervals as to exactly what values are included, though such a solution is sometimes limited based on the availability of space.

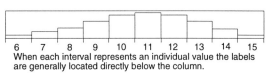

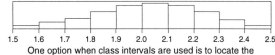
When each interval represents an individual value the labels are generally located directly below the column.

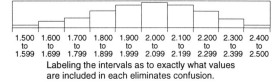
One option when class intervals are used is to locate the labels at the boundaries between the class intervals. This method leaves some doubt as to where data points with values the same as the boundary values are counted.

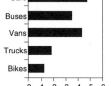
Labeling the intervals as to exactly what values are included in each eliminates confusion.

Examples of possible variations in histogram labels

Histograph

Many times referred to as histogram. See Histogram.

Horizontal Axis

In a two-dimensional graph, the horizontal axis is the one that runs left and right. It is also frequently called the X-axis. Positive values generally proceed from left to right. In a three-dimensional graph, typically none of the axes are referred to as a horizontal axis.

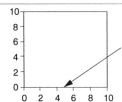
Generally referred to as the horizontal or X-axis in a two-dimensional graph. Positive values typically increase from left to right.

Horizontal Bar Graph or Horizontal Column Graph

Sometimes referred to as a rotated column graph or bar graph. See Bar Graph.

Horizontal bar graph

"How-To" Chart

Many how-to charts are variations of diagrams and/or illustration charts. How-to charts are graphical representations of the steps required or recommended to accomplish a given task. In some cases actual representations such as pictures, sketches, drawings, etc., are used. In other cases only text or abstract symbols are used. How-to charts can be very elaborate or very simple. They might consist of a single chart or a series of charts, as shown below. The tasks they deal with range from very practical, such as how to assemble a pencil sharpener, to very complex and abstract, such as how to solve complicated equations.

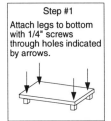
Step #1
Attach legs to bottom with 1/4" screws through holes indicated by arrows.

Step #2
Attach sides to bottom with screws provided

Step #3
Apply adhesive with applicator provided.
Adhesive

Step #4
Position top and let dry for 24 hours.

Example of a series of how-to charts for assembling a small table

Hydrographic Map

A map used for navigation on water such as oceans, lakes, rivers, etc.

Hypsometric Map

Sometimes called a relief, or contour map. A hypsometric map shows the surface features of the terrain such as locations, configurations, and elevations, with particular emphasis on elevations. This is accomplished largely through the use of contour lines which connect points of equal elevation. Shades of color, called hypsometric tints, are frequently applied between the contour lines. See Topographic Map.

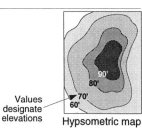
Values designate elevations
Hypsometric map

Icon

Miniature (1" square or less) graphs or images. When graphs are used they have no titles, labels, tick marks, or grid lines. The purpose of icons is not to convey specific quantitative information. Instead, they are used to show such things as:

- Relative sizes, values, ratings, etc., of specific items, characteristics, or attributes.
- Overall comparisons of multiple entities or the same entity at different times
- Trends of a single entity or a comparison of trends for multiple entities
- Unusual information patterns in one or multiple entities

There are a number of different types of icons. Examples of some of the more straightforward designs are shown below.

Icon / Symbol	Names sometimes used to describe	Features
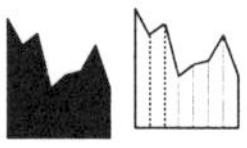	Area or profile	Each point represents a different variable or the same variable over time. The higher the point, the more favorable the entity in that characteristic. The greater the area of a symbol, the higher the overall rating for that entity. Fill and/or drop lines may or may not be used.
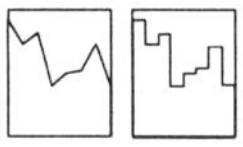	Line	Similar to an area icon except fill is removed and a border is added as a frame of reference for estimating relative values. When a scale of zero to one is used, the bottom of the frame is zero and the top is one. A segmented or stepped line format can be used.
	Column or histogram	Each column may represent a different variable or the same variable over time. The heights of the columns are proportional to the values they represent.
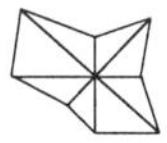	Polygon, star, snowflake, or profile	Each point on the polygon represents a different variable. The length of the spoke leading to the point is proportional to the value it represents. This type of icon is often generated using a radar graph template. Polygons have an advantage over other icons because it is easier to locate the same characteristic on each icon. For instance, it is easier to compare the characteristics at the six o'clock position on each star than the fifth one from the left on each line icon.
	Chernoff faces or faces	In Chernoff faces, each facial feature represents a different characteristic. The size, shape, and appearance of the facial features relate to values they represent. For example, eyes to the left might mean a large value and eyes to the right a small value.

Examples of some of the more widely used icons

Icons are used as symbols on charts, graphs, and maps. Groups of icons are also used to compare multiple entities with regards to three or more variables. This latter application is sometimes called an icon comparison display. See Icon Comparison Display.
• When icons are used, a legend is required for decoding purposes. Three applications using icons are shown here.

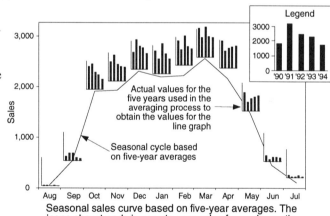

Seasonal sales curve based on five-year averages. The icons show trends in year-to-year sales for each month.

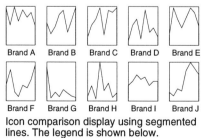

Icon comparison display using segmented lines. The legend is shown below.

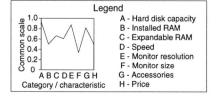

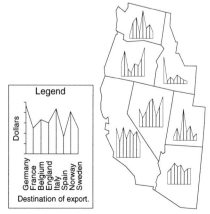

Icons used on a map to compare exports to eight different countries from six different states.

Icon Comparison Display

Sometimes called a symbol display or symbolic display. An icon comparison display is a group of icons assembled for the purpose of simultaneously comparing or screening a sizable number of entities with regards to three or more variables. In this type of display, a separate icon/symbol is generated for each entity under review. The icons are then assembled into a matrix similar to the one shown at the right. In this example, 24 brands of desktop computers are compared with regards to eight different characteristics. The arrangement of the icons in the matrix is arbitrary, though an effort is generally made to group them in some meaningful way such as by size, market, manufacturer, etc. • The icon used in this example is called an area or profile icon. Some of the other icons used in icon comparison displays are shown under the heading Icon. • Sometimes values and units of measure for the different characteristics vary significantly. In addition, characteristics might be quantitative or qualitative. To plot all of these on the same icon, a common scale of 0 to

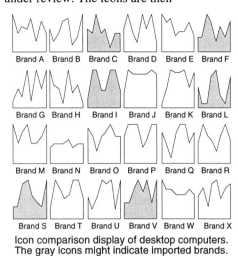

Icon comparison display of desktop computers. The gray icons might indicate imported brands.

1.0 is frequently used for all variables. The highest value or rating in each variable is assigned the value of 1.0. The lowest value or rating in each variable is assigned some value less than 1.0, often zero. In some cases, such as the polygon icon used in radar graphs, the graph can accommodate different scales on each axis and the conversion of values to a common scale is not necessary. Regardless of which type of scale is used in the construction of the icons, the titles, labels, and tick marks are eliminated when the icons are assembled into the matrix. • During the generation of the icons, a template similar to the one shown at the right is used to assure that the same characteristic is always in the same location on every icon and that the same scale is used for all icons. The order of the characteristics along the horizontal axis is arbitrary, though they are typically arranged

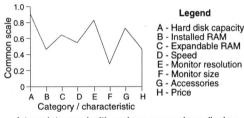

A template used with an icon comparison display

in meaningful groupings such as functional, appearance, price, quality, etc. Other than legibility, there is no limit to the number of variables that can be included in each icon. There also is no limit as to how many icons can be included in the same display.

Illustration Chart

A chart consisting solely or mainly of a sketch, picture, photograph, drawing, etc. When text is included, its primary function is generally to explain or give additional insight into the subject illustrated on the chart, usually in the form of notes, labels, or a legend. The illustration might be very general for a large audience or very detailed for a focused audience such as a training class. The illustrations may be symbolic, pictorial, or photographic, accurately drawn to some scale, or hand sketched. Enlargements and cross-sections are often used. The illustration might serve to describe a person, place, or thing; show how to do something; graphically indicate how things move, grow, change, etc.; or orient an audience during a presentation. An illustration chart may by used by itself or as part of a series of charts. Two examples are shown at the right.

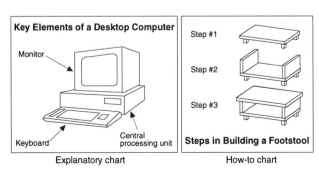

Examples of illustration charts

Independent Variable

A term used to describe certain types of variables. See Variable.

Index Graph

An index graph is a graph that displays values in terms of a percent, fraction, or ratio of some reference value. For example, a graph might show the value of a house each year as a percent of its purchase price. If a house purchased in 1985 for $20,000 is now worth $40,000, it would be plotted at the 200% value. The 200% is considered an index value. A comparison of an index graph and a graph showing the corresponding actual values is shown at the right. If an index includes several factors in the values, such as the consumer price index (food, housing, fuel, etc.), it is sometimes referred to as a composite index.

Graph of actual data — Index graph using 1985 as a reference

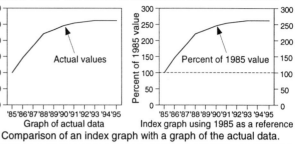

Comparison of an index graph with a graph of the actual data.

Multiple data series

There is no limit on the number of data series that can be plotted on a single graph. Values smaller and larger than the reference can be plotted. The relative positions of curves frequently change when plotted on an index graph.

Graph of actual data — Index graph using 1985 as a reference

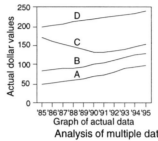

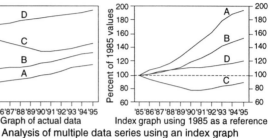

Analysis of multiple data series using an index graph

Index type comparative graphs

Sometimes only the beginning and ending or current index values are plotted. The graph at the right contains the same values as the graph immediately above it. This format sometimes makes the data stand out more crisply, particularly when there are many values. This format is many times called a comparative graph or fan chart.

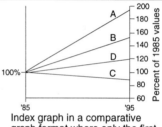

Index graph in a comparative graph format where only the first and last years are shown.

Different units of measure

In addition to comparing data series with different values, index graphs are used to compare data series using different units of measure. For instance, in the example at right, the units produced increased at a much greater rate than the sales dollars.

Actual values — Index values

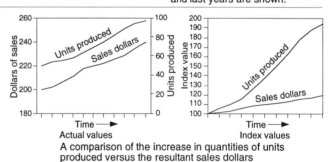

A comparison of the increase in quantities of units produced versus the resultant sales dollars

Options in selecting the reference value

Often the first value in a data series is used as the reference against which all other values are measured. However, the reference value may be any one of the data elements in the series. Three variations are shown below.

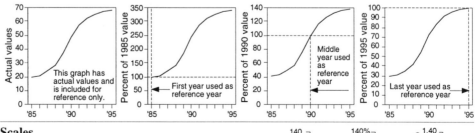

Scales

The scale on the horizontal axis is almost always sequential. The scale on the vertical axis is quantitative. Three examples of how the values sometimes appear are shown at the right. The values are referred to by several different terms, including index value, percent, and fraction or ratio.

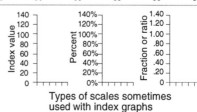

Types of scales sometimes used with index graphs

Indicators	Graphic indicators are used in many fields and for many different purposes. An indicator generally foretells or infers something about another entity, event, data series, performance, etc. For example, interests rates are many times used as an indicator of housing starts (e.g., as interest rates go up, housing starts go down). Most indicators have a sequential scale on the horizontal axis. A large percentage of the indicators use time scales. • In many cases the shape or general pattern of an indicator is similar to or inverse to the curve of the thing it is indicating. The curve of the indicator and the thing it is indicating might be in time phase (coincident indicator), the indicator might be shifted ahead (leading indicator), or the indicator might be shifted behind (lagging indicator). Examples of these three time options are shown at the right. • In another major type of graphic indicator, the shape of the indicator curve looks nothing like the curve of the thing it is indicating. In this type of indicator, observations focus on such things as when the indicator exceeds a particular value, when the direction of the indicator changes, when two data series cross, when the indicator repeats itself, etc. These types of indicators are widely used in the investment field. Several examples are shown below.

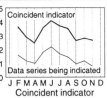
Coincident indicator

Leading indicator

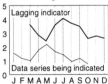
Lagging indicator

Examples where the curve of the indicator is similar to the thing it is indicating

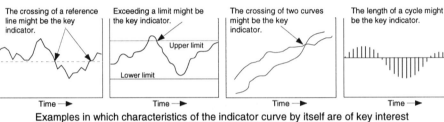

Examples in which characteristics of the indicator curve by itself are of key interest

Indiscrete Data	Sometimes referred to as continuous data. See Continuous Data.
Information	In this book, the word information is used interchangeably with the word data.
Information Graphics	Charts, graphs, maps, diagrams, and tables whose primary function is to consolidate and display information graphically in an organized way so a viewer can readily retrieve the information and make specific and/or overall observations from it. Information graphics may be contrasted with graphics whose primary functions are artistic or for purposes of entertainment, promotion, identification, etc. Such things as engineering and architectural drawings are not included under the classification of information graphics. See Chart.

Insert or Inset	An inset is a small graph, chart, map, or table that is displayed inside or alongside a larger chart, graph, or map to provide supplemental information about the larger document. Insets are used for expanding on a specific part of the larger document, looking at the data in a different format, or orienting the viewer. The same graphic principles that apply to other forms of charts also apply to insets. Insets on maps are sometimes referred to as supplementary or ancillary maps. Multiple insets can be used on the same chart.

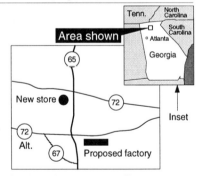

Inset, supplementary, or ancillary map used to help orient the viewer

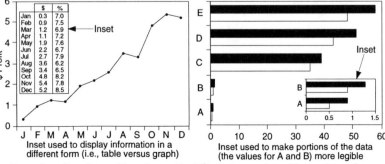

Inset used to display information in a different form (i.e., table versus graph)

Inset used to make portions of the data (the values for A and B) more legible

Examples of insets used to enhance the information contained in graphs and maps

Intercept	The point or value at which a line or axis joins or crosses an axis, – for example the point or value at which the X-axis joins or crosses the Y-axis.

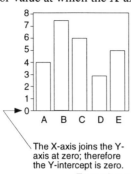 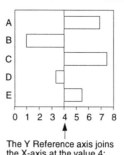 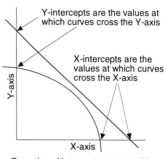

The X-axis joins the Y-axis at zero; therefore the Y-intercept is zero.

The Y Reference axis joins the X-axis at the value 4; therefore the X-intercept is 4.

Examples of intercepts generated by curves crossing the X and Y axes.

Examples of intercepts formed by axes joining and curves crossing axes

Interior Label	Labels that are placed within the plot area, on or immediately adjacent to the data graphic they apply to are called interior labels, data graphic labels, or data point labels.

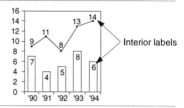

Interpolation	Interpolation is the process of estimating values between known values. When two known data points are to be connected with a line, a visual or mathematical interpolation process takes place, and the shape of the connecting line is the result. For instance, in the example at the right, points A and B are common to data series #1 and #2. The lines connecting points A and B are different for the two series. This is because, based on the other data points in the series, it was estimated (interpolated) that in series #1 the points between A and B formed a straight line, while in data series #2 it was estimated (interpolated) that the points between A and B formed a curved line.

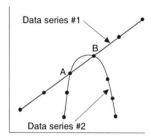

Examples of lines connecting data points based on interpolation

Intersecting Band Graph or **Intersecting Silhouette Graph**	Sometimes referred to as a curve difference or difference line graph. An intersecting band graph is a graph on which the curves representing two data series intersect one another and the areas between the two curves are highlighted by means of shading or color. For example, one curve might represent the dollar value of exports each year, and the other the dollar value of imports, as shown in the example at the right. When exports have a higher value than imports, one shade or color is used. When exports are lower than imports, another shade or color is used.

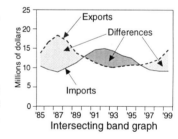

Intersecting band graph

Interval Axis	An axis with an interval scale on it.

Interval Scale	Sometimes referred to as a quantitative, value, amount, or numeric scale. An interval scale consists of numbers organized in an ordered sequence with meaningful and uniform spacings between them. The numbers are used for their quantitative value, as opposed to being used to identify entities. See Scale.

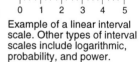

Example of a linear interval scale. Other types of interval scales include logarithmic, probability, and power.

Ishikawa Diagram	Sometimes referred to as a cause-and-effect, CE, fishbone, or characteristic diagram. The purpose of a Ishikawa diagram is to provide a method for systematically reviewing all factors that might have an effect on a given quality objective or problem. This is accomplished by assigning each factor to a line or arrow and arranging the lines and arrows into meaningful clusters in a hierarchal fashion. The lines and clusters are arranged so that they all feed into a central line that, in turn, feeds into a symbol representing the overall object or problem. See Cause-and-Effect Diagram.

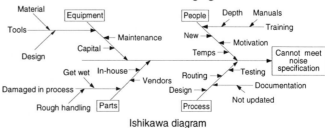

Ishikawa diagram

199

Isobath	A line connecting points of equal depth under the surface of a body of water. For example, the points on the ocean floor that are at the same depth may be connected by an isobath line. Similar lines above water are called contour lines.
Isogram Map	A map that displays lines or bands along which all points are of equal value. When the lines are of equal value (left), the map is sometimes referred to as an isoline map. When it is bands of equal values (right), the map is sometimes referred to as an isopleth map.

Isoline map
A variation of isogram map in which points of equal value are connected by lines

15¢/lb.
20¢/lb.
25¢/lb.
30¢/lb.
35¢/lb.

Isopleth map
A variation of isogram map in which areas of equal value are connected by bands

Isoline	Sometimes referred to as an isoquantity or isoquant line. Isolines are lines of equal value. For example, on a contour map a given isoline or contour line designates all points that have the same elevation. In the map at the right, all points on the line marked 400 have an elevation of 400 feet above sea level. All points on the line marked 380 are 380 feet above sea level, and so on. • On graphs, all points on a given isoline have some common value, which normally implies equal values along either the X, Y, or Z axis. In the graph below all points on line A have the same tensile strength of 75 lbs. All points on line B have the same tensile strength of 50 lbs. If a tensile strength of 25 lbs. is desired, one can use this type of graph to determine which combination of hardener and cure

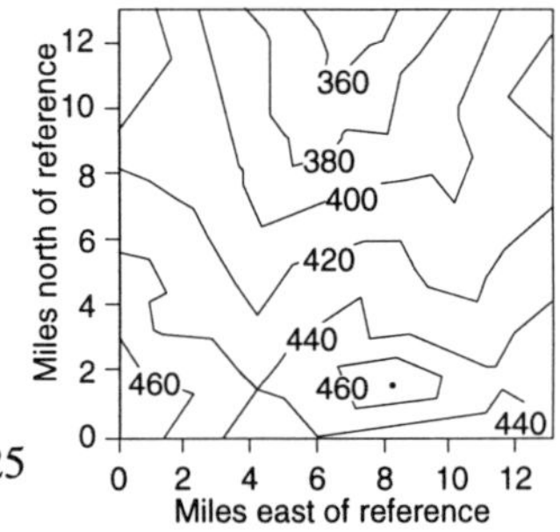

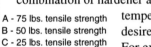

Contour map in which isolines indicate points of equal elevation

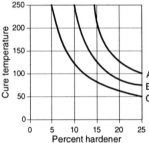

A - 75 lbs. tensile strength
B - 50 lbs. tensile strength
C - 25 lbs. tensile strength

temperature are required to obtain the desired 25 pounds of tensile strength. For example, if a cure temperature of 50 degrees is used, the graph indicates 25 percent of hardener is required. If a temperature of 250 degrees is used, the amount of hardener can be reduced to 5 percent to achieve the same 25 pounds of tensile strength. • Isolines can be displayed on two- or three-dimensional graphs. To illustrate, the three-dimensional graph at the left was generated using the same data as the two-dimensional graph above it. In this three-dimensional graph, instead of all three curves being shown in the same X-Y plane, each is displayed on the Z-axis at its respective tensile strength value. In addition, for illustrative purposes, the three lines are projected onto the floor of the graph to show how the two- and three-dimensional graphs are related. Specifically, it can be noted that the graph on the floor of the three-dimensional graph is the same as the two-dimensional graph above, except the scales progress in different directions. In an actual application either the two- or three-dimensional

Isolines are used to indicate all combinations of hardener and cure temperature that yield the same tensile strength.

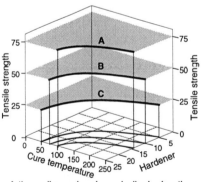

A three-dimensional graph displaying the same data as above. In this graph the three isolines are shown on the Z-axis at the applicable tensile strength value. They are also projected onto the floor of the graph to illustrate how they relate to the two-dimensional graph above.

variation would be shown, not both. Generally, the two-dimensional graph is easier to read. • All three examples shown here use lines of equal value along the Z-axis. Isolines can also be drawn with equal values along the X or Y axes.

Isoline Graph	A graph on which data points of equal value are connected by lines. If equal values along the Z-axis are displayed, it is sometimes called a contour graph. If lines of equal value along both the X and Y axis are used, it is sometimes called a fishnet graph. See Isoline, Surface Graph, and Contour Graph.

Isoline Map	A map on which lines connect points of equal value. Two examples are shown at the right.	

Lines of equal pressure (isobar map)   Lines of equal elevation (contour map)
Examples of isoline maps with lines connecting points of equal value

Isometric Projection (View)	To give the appearance of three dimensions, a graph may be rotated and tilted forward so the viewer can see the top and two sides at the same time. Such views, or projections, as they are frequently called, have the general name of axonometric. When the angles between each pair of axes (i.e., X, Y, and Z) is 120°, the view is called isometric. The same names and principles apply to graphs, maps, or other visual representations. In an isometric view, no major plane (i.e., X-Z, Y-Z, or X-Y) is parallel to the plane of the paper and with rare exceptions, all surfaces are somewhat distorted. Since actual values are difficult to read using axonometric or isometric views, the major function of such graphs is generally to give an overview of the data and show relative values, trends, correlations, patterns, unusual data points, etc. Exact values are typically determined some other way.

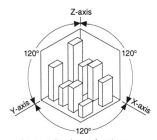

Isometric view of a three-dimensional column graph

Isopleth Map	A map with bands representing areas of equal values or ranges of values such as equal crime rates, equal rain fall, equal temperature, etc. An isopleth map displaying areas of equal temperature, as shown at the right, is sometimes called an isotherm map.

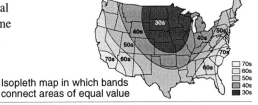

Isopleth map in which bands connect areas of equal value

Isoquantity Line **or** **Isoquant Line**	Sometimes referred to as an isoline. Isoquantity or isoquant lines are lines used on charts and maps to connect points of equal value. They can be used with two- or three-dimensional charts and maps. See Isoline.
Italics	*The term italics refers to a style of type in which all of the characters slant slightly to the right. These two sentences have italic type.*
Japanese Candlestick Chart	Sometimes referred to as candlestick chart. See Candlestick Chart.

Jittering	Jittering addresses the problem of overlapping data symbols on graphs. With this technique, data points on a graph are randomly shifted a slight amount so no symbol obscures another. When there is only one quantitative scale, the symbols are shifted perpendicular to the quantitative scale so accuracy is not reduced. When there are two quantitative scales, the symbols might be shifted with regards to one or both axes. Jittering causes a slight degradation in accuracy on a graph with two quantitative scales. • The points are shifted just enough to assure that all the data symbols are visible, even though there might still be some slight overlapping. The amount of shift may be determined visually or by a mathematical formula. When jittering has been employed, the viewer is normally alerted, though no indication is given as to how much the symbols have been shifted.

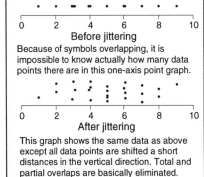

Joined Bar Graph **and** **Joined Column Graph**	When the rectangles on a bar or column graph become so wide there are no spaces between them, the graph is sometimes called a joined bar/column graph, connected bar/column graph, stepped bar/column graph, or histogram. See Bar Graph and Column Graph.

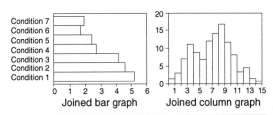

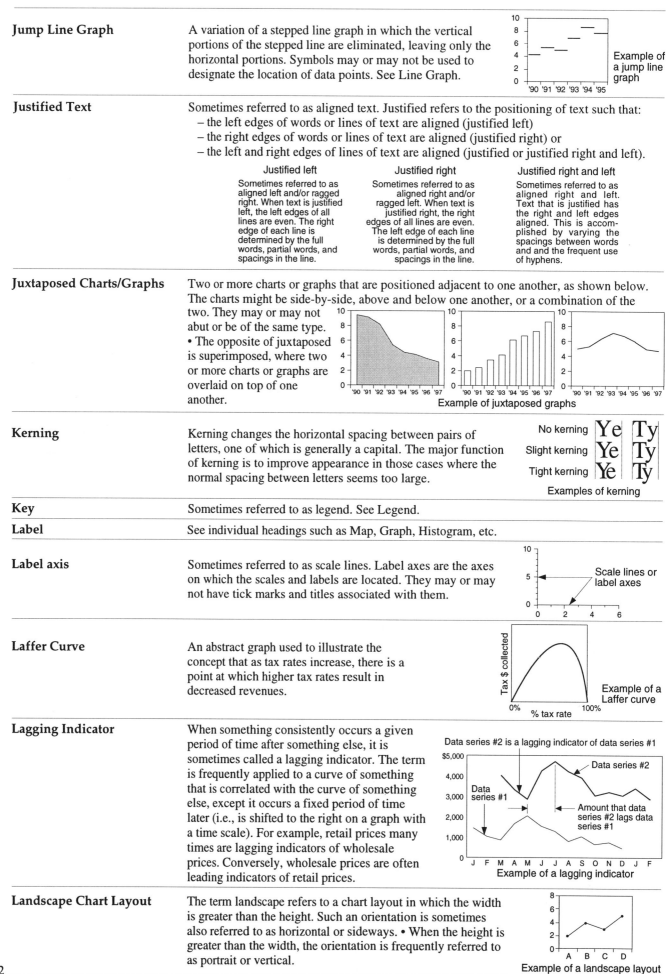

Jump Line Graph

A variation of a stepped line graph in which the vertical portions of the stepped line are eliminated, leaving only the horizontal portions. Symbols may or may not be used to designate the location of data points. See Line Graph.

Example of a jump line graph

Justified Text

Sometimes referred to as aligned text. Justified refers to the positioning of text such that:
– the left edges of words or lines of text are aligned (justified left)
– the right edges of words or lines of text are aligned (justified right) or
– the left and right edges of lines of text are aligned (justified or justified right and left).

Justified left

Sometimes referred to as aligned left and/or ragged right. When text is justified left, the left edges of all lines are even. The right edge of each line is determined by the full words, partial words, and spacings in the line.

Justified right

Sometimes referred to as aligned right and/or ragged left. When text is justified right, the right edges of all lines are even. The left edge of each line is determined by the full words, partial words, and spacings in the line.

Justified right and left

Sometimes referred to as aligned right and left. Text that is justified has the right and left edges aligned. This is accomplished by varying the spacings between words and and the frequent use of hyphens.

Juxtaposed Charts/Graphs

Two or more charts or graphs that are positioned adjacent to one another, as shown below. The charts might be side-by-side, above and below one another, or a combination of the two. They may or may not abut or be of the same type.
• The opposite of juxtaposed is superimposed, where two or more charts or graphs are overlaid on top of one another.

Example of juxtaposed graphs

Kerning

Kerning changes the horizontal spacing between pairs of letters, one of which is generally a capital. The major function of kerning is to improve appearance in those cases where the normal spacing between letters seems too large.

No kerning
Slight kerning
Tight kerning

Examples of kerning

Key

Sometimes referred to as legend. See Legend.

Label

See individual headings such as Map, Graph, Histogram, etc.

Label axis

Sometimes referred to as scale lines. Label axes are the axes on which the scales and labels are located. They may or may not have tick marks and titles associated with them.

Scale lines or label axes

Laffer Curve

An abstract graph used to illustrate the concept that as tax rates increase, there is a point at which higher tax rates result in decreased revenues.

Example of a Laffer curve

Lagging Indicator

When something consistently occurs a given period of time after something else, it is sometimes called a lagging indicator. The term is frequently applied to a curve of something that is correlated with the curve of something else, except it occurs a fixed period of time later (i.e., is shifted to the right on a graph with a time scale). For example, retail prices many times are lagging indicators of wholesale prices. Conversely, wholesale prices are often leading indicators of retail prices.

Data series #2 is a lagging indicator of data series #1

Data series #2

Data series #1

Amount that data series #2 lags data series #1

Example of a lagging indicator

Landscape Chart Layout

The term landscape refers to a chart layout in which the width is greater than the height. Such an orientation is sometimes also referred to as horizontal or sideways. • When the height is greater than the width, the orientation is frequently referred to as portrait or vertical.

Example of a landscape layout

Latitude and Latitudinal Lines	Latitude is a distance north or south of the equator. It is measured in degrees with zero at the equator and 90° at the North and South Poles. Latitudinal lines are the grid lines that connect all points of equal latitude. They are also called parallels. • East-west distances are also measured in degrees with zero at Greenwich, England, and 180° on the opposite side of the earth. Using values of latitude and longitude, places and things can be precisely located anywhere on the globe. See Map. Illustration of latitude and longitude lines
Layer Area Graph	Sometimes called a divided, multiple-strata, stacked area, strata, stratum, subdivided area, or subdivided surface graph. An area graph in which multiple data series are stacked on top of one another. This is the most frequently used area graph format. See Area Graph. Layer area graph
Layer Graph	Sometimes called a slice graph, spectral plot, ridge contour, strata graph, or partition plot. A layer graph is a graph in which the data points of a three-dimensional scatter graph are condensed onto a limited number of planes to aid in the interpretation of the data. See Slice Graph. Layer graph
Layer Line Graph	Sometimes called a stacked line graph. A layer line graph is a graph in which multiple data series are stacked on top of one another. In this type of graph the top curve represents the total of all the components or data series below it. This format of line graph should be used cautiously because of potential confusion with the more familiar grouped line graph format, in which each curve is referenced from the same zero axis. See Line Graph. Layer line graph
Leader or Leader Dots	A dotted line such as a row of periods to assist the viewer in visually relating material on one side of a page or chart with material on the other side. For example, a heading on this side...with a number on this side.
Leading	Sometimes referred to as line space. Leading (pronounced "ledding") is the vertical distance between lines of type. It is measured from the base line of one line of type to the base line of the next line of type. Leading is generally specified in the same point system used to specify type size (i.e., 72 points to the inch). The examples below show three different leadings using the same nine-point type size in each.

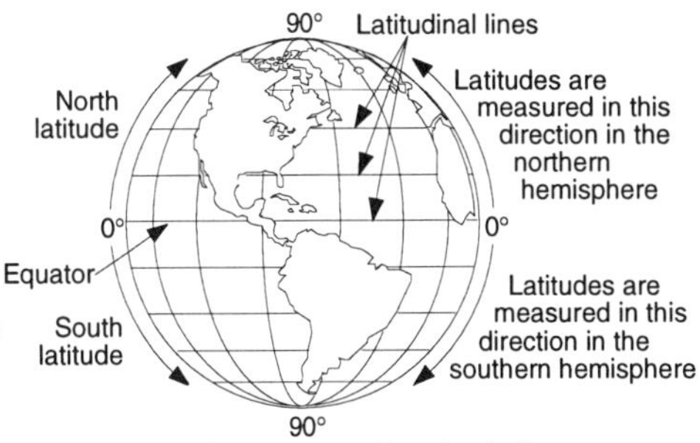

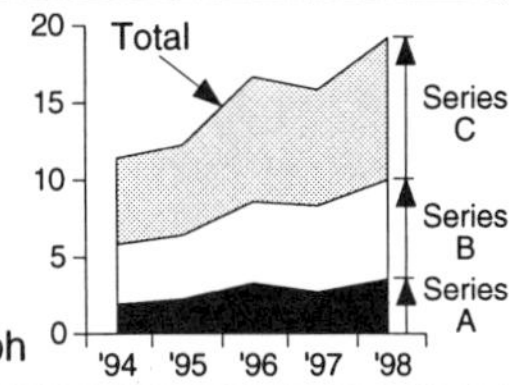

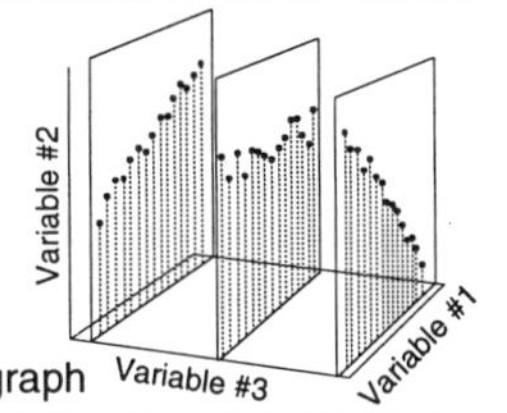

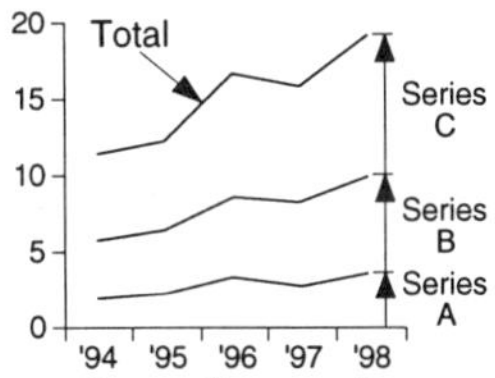

#9 size type and #9 leading. Sometimes designated as 9/9	#9 size type and #11 leading. Sometimes designated as 9/11	#9 size type and #13 leading. Sometimes designated as 9/13
Brief paragraphs sometimes appear below charts and graphs to explain the contents of the chart. When the type is small, the clarity can sometimes be improved by increasing the leading. 9-point leading	Brief paragraphs sometimes appear below charts and graphs to explain the contents of the chart. When the type is small, the clarity can sometimes be improved 11-point leading	Brief paragraphs sometimes appear below charts and graphs to explain the contents of the chart. When the type is small, the clarity can 13-point leading

Leading Indicator	When something consistently occurs a given period of time before something else, it is sometimes called a leading indicator. The term is frequently applied to a curve of something that is correlated with the curve of something else, except it occurs a fixed period of time before (i.e., is shifted to the left on a graph with a time scale). For example, wholesale prices often are leading indicators of retail prices. Conversely, retail prices are often lagging indicators of wholesale prices. Example of a leading indicator
Least Squares Line	Sometimes referred to as linear regression line. See Linear Regression Line.

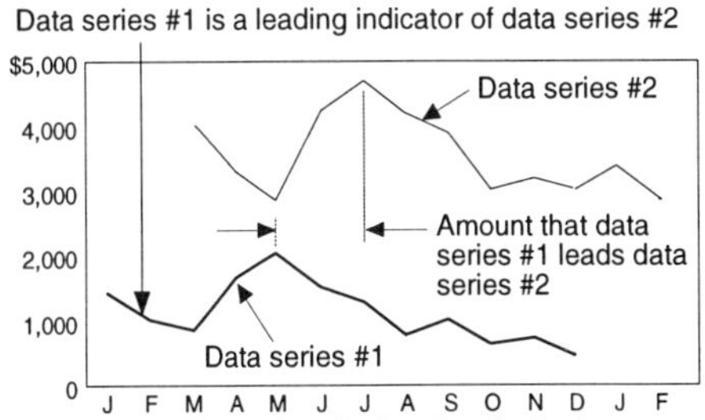

Legend

Sometimes referred to as a key. Legends are often critical to the understanding of a chart because they frequently contain information necessary for decoding the data graphics. Three of the major functions of legends are: 1) to identify what the data graphics represent, 2) to indicate certain characteristics of the thing represented, and 3) when quantitative information is encoded, to enable the viewer to estimate the values that individual data graphics represent. Examples of each usage are shown below.

Examples of legends that identify and/or describe what the data graphic represents

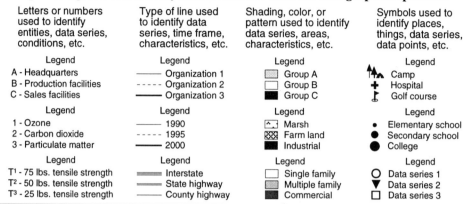

Letters or numbers used to identify entities, data series, conditions, etc.
Legend
A - Headquarters
B - Production facilities
C - Sales facilities
Legend
1 - Ozone
2 - Carbon dioxide
3 - Particulate matter
Legend
T1 - 75 lbs. tensile strength
T2 - 50 lbs. tensile strength
T3 - 25 lbs. tensile strength
Type of line used to identify data series, time frame, characteristics, etc.
Legend
Organization 1
Organization 2
Organization 3
Legend
1990
1995
2000
Legend
Interstate
State highway
County highway
Shading, color, or pattern used to identify data series, areas, characteristics, etc.
Legend
Group A
Group B
Group C
Legend
Marsh
Farm land
Industrial
Legend
Single family
Multiple family
Commercial
Symbols used to identify places, things, data series, data points, etc.
Legend
Camp
Hospital
Golf course
Legend
Elementary school
Secondary school
College
Legend
Data series 1
Data series 2
Data series 3

Examples of legends that enable the viewer to decode quantitative information

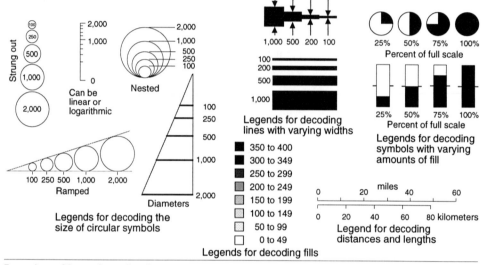

Strung out
100
250
500
1,000
2,000
2,000
1,000
0
Can be linear or logarithmic
Nested
2,000
1,000
500
250
100
100 250 500 1,000 2,000
Ramped
100
250
500
1,000
2,000
Diameters
Legends for decoding the size of circular symbols
1,000 500 200 100
100
200
500
1,000
Legends for decoding lines with varying widths
350 to 400
300 to 349
250 to 299
200 to 249
150 to 199
100 to 149
50 to 99
0 to 49
Legends for decoding fills
25% 50% 75% 100%
Percent of full scale
25% 50% 75% 100%
Percent of full scale
Legends for decoding symbols with varying amounts of fill
miles
0 20 40 60
0 20 40 60 80 kilometers
Legend for decoding distances and lengths

Location of legend on the chart

Legends can be located anywhere on the chart. The closer they are to the information they explain, the more convenient it is for the viewer and the less chance for errors to occur in the decoding process. The variations in location of legend shown in the examples below are applicable to almost any type of chart. When possible, the information in the legend appears in the same order as it appears on the chart.

20
10
0
A B C D E F
Car #3
Car #2
Car #1

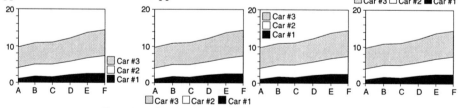

Examples of potential locations for legends on charts

"Less Than" Graph

A "less than" graph is a variation of a cumulative frequency graph. It is constructed so that one can read directly from the graph the percentage of data elements in a data set having values less than a certain value. For example, in the graph at the right, 63% of the data elements in the data set are equal to or less than 2.1.

- The cumulative histogram is included for reference only. See Cumulative Frequency Graph.

Percent of data elements "less than" the corresponding value on the X-axis
100%
80%
60%
40%
20%
0%
1.5 1.6 1.7 1.8 1.9 2.0 2.1 2.2 2.3 2.4 2.5

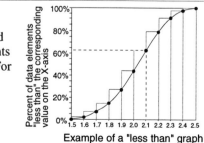

Example of a "less than" graph

Line

Lines are one of the most versatile building blocks of information graphics. Shown below are six major ways lines are modified to achieve their high degree of flexibility. The specific variations shown are meant to be representative and are not all-inclusive. See Line Graph for terminology specifically oriented to curves on graphs.

Weight (thickness or width)

The weight (sometimes referred to as thickness or width) of lines is specified several different ways. Shown here is a comparison of four of the more widely used systems. The line samples are of the approximate width indicated. Technically, there is no upper or lower limit on the size of lines that can be drawn.

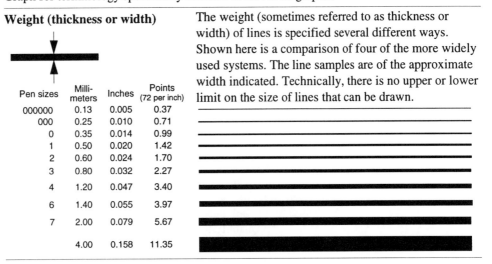

Pen sizes	Milli-meters	Inches	Points (72 per inch)
000000	0.13	0.005	0.37
000	0.25	0.010	0.71
0	0.35	0.014	0.99
1	0.50	0.020	1.42
2	0.60	0.024	1.70
3	0.80	0.032	2.27
4	1.20	0.047	3.40
6	1.40	0.055	3.97
7	2.00	0.079	5.67
	4.00	0.158	11.35

Pattern

Variations in patterns are often used to differentiate one line from another. Patterns are sometimes grouped and given a common name. For example, ball & line and circle & line could be combined into the common heading of symbol & line. They are shown separately here for informational purposes since they are so widely used.

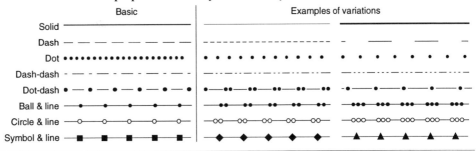

Color, shade, and fill

Different colors and shades can be effective with all line weights. Pattern, fill, and outlines generally require thicker lines.

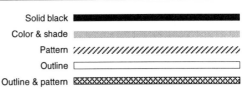

Shape

Unusually shaped lines are most widely used in maps and illustrations. They might be used for such things as indicating walls, streams, varying width paths, road type, etc.

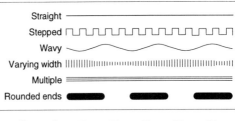

Special applications

Considerable additional information can be encoded into lines by means of special symbols. For example, small cross lines can indicate a railroad. By dashing the line, another level of information is added.

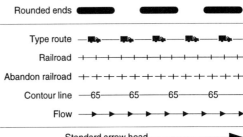

Arrows

The shape of an arrow generally has more to do with appearance than function, although on occasion its shape or color is used to encode additional information.

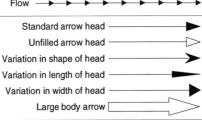

Linear	A relationship between two sets or series of data such that a given change in one set is always accompanied by a uniform amount of change in the other. For example, assume an increase of ten from 20 to 30 in one data set is accompanied by an increase of five in the other set of data. If the two sets have a linear relationship, any increase of ten in the first set, (for instance, from 43 to 53 or 81 to 91) will be accompanied by an increase of five in the second data set. When data sets with linear relationships are plotted on a standard linear graph, the data points cluster around a straight line.

Linear Regression Line

Sometimes referred to least squares line. A linear regression line is a variation of a fitted curve. It is a straight line superimposed over the data points of a data series and is considered the straight line that best approximates the data series. When sequential information is plotted on the horizontal axis, a linear regression line of the data is sometimes referred to as a trend line. The regression line can be drawn by visually estimating its location and slope; however, its location and slope are generally determined using a mathematical method known as least squares. In determining the location of the line, no attempt is made to have the line pass through all of the data points and, in fact, many times the curve passes through few or none of the points. Linear regression lines are used for several purposes including:

- Quantify the slope of a data series or the relationship between two sets of data;
- Compare multiple data series and make forecasts and projections;
- Determine how closely the data series follows a straight line; and
- Analyze the data for unusual patterns such as seasonal and cyclical fluctuations.

Linear regression lines used to compare multiple data series

Regression lines can be fitted to multiple data series on the same graph to compare the slopes of the two series. This is particularly advantageous when data series are interspersed, as in the example at the right.

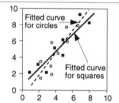

Linear regression lines used to make projections

When data is closely approximated by a linear regression line, the line can sometimes be extended to make projections beyond the actual data available. This can be done with point, line, area, or column graphs.

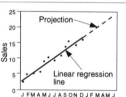

Residuals

The distances between the linear regression line and the data points are called residuals. The mathematical technique for establishing the linear regression line makes the sum of the squares of these distances the smallest value possible; thus the name least squares line. The residuals are sometimes plotted by themselves to look for patterns within the residuals. See Residuals.

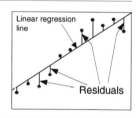

Linear Scale

A linear scale is sometimes referred to as an arithmetic scale. A linear scale is arranged such that the distance and value between any two major tick marks (major interval) is always the same on a given scale. For example, the distance between any two major tick marks might be one inch and represent ten units of measure. The distance between any two minor tick marks on the same scale might be one-tenth of an inch and represent one unit of measure. On a linear scale, these two criteria would apply at any point along the entire length of the scale.

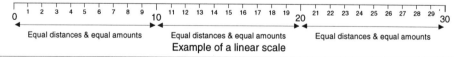

Example of a linear scale

Lined Surface Graph

Lined surface graphs are three-dimensional graphs on which the data points form a surface or surfaces. These surfaces have lines drawn on them to assist the viewer in visualizing the shape, and to some extent the dimensions, of the surfaces. See Surface Graph [1].

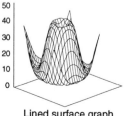

Lined surface graph

Line Graph

Sometimes referred to as a curve graph. Line graphs are a family of graphs that display quantitative information by means of lines. They are extremely versatile and therefore are used extensively.

Major types of lines used with line graphs

Shown at the right are examples of the three major types of lines used to connect data points to form line graphs. The segmented variation is sometimes referred to as a broken, fever, thermometer, or zigzag graph. In actual graphs the plotting symbols may or may not be shown.

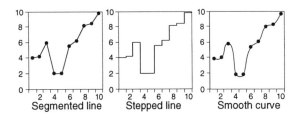

Segmented line   Stepped line   Smooth curve

Simple line graph

A simple line graph displays a single data series. It typically has a quantitative scale on the vertical axis and a category, quantitative, or sequence scale on the horizontal axis. Examples of all three are shown below. With the category scale, a data point is located directly above every label (category) on the horizontal scale. All categories must be shown on the scale in order for the viewer to properly interpret the graph. With sequence and

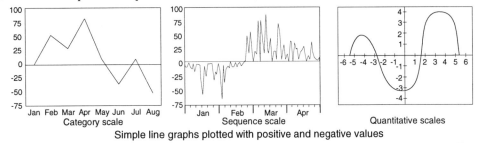

Simple line graphs with one data series per graph

quantitative scales, data points are frequently not located over labels and in many cases labels are not displayed for every value. Linear or nonlinear scales can be used on quantitative scales. When all scales are quantitative, the two-axis graph is sometimes called a simple XY line graph, and the three-axis a simple XYZ line graph. Positive and negative values can be plotted on quantitative scales on either axis (below).

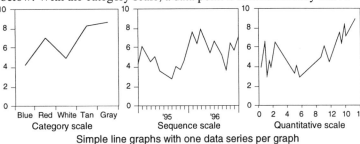

Simple line graphs plotted with positive and negative values

Simple scatter line graph

On most line graphs, the data points are connected sequentially from left to right. When quantitative scales are used on both axes, similar to a scatter graph, the data points are occasionally connected some other way such as the order in which the data points were collected or plotted. When this is done the graph is sometimes referred to as a simple scatter line graph. As shown below, the shape of the curve can be significantly different, depending on which method is used. Only a person familiar with the data can decide which is the proper sequence for a given data series. The same principle applies to three-axis graphs.

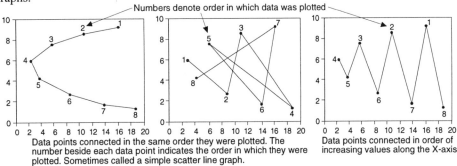

Numbers denote order in which data was plotted

Data points connected in the same order they were plotted. The number beside each data point indicates the order in which they were plotted. Sometimes called a simple scatter line graph.

Data points connected in order of increasing values along the X-axis

Line graphs on which the same data points are connected three different ways

Grouped line graph - multiple data series all referenced from the zero axis

Sometimes referred to as a dual line, multiple line, overlapped line, or compound line graph. This is the format generally used when multiple data series are shown on the same line graph. Grouped line graphs generally have a quantitative scale on the vertical axis and either a category, quantitative, or sequence scale on the horizontal axis. All values along the vertical axis are referenced/measured from a common zero base line axis which typically is the horizontal axis. An example is shown at the right. When a category scale is used, all data series use the same categories. When the horizontal axis has a quantitative or sequential scale, there are two potential variations.

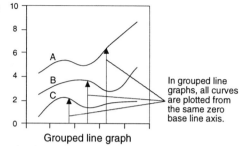

In grouped line graphs, all curves are plotted from the same zero base line axis.

Grouped line graph

– All data series use the same values along the horizontal axis, but each data point has a unique value along the vertical axis.

– All data points have a unique pair of values along both the horizontal and vertical axes.

These variations are of most interest/concern when the graphs are being generated on a computer since some programs require the two types of data be entered in different ways. Examples of graphs plotted with these two types of data are shown below.

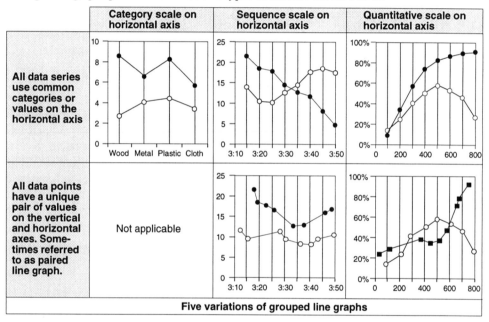

Five variations of grouped line graphs

Technically there is no limitation on the number of data series that can be shown on a grouped line graph. Practically, when there are more than four or five data series, the graph becomes confusing. • There are many ways to differentiate lines for identification and encoding purposes. See Line and Fill elsewhere in this book for examples of potential variations. • Segmented and smooth curve lines are most frequently used for graphs with multiple data series. Stepped lines are sometimes used but can be confusing if the lines intersect. • With all quantitative scales, the two-axis graph is sometimes called a grouped XY line graph, and the three-axis variation a grouped XYZ line graph. • Positive and negative values can be plotted on axes with quantitative scales.

Grouped scatter line graph

On most line graphs, the data points are connected sequentially from left to right. When quantitative scales are used on both axes, similar to a scatter graph, the data points are occasionally connected some other way, such as the order in which the data points were collected or plotted. When this is done on a graph with multiple data series, it is sometimes referred to as a grouped scatter line graph. The shapes of the curves can be significantly different depending on which method is used. (See previous page for an example). Only a person familiar with the data can decide which is the proper sequence for a given set of data. Whichever order of connecting the data points is decided on, it is used with all data series on a given graph. The same options for connecting data points applies to three-axis line graphs.

Line Graph (continued)

Stacked line graph - multiple data series positioned on top of one another

Sometimes referred to as a layer line graph. In a stacked line graph, one data series is plotted with all data points measured from the zero axis. The data points for the second data series are determined by adding the values of the second series to the values of the first series. For the third data series the data points are determined by adding the values for the third data series to the values for the first two data series, etc. Thus, the top curve on a stacked line graph represents the sum of all of the data series plotted, and the values for any particular data series is represented by the difference between two curves or a curve and an axis. For example, if a company sells three products and the sales for each of the three product lines are plotted, the difference between each pair of lines represents the sales for an individual product, and the top curve represents the total sales of all three products. (right). The stacked variation of

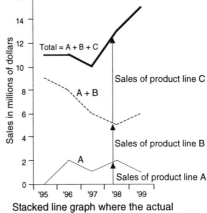

Stacked line graph where the actual values for each data series are added to the data series below it.

line graph is used with caution because of potential confusion with the more widely used grouped line graph. Placing titles of the curves directly on the data graphics can help diminish the potential for misunderstanding. Conversely, filling the areas between the lines increases the chance for misunderstanding. When stacking data series, a stacked area graph is most frequently used.

• Segmented, smooth, or stepped curves work equally well in most applications. An example of a stepped stacked line graph is shown at the left.

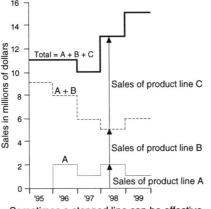

Sometimes a stepped line can be effective when used as a stacked line graph

100% stacked line graph

The example at right works on the same principle as the other two examples, except percent-of-the-whole values are used instead of actual values. In other words, the lower curve indicates the percent of the total company sales that product A represented each year. The middle curve is arrived at by adding the percents that product B represented to the percents for product A. The percents for product C are added to the middle curve in the same fashion, arriving at the top line, which is 100% of the company's sales for each year.

This variation is referred to as a 100% stacked line graph. Stacked and 100% stacked graphs are generally used with categorical and sequential data. All positive or all negative values can be plotted; however, combinations of positive and negative values create very confusing graphs and are therefore generally avoided. Features such as multiple scales, fitted curves, and error bars are generally not used with stacked line graphs.

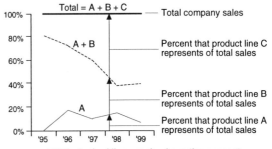

100% stacked line graph where the percent that each data seres represents of the whole are added to the data series below it.

Three-dimensional stacked line graph

Three-dimensional stacked line (sometimes called ribbon) graphs are sometimes used for presentation purposes. An example is shown at the left.

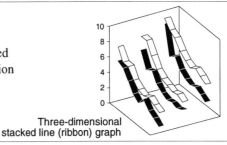

Three-dimensional stacked line (ribbon) graph

Range and high-low line graphs

There are several variations of line graphs that designate ranges of values. The ranges designated might be maximums and minimums, plus and minus three standard deviations, upper and lower confidence limits, upper and lower standard errors, etc. In addition to the upper and lower values, a series of midvalues may also be shown such as average or median. When midvalues are included, the graph is frequently referred to as a high-low graph. When the data is continuous, the common data points such as all of the uppers, all of the lowers, and/or all of the midvalues are sometimes

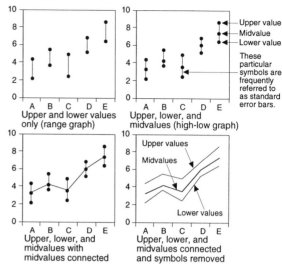

connected by lines running from left to right. When this is done, the vertical lines or symbols designating the ranges are sometimes eliminated. Examples of four widely used variations are shown above. Range and high-low symbols can be added to any of the three types of curves, as shown below. See Range Symbols and Graphs.

Shown at the right are examples of range symbols added to each of the three different types of line curves.

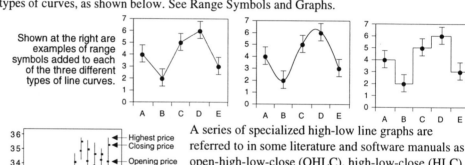

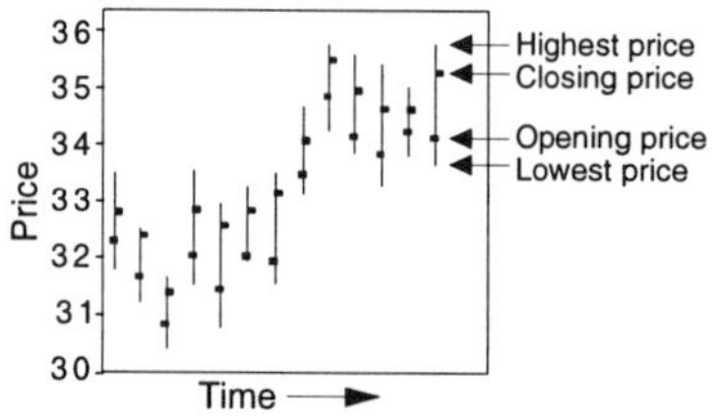

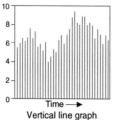

Open-high-low-close graph,. frequently called bar graph

A series of specialized high-low line graphs are referred to in some literature and software manuals as open-high-low-close (OHLC), high-low-close (HLC), etc. In investment literature and software, the same graphs are referred to as bar, price, or vertical line charts. These graphs are generally used to record the price of stocks, securities, commodities, etc., over time. An example is shown at the left. See Bar Chart [2].

Vertical line graph

Sometimes referred to as a needle or spike graph. Vertical line graphs are graphs in which a vertical line extends from each data point down to the horizontal axis or plane. They are sometimes thought of as column graphs where the columns have the widths of lines, or as point graphs with drop lines in which the plotting symbols are eliminated. Vertical line graphs are generally used when many successive data points are to be plotted or where it is desirable to enable the viewer to a identify the exact value or time associated with each data point. When the lines are close together, they sometimes resemble an area graph in which the area under a line is filled with a pattern. • Sometimes a conventional line and the vertical lines are used together for emphasis. Selected vertical lines can be darkened, dashed, or made a different color to designate intervals such as weeks, months, etc. The lines may or may not have symbols at the top to designate the exact data points. Vertical lines can be used on two- or three-dimensional graphs.

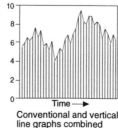

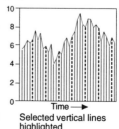

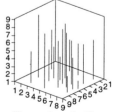

Vertical line graphs with and without the conventional line curve

Line Graph (continued)

Difference line and deviation line graphs

• The term difference graph is occasionally used when referring to a graph with intersecting data curves in which the areas or differences between the curves are filled. For example, one curve might represent the dollar value of exports each year, and the other the dollar value of imports. The areas between the two would represent the differences, as shown at the right. This type of graph is also sometimes referred to as a curve difference, intersecting silhouette, or intersecting band graph.

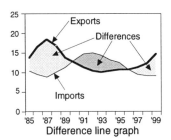
Difference line graph

• The term deviation graph is used to describe a graph on which the values of the differences between a data series and a known reference are plotted. For example, if profit is being compared against budget, only the differences might be plotted, as shown at right. If actual profits are over budget, the values plotted are positive. If actual profits are under budget, the values plotted are negative. The reference or data series against which deviations are calculated might be constant (e.g., the same value each month) or variable (e.g., a different value each month).

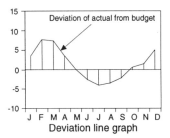
Deviation line graph

• A cumulative deviation graph plots the cumulative sum of all of the prior deviations plus the current deviation. For example, a regular deviation graph might show how profit compared to budget each month. A cumulative deviation graph shows the difference between actual profit and budget on a year-to-date basis. The graph at the right is a cumulative variation of the deviation graph immediately above it.

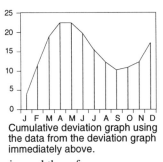
Cumulative deviation graph using the data from the deviation graph immediately above.

• The areas between the two data series or between the data series and the reference are sometimes filled or have drop lines added for emphasis. When fill is used the resulting graph is sometimes called an area or filled line graph.

Isoline graphs

Isolines are lines of equal value. Put another way, every point on an isoline has some value in common. One of the most familiar examples is a contour graph or map on which lines represent points of equal value or elevation. For instance, a line labeled 400 designates all points on the graph that have a value of 400. This applies whether the graph is two- or three-dimensional. Isolines are also sometimes shown on graphs to indicate all possible combinations that yield the same result – for example, all the combinations of hardener and cure temperature that yield the same tensile strength, or all the combinations of sales of two differently priced products that result in the same total sales dollars. Three examples are shown below.

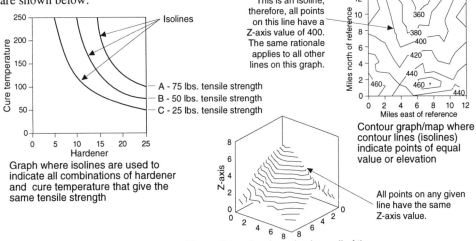

Graph where isolines are used to indicate all combinations of hardener and cure temperature that give the same tensile strength

This is an isoline; therefore, all points on this line have a Z-axis value of 400. The same rationale applies to all other lines on this graph.

Contour graph/map where contour lines (isolines) indicate points of equal value or elevation

All points on any given line have the same Z-axis value.

Three-dimensional graph where all of the points on a given line (isoline or contour) have the same value on the vertical (Z) axis.

Stepped line graph

Stepped lines are sometimes used to differentiate data series more clearly, to highlight differences and comparisons rather than trends, and to emphasize the nature of the data such as abrupt changes and then constant values. These graphs are especially suitable for things such as prices, interest rates, plant capacity, etc., that change abruptly and then remain level for a period of time. The example at the right gives the viewer a clear understanding of when the percents changed, by how much, and how long each percent was in effect

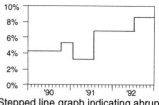

Stepped line graph indicating abrupt changes in something such as interest rates and then a constant rate until a subsequent change

before the next change occurred. When a category scale is used on the X-axis, the horizontal portions of a stepped line are as wide as the space allotted to each category. When a plotting symbol is used it is frequently located in the center of the horizontal portion. When a sequential or quantitative scale is used, the horizontal lines extend from point to point regardless of how near or far apart the points are. Plotting symbols are sometimes shown each time the direction of the line changes, as on a segmented line graph. Examples with a quantitative scale and a category scale are shown at the left. • Sometimes, in order to emphasize certain characteristics of the data, plotting symbols will be shown only at the left or right ends of the horizontal portions of the lines (below). If only the symbols on the left

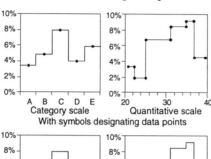

With symbols designating data points

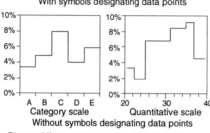

Without symbols designating data points

Stepped line curves on graphs with category and quantitative scales on the horizontal axis.

ends are displayed, the graph is sometimes referred to as left stepped graph. If the symbols are shown on the right ends it is called a right stepped graph, and in the center, a center stepped graph. • In some category graphs, symbols are located on the right or left ends of the horizontal line

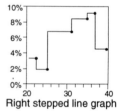

Left stepped line graph Right stepped line graph

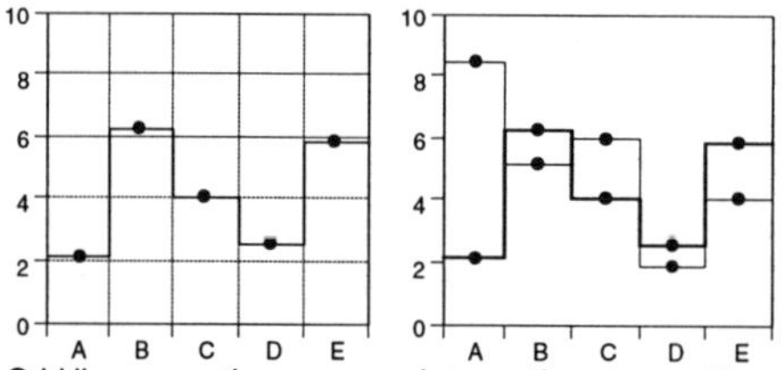

segments, even though data points typically do not exist there. • Stepped line graphs have limitations that segmented and smooth curve line graphs do not. For example, when grid lines are used or when multiple stepped curves intersect one another, the legibility of the data can sometimes be seriously degraded, as shown at the left.

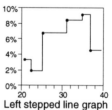

Grid lines sometimes interfere with stepped lines.   Intersecting stepped lines can cause confusion.

Some combinations of stepped line graphs and grid lines can cause confusion.

Jump line graph

A variation of the stepped line graph called a jump curve eliminates the vertical portions of a stepped line, leaving only the horizontal portions. This type of curve sometimes more correctly reflects what actually occurs. For example, when a price is changed, it does not pass through all the values between the old and new price, but instead, simply jumps from one level to another. Four variations of jump graphs are shown below.

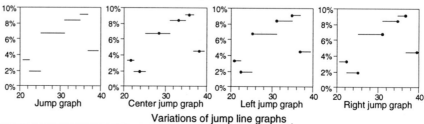

Jump graph   Center jump graph   Left jump graph   Right jump graph

Variations of jump line graphs

Line Graph (continued)

One-axis line graph

What is commonly referred to as a two-dimensional graph typically has data plotted on the vertical and horizontal axes. In some situations the data for both axes can be combined onto a single axis, either the vertical or horizontal, to create what is sometimes called a one-axis graph. Shown below are examples of the same data plotted on horizontal and vertical one-axis line graphs.

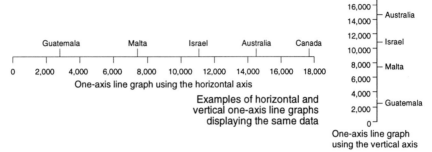

One-axis line graph using the horizontal axis

Examples of horizontal and vertical one-axis line graphs displaying the same data

One-axis line graph using the vertical axis

For comparison, two conventional two-axis graphs are shown below displaying the same data used in the one-axis graphs above. Many times, one of the scales on a one-axis graph is of the category type; however, this is not a prerequisite.

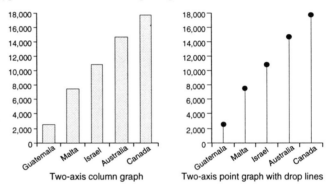

Two-axis column graph   Two-axis point graph with drop lines

Two-axis graphs for comparison. Both display the same data as the one-axis graphs above.

Other line graphs

The examples shown here illustrate some of the less frequently used variations of line graphs. See individual headings for more information.

Radial line graph

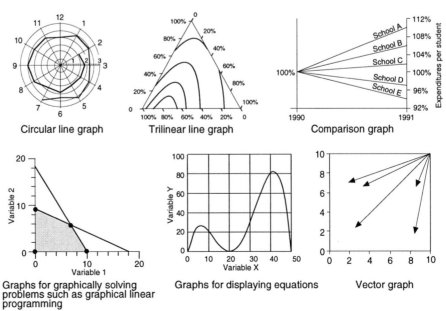

Circular line graph   Trilinear line graph   Comparison graph

Graphs for graphically solving problems such as graphical linear programming

Graphs for displaying equations

Vector graph

Line Graph (continued)

Three-axis (frequently called three-dimensional) line graph

When a third axis is added to a two-axis grouped line graph, the data curves are uniformly distributed along the third axis as shown below. Three-axis variations of two-axis line graphs typically yield no additional information over the two-axis variation. The data graphic may be a conventional line or, as is sometimes done for presentation purposes, the line may be given width and depth and called a ribbon or ribbon graph. The ribbon may simulate a segmented or a smooth curve. • There are several

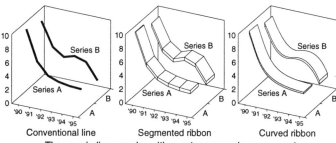

Conventional line   Segmented ribbon   Curved ribbon
Three-axis line graphs with a category scale on one axis

additional types of three-axis graphs, most of which provide insights that would not be possible, or at best would be difficult to observe using only two-axis graphs. On three-axis graphs there is always a quantitative scale on the vertical axis. Quantitative, sequential, or category scales can generally be used on the other two axes depending on the nature of the data and the type of graph. Some variations accept only quantitative scales on all three axes. Several examples of three-axis, three-dimensional graphs are shown below.

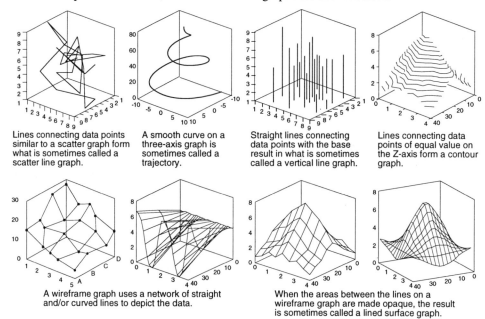

Lines connecting data points similar to a scatter graph form what is sometimes called a scatter line graph.

A smooth curve on a three-axis graph is sometimes called a trajectory.

Straight lines connecting data points with the base result in what is sometimes called a vertical line graph.

Lines connecting data points of equal value on the Z-axis form a contour graph.

A wireframe graph uses a network of straight and/or curved lines to depict the data.

When the areas between the lines on a wireframe graph are made opaque, the result is sometimes called a lined surface graph.

Variations of three-axis (frequently called three-dimensional) graphs

Line graph matrix

A line graph matrix consists of line graphs arranged in rows and columns in some organized fashion. The matrix simplifies the analysis of large quantities of data by enabling the viewer to study multiple graphs at one time. For instance, the example at the right shows the trends of six different measures in four divisions on a single page. See Matrix Display.

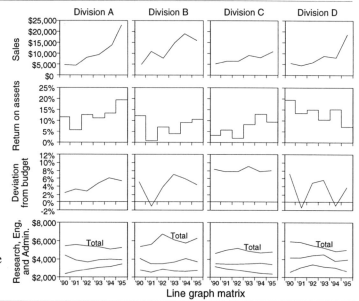

Line graph matrix

214

Line Graph (continued)

Line graph icons

Miniature (1" square or less) line graphs, frequently without titles, labels, tick marks, or grid lines, are sometimes called icons or symbols. Icons do not convey specific quantitative information. Instead, they show such things as relative sizes, values, ratings, etc., of specific things; overall comparisons of multiple entities; trends; or unusual patterns of information. Icons might use segmented, stepped, or smooth curves as the data graphic. Three examples of how line graph icons are used are shown below. See Icon.

——— As symbols on graphs ———

Icons can be used on graphs to give detailed information about specific data points. For instance, in the example at the right, the dashed line connects data points that represent five-year average values. The icons show trends in the data from which the averages were calculated. A typical observation from a graph such as this would

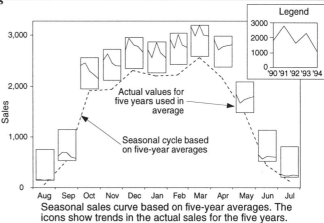

Seasonal sales curve based on five-year averages. The icons show trends in the actual sales for the five years.

be the fact that even though October has a higher average value than May, for the last four years sales in October have gone down while in May they have gone up.

——— To compare multiple entities with regards to three or more characteristics ———

To simultaneously compare a sizable number of entities with regards to three or more variables (sometimes called multivariate), a display of line icons is sometimes used. The example at the right compares ten desktop computers with regards to eight different characteristics. A template, as shown in the legend, is used when generating the icons to assure that the same characteristic is always in the same location on every icon and that the same value scale is used. Since values and units of measure can vary significantly and some characteristics might be qualitative, a common scale of 0 to 1.0 is generally used for all characteristics. The highest value or rating in each characteristic is assigned the value of 1.0. The lowest value or rating is assigned some value less than 1.0, often zero. Other than legibility, there is no limit to the number of variables that can be included in each icon. There also is no limit as to how many icons can be included in the same display. Icons used in this way are sometimes referred to as icon comparison displays. See Icon Comparison Display.

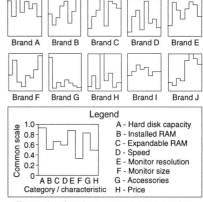

Example of an icon comparison display

——— As symbols on maps ———

Line graph icons are used on maps to convey two major types of information. One usage is to show how things change over time. In this kind of application, a time scale is used on the horizontal axis of the template/legend. The other major usage is for comparing multiple variables or characteristics between areas on the map. For example, the icons may show the values of exports for each state to ten different countries, the production of six different crops in each state, population by ethnic background, etc. When used in this way the template/legend has a category scale on the horizontal axis. A legend is always required when icons are used.

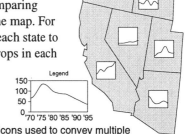

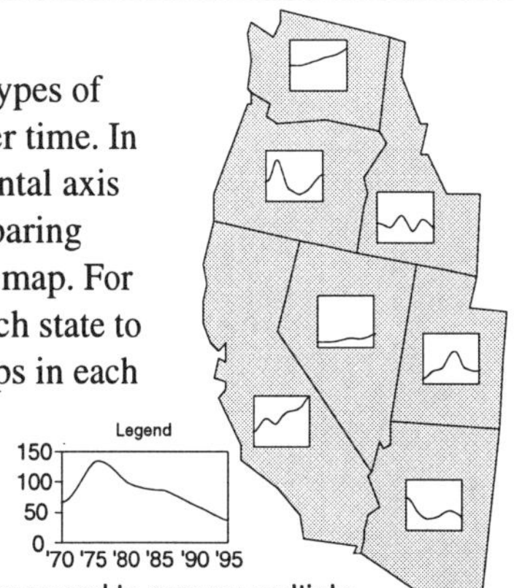

Line icons used to convey multiple bits of quantitative information by area.

Curve fitting with line graphs

Sometimes the irregularities and fluctuations of curves on line graphs make it difficult to accurately discern the nature of an overall trend, make meaningful projections, or compare multiple data series. Straight or gently curving lines that best approximate the data series are sometimes fitted to data points to aid in the interpretation of the data. The curves might be fitted visually or by well-established mathematical techniques. See Curve Fitting.

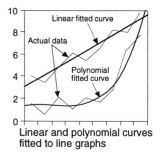

Linear and polynomial curves fitted to line graphs

Other names of line graphs when areas are filled

When the areas under or between curves on a line graph are filled, the graph sometimes takes on an additional name or assumes a new name. There are no guidelines as to which names should be used. To a large extent the choice of name is dependent on individual circumstances and customs. Shown below are four examples, along with alternate names they are sometimes called.

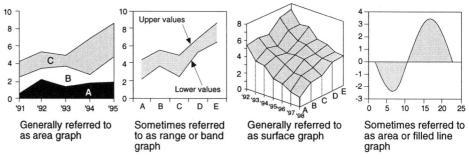

Generally referred to as area graph

Sometimes referred to as range or band graph

Generally referred to as surface graph

Sometimes referred to as area or filled line graph

Connecting dissimilar information with lines

Most people agree that data points representing continuous data can be connected with lines to form a line graph. An example of continuous data is the temperature in an office. If the office temperature was measured every hour and the data points plotted, they would typically be connected by lines, since it is known that the temperature at one hour gradually and continuously transitions to the temperature the next hour.
• There is disagreement as to whether discrete data should be connected by lines to form a line graph, as in, for example, data points for the populations of Paris, Rome, London, and Moscow. Some feel that since the four data points represent distinctly different things and do not represent data elements that transition smoothly from one to another, they should not be represented by a line graph, but instead, by a point, column, or bar graph. Others feel it is OK to connect the data points to form a line graph, reasoning that it helps the viewer interpret the information and the viewer will know that the line is simply a visual aid. Several variations of the population example are shown at the right. The column graph is included for comparison purposes. When the points are connected, stepped line graphs are sometimes used to emphasize the fact that the graph is showing values for comparison purposes and not to establish a trend. The horizontal portion of the line can be darkened to emphasize the values instead of the transitions. The variation selected is an individual decision.

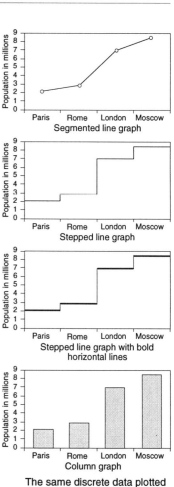

Segmented line graph

Stepped line graph

Stepped line graph with bold horizontal lines

Column graph

The same discrete data plotted on various types of graphs

Line Graph (continued)

Scales on line graphs

Line graphs are one of the most versatile types of graphs with regards to scales. The following are some of the key features regarding scales on line graphs:

 – Typically have a quantitative scale on the vertical axis
 – Can have a category, sequence, or quantitative scale on the horizontal axis
 – Quantitative scales can be linear or nonlinear *
 – Both positive and negative values can be plotted on the quantitative scales*
 – Multiple scales are often used on the vertical axis*

* Except on stacked line types.

As a general rule it is recommended that quantitative scales include zero and be continuous. In situations where the actual values are large and the differences between values are small, these restrictions sometimes make it difficult or impossible to discern differences between values. In these cases the scale may be enlarged to make the differences stand out. Several methods for accomplishing this are shown below.

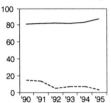

For reference, the scale starts at zero on this example. Actual values and changes from year to year are difficult to estimate.

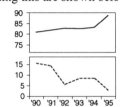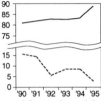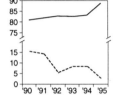

These three examples expand the scale and then take a section out of the middle (sometimes referred to as a scale break) to retain the original size of the graph. The center example tends to make the scale break most obvious. The example on the right makes the scale break least obvious.

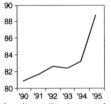

An alternative to a scale break when there is only one data series or multiple data series are clustered at the top, is to eliminate the lower portion of the scale. This variation presents the danger of the viewer overlooking the fact that the lower portion of the scale is missing.

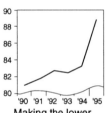

Making the lower portion of the graph uneven helps to call the viewer's attention to the fact that the lower portion of the scale is missing.

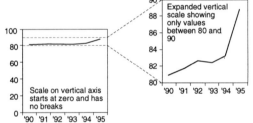

Scale on vertical axis starts at zero and has no breaks

Expanded vertical scale showing only values between 80 and 90

An alternative is to use two graphs. One with a scale starting at zero with no breaks which lets the viewer put the data in perspective (above left). The other, an enlargement of the area with the data (called an expanded scale), which lets the viewer better determine exact values (above right).

• Scales are generally located on the left side and bottom. It is common to have additional scales on the right side and occasionally on the top. Three of the major reasons for additional quantitative scales on the right side are:

 – Identical scales on both sides improves the ease and accuracy of reading the graph, particularly when it is wide, when there are multiple data series plotted, and/or when grid lines are not used
 – The right-hand scale can specify the same thing as the left scale, except in different units such as kilometers and miles.
 – When multiple data series use different units of measure or significantly different values, a second scale is often required.

Examples of these three types of multiple quantitative scales are shown below.

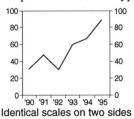

Identical scales on two sides

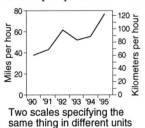

Two scales specifying the same thing in different units

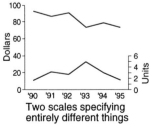

Two scales specifying entirely different things

When a second scale is added to the top it is generally a repeat of the scale on the bottom. Occasionally, different units of measure or time intervals might be used on the top.

Vertical versus horizontal orientation

The curves on line graphs almost always progress from left to right. Although it is technically possible to have them run up and down, it is only occasionally done, primarily because line graphs are so firmly associated in people's minds with time series which typically progress from left to right. The examples at the right display the same data both ways. Most people feel more comfortable with the example on the right.

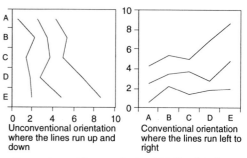

Unconventional orientation where the lines run up and down

Conventional orientation where the lines run left to right

Comparison of line graphs in which the data lines have horizontal and vertical orientations

Grid lines on line graphs

The examples below illustrate the variations of grid lines commonly used on line graphs. See Grid Lines.

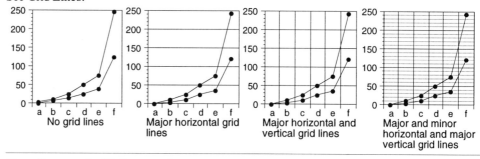

No grid lines

Major horizontal grid lines

Major horizontal and vertical grid lines

Major and minor horizontal and major vertical grid lines

Drop lines used with line graphs

Faint lines called drop lines are sometimes used with both two- and three-dimensional line graphs to help the viewer's eye more easily relate a data point to a label or value on the axis. Vertical drop lines, horizontal drop lines, or a combination of the two might be used on a given graph. Vertical drop lines are most widely used. See Drop Line.

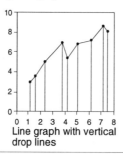

Line graph with vertical drop lines

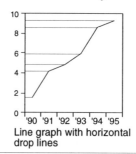

Line graph with horizontal drop lines

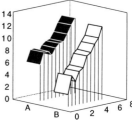

Three-dimensional line graph (ribbon) with vertical drop lines

Missing or irregular data

Line graphs are many times used to plot sequential data, particularly time series. It is generally recommended that there be no breaks in a time scale, since this can distort the pattern or trend of the data. Even though data might be irregular and in some cases missing, the time scale is generally continuous and uniform. The examples shown here illustrate several ways to call missing data to the viewers attention.

The presence of data symbols indicates where actual data was used.

Dashed lines sometimes bridge gaps where data is missing.

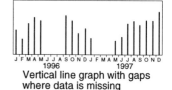

Vertical line graph with gaps where data is missing

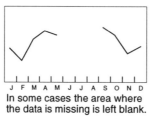

In some cases the area where the data is missing is left blank.

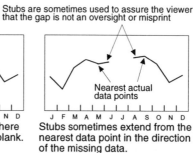

Stubs are sometimes used to assure the viewer that the gap is not an oversight or misprint

Nearest actual data points

Stubs sometimes extend from the nearest data point in the direction of the missing data.

Line Graph (continued)

Curve variations

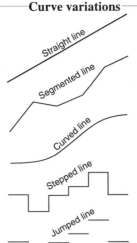

- Straight lines on a linear grid generally indicate linear relationships between data sets.

- A segmented line is made up of a series of straight lines connecting adjacent data points of a data series. Segmented lines may represent linear or nonlinear data.

- A curved line is a continuous line with gentle bends and no sharp corners. Certain curved lines are referred to as curilinear and represent nonlinear data.

- Stepped lines result from connecting data points with a series of horizontal and vertical lines.

- Jumped lines are stepped lines without the vertical portions.

Slope

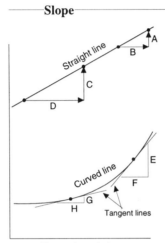

Examples of slopes

- The slope of a straight line is the same at any point along the line and is equal to the vertical distance between any two points on the line divided by the horizontal distance between the same two points. In the example at the left, the slope of the straight line is A divided by B, or C divided by D. Both yield the same value, fraction, proportion, and slope.

- The slope of a curved line varies from point to point. At any given point, the slope of the curve is equal to the slope of a straight line drawn tangent to the curve at that point. A tangent line is a line that just touches the curve and is drawn perpendicular to the radius of the curve at that point. In this example the two slopes illustrated are equal to E divided by F and G divided by H.

- The larger the ratio (fraction) between the vertical distance and the horizontal distance, the steeper the slope.

- Slopes can be positive or negative, as shown below.

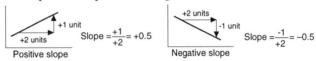

Intercept

A y-intercept is the value at which a curve and/or the X-axis touches or crosses the Y-axis.

An x-intercept is the value at which a curve and/or the Y-axis touches or crosses the X-axis.

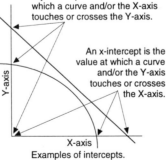

Examples of intercepts.

Extrapolation

Extrapolation is the process of projecting or estimating values based on known information. In this example, the values between B and C are extrapolated from the known values between A and B.

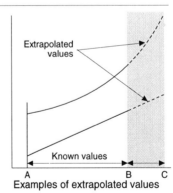

Examples of extrapolated values

Interpolation

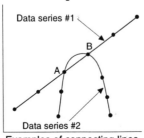

Examples of connecting lines based on interpolation

Interpolation is the process of estimating values between known values. When data points representing continuous data are connected with lines, an interpolation process takes place and the shapes of the connecting lines are the result of that process. For instance, in the example at the left, points A and B are common to data series #1 and #2. The lines connecting points A and B are different for the two series. This is because it was estimated (interpolated) that in series #1 the points between A and B were linear, while in series #2 it was estimated (interpolated) that the points were nonlinear.

Line Map	A map generated by hand or computer using lines, symbols, shading, etc., as opposed to a map generated by means of photographs.

Line-of-Best-Fit	Sometimes referred to as curve fitting or smoothing. Line-of-best-fit is a process in which a curve that most closely approximates a data series is superimposed over a plot of the data points of that data series. There are many different types of curves that can be fitted to a given set of data points. Two examples are shown at the right. See Curve Fitting.	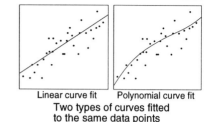

Linear curve fit   Polynomial curve fit

Two types of curves fitted
to the same data points

Line Space	Frequently referred to as leading. Line space is the vertical distance between lines of type. Line space is measured from the base line of one line of type to the base line of the next line of type. See Leading.

Link[1]	The short lines connecting the boundaries between data series on stacked bar or column graphs.	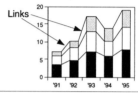

Link[2]	The lines connecting the boxes or symbols on diagrams such as flow charts and PERT charts.	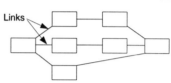

Linkage Tree	Sometimes referred to as a tree diagram, cluster map, or dendrogram. A linkage tree is a graphical means of organizing information for the purpose of establishing groupings and/or categorizing individual elements. For example, one might study different groups of consumers to determine how they perceive a product, or a linkage tree might be used to look for relationships in a group of archeological specimens. See Dendrogram.	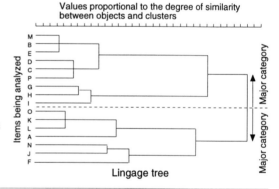

Lingage tree

Linked Stacked Graph

In an attempt to make relationships easier to see on stacked bar and column graphs, lines are sometimes drawn connecting the boundaries between the data series. When this is done the graph is referred to as a linked or connected stacked graph. Connecting lines can be used with simple or 100% stacked graphs. The area between the links may or may not be filled. See Bar Graph and Column Graph.

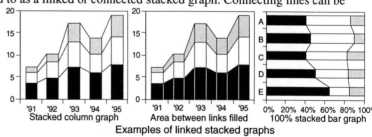

Stacked column graph   Area between links filled   100% stacked bar graph

Examples of linked stacked graphs

Loading Chart

Loading charts are used to plan and schedule resources, frequently with the objective of optimizing their utilization. Key objectives are to assure resources are not overloaded, while at the same time assuring they are idle a minimum amount of time. Time is generally shown along the horizontal axis and resources along the vertical axis. Color, shading, symbols, and text are used to encode the information.

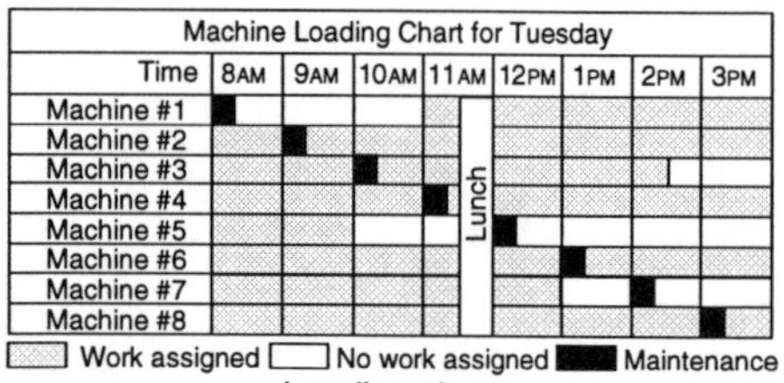

Machine Loading Chart for Tuesday								
Time	8AM	9AM	10AM	11AM	12PM	1PM	2PM	3PM
Machine #1								
Machine #2								
Machine #3								
Machine #4								
Machine #5								
Machine #6								
Machine #7								
Machine #8								

☐ Work assigned ☐ No work assigned ■ Maintenance

Loading chart

**Logarithmic Graph
(Full Log or Log Log)
and
Semilogarithmic Graph
(Semilog or Semi-Log)
and
Logarithmic Scale**

Sometimes referred to as a ratio or rate-of-change graph. A graph is referred to as a logarithmic graph if it has one or more logarithmic scales. • For reference: a logarithm is the power to which a base number must be raised to equal a given value. For example, the logarithm of 100 using a base of 10 is two since 10 must be raised to the power of two (10^2) to equal 100.

Types of logarithmic graphs

If a graph has one logarithmic scale and one nonlogarithmic scale (generally quantitative or sequential), it is referred to as a semilogarithmic or semilog graph. On semilog graphs, the logarithmic scale is generally on the vertical axis; however, it can be on the horizontal axis. If the graph has two logarithmic scales, it is referred to as a full logarithmic or log-log graph. Examples of linear, semi-logarithmic, and full logarithmic grids are shown at the right. • When a data series forms a straight line on a linear grid, the data along both axes are increasing

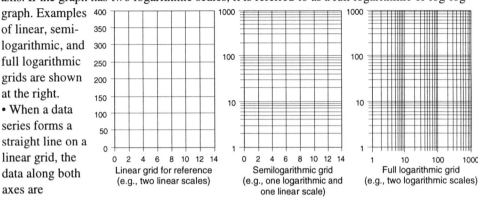

Examples of linear, semilogarithmic, and full logarithmic grids

linearly. When a data series forms a straight line on a semilogarithmic grid, the data along the logarithmic axis is increasing at a constant percentage rate while the variable on the other axis is increasing linearly. A straight line on a full logarithmic grid means the data along both axes are increasing at constant percentage rates. The matrix below shows examples of each type of data plotted on each type of grid.

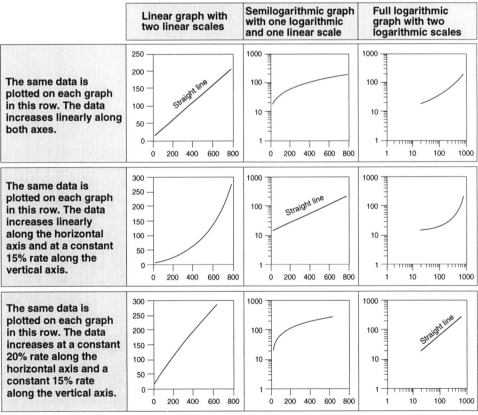

A comparison of linear, semilogarithmic, and full-logarithmic graphs

Scatter and line type graphs are almost always used with logarithmic graphs. It is possible to use other types such as column, bar, and area; however, these types are seldom used because viewers tend to equate the size or areas of the data graphics with the values represented, which, in the case of logarithmic graphs, gives an incorrect impression.

**Logarithmic Graph
(Full Log or Log Log)
and
Semilogarithmic Graph
(Semilog or Semi-Log)
and
Logarithmic Scale**
(continued)

Scales

Logarithmic scales are always quantitative. Because of the nature of logarithms from which logarithmic scales are derived, a logarithmic scale does not have a zero. If the variable being plotted approaches zero, values such as 0.1, 0.01, 0.001, etc., are used as the lower value. • For all practical purposes, negative numbers are not plotted on logarithmic graphs. • The major scale intervals on logarithmic scales are called cycles (also referred to as banks, decks, phases, or tiers).

Major scale intervals on logarithmic scales are called cycles, banks, decks, phases, or tiers

When intervals of equal physical length are marked off on a logarithmic scale, the upper and lower values of each interval are in the same ratio to one another as their counterparts in any other interval of equal physical length on the same scale. For example, the physical distance between 1 and 10 (a ratio of 10 to 1) is the same as the physical distance between 100 and 1,000 (also a ratio of 10 to 1). This principle applies regardless of what length of interval is used, as shown at the right. • By contrast, when intervals of equal physical length are marked off on a linear scale, each interval encompasses an equal amount of the variable being plotted regardless of where the interval is marked off. For example, the physical distance between 1 and 2 (a difference of 1) is the same as the physical distance between 99 and 100 (also a difference of 1).

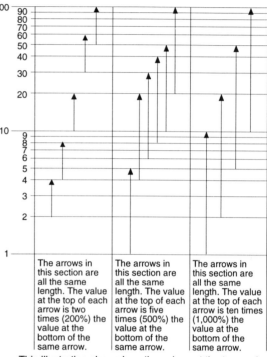

The arrows in this section are all the same length. The value at the top of each arrow is two times (200%) the value at the bottom of the same arrow.

The arrows in this section are all the same length. The value at the top of each arrow is five times (500%) the value at the bottom of the same arrow.

The arrows in this section are all the same length. The value at the top of each arrow is ten times (1,000%) the value at the bottom of the same arrow.

This illustration shows how the values at the two ends of all intervals of equal length along a given logarithmic scale are always in the same proportion/ratio.

Labels

In most cases the actual values and units of measure of the thing being plotted are used on the logarithmic scale (e.g., dollars of sales, number of participants, prices of stock). This enables the people plotting and reading the graph to deal with the same terms as in a linear graph. Occasionally, especially in scientific uses, the logarithms of the values being plotted are displayed on a second scale, as shown at the right. In a few cases only the scale showing the logarithms is used.

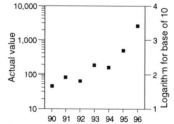

Semilog graph with a scale showing actual values on the left and a scale showing logarithms on the right

Tick marks

Major tick marks designate the cycles on a logarithmic graph. The number of minor tick marks varies depending on the values on the scale and the logarithmic base used. For instance, if a base of five is used, there may be only three minor tick marks between major tick marks. With a base of ten, eight minor tick marks are common. • Minor tick marks are sometimes eliminated, as shown at the right; however, there is the potential hazard that the viewer will inadvertently estimate values based on a linear extrapolation instead of logarithmic extrapolation.

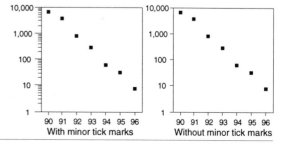

Logarithmic Graph (Full Log or Log Log) and Semilogarithmic Graph (Semilog or Semi-Log) and Logarithmic Scale (continued)

Logarithmic base value

Logarithms, from which logarithmic graphs are generated, are calculated with reference to what is called a base number. The number ten is frequently used; however, it can be almost any value. With regards to graphs, the major effect of using various bases is a change in the values and the number of tick marks shown on the scale. The choice of the base value has no effect on the shape of the curves being plotted, as shown below. In this illustration, base values of 5, 10, and 22 are used. The same data is plotted in each graph as well as in the graph at the right, which has linear scales for reference.

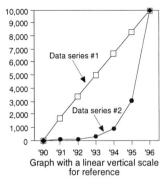

Graph with a linear vertical scale for reference

Vertical scale uses a base of 5 | Vertical scale uses a base of 10 | Vertical scale uses a base of 22

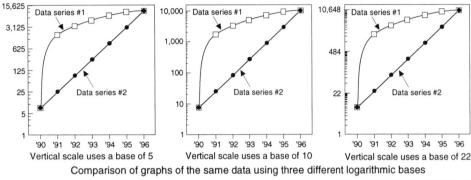

Comparison of graphs of the same data using three different logarithmic bases

Interpreting curves on logarithmic graphs

The steepness of the curve on a semilog graph at any point or overall is proportional to the actual rate of change of the thing being plotted. The steeper the slope, the greater the rate of change, either positive or negative. If the line slopes up, it indicates a positive rate of change. A downward slope indicates a negative rate of change.

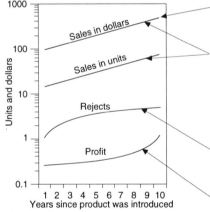

Typical application of a semilogarithmic graph

- When the curve is a straight line, change is taking place at a constant rate, such as 20% per year for each successive year.
- When curves are parallel, the rates of change are the same for the two data series, even thought the actual values are different. In this example the rates of growth for the product are the same (20% per year), both in terms of units and dollars.
- When the slope of the curve becomes shallower as it progresses, the rates of change are becoming smaller and smaller.
- When the slope of the curve becomes steeper and steeper as it progresses, the rates of change are becoming greater and greater.

- The absolute slope (ignoring plus and minus signs) of the curve is not the same for equal positive and negative rates of change (e.g., +40% and -40%). As illustrated in the example at the right, the slope of a curve with a constant plus 40% rate of change is less steep than the slope of a curve with a constant minus 40% rate of change. The same data is plotted on a semilog and linear graph for comparison purposes.

Semilog graph | Linear graph for reference

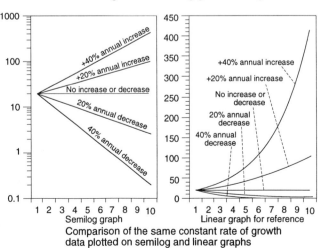

Comparison of the same constant rate of growth data plotted on semilog and linear graphs

**Logarithmic Graph
(Full Log or Log Log)
and
Semilogarithmic Graph
(Semilog or Semi-Log)
and
Logarithmic Scale**
(continued)

Supplementary scale

Daily stock prices are often plotted on semilog graphs, as shown in the example below. If the stock prices tend to fluctuate around a straight line with an upward slope, the price of the stock, on the average, is increasing at a constant rate comparable to the percent represented by the straight line it fluctuates around. Since actual rates of change cannot be read directly from the graph, a supplementary scale is sometimes used, as shown in the example. From such a graph, one can estimate short-term growth rates (e.g., between 1980 and 1982 the stock price increased at a rate of about 20%), as well as long-term growth rates (e.g., between 1980 and 1990 the stock price increased at an average rate of about 15%). Although stock prices are used in the example, the concept is applicable to any type of data.

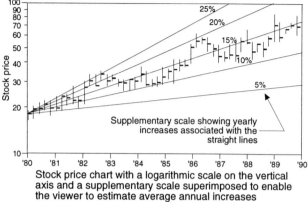

Stock price chart with a logarithmic scale on the vertical axis and a supplementary scale superimposed to enable the viewer to estimate average annual increases

Use of semilog graph to compare rates of growth of various entities

A logarithmic scale can be helpful when analyzing data to see what is happening to the rates of growth of various entities. For example, if managers want to compare the growth rates of various divisions in a company, they might plot the annual sales for each division on a semilog graph, as shown below on the left. The decreasing slope of the curve for division A indicates that the division's rate of growth decreased each successive year. The fact that division B's curve is a straight line indicates that this division's annual growth was constant and positive for the entire time. The increasing slope of the curve for division C indicates that this division grew faster each year than the previous year for all 12 years.

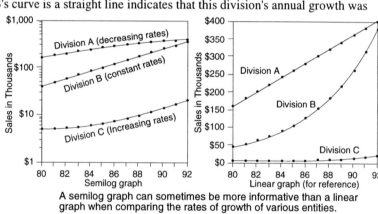

A semilog graph can sometimes be more informative than a linear graph when comparing the rates of growth of various entities.

Use of semilog graph to enhance the readability of individual data points

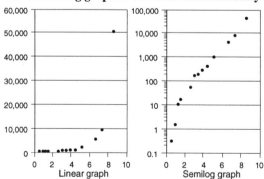

Illustration of how a logarithmic scale can sometimes provide better resolution when the value of data points differ significantly. The same data are plotted on both graphs.

A logarithmic scale can be useful when displaying data with large differences in numeric values and/or large differences between multiple data series. For example, if values range from 0.3 and 1.5 to 8,719 and 50,444, it would be impossible to display these on a linear scale so that the viewer could identify values and differences with a reasonable degree of accuracy. The use of a logarithmic scale improves the resolution of the individual data points in such a series, as shown at the left.

Preprinted logarithmic graph paper

Although the grid for a logarithmic graph can be prepared by hand, such graphs are usually generated using computers or preprinted graph paper. Preprinted graph paper typically comes with the numbers 1 to 9 on the logarithmic axis. These can be converted to the scale numbers needed by the addition of zeros and/or decimal points.

Logarithmic Map	Sometimes referred to as a map with exaggerated perspective. This type of map places emphasis on those things towards the center or foreground and reduces the size and amount of detail of things further away at the edges of the map.

Logarithmic Map: Sometimes referred to as a map with exaggerated perspective. This type of map places emphasis on those things towards the center or foreground and reduces the size and amount of detail of things further away at the edges of the map. More space is provided for items in the foreground by reducing the space required for distant items. Because of the distortion introduced, these types of maps are generally nonquantitative. Rental car agencies sometimes use logarithmic maps, since the majority of the renters use the cars close to where they are rented and that is where the greatest amount of detail and the least amount of distortion appears. The outlying information tends to be used primarily for orientation purposes.

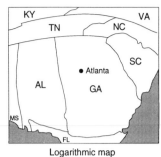

Logarithmic map

Longitude and Longitudinal Line

Longitude is a distance measured in degrees east or west of a reference called the prime meridian. The prime meridian, which passes through Greenwich, England, designates zero, with 180° on the opposite side of the earth. Longitudinal lines are the grid lines that connect all points of equal longitude. They are also called meridians.
• Latitude is a distance north or south of the equator. It is measured in degrees with zero at the equator and 90° at the North and South Poles. Using values of latitude and longitude as coordinates, places and things can be precisely located anywhere on the globe. See Map.

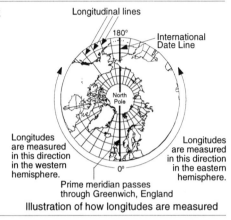

Illustration of how longitudes are measured

Lorenz Curve or Lorenz Graph

A graph with cumulative frequency percents plotted on both the vertical and horizontal axes. Lorenz graphs are used primarily for comparing two or more variables to see if their distributions are similar. For example, Lorenz graphs are used extensively to compare the distribution of income versus the distribution of population, i.e., how the total income received by a certain population is distributed across that population. The population might be that of a country, state, profession, company, etc. The graphs are normally square. The scales on both axes range from zero to 100%, with the zero for both axes at the lower left-hand corner. A diagonal line is drawn from the lower left-hand corner to the upper-right hand corner to represent the curve that would result if the distributions on the two axes were exactly the same. This diagonal is referred to as the line of perfect equality, absolute equality, curve of complete equality, or line of equal distribution. The graph at the left is an example of a Lorenz curve with cumulative percent of total income plotted on the vertical axis and cumulative percent of the population on the horizontal axis. The shaded area indicates the difference or deviation from the diagonal. The dashed lines have been added to illustrate how the graph is interpreted. For example, in this hypothetical situation, 40% of the population earned about 22% of the total income. At the other end of the curve the reference lines indicate that the upper 10% of the population earned about 21% of the total income.

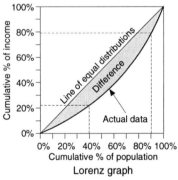

Lorenz graph

• Sometimes Lorenz graphs are used to study such things as changes in the distribution of income over time, under different sets of conditions, or in different populations. Examples of several of these applications are shown below.

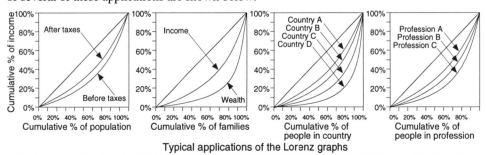

Typical applications of the Lorenz graphs

Map

Maps provide the unique ability to display information visually in relationship to its physical (spatial) location. Tables and graphs provide information on the quantitative aspect of data. For instance, for quick reference and precision, a table is usually the best choice. To make rapid comparisons of individual values or to look at the overall quantitative nature of the data, graphs provide a good option. To see how the data is distributed geographically, there is no substitute for a map. Examples of all three types of graphics displaying the same data are shown at the right. • Maps used as charts can be categorized into six major classifications, shown below with examples of each. Statistical, descriptive, flow, topographic, and weather maps are discussed under those family headings. The maps categorized as special purpose are discussed under their individual headings. This section discusses the general aspects of maps that apply to multiple families and types. • Although the various types of maps are best discussed separately, in practice, multiple types are frequently combined into a single map.

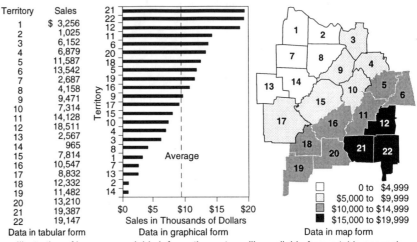

Territory	Sales
1	$ 3,256
2	1,025
3	6,152
4	6,879
5	11,587
6	13,542
7	2,687
8	4,158
9	9,471
10	7,314
11	14,128
12	18,511
13	2,567
14	965
15	7,814
16	10,547
17	8,832
18	12,332
19	11,482
20	13,210
21	19,387
22	19,147

Data in tabular form

Data in graphical form

Data in map form

Illustration of how a map yields information not readily available from a table or graph

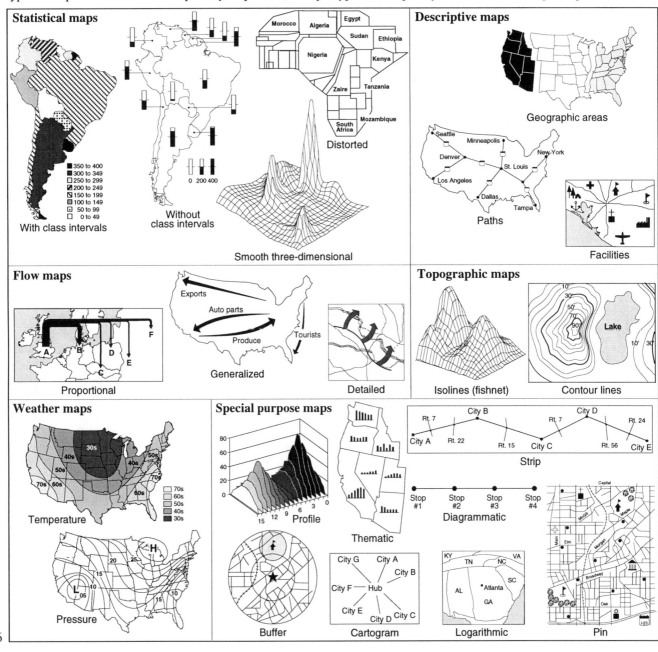

Statistical maps

With class intervals

Without class intervals

Smooth three-dimensional

Distorted

Descriptive maps

Geographic areas

Paths

Facilities

Flow maps

Proportional

Generalized

Detailed

Topographic maps

Isolines (fishnet)

Contour lines

Weather maps

Temperature

Pressure

Special purpose maps

Profile

Thematic

Strip

Diagrammatic

Buffer

Cartogram

Logarithmic

Pin

Map (continued)

Grid lines, meridians, and parallels

Parallels and meridians

Two of the primary functions of grid lines on maps are to locate places and things and to orient the viewer. When precise locations on the earth's surface are required (sometimes referred to as absolute location), a coordinate system of latitudes and longitudes is used (sometimes referred to as geographic grid or graticule). The latitude and longitude of a particular location are sometimes called geodetic or geographic coordinates. The latitude is frequently considered the Y coordinate and the longitude the X coordinate. The grid lines running north and south are called meridians or longitudinal lines. The grid lines running east and west are called parallels or latitudinal lines. The example at the right shows how parallels and meridians appear on a globe. The examples below show various way they appear on maps.

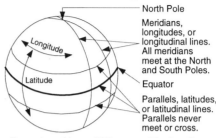

North Pole

Meridians, longitudes, or longitudinal lines. All meridians meet at the North and South Poles.

Equator

Parallels, latitudes, or latitudinal lines. Parallels never meet or cross.

Examples of key grid lines used for locating places or things on the earth's surface

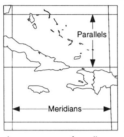

In many maps of small areas, parallels appear almost as horizontal straight lines and meridians as vertical straight lines

As the area becomes larger, parallels and meridians frequently appear as gently curving lines.

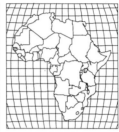

When a large area is viewed perpendicular to the north-south axis, parallels approach straight horizontal lines and meridians become curved vertical lines.

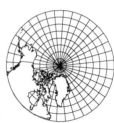

When viewed along the north-south axis, parallels appear as circles and meridians as straight lines radiating from the center.

Examples of how parallels and meridians appear on maps, depending on the size of the area involved and the angle from which the view is taken

Parallels are used to measure distances north and south of the equator. They start with zero degrees at the equator and increase to 90° at either pole. The letter N or S is appended to the values (e.g., 17°N, 47°S) if it is not obvious whether they apply to the northern or southern hemisphere. • Meridians are used to measure distances east and west from a generally accepted reference line called the prime meridian, which passes through Greenwich, England. The meridians start with zero at the prime meridian and increase to the east and west until the two sets meet at 180° on the opposite side of the world. The point where they meet is called the International Date Line. The letters E and W are appended (e.g., 27°E, 115° W) if it is not obvious which hemisphere is meant. Both longitudinal and latitudinal distances are measured in degrees, minutes, and seconds. • Generally only major grid lines are used. • See illustrations below and Scales in this section on maps.

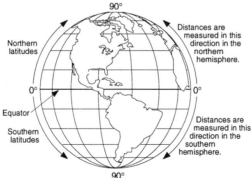

Illustration of how latitudes are measured north and south from the equator

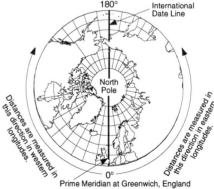

Illustration of how longitudes are measured east and west from the prime meridian

Quadrangle, quadrilateral, or quad

The area bounded by two lines of latitude and two lines of longitude is sometimes called a quadrangle, or quad, and is frequently designated by the angles included in the enclosure. For example, a seven-and-one-half minute quadrangle is a quadrangle encompassing 7.5 minutes of latitude and 7.5 minutes of longitude. The U.S. Geological Survey issues a series of maps referred to as 7.5-minute maps or quadrangles.

Grid lines, meridians, and parallels (continued)

Grid lines for determining size and location

With many maps, such as statistical and descriptive maps, precise locations with regards to latitude and longitude are not important. Of more importance are things like the relative location of a place or thing with respect to other places or things; distances between various locations in units of feet, meters, miles, or kilometers; statistics with regards to various areas; the size of an area; etc. As a result, many maps do not show parallels and meridians and in fact, may have no grid lines at all. Those that do, often use them to assist in finding specific entities on the map. For this purpose the map is generally divided in to rows and columns by means of horizontal and vertical grid lines, though these grid lines are generally unrelated to lines of latitude and longitude. The number of rows and columns is optional, depending on the size of the map and the preciseness with which places and things are to be referenced. The lines are generally faint and straight. Such grid lines frequently have no particular mileage distance between them. Each vertical column is designated by a number or letter and each horizontal row by a letter or number, respectively, as shown below. A table accompanying the map cross references each major item on the map with its appropriate letter and number combination so it can readily be located on the map.

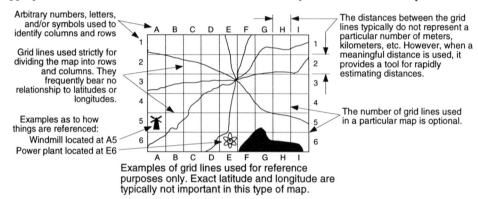

Arbitrary numbers, letters, and/or symbols used to identify columns and rows

Grid lines used strictly for dividing the map into rows and columns. They frequently bear no relationship to latitudes or longitudes.

Examples as to how things are referenced:
Windmill located at A5
Power plant located at E6

The distances between the grid lines typically do not represent a particular number of meters, kilometers, etc. However, when a meaningful distance is used, it provides a tool for rapidly estimating distances.

The number of grid lines used in a particular map is optional.

Examples of grid lines used for reference purposes only. Exact latitude and longitude are typically not important in this type of map.

Supplementary grid lines and scale

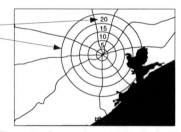

Supplementary scale
Supplementary grid lines

In some instances supplementary grid lines are added, such as a series of circles or lines bracketing a specific point to indicate distances from that point. The scale for this additional grid is generally shown directly on the grid.

Example of a supplementary set of grid lines used on some maps for determining distances, designating zones, etc.

Location of grid lines

Grid lines may be drawn in front of or behind the land masses on maps. They frequently are drawn in front of. Whether they are in front or back or whether a grid is used at all depends largely on the purpose of the map and the degree of accuracy desired in decoding the information. Examples are shown below.

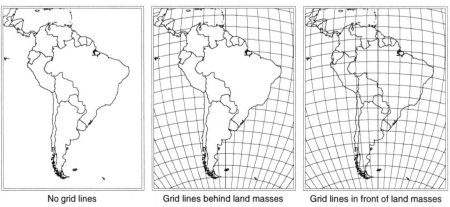

No grid lines
Grid lines behind land masses
Grid lines in front of land masses
A comparison of three variations of grid lines

Map (continued)

Scales

--- **Scales for establishing absolute locations using latitudes and longitudes** ---

Latitudes and longitudes are imaginary lines that locate places and things on the earth's surface. These are sometimes referred to as absolute locations. The angular scales used for identifying latitudes and longitudes are referenced from imaginary planes passing through the center of the earth. Degrees of longitude start at what is called the prime meridian and increase in both the easterly and westerly directions, meeting at 180° on the opposite side of the earth. • Degrees of latitude start at the equator and increase in both the northerly and southerly directions to 90° at the two poles. The angles are generally stated in terms of degrees, minutes, and seconds (e.g. 35° 47' 12"). Sometimes fractions of degrees or minutes are used, such as tenths of a minute (35° 47.2') or hundredths of a degree (e.g. 35.79°). Illustrations of how the angles are measured are shown below.

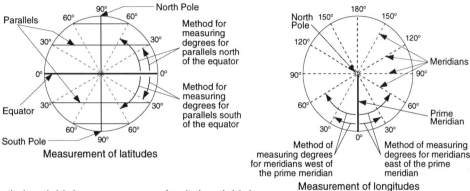

Measurement of latitudes

Measurement of longitudes

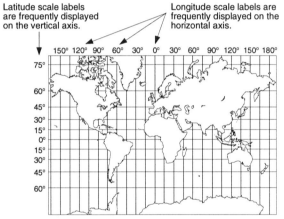

Example of a map with latitude and longitude scales

When displayed on a flat map with a view from the side, the scale for longitudes and their labels are generally on the horizontal axis. East values increase to the right and west values to the left. Scales for latitudes and their labels are generally on the vertical axis. Northern values increase upwards and southern values downward. Examples of both are shown at the left.

--- **Placement of scale labels** ---

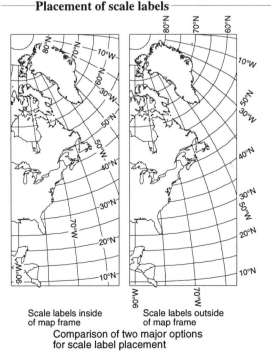

Scale labels inside of map frame

Scale labels outside of map frame

Comparison of two major options for scale label placement

Sometimes it is convenient to display scale labels outside the map frame or neat line. In other cases it is inconvenient and/or confusing to show the labels outside. In these cases it may be clearer if the labels are located directly on their respective grid lines. Examples of both placements are shown at the left. Even when the labels could be shown outside the frame, the option of placing them on the grid lines is often preferred. When the map is large, the labels are sometimes repeated two or more times at opposite ends of the map and in some cases in the middle. Adding letters to the labels indicating the hemisphere (e.g., N, S, E, W), as shown in the example at the left, can also aid the viewer.

Scales (continued)

─────**Surface distance scales**────────────────────

This type of scale is used primarily for determining the size, distance, or relative location of places and things, — for example, to determine how wide Kansas is or how far it is from Paris to Moscow. The following three methods are widely used for indicating surface distances. Scales such as these are typically included on the map or in the margin. It is common for two or more different types of scales to be included on the same map.

- **Simple fraction or ratio** - Sometimes referred to as representative fraction or RF. This type of scale relates a unit distance on a map to the distance it actually represents on the ground or earth. For example, the fraction or ratio of 1/1,000,000 or 1:1,000,000 means that 1 unit on the map represents 1,000,000 of the same units on the ground. If the units are inches then one inch on the map represents 1,000,000 inches on the ground. The same technique is sometimes used to scale areas in which case the units are squared (e.g. $1^2 : 10,000^2$). This type of scale will be incorrect if the map is resized by a photocopying process.

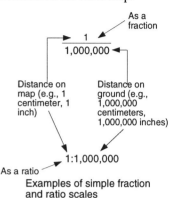

Examples of simple fraction and ratio scales

- **Written statement or verbal -** This type of scale will generally be incorrect if the map is enlarged or reduced by a photocopying process.

One inch equals one mile
One centimeter equals one kilometer

Examples of written statement scale

- **Graphic representation** - Graphic scales are unaffected by enlargements or reductions of the map by photocopying processes.

  - **Bar scales** are frequently used when distances are not distorted. The examples shown here all display miles and kilometers. When the distances are shorter, the labels are in smaller units such as feet or meters.

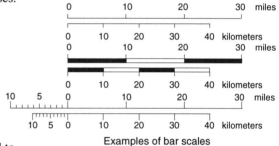

Examples of bar scales

  - **Special scales** are sometimes used to compensate for distortion. This example takes into consideration the fact that the further one gets from the equator, the shorter the actual distance represented by a unit length on the map.

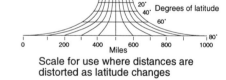

Scale for use where distances are distorted as latitude changes

─────**Elevation or vertical axis scales**────────────────

Elevations might be noted directly on a map or, as is often done, various colors or shading are used to indicate ranges of elevations. When color or shading is used, a scale similar to those at the right is frequently employed. The color/shading concept can be applied to surfaces above or below water.

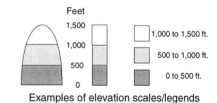

Examples of elevation scales/legends

─────**Large scale, intermediate scale, and small scale maps**────────────

Maps are sometimes grouped into the following three categories based on scales. The ranges of scale values for the categories are approximate and are not industry standards.

Category	Description	Scale
Large scale	A map that shows a small area in great detail (e.g., a city map)	1 / 24,000 and larger fractions (e.g., 1 / 10,000)
Intermediate scale	Maps with areas and detail somwhere between large and small scale	1 / 25,000 to 1 / 29,000,000
Small scale	A map that shows a large area with relatively little detail (e.g., a world map)	1 / 30,000,000 and smaller fractions (e.g., 1 / 50,000,000)

Map (continued)

Symbols

In addition to designating the location of things on maps, symbols are used extensively to encode quantitative and descriptive (qualitative) information. The table below shows examples of some of the diverse ways symbols are used for these purposes. Additional information is available under the heading Symbol elsewhere in this book.

Type of symbol	Classification of information	Descriptive information — Nominal (qualitative)	Descriptive information — Ordinal (Ranked nonquantitatively, e.g., first, second, third, etc.)	Quantitative information — Interval (quantity, value)	Quantitative information — Proportional (e.g., ratio, relative, etc.)
Literal (e.g., words, numbers, etc.)	Standard	library, coal mine, plant	L, M, S / large, medium, small	10 ton / 20 passengers / 50 births	2.7 tons per acre / 2,080 passenger-miles / 1.2 births per family
Literal	Multiple bits of information encoded into symbol	**Portland** = In the west / *New York* = In the east / Tulsa = In the south	**NEW YORK** = Largest / Portland = Medium / TULSA = Smallest	Portland = 100k to 500k / New York = 50k to 100k / Tulsa = .5k to 1k	2.7 = current year / 2.1 = previous year / 1.2 = 5yr. average
Point (geometric or pictorial)	Size	unincorporated city / incorporated city	= minor / = average / = major	= $100 Sales / = $200 Sales / = $300 Sales	= X$ purchases per household / = Y$ purchases per household / = Z$ purchases per household
Point	Shape	airport has feature / church tin deposit	= third place / = second place / = first place	= 10 to 99 employees / = 100 to 499 employees / = 500 to 999 employees	= 1/2 depleted / = 3/4 depleted / = fully depleted
Point	Multiple	NA - if used in multiples they become quantitative	= least / = mid-value / = most	= 100 / = 200 / = 500 etc.	size of A / B twice size of A / C four times size of A
Point	Multiple bits of information encoded into symbol	Bridge / Bridge closed / ● Existing ○ Proposed	minor airport / average airport / major airport	62% / 38% / = 100 kilograms / = 200 kilograms / ○ = 300 kilograms	size of A / B twice size of A / C three times size of A / ○ seasoned ● not seasoned
Line	Thickness (weight)	original contours / natural changes / man-made changes	unimproved road / light duty road / primary highway	10 messages / 100 messages / 1000 messages	2 messages per hour / 4 messages per hour / 8 messages per hour
Line	Solid or dashed	property boundary / township boundary / county boundary	highest elevation / middle elevation / lowest elevation	300 ft. elevation / 200 ft. elevation / 100 ft. elevation	10 X elevation / 5 X elevation / 1 X elevation
Line	Patterned	wall / stream / power line	most curves / medium / least curves	100 hertz / 200 hertz / 300 hertz	100% / 200% / 300%
Line	Multiple	walking trail / bike tail / walking & bike trail	narrowest road / average width road / widest road	10 ton capacity / 20 ton capacity / 30 ton capacity	1 ton per axle / 2 tons per axle / 3 tons per axle
Line	Symbols included	fence line / truck route	densest / medium / sparsest	l or ▮ = width of road in meters	3 X base / 2 X base / base
Line	Multiple bits of information encoded into symbol	railroad / abandoned railroad	steepest incline / intermediate incline / no incline	50 to West by truck / 60 to East by air	speed of A / 2 X speed of A / 3 X speed of A
Area and volume (includes geometric and nongeometric shapes)	Size	children's park / adult's park	smallest / medium / largest	500 1,000 2,000	all products / product A / product B
Area and volume	Shape	foot print plot / oil reserves	#1 rank #2 rank #3 rank	10 10² 10³	50% complete / 75% complete / 100% complete
Area and volume	Multiple bits of information encoded into symbol	patterns designate features	by class, category, feature, etc.	500 1,000 2,000	by type, category, characteristic, etc.

The table above illustrates how relatively standard forms of symbols might be used to represent both quantitative and descriptive information. In most of the examples, the symbol represents a single bit of information. With each basic form (i.e., literal, point, line, area, and volume) a row of examples is included to show how multiple bits of information might be incorporated into that type of symbol. Some symbols are applicable to only one type of information. In most cases the same symbol can be used with several or all types of data.

231

Map projections

Geographical information transferred from a spherical surface (earth or globe) to a flat surface (map) is called a map projection. The transfer process is done mathematically but might be visualized as the process of projecting each point on a globe onto a simple geometric shape such as a cylinder, cone, or plane. When the geometric shape is spread out flat and trimmed, the result is a map as we normally think of it. Shown at the right is an illustration of three major methods for making projections, plus examples of maps resulting from each method. • All map projections produce distortion, the amount and type depending largely on the method of projection used and the amount of the earth's surface displayed on the map. World maps often

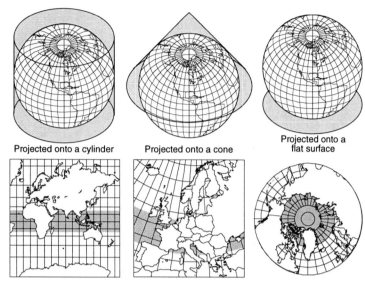

Projected onto a cylinder   Projected onto a cone   Projected onto a flat surface

Illustration of three major projection methods with examples of the types of maps that result from each. The gray areas on the maps indicate the shape of the area on the map where the minimum distortion tends to occur.

have significant distortion in certain areas, particularly in the upper and lower latitudes, while local maps covering much smaller areas have negligible distortion. • Before maps of large areas are generated, it is decided which is the most important, true areas, true shapes, true distances, and/or true directions. Reasonable accuracy can be achieved in one or more of these measures, but never in all four. The type of projection is then selected to minimize distortion in the characteristics considered most important. The gray areas in the map examples above indicate the shape of the areas where the distortion tends to be least using that specific projection. • When looking at a map it is generally difficult to know what type of projection was used and, more importantly, where and how much distortion exists. In some cases the type of projection used is noted on the map. In other cases, observing the parallels and meridians can help in estimating the areas of greatest distortion. • The maps at the left, which are made from three different projections, illustrate the concept of distortion. In these three maps, Greenland appears to have three different shapes, three different sizes compared to the 48 contiguous states of the US, and three different orientations to, for instance, New England. On the other hand, Greenlands position

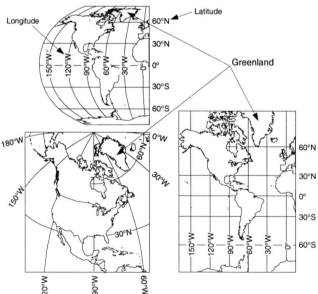

An illustration of how differently the same land mass can appear, depending of the projection used to generate the map. Since each map encompasses different amounts of the earth's surface, only relative shapes, sizes, distances, and orientations are meaningful.

with regards to parallels and meridians is the same on all three projections. The apparent differences do not mean that one map is right and another wrong; rather, it is a matter of one type of projection being better than another for a given task.

Map (continued)

Base map

Most statistical and descriptive maps are generated by superimposing what is sometimes called attribute information on a base map, as shown below.

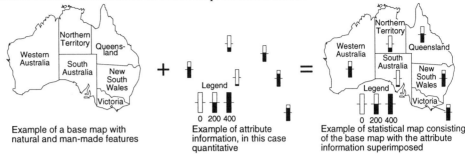

Example of a base map with natural and man-made features

Example of attribute information, in this case quantitative

Example of statistical map consisting of the base map with the attribute information superimposed

An illustration of how attribute data is superimposed on a base map to form a statistical map

There are no clear-cut guide lines as to what information should be included on a base map and in fact, what is considered part of the base map in one map might be considered part of the attribute data in another. As a general rule, the amount of information on the base map is kept to a minimum so as not to detract from the attribute data. If information is not germane to the purpose of the map or the viewer's understanding, it most times is omitted.

• Most base maps contain a combination of two major types of information: natural geographic features and man-made geographic features. Examples of both are shown below. In some cases man-made structures are considered a third category. In other cases they are considered part of the attribute data.

Base map with natural geographic features

Included in the definition of natural geographic features are such things as the outlines of natural land areas (coastlines) and the location of rivers, mountain ranges, lakes, islands, etc. In the example at the right, all lines and symbols represent natural geographic features.

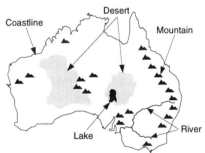

Base map with natural geographic features

Base graph with man-made geographic features

Included in the definition of man-made geographic features are such things as political boundaries (cities, states, countries, etc.), postal zones (ZIP codes), trading areas, census regions, school districts, etc. Sometimes roads, streets, canals, or parks are also classified as man-made geographic features, depending on the purpose of the map. The map at the right shows a combination of man-made features (political boundaries of states/territories) and natural geographic features (coast lines). For certain types of information, this would be considered an adequate base map.

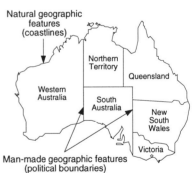

Base map with man-made geographic features superimposed over natural geographic features

Polygon

Many of the representations of man-made features such as countries, states, territories, or counties are designated by outlining an area with lines. The graphic unit that is formed by this process is called a polygon. An example is shown at the right. A polygon can be made of straight or curved lines with any number of sides .

The closed outline of an area such as the bold line around Queensland is called a polygon.

Map (continued)

Terminology and locations of key elements

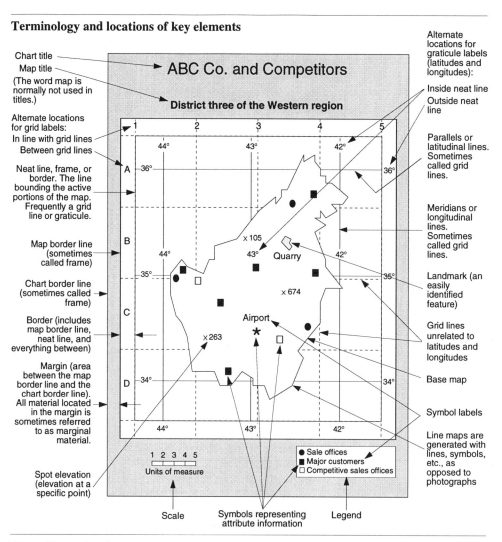

Chart title

Map title
(The word map is normally not used in titles.)

Alternate locations for grid labels:
In line with grid lines
Between grid lines

Neat line, frame, or border. The line bounding the active portions of the map. Frequently a grid line or graticule.

Map border line (sometimes called frame)

Chart border line (sometimes called frame)

Border (includes map border line, neat line, and everything between)

Margin (area between the map border line and the chart border line). All material located in the margin is sometimes referred to as marginal material.

Spot elevation (elevation at a specific point)

Alternate locations for graticule labels (latitudes and longitudes):

Inside neat line
Outside neat line

Parallels or latitudinal lines. Sometimes called grid lines.

Meridians or longitudinal lines. Sometimes called grid lines.

Landmark (an easily identified feature)

Grid lines unrelated to latitudes and longitudes

Base map

Symbol labels

Line maps are generated with lines, symbols, etc., as opposed to photographs

ABC Co. and Competitors

District three of the Western region

× 105

Quarry

× 674

Airport

× 263

● Sale offices
■ Major customers
□ Competitive sales offices

1 2 3 4 5
Units of measure

Scale

Symbols representing attribute information

Legend

Three-dimensional map

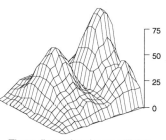

Three-dimensional map utilizing a third axis capable of displaying elevations or Z-axis values

There are two major categories of three-dimensional maps. One has a third axis along which elevations or quantitative Z-axis values can be plotted (example at left). The other category is simply a two-axis map that has been tilted and/or rotated with depth added for aesthetic purposes (example at right). See Three-Dimensional Map.

Three-dimensional map without a third axis. The depth is added for cosmetic purposes only.

Exploded map

An exploded map is one in which portions of the map have been graphically separated. The separations are typically made along man-made boundaries such as countries, states, territories, or counties. Maps are exploded to emphasize certain areas or groups of areas or to more clearly define where boundaries are located.

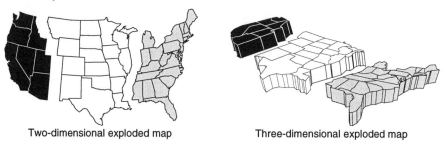

Two-dimensional exploded map

Three-dimensional exploded map

Map (continued)

Inset, ancillary, or supplementary map

If a viewer is not familiar with the area shown on a detailed map, an inset or small supplementary map is often used to help with orientation. For example, the detailed map at the right shows a hypothetical area in the northern part of the state Georgia. If shown only the detailed map, most people would not know what state it was located in, whether it was in the northern, southern, eastern, or western portion of the state, and what is close to it that they might recognize. A supplementary map, as shown in the upper right-hand corner of the detailed map, supplies answers to these unknowns.

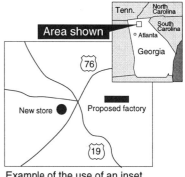

Example of the use of an inset, ancillary, or supplementary map

Orientation

Maps are normally oriented with their tops in a northerly direction. If a different orientation is used, an arrow or symbol is sometimes included to orient the viewer. If lines of latitude and longitude are present on the map, arrows may not be included. On large maps, multiple directional arrows are sometimes used since the angles of the north-south meridians, as drawn on a two-dimensional map, change as the longitude

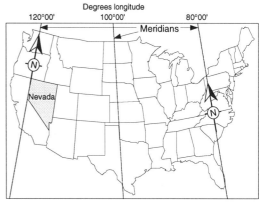

An example illustrating how the angles of the north-south meridians on a two-dimensional map change as the longitudes (horizontal distances) vary.

Same orientation as in the larger map

Rotated to a true north-south orientation

Two options frequently used when displaying a portion of a larger map

(horizontal distance) varies. An example is shown above.
• When a small portion of a large map is shown by itself, it sometimes is reoriented into a true north-south position. In other cases, it is left in the same orientation as in the larger map since that might be the more familiar orientation. The two examples of the state of Nevada at the left illustrate the point.

Contrast between adjacent entities

Sometimes it is difficult for viewers to differentiate adjacent entities that are unfamiliar and have no colors or shading. An example is coast lines where it is not clear which is water and which is land as illustrated at the right. Colors, shading, and sufficient labeling normally overcome such problems.

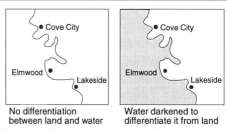

No differentiation between land and water

Water darkened to differentiate it from land

Examples illustrating the value of fill to differentiate adjacent entities from one another

Frame, border, and neat line

There are three major terms used to describe the lines around maps: frame, border or border line, and neat line. The term neat line is generally restricted to the line that abuts the active map area. The other terms vary somewhat in their definitions. The examples below indicate how the other terms are sometimes used.

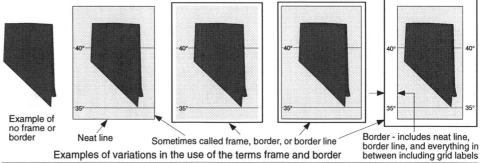

Example of no frame or border

Neat line

Sometimes called frame, border, or border line

Border - includes neat line, border line, and everything in between including grid labels

Examples of variations in the use of the terms frame and border

| **Margin** | Sometimes referred to as a border. The margin is the area immediately outside of and around a graph, map, or table as shown in the example at the right. The material shown in these areas (notes, legends, data, etc.) is sometimes referred to as marginal data or marginal information. |

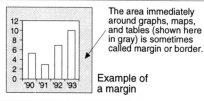

The area immediately around graphs, maps, and tables (shown here in gray) is sometimes called margin or border.

Example of a margin

| **Marginal Frequency Distribution Graph** | Sometimes referred to as a border plot. Marginal frequency distribution graphs provide a means of displaying the distribution of data along one axis of a two-axis scatter graph. This is accomplished by placing a one-axis data distribution graph of the data in the margin or border of a two-axis graph. This in essence condenses all the data points on the two-axis graph into a single line of data points. The technique can be applied to one or both axes. The graphic displays might be one of a number of different types, including one-axis point graphs, stripes, boxes, or histograms. Examples are shown below. |

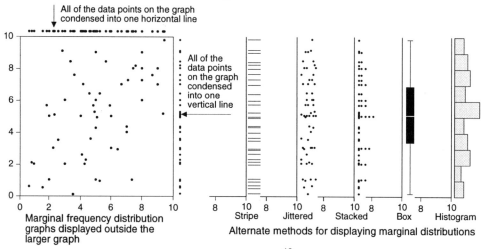

Marginal frequency distribution graphs displayed outside the larger graph

Alternate methods for displaying marginal distributions

The one-axis point graphs may have the data points in a single line or displaced by means of jittering or stacking to avoid overlapping (see Symbol). The marginal frequency distribution graphs can be displayed outside the frame, as in the example above, or inside the frame of the two-axis graph, as shown at the right. When this is done the frame and scales sometimes have to be shifted so the marginal graphs do not interfere with the data. The one-axis data distribution graphs might be located at the top, bottom, left, right, or some combination of these. They can touch the frame or a space might be provided.

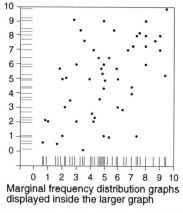

Marginal frequency distribution graphs displayed inside the larger graph

| **Marginal Information and Marginal Distribution** | Information that is shown in the margin of a graph, table, map, or chart such as notes, legends, titles, and data. When overall percentages are shown in conjunction with tables, the information is sometimes called marginal distribution. |

| **Marker[1]** | Sometimes referred to as a data graphic, data measure, or data marker. Marker refers to the dots, lines, bars, columns, symbols, or other graphic configurations used to represent quantitative information on charts, graphs, and maps. For example, in a bar graph the bars are referred to as markers. In a pie chart, the markers are the wedges. In a point graph the dots or symbols are the markers. In a map the markers might be framed rectangles. |

Examples of marker, data marker, data graphic, or data measure

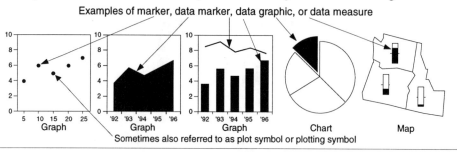

Graph   Graph   Graph   Chart   Map

Sometimes also referred to as plot symbol or plotting symbol

| Marker[2] | A marker is a notation along the axis of a graph designating a point on the scale that may be significant with regards to the analysis or understanding of the data. For example, markers may be used to designate the time a significant event happened. Markers might be located inside or outside the plot area and are sometimes used in conjunction with a reference line. | 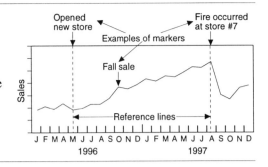 |

Master Chart/Graph

Sometimes referred to as a template chart or graph. A master chart or graph contains all or most of the common elements of a series of charts or graphs. Detailed information is then added to copies of the master to form specific variations. For example, if the sales performance of three different salespersons were to be looked at individually, a master graph might be generated and the sales data for the individual salespersons added to three different copies of the master graph, as shown below. In the example, the three specific charts generated from the master are all column graphs. It is not required that they all be the same, nor is there a restriction as to the type of chart or graph that can be generated using masters. Using masters allows more rapid construction and revision of repetitive types of charts and graphs and the potential for improved accuracy and consistency in a series of charts. Masters are frequently used as the background for presentation charts to provide a uniform appearance throughout the presentation.

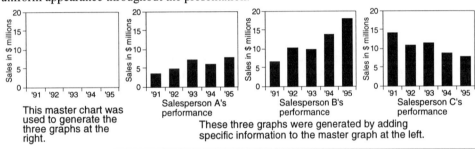

This master chart was used to generate the three graphs at the right.

Salesperson A's performance — Salesperson B's performance — Salesperson C's performance

These three graphs were generated by adding specific information to the master graph at the left.

Matrix

The term matrix is used two different ways in the context of information graphics. Some people use the term matrix interchangeably with the term table to refer to any meaningful arrangement of information in rows and columns. Others restrict the term matrix to only those tables in which the row headings and column headings are identical (sometimes called a bidirectional table). All three examples shown here qualify as matrixes under the first definition. Only the example at the right qualifies as a matrix under the second definition. The first, broader definition is used in this book. Under both definitions, the information in the rows and columns might be in any form including words, numbers, symbols, graphs, icons, maps, etc. See Table, Matrix Display, and Icon Comparison Display.

	City A	City B	City C	City D	City E	City F	City G	City H	City I
City A		15	19	40	37	5	24	43	19
City B	15		24	13	4	10	11	30	10
City C	19	24		28	24	15	15	34	15
City D	40	13	28		10	36	15	34	15
City E	37	4	24	10		33	12	31	11
City F	5	10	15	36	33		30	39	16
City G	24	11	15	15	12	30		19	2
City H	43	30	34	34	31	39	19		21
City I	19	10	15	15	11	16	2	21	

Example of a matrix in which the row and column heading are the same. Sometimes called a bidirectional table.

		Sales			
		1995 Estimate		1999 Forecast	
Product		Dollars (000)	% of total	Dollars (000)	% of total
Drills	Large	14.4	8.6	25.6	12.2
	Medium	28.2	16.9	45.6	21.7
	Small	40.8	24.5	76.4	36.4
Nail Drivers	Large	8.7	5.4	13.2	6.3
	Small	11.4	6.8	15.1	7.2
Torque Wrench	Large	6.2	3.7	6.1	2.8
	Small	8.9	5.3	11.3	5.4
	Pneumatic	4.3	2.6	7.1	3.4
	Electric	5.9	3.5	5.0	2.4
Saws		37.9	22.7	4.6	2.2

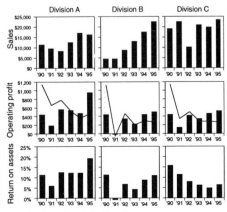

Two examples of matrixes included in the broader definition in which the terms table and matrix are used interchangeably

237

Matrix Display

A matrix display consists of multiple charts arranged in rows and columns in some organized fashion. The display's major purpose is to simplify the analysis of large quantities of data by enabling the viewer to study multiple charts at one time. Examples of two of the more widely used matrix display formats are shown below.

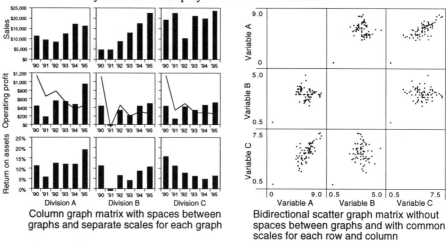

Column graph matrix with spaces between graphs and separate scales for each graph

Bidirectional scatter graph matrix without spaces between graphs and with common scales for each row and column

General observations regarding matrixes

- Matrixes are used for analysis, monitoring, and communication purposes.
- They have potential applications in most fields.
- Individual scales can be used for each graph, or a single scale can be used for an entire row or column (see examples above). When one is simply looking for correlations or deviations, scales are sometimes not used at all.
- Other than legibility, there is no limit on the number of rows and columns that can be incorporated into a single display.
- Derived values such as deviations can be used as effectively as actual values.
- Individual charts may abut one another or have spaces between them (see examples above).
- Most of the enhancement techniques used with individual charts can be used with charts in matrixes (e.g., curve fitting, color coding, confidence limits, the use of symbols, etc.).
- In many cases, such as monitoring for deviations and trends or when looking for correlations, the individual charts can be very small without adverse effects.
- Almost any type of chart can be used. A matrix might be made up of all the same type of charts or a combination of several types of charts. See example below.

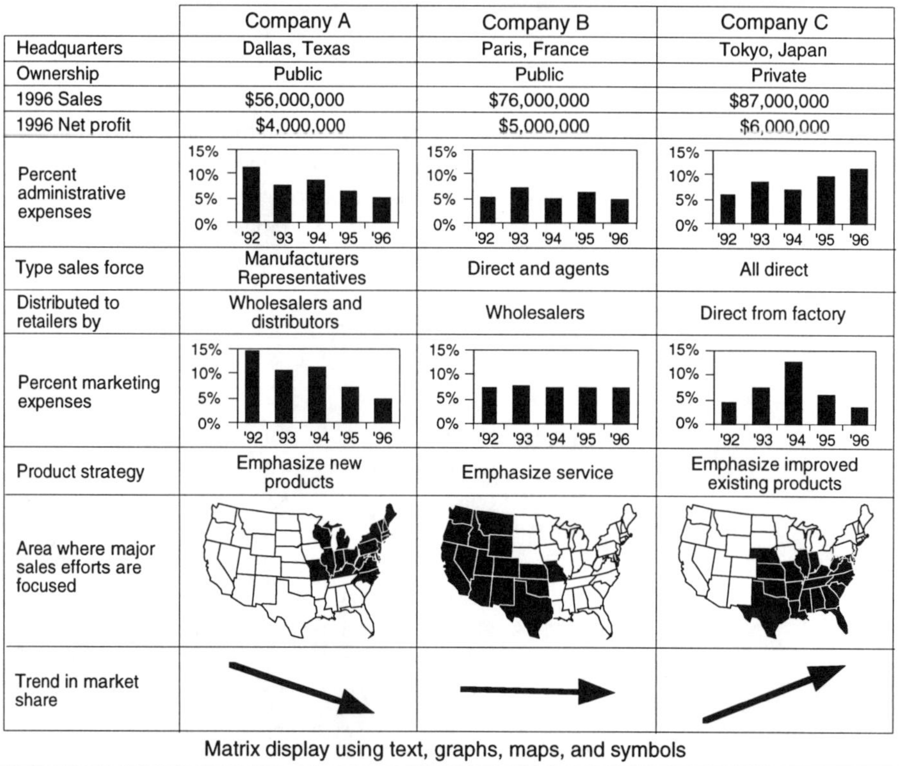

	Company A	Company B	Company C
Headquarters	Dallas, Texas	Paris, France	Tokyo, Japan
Ownership	Public	Public	Private
1996 Sales	$56,000,000	$76,000,000	$87,000,000
1996 Net profit	$4,000,000	$5,000,000	$6,000,000
Percent administrative expenses			
Type sales force	Manufacturers Representatives	Direct and agents	All direct
Distributed to retailers by	Wholesalers and distributors	Wholesalers	Direct from factory
Percent marketing expenses			
Product strategy	Emphasize new products	Emphasize service	Emphasize improved existing products
Area where major sales efforts are focused			
Trend in market share			

Matrix display using text, graphs, maps, and symbols

238

Matrix Display (continued)

Scatter graph matrix

When all the charts in a matrix are scatter graphs, the display is sometimes called a scatter graph matrix or stacked scatter graph. This type of display is frequently used to analyze a series of variables for correlations. For example, if four sets of data are being studied for correlations, they could be put in a matrix form, as shown at the right, and all combinations of the four sets could be observed at one time. If only correlations are being studied, scales may not be necessary. The graphs in the examples shaded gray are

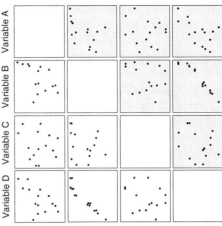

Full scatter graph matrix. The graphs shaded gray are duplicates of the unshaded graphs.

actually duplicates of the unshaded graphs, except for a different orientation. The duplicates can be eliminated to conserve space, resulting in the half-matrix shown at the left. Although space is conserved, the half-matrix is a little more difficult

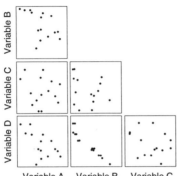

Half-matrix with the duplicate graphs (shown in graph above) eliminated

to analyze because the viewer must visually follow along both rows and columns to check for possible correlations with a particular variable.

• Distribution graphs are sometimes placed in the blank cells in scatter graph matrixes to display the distribution of one set of data. For instance, in the example at the right, there are three small histograms displaying the distributions of the three sets of data represented in the display.

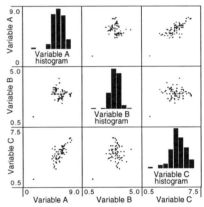

Scatter graph matrix with individual distribution graphs

Icon comparison display

Another type of matrix display assembles icons or symbols for simultaneously comparing or screening a sizable number of entities with regards to three or more variables (sometimes referred to as multivariate). In this type of display, a separate icon is generated for each

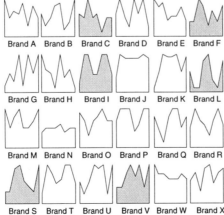

entity under review. The icons are then assembled into a matrix for comparison, as shown at the left. A template (as shown below) is required for encoding and decoding the information. See Icon Comparison Display.

Example of how icon symbols might be used to compare 24 brands of desktop computers with regards to 8 different characteristics. The gray plots might represent foreign-built brands.

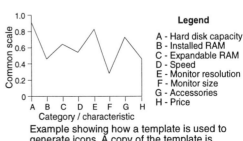

Example showing how a template is used to generate icons. A copy of the template is generally included in a legend by the display.

Mean	Also called average. The mean of a set of data is a value equal to the sum of the values of all the data elements in the set of data, divided by the number of data elements in the set. For example, the mean of the seven-element data set 1, 2, 2.5, 3, 4, 4.5, and 5 is 3.14.
Mean Graph	A graph that plots the mean (average) of multiple sets of data. For example, a graph that displays the average heights of individuals by country for ten different countries might be called a mean graph. Any of the basic types of graphs can be used to construct a mean graph.
Median	The number in a set of data that has half the data elements in the set with values less than it and half with values greater than it. There are no mathematical calculations required to determine the median. All that has to be done is to arrange the values of a data set in ascending or descending order. This can be done in tabular or graphical form. The graphical form is shown below. The value of the data element in the center is the median value. In the

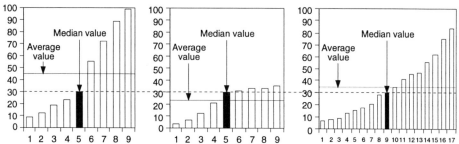

An illustration of how median values are independent of the exact values of the other data elements in a data set, the average of the data set, or the number of data elements in the set. Medians are determined strictly by their position in the set.

three examples shown here, thirty is the median since that is the value of the data element having the same number of data elements above it as below it. The median value is independent of the exact values of the other data elements in the set, the average of the data set, or the number of data elements in the set.

Mercator Projection Map	A well known and widely distributed world map. One of its key features is that directions are correct between any two points within a reasonable distance of each other. This feature makes the Mercator map particularly valuable to navigators. Shapes of land masses are fairly representative at all points on the map. Distances and areas are somewhat accurate near the equator but get progressively more distorted the further they are from the equator. The distortion is extreme in the polar areas.

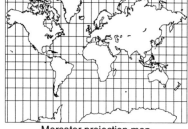

Mercator projection map

Meridian	Meridians are the north-south grid lines (graticules) on maps and globes. On a globe, all meridians meet at the North and South Poles. The meridian that passes through Greenwich, England, is called the prime meridian and is the meridian from which all other meridians are referenced in terms of degrees. Since meridians are used to measure longitudinal distances (east and west), they are also sometimes referred to as longitudinal lines. See Map.

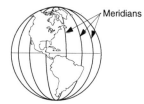

Examples of meridians

Milestone Chart	A variation of a time and activity bar chart in which major projects are broken into smaller tasks that have identifiable activities at their beginning and end. For example, when building a house, pouring the foundation, putting the roof on, and installing the plumbing are examples of subtasks with identifiable activities at the beginning and end. When a milestone chart is generated, each subtask is assigned a bar with a symbol (milestone) at both ends. When a subtask has been started, the symbol (milestone) at the left is filled or modified in some way to communicate to the viewer that the task has been started. When the subtask is completed the milestone at the right end is modified for the same purpose. This procedure is repeated until all milestones indicate that all subtasks are finished and the overall project is complete. See Time and Activity Bar Chart.

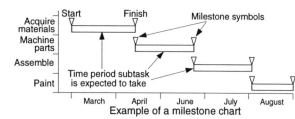

Example of a milestone chart

Minimal Spanning Tree
or
Minimum Spanning Tree

A planning diagram consisting of a series of points called nodes and a network of lines called arcs, branches, paths, or links that connect (span) the points. The points (coded circles) represent the locations of things such as cities, computer terminals, pumping stations, communications centers, etc. The lines generally represent the things that will connect the locations, such as highways, cables, pipe lines, and microwave corridors. The key objective when determining where the lines should go is to assure that every point (location) is connected to the network and that the total length of all the paths (connecting lines) is the shortest possible. For example, when building a subway system, a minimal spanning tree might be constructed to display the shortest length of subway tunnels that can be built and still provide a way to get from every community (node) to all other communities. A basic characteristic of a minimal spanning tree is that there is a single path to get from one node to any other node. Other than the single path shown in the minimal spanning tree, there is no other way to get from one point to another even, by taking branches that pass through other nodes. For instance, in the example at the right, there is one and only one path to get from point A to point K. Minimal spanning trees might use arrows if the movement is in only one direction, such as the flow of water through water mains. The lines between the nodes may or may not be drawn to scale. There seldom is any effort to show the exact path the branch takes. For instance, there may be many bends and turns in the path between points C and E, but on the diagram the path is represented as a straight line.

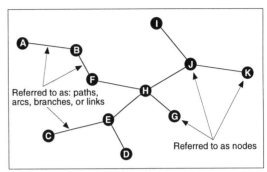

Example of a minimal spanning tree used for planning purposes in which all points are connected to a proposed network in such a way that there is only one path between any two points and the total of all the paths is the shortest possible distance.

Mixed Graph

Sometimes referred to as a combination, composite, or overlay graph. A mixed graph displays multiple data series, using two or more types of data graphics to represent them. See Combination Graph.

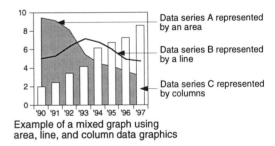

Example of a mixed graph using area, line, and column data graphics

Mode

The mode is the value or class interval that occurs most frequently in a set of data. The highest point in a histogram or frequency polygon designates the mode. See Histogram and Frequency Polygon.

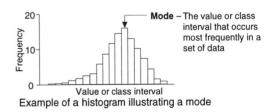

Example of a histogram illustrating a mode

"More Than" Graph

A "more than" graph is a variation of a cumulative frequency graph. It is constructed such that one can read directly from the graph the percentage of data elements in a data set that have values more than a certain value. For example, in the graph at the right, one can determine that approximately 37% of all of the data elements in the data set have values equal to or more than 2.1. A cumulative histogram is included in the example for reference purposes only. In an actual application, the cumulative histogram would not be shown. See Cumulative Frequency Graph.

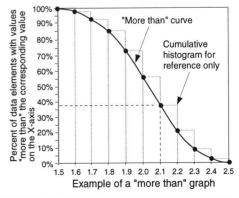

Example of a "more than" graph

241

Mosaic Graph

A mosaic graph is made up of a series of interspersed 100% stacked column graphs and 100% stacked bar graphs. Its major function is to display a system of interrelated values in such a way that groupings and relative sizes of the many elements can be seen at the same time. The steps involved in constructing such a graph are used to describe the nature of the graph. A manufacturing company is used in this example; however, a mosaic graph can be used in many different applications.

Step 1

In this example, how a company's overall sales dollars are allocated is analyzed. Therefore, the full length of the mosaic graph, which is a 100% stacked bar graph, represents total sales dollars. The components of the bar graph are the five major categories for which the sales dollars are used. A zero to 100% scale is displayed at the bottom of the bar, so the viewer can determine the percent that each component represents of total sales. For example, 50% ($25 million) of the revenue was used to manufacture the product. Actual sales values may or may not be shown at the top.

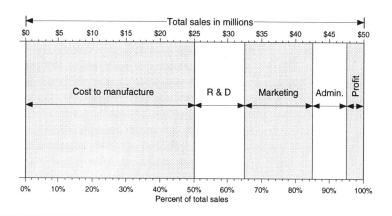

Step 2

Each segment of the horizontal bar graph is made into a 100% stacked column graph showing how the dollars are allocated within each major component. For example, in the cost to manufacture, it can be seen that there were three major types of costs: material, labor, and overhead. It can also be seen that material alone accounted for 55% of the total manufacturing costs. The same types of observations can also be made for R&D, Marketing, and Administrative.

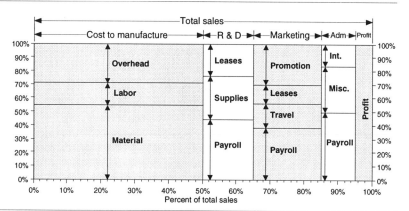

Step 3

In this step, each of the components of the column graphs are made into 100% bar graphs. When this is done with material costs, for example, it can be seen that about 37% went for raw material, 35% for fabricated parts, and about 28% for small parts. If one is only interested in a general feel for the data, additional scales are sometimes not required. If specific values are preferred, scales can be included with each group of smaller bar and column graphs, as shown in the cost to manufacture and marketing sections of the example.

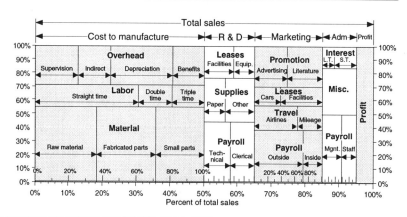

Final step

Additional sets of 100% graphs can be added; however, the graph begins getting very cluttered. In this final variation, the reference arrows have been removed since in practice they are generally not used. Shading or color may be added to highlight certain information, call attention to numbers that are related, aid the viewer in differentiating between various segments and subsegments, etc. Mosaic graphs can be made even more useful by noting the actual dollars and percents in each component and subcomponent.

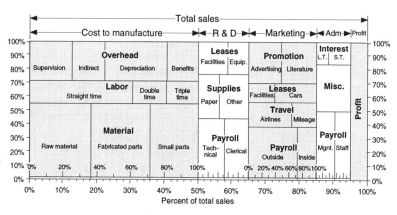

| **Moving Average** | Sometimes referred to as a rolling average or trend line. A method used to smooth the curve of a data series and make general trends more visible. The method involves generating a second curve of a data series with the short-term peaks and valleys smoothed out, as shown |

at the right. The degree to which the peaks and valleys are smoothed depends on the type of moving average and the number of intervals used. Each point on a moving average curve is generally calculated by averaging the value for the current period plus a fixed number of prior periods. Each time the value for a new period is added, the value for the oldest period in the previous calculation is dropped. For example, if monthly sales data were being tracked, a

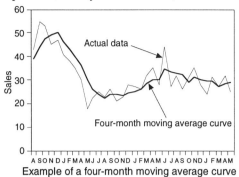

Example of a four-month moving average curve

three-month period might be used for the average. Thus, in March, the values for January, February, and March would be averaged and that point plotted. In April, the values for February, March, and April would be averaged and that point plotted. The number of prior time periods included in the average varies significantly. Three to 200 are commonly used, though there is no limit on the number of periods that can be included in the averaging process. • Occasionally, averages are calculated using what is called a centered moving average. With this process, the average is based on a given period plus an equal number of periods on either side, such as three in front and three behind. Moving average curves are primarily used with sequential data. The curves are generally superimposed over a graph of the actual data and in time phase with the actual data.

Effect of the number of intervals used in the average

As a general rule, the fewer the time intervals used in the averaging process, the more closely the moving average curve resembles the curve of the actual data. Conversely, the greater the number of intervals, the smoother the moving average curve. This is illustrated in the examples below which show three moving average curves for the same data series, each based on a different number of time periods used in the averaging process. Moving average curves tend to have a delayed reaction to changes.

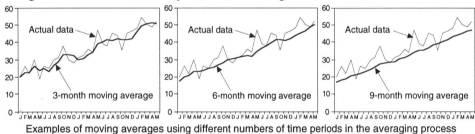

Examples of moving averages using different numbers of time periods in the averaging process

Three major types of moving average curves

There are three major types of moving average curves. They are:

– **Simple average** - Values plotted are based on averaging the actual values for a uniform number of periods.

– **Weighted average** - Values are calculated the same as for the simple average except each period used in the average is given a different weighting with the most recent value receiving the highest weighting.

– **Exponential average** - Similar to the weighted average variation, except that the weighting values decrease exponentially as the age of the data increases.

• Examples of a simple and a weighted six-month moving average curve are shown at the left. An example of an exponential is not included since exponential moving average curves often look like weighted moving average curves, depending on the weighting and exponential multipliers. Weighted and exponential curves generally are more responsive to short time fluctuations than simple moving averages because of the greater emphasis placed on the most current values.

Comparison of a simple and weighted moving average curve using the same number of periods in the averaging process

Moving Average (continued)

Use of moving averages by technical stock analysts

Technical stock market analysts sometimes use moving average graphs in ways not generally done in other fields. Three examples are shown below.

Moving average curves with different periods

Some analysts plot a fast moving average curve (e.g., 3 to 5 days) and a slow moving average curve (e.g., 7 to 20 days) on the same graph. When the two curves cross is sometimes considered as an indication to take some action. An example is shown at the right.

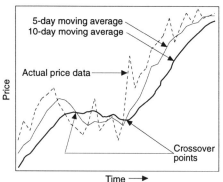

Example of multiple moving average curves with different time periods used in the averaging process.

Band or envelope formed by moving averages

Sometimes a moving average envelope or band is formed by generating two additional curves at a prescribed amount, normally a percent of the moving average, above and below a standard moving average curve. The percent varies depending on the analyst. Some analysts feel that when the stock price crosses one of the boundaries of the envelope, an appropriate action should be initiated. An example is shown at the left.

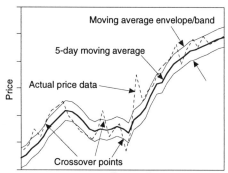

Example of a moving average curve with envelope/band

Shifted moving average curve

Sometimes one or more moving average curves are shifted or displaced horizontally. If only one displaced moving average curve is used, the analyst might look for the point where the stock price crosses the moving average curve. In other cases the single moving average curve might serve as a trend line. If two moving average curves are used, one shifted and one not shifted, the analyst might look for where the two cross as an indication to take some action. An example is shown at the right.

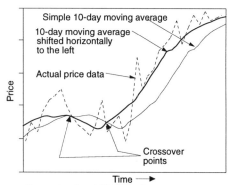

Example of a shifted moving average curve

Special terminology

Depending on the analyst, moving average curves might be based on open, high, low, or close stock prices or some combination of two or more of them. In addition to several unique applications of moving averages, technical stock analysts also use some specialized terminology. For example, moving averages with just a few periods used in the averaging process are referred to as fast moving averages. Accordingly, greater numbers of periods are referred to as medium or slow moving averages. Weighted and exponential moving averages are sometimes referred to as front loaded. Standard weighted moving average curves where each weighting factor is applied to two or more successive values of the original data are sometimes referred to as stepped weighted curves.

Moving Total Curve

Sometimes referred to as a rolling total curve. On a moving total curve, the value plotted for each interval is equal to the sum of the incremental value for that interval, plus the values for a fixed number of preceding intervals. For instance, if one is preparing an annual moving total curve, the value plotted in May, for example, will equal the actual value for May plus the values for the preceding eleven months, making it a 12-month or annual total. In June, the plotted value will be equal to the actual value for June, plus the values for the preceding eleven months. Any number of intervals can be used in the total. Whatever number of intervals are selected, that same number is used throughout the entire graph. Incremental values are many times plotted on the same graph with the moving total, as shown at the right. The purpose of moving total curves is

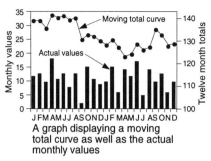

A graph displaying a moving total curve as well as the actual monthly values

to smooth out the interval-to-interval fluctuations and provide a graphic indication of the overall trend of the data. In the example shown here, the moving total indicates a general decline in the annualized sales until the first quarter of the second year when they seem to start up again. Such a trend is not as obvious in the monthly values.

Multiple Area Graph

Sometimes referred to as an overlapped or grouped area graph. A multiple area graph displays multiple data series, all of which are measured from the same zero base line, usually the horizontal axis. As a result of the data series extending down to the zero line, the values for each data series can be read directly from the graph. This type of area graph should be used with caution because it can easily be confused with the more familiar stacked area graph. See Area Graph.

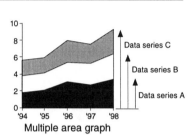

Multiple area graph

Multiple Bar Graph

When two or more data series are plotted side-by-side on the same bar graph it is called a clustered bar, grouped bar, multiple bar, or side-by-side bar graph. The bars for a given data series are always in the same position in each group throughout a given graph. Each data series typically is a different color, shade, or pattern. See Bar Graph.

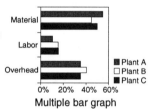

Multiple bar graph

Multiple Choice Decision Diagram

If more than two options are possible at decision points on a decision diagram, the chart is sometimes called a multiple-choice decision diagram. See Decision Diagram.

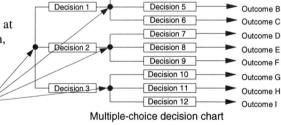

Multiple-choice decision chart

Multiple Column Graph

Sometimes called a clustered, grouped, or side-by-side column graph. A column graph in which two or more data series are plotted side-by-side. The columns for a given data series are always in the same position in each group throughout a given graph. Each data series typically is a different color, shade or pattern. See Column Graph.

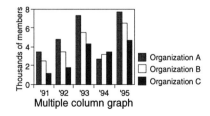

Multiple column graph

Multiple Line Graph

Sometimes referred to as a compound line, dual line, grouped line, or overlapped line graph. A multiple line graph has multiple data series, all of which are measured from the same zero base line axis. See Line Graph.

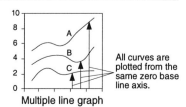

Multiple line graph

245

Multiple Point Graph
or
Multiple Scatter Graph

A point graph with multiple data series, all of which are measured from the same zero base line axis, is referred to as a grouped or multiple point graph if one of the axes has a category scale. If there are quantitative scales on both the vertical and horizontal axes, the graph is frequently called a grouped or multiple scatter graph. See Point Graph and Scatter Graph.

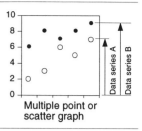

Multiple point or scatter graph

Multiple Strata Graph

Sometimes called a layer area, divided, stacked area, strata, stratum, subdivided area, or subdivided surface graph. A multiple strata graph is an area graph in which multiple data series are stacked on top of one another. This is the most frequently used area graph format. See Area Graph.

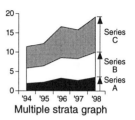

Multiple strata graph

Multiple Time Series Graph

A multiple time series graph displays two or more time series curves, each of which has its own time scale. The major purpose of such graphs is generally to compare things during similar time periods. For example, the graph might compare population growth in two post-war periods or prices during two periods of inflation. See Time Series Graph.

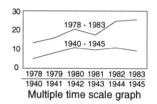

Multiple time scale graph

Multivariate
or
Multivariate Chart

Multivariate means two or more variables, thus, a multivariate chart is a chart that displays two or more variables. With two variables, the term bivariate is sometimes used. With three variables the term trivariate is sometimes used.

Multiway Table

A table displaying four or more variables distributed between the two axes and the body of the table. Two examples of three-way tables with four variables are shown at the right. In one example, two variables are displayed along the horizontal axis, one on the vertical axis, and the fourth in the body of the table. In the other, two variables are displayed on the vertical axis and one on the horizontal. In this type of multi-way table, an additional set of columns or rows is introduced for every additional variable over three. The data for both three-way tables was taken from the one-way table below.

Year	Sex	Age	Value
1980	Male	Under 40	69
1980	Male	Over 40	49
1980	Female	Under 40	21
1980	Female	Over 40	27
1990	Male	Under 40	41
1990	Male	Over 40	66
1990	Female	Under 40	43
1990	Female	Over 40	52

A one-way table of the same data displayed in the multiway tables at the right

• An alternative multiway table format uses the equivalent of a series of two-way tables, as shown at the right. In this design, an additional two-way table is generated for each additional variable over three. In this way large numbers of variables can be displayed. As the number of variables increases, however, it becomes more and more difficult to spot trends and relationships within the data.

Two variables along horizontal axis

	Male		Female	
	Under 40 years old	Over 40 years old	Under 40 years old	Over 40 years old
1980	69	49	21	27
1990	41	66	43	52

One variable along vertical axis

Fourth variable in body of table (shaded area)

Two variables on the horizontal axis, one on the vertical axis and one in the body of the table

Fourth variable in body of table (shaded area)

One variable along horizontal axis

		1980	1990
Male	Under 40 years old	69	41
	Over 40 years old	49	66
Female	Under 40 years old	21	43
	Over 40 years old	27	52

Two variables along vertical axis

Two variables on the vertical axis, one on the horizontal axis and one in the body of the table

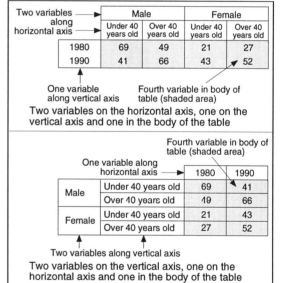

Examples of a multiway table design in which an additional set of columns or rows is introduced for each additional variable over three.

One variable on the vertical axis

One variable on the horizontal axis

	Under 40 years old	Over 40 years old
Male		
1980	69	49
1990	41	66
Female		
1980	21	27
1990	43	52

One variable as spanner headings

One variable in the body of the table

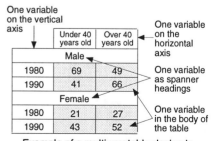

Example of a multiway table design in which an additional two-way table is generated for each additional variable over three.

Neat Line	Sometimes called a frame or border. A neat line is is a series of lines that form the boundary of the active portion of a map. Neat lines frequently are grid lines or graticules.

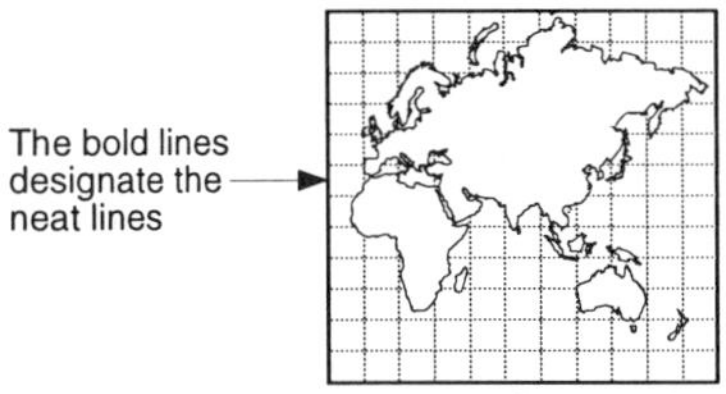

The bold lines designate the neat lines →

Example of a neat line

Needle Graph	Sometimes referred to as a spike or vertical line graph. On this type of graph, a separate vertical line is used to designate each individual data point. The top of the vertical lines designate the actual data points. Such graphs are often used when many successive data points are to be plotted at uniform intervals. Needles (vertical lines) can be used with two- and three-dimensional graphs. See Line Graph.

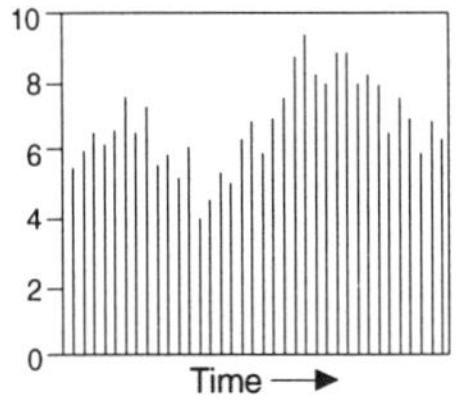

Needle graph

Net Deviation Graph **or** **Net Difference Graph**	A variation of a difference graph. See Difference Graph.

Network **or** **Network Diagram**	A diagram or map that shows how things are interconnected. A network diagram may show interconnected roads, or several pieces of computer equipment connected by wires. The definition is sometimes extended to include how such things as people and events are interrelated, – for example, how lines of responsibility connect people in an organization chart, or how events are related in PERT and CPM charts. In many cases, little or no attempt is made to show the elements of the network in their proper physical locations or relative sizes. The things connected might be identified by words, symbols, or pictorial representations. Three examples are shown here.

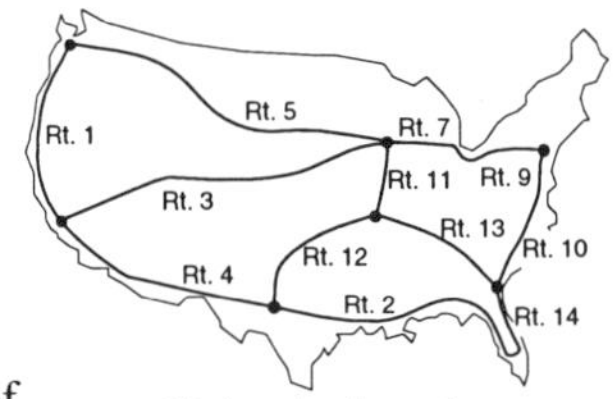

Network of roads.

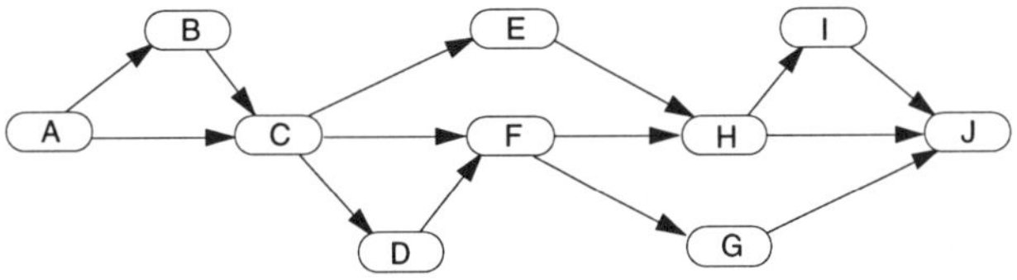

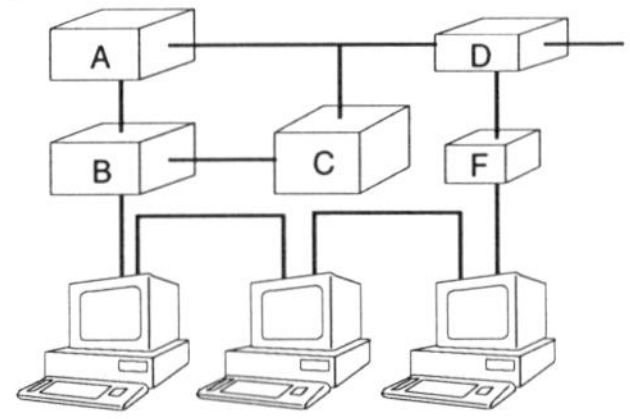

Network of computer equipment

PERT or CPM network

Nightingale Rose	Sometimes referred to as a sector graph. See Sector Graph.
No Class Map	Sometimes referred to as an unclassed or classless map. A map that does not use class intervals to encode statistical information. See Statistical Map.
Node	As one of the major building blocks in many types of diagrams, nodes denote the junction of two or more lines, arrows, links, etc. Nodes are graphical representations of events, activities, individuals, pieces of equipment, functions, decision points, etc. A typical node is made up of a symbol, such as a square, circle, or rectangle with words and/or numbers inside for identification or description. The shape of the symbol sometimes conveys information as to the type of thing or event it represents, such as a decision, city, data input, etc. Nodes are typically connected by links (lines or arrows) which show the relationships, directions of movement, channels of communication, etc., between nodes.

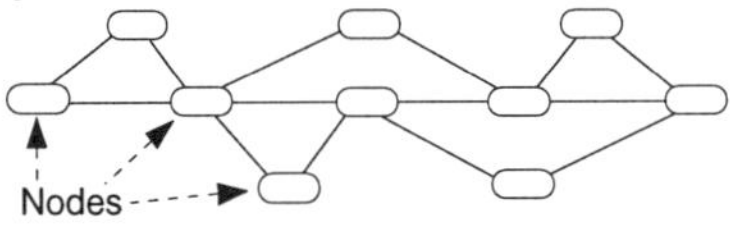

Nodes - - - -
Examples of nodes used in a diagram

Node Diagram	Sometimes referred to as PERT chart. See PERT Chart.
Nominal Axis	An axis with a nominal scale on it.
Nominal Scale	Sometimes referred to as a category or qualitative scale. A scale consisting of a series of words and/or numbers that name, identify, and/or describe people, places, things, or events. The items on the scale do not have to be in any particular order. When numbers are used on a nominal scale, they are for identification purposes only, since nominal scales are not quantitative. Each word or number defines a distinct category containing one or more entities. In other words, a category might refer to one man or many men. See Scale.

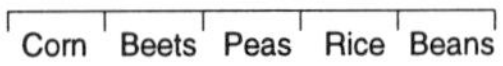

Corn | Beets | Peas | Rice | Beans
Example of a nominal scale. Sometimes called category or qualitative scale.

247

**Nomograph
or
Nomogram**

Sometimes referred to as a calculation graph or alignment graph. Nomographs are sometimes thought of as special purpose calculators or reference tables. They are particularly useful in situations where a given calculation is done repetitively or where a convenient, quick reference document is desired. A nomograph is designed to solve an

equation involving three or more variables. It consists of three or more scales arranged so that a straight line crossing each of the scales intersects the scales at values that satisfy the equation. The equation might be complex, or as simple as A + B = C. An example of a nomograph for this simple equation is shown at the right. The three parallel vertical scales are labeled to correspond to the variables in the equation (e.g., A,B, and C). The dashed lines crossing the three scales represent solutions. For example, the lower line crosses C at 5, which is the sum of 1 (on scale A) and 4 (on scale B).

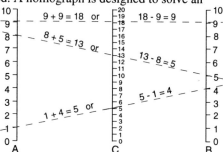

Example of a nomograph for addition or subtraction. A straight line connecting numbers in scales A and B crosses scale C at a number equal to the sum of the two numbers connected in scales A and B. Alternately, a straight line connecting numbers in scales A and C crosses scale B at a number equal to the difference between the numbers in C and A.

• The two examples below illustrate several features of monographs. For example:
  – Multiplication and division can be performed as easily as addition and subtraction
  – Linear and nonlinear scales are used
  – Nomographs can be arranged multiple ways to perform the same function

The graph on the left also demonstrates a unique feature of nomographs that allows the user to pivot a straight line around a fixed value for one of the variables to see all possible combinations of the other two variables that satisfy the equation. In this example, the lines show combinations of A and B that when multiplied together equal 50. In practice, a line might be pivoted around a desired liquid flow rate to see what combinations of pipe diameter and liquid velocity yield that flow rate.

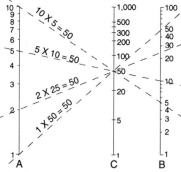
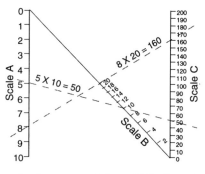

Example of a nomograph for multiplication using parallel nonlinear scales. A straight line connecting numbers in scales A and B crosses scale C at a number equal to the product of the two numbers connected in scales A and B. A reverse procedure is used for division.

Example of a nomograph for multiplication using parallel linear scales. A straight line connecting numbers in scales A and B crosses scale C at a number equal to the product of the two numbers connected in scales A and B. A reverse procedure is used for division.

Two different types of nomographs used to perform multiplication

The upper and lower limits on the scales are generally based on what is realistic for the particular situation. Many scales do not start at zero. • The precision of the solutions is proportional to the size of the graph and detail of the scales. • The configuration of the graphs and the complexity of equations they can solve is very diverse (examples below).

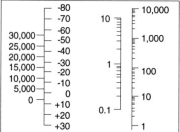
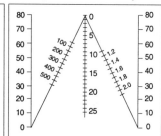
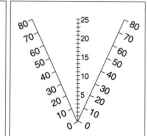

Example of a chart with plus and minus values, log and linear scales and two scales on one axis

Example of a double N or Z chart for situations with four variables

Example of a V-shaped chart for special applications

Nonlinear	Any relationship between two sets of data that is not linear. See Linear.

Nonquantitative Map	Sometimes referred to as an explanatory, descriptive, or qualitative map. This type of map is generally used to show nonquantitative information such as where things are located, how areas are organized or subdivided, and where the routes are located that get people and things from one place to another.

Nonquantitative maps may include grain fields, buildings, rivers, cities, roads, land masses, power lines, sewer lines, sales areas, public facilities, voting precincts, schools, political divisions, resorts, pet stores, sales areas, etc. See Descriptive Map.

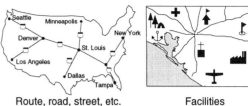

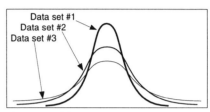

Route, road, street, etc.   Facilities

Examples of nonquantitative maps

Normal Distribution Curve
or
Normal Frequency Curve

A normal distribution curve is a frequency polygon for a set of data that has a normal distribution. The exact shape of normal distribution curves differ from data set to data set; however, all normal distribution curves have a similar bell-shaped appearance, as shown at the right. A normal distribution is one of the most common types of distributions. They often occur in nature. All data sets with normal distributions have certain distinguishing characteristics including:

Examples of three normal distribution curves with the characteristic bell shape. The shape of the curve reflects how spread out the data elements are within each set.

– The average, median, and mode are all the same and are located at the center of a curve of the distribution;
– The curve is symmetrical about the average value; and
– The distribution of data elements follow well-defined patterns, as described below.

Normal distribution curves are broken into the following eight segments along the horizontal axis:

– Three equal segments to the left of the average, called standard deviations, plus one segment that includes all values below the three standard deviations.
– Three equal segments to the right of the average, called standard deviations, plus one segment that includes all values above the three standard deviations.

Each data set has a unique standard deviation based on the values of the data elements and how spread out they are. All standard deviations (sometimes designated with the symbol σ) for a given data set have the same value. Data elements, for all data sets with normal distributions, are distributed approximately as shown in the illustration below.

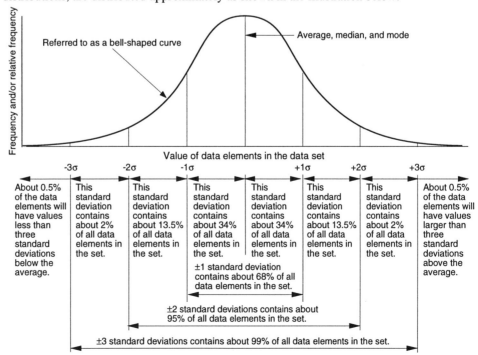

Illustration of the theoretical distribution of data elements in a data set with a normal distribution

249

Normal Distribution Curve
or
Normal Frequency Curve
(continued)

When the cumulative percent frequency data for a data set with a normal distribution is plotted on a linear grid, the curve resembles an elongated S, as shown at right. This type of graph is sometimes called a cumulative frequency or ogive graph.

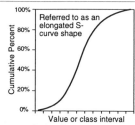

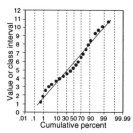

When the cumulative frequency data for a data set with a normal distribution are plotted on a normal probability grid, the data points cluster around a straight line.

When the cumulative frequency data for a data set with a normal distribution are plotted on a graph with linear scales, the data points tend to form an elongated S.

When the cumulative percent frequency data for a data set with a normal distribution is plotted on a normal probability grid, the curve approximates a straight line, as shown at left. This type of graph can be used to determine if a data set has a normal distribution.

Normal Probability Graph

Normal probability graphs are special applications of scatter graphs. This type of graph is used to show how closely the elements of a set of data approach a normal distribution. Two widely used variations are the cumulative frequency and quantile type graphs. In both cases the data points of a set of data with a normal distribution approach a straight line when plotted on the special sets of grids and scales associated with these graphs. See Cumulative Frequency Graph and Quantile Graph.

———— **Cumulative frequency graph** ————

To determine whether a data set has a normal distribution using this type of graph, the cumulative relative frequency data are plotted on a normal probability grid. The special grid has a non-linear percent scale on the horizontal axis; therefore, a computer or special graph paper is generally used to generate these graphs. If the distribution is normal, the data points tend to cluster around a straight line.

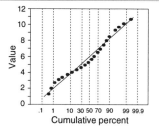

Cumulative frequency data plotted on a normal probability grid

———— **Quantile graph** ————

When using a quantile graph to determine whether a data set has a normal distribution, the data set under study is compared against a theoretical data set with a known normal distribution, an average value of zero, and a standard deviation (and variance) of one. If the distribution of the data set under study is normal, the data points will cluster around a straight line.

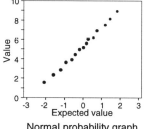

Normal probability graph using quantile data

Number Axis
or
Number Line

Sometimes referred to as a one-axis distribution graph, or univariate graph. The major purpose of this type of graph is to show graphically how the data elements of a data set are distributed. Based on such charts, one might observe such things as the largest and smallest values, whether the points are clustered and if so where, whether the points are skewed towards one end or the other, whether there are many or few data elements in the set, and whether there are any points at the extremes that look like they don't belong with the series. See One-Axis Data Distribution Graph.

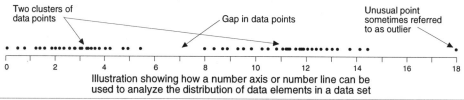

Illustration showing how a number axis or number line can be used to analyze the distribution of data elements in a data set

Aligning columns of numbers

Columns of numbers are typically aligned on the decimal point when one or more of the numbers in the column has a decimal point. When no decimal points are shown, a column of numbers is generally aligned to the right side.

5,869.12	2,578
18.5234	450,850
4,215,632.	289,352
18.65782	14
-254.123	369
4,521.2565	14,723
.0047	9
-127.58	36,251
Column of numbers aligned on the decimal points	Column of numbers without decimal points aligned to the right

Commas

Commas after every third digit to the left of the decimal generally make large numbers easier to read. Commas are not used to the right of the decimal.

200000000	200,000,000
100000	100,000
-1000	-1,000
-200000000	-200,000,000
Without commas	With commas

Currency

– When numbers representing positive dollar values are shown, they generally assume one of the formats shown at the right.

$200,000	$200,000.00	200,000	200
$100,000	$100,000.00	100,000	100
$50,000	$50,000.00	50,000	50
$5,000	$5,000.00	5,000	5
$0	$0.00	0	0
Dollar signs and no decimals to designate cents	Dollar signs and two decimal places for cents	Units of measure (dollars) shown in title [Dollars]	Units of measure and multiplier shown in title [Dollars (000)]

– Negative dollar values introduce additional variations, as shown at the right. Financial and accounting personnel many times prefer parentheses, but there is no general agreement whether the front parentheses should be before or after the dollar sign.

$–200.00	–$200.00	$(200.00)	($200.00)
$–100.00	–$100.00	$(100.00)	($100.00)
$–50.00	–$50.00	$(50.00)	($50.00)
Minus sign between dollar sign and number	Minus sign in front of dollar sign	Parentheses between dollar sign and number	Parentheses in front of dollar sign

– For values less than one dollar, the units of measure are sometimes noted in cents rather than dollars. Whichever method is used it should remain consistent throughout a given chart.

$0.97 .... 97¢	
$0.23 .... 23¢	
$0.07 .... 7¢	
Values in dollars	Values in cents

Fractions

Whether a fraction or decimal is used depends on the particular situation. On the whole, decimals are used much more widely than fractions. When the same size type is used for a whole number and a fraction, a hyphen is frequently placed between the two to avoid misunderstanding. Hyphens are not used when the type size used for the fraction is noticeably smaller than the type size of the whole number.

Examples in which the type size of the whole number and the fraction are the same	Examples in which the type size of the fraction is noticeably smaller than that of the whole number	Decimal equivalent for reference only
1-1/4	$1\frac{1}{4}$ $1\frac{1}{4}$	1.25
3-1/2	$3\frac{1}{2}$ $3\frac{1}{2}$	3.50
7-3/4	$7\frac{3}{4}$ $7\frac{3}{4}$	7.75

Negative numbers

The two generally accepted ways of expressing negative numbers are with a minus sign in front of the number or with parentheses around the number. Accounting and financial personnel tend to prefer parentheses. When a negative number is considered undesirable, as in many financial applications, the number is sometimes also shown in red.

Minus sign	–50,000
	$–50,000
	–$50,000
Parentheses	(50,000)
	$(50,000)
	($50,000)
Examples of negative numbers	

Percent

Numbers that represent percents are typically shown in one of three ways:
– With percent sign behind each value. If in column form, a percent sign may be shown only with the top value. If there is a total at the bottom of the column, a percent sign may be shown with the total also.
– Specified in the scale title, either in word form or as a percent sign
– As a decimal equivalent determined by dividing the percent value by 100.
All three methods are applicable with positive and negative values as well as decimals.

200%	138.20%	209	2.09
100%	68.06%	140	1.40
50%	14.75%	52	0.52
0%	0.00%	0	0.00
-50%	-50.10%	-23	-0.23
-100%	-75.25%	-115	-1.15
Percent signs and no decimals	Percent signs and two decimal places. Any number of decimal places can be used but generally no more than the largest number of significant digits after the decimal in any value.	Units of measure (percent) shown in title [Percent]	Sometimes decimal equivalents are used, in which case no percent sign is used.

251

Roman and Arabic numerals

The two major types of numerals used on information graphics are Roman and Arabic. For showing measured, counted, or calculated values, Arabic numerals are used almost exclusively. Roman numerals are used primarily for indicating headings, groupings, subdivisions, phases, quadrants, etc.

Roman	Arabic	Roman	Arabic	Roman	Arabic
I	1	VIII	8	LX	60
II	2	IX	9	LXX	70
III	3	X	10	LXXX	80
IV	4	XX	20	XC	90
V	5	XXX	30	C	100
VI	6	XL	40	D	500
VII	7	L	50	M	1,000

Examples of numbers made up of
Roman and Arabic numerals

Zeros before and after the decimal

0.190	.190	78.690	78.69
0.002	.002	8,521.002	8,521.002
0.174	.174	9.000	9.
0.400	.400	642.400	642.4
0.010	.010	34.011	34.011
0.700	.700	0.700	0.7

A zero is frequently placed in front of the decimal for values that have numerals only after the decimal.	No zero in front of the decimal is an acceptable alternative.	An equal number of digits after the decimal sometimes makes it easier for the viewer to compare values.	An unequal number of digits after the decimal is an acceptable alternative, particularly when there is a large difference in the number of significant digits after the decimal.

Zeros before the decimal   Zeros after the decimal

Very large numbers

Each column of numbers express the same values in different format.

General format	Multiplier shown in title (Value times 1,000)	Roman numerals or international symbols		(See note)	(See note)	
200,000	200	200M	200K	2.0E+5	200E+3	2.0x10^5
150,000	150	150M	150K	1.5E+5	150E+3	1.5x10^5
100,000	100	100M	100K	1.0E+5	100E+3	1.0x10^5
50,000	50	50M	50K	5.0E+4	50E+3	5.0x10^4
0	0	0	0	0.0E+0	0E+0	0.0x10^0

General format	Multiplier shown in title	Roman numerals or international symbols can be substituted for multiples of three zeros.	In this format all values have the same number of digits (1) left of the decimal, plus an indication of how many places the decimal point should be moved to the right.	In this format there are varying numbers of digits left of the decimal point, plus an indication as to how many places the decimal point should be moved to the right.	The number to the left of the times sign is to be multiplied by the number to the right of the times sign. (e.g., 5.0 x 10^4 means 5 should be multiplied by 10 four times.).

Note:

The capital E with a number behind it is one of several methods technical personnel use to designate large and small numbers. The number behind the E with a plus sign in front of it indicates how many places the decimal point should be moved to the right. For example, E+3 means the decimal point should be moved three places to the right.

Very small numbers

Each column of numbers express the same values in a different format.

General format	Multiplier shown in title (Value times 0.001)			
0.010000	10.000	1.00E-2	10.0E-3	1.00x10^{-2}
0.008170	8.170	8.17E-3	81.7E-4	8.17x10^{-3}
0.006000	6.000	6.00E-3	60.0E-4	6.00x10^{-3}
0.004070	4.070	4.07E-3	40.7E-4	4.07x10^{-3}
0.000205	0.205	2.05E-4	20.5E-5	2.05x10^{-4}
0.000005	0.005	5.00E-6	50.0E-7	5.00x10^{-6}

General format	Multiplier shown in title	The numbers in the column on the left have one digit to the left of the decimal, while those in the column on the right have two. The letter E with the negative number behind it indicates how many digits the decimal should be moved to the left (see note below).		The number to the left of the times sign is to be multiplied by the number to its right. (e.g., 5.00 x 10^{-6} means 5 should be multiplied by 0.1 six times)

Note:

The capital E with a number behind it is one of several methods technical personnel use to designate large and small numbers. The number behind the E with a minus sign in front of it indicates how many places the decimal point should be moved to the left. For example, E-3 means the decimal point should be moved three places to the left.

Angular numbers

Angular values are generally expressed in degrees or radians. There are three major variations as to how degrees might be expressed, depending on whether fractions of a degree are expressed as a decimal of a degree or as minutes and seconds and decimals thereof. (There are 60 minutes to a degree and 60 seconds to a minute in angular terminology). There are two major variations for expressing radians. One is to use the actual value of radians; one radian equals 57.296 degrees. The other is to express the radians in terms of a constant called pi (π), which has a value of 3.142. Examples of all five variations are shown below.

Degrees and decimals of degrees	Degrees, minutes, and decimals of minutes	Degrees, minutes, seconds, and decimals of seconds	Radians	Radians in terms of the constant pi (π)
57.296°	57°17.743'	57°17'44.6"	1.0	0.318π
90°	90°00'	90°00'00"	1.571	0.5π
114.591°	114°35.487'	114°35'29.2"	2.0	0.637π
171.887°	171°53.23'	171°53'13.8"	3.0	0.955π
180°	180°00'	180°00'00"	3.142	1.0π
229.183°	229°10.973'	229°10'58.4"	4.0	1.273π
270°	270°00'	270°00'00"	4.713	1.5π
360°	360°00'	360°00'00"	6.284	2.0π

Elapsed time numbers

There are four widely used methods for denoting elapsed time. The major difference between the four is the combination of hours, minutes, seconds, and decimal or fractions of each that are used. Examples of the four using decimals are shown below. Occasionally fractions are substituted for the decimals. • Because the hour and minute numbers (column two) and the minute and second numbers (column four) occasionally look alike, if not obvious, it is sometimes noted on the chart what times are being displayed to avoid confusion (e.g., 8:16 might be 8 hours and 16 minutes, or 8 minutes and 16 seconds).

Hours and decimals of hours	Hours, minutes, and decimals of minutes	Hours, minutes, seconds, and decimals of seconds	Minutes, seconds, and decimals of seconds
0.200	0:12	0:12:00	12:00
12.000	12:00	12:00:00	720:00
0.327	0:19.620	0:19:37.22	19:37.22
0.702	0:42.133	0:42:08	42:08
0.950	0:57	0:57:00	57:00
1.767	1:46	1:46:00	106:00
2.454	2:27.254	2:27:15.23	147:15.23
34.117	34:07	34:07:00	2,047:00

Actual time numbers

The major variable with regards to numbers used for time is whether a 12-hour or 24-hour clock is used. Examples of both are shown below.

Word description	12 hour clock	24 hour clock	
		Option #1	Option #2
One-twenty-seven in the morning	1:27 A.M.	1:27:00	0127
Fifteen minutes and seven seconds after nine in the morning	9:15:07 A.M	9:15:07	0915:07
Noon	12:00 P.M., 12:00M, or 12:00 N	12:00:00	1200
Eight minutes and six seconds after two in the afternoon	2:08:06 P.M	14:08:06	1408:06
Five-nineteen in the afternoon	5:19 P.M.	17:19:00	1719
Nine-forty-five in the evening	9:45 P.M.	21:45:00	2145
Midnight	12:00 P.M., 12:00 A.M., or 12:00 M	24:00:00 or 00:00:00	2400 or 0000

Numbers, Times, and Dates
(continued)

On many charts and graphs, space considerations require that dates and related information such as days and quarters be abbreviated. The abbreviations shown below are among the more widely used.

Years

	1990	1991	1992	1993
Option 1	'90	'91	'92	'93
Option 2	90	91	92	93

Quarters

Time periods, particularly years, are often broken into quarters. When quarters are applied to years, the first quarter might be the first three months of the calender year. If the fiscal year is different from the calendar year, the quarters frequently apply to the fiscal year, not the calendar year. If not obvious, the meaning is often specified in a note.

	First Quarter	Second Quarter	Third Quarter	Fourth Quarter
Option 1	1st Qtr	2nd Qtr	3rd Qtr	4th Qtr
Option 2	1st Q	2nd Q	3rd Q	4th Q
Option 3	Q1	Q2	Q3	Q4
Option 4	1st	2nd	3rd	4th

Months

Option 1	Jan.	Feb.	Mar.	Apr.	May	June	July	Aug.	Sept.	Oct.	Nov.	Dec.
Option 2	Jan	Feb	Mar	Apr	May	Jun	Jul	Aug	Sep	Oct	Nov	Dec
Option 3	Ja	F	Mr	Ap	My	Je	Jl	Ag	S	O	N	D
Option 4	J	F	M	A	M	J	J	A	S	O	N	D

Days

	Sunday	Monday	Tuesday	Wednesday	Thursday	Friday	Saturday
Option 1	Sun.	Mon.	Tues.	Wed.	Thurs.	Fri.	Sat.
Option 2	Sun	Mon	Tue	Wed	Thu	Fri	Sat
Option 3	Su	M	Tu	W	Th	F	Sa
Option 4	S	M	T	W	T	F	S

Dates

Some of the variations shown below are the result of different standards used in America, Europe, the military, the International Standards Organization, etc.

		The first day of the year in 1999	The twelfth day of June in the year 2000	The eighth day of November in the year 2005
Sequence of information: Month, day, year	Option 1	Jan 1, 1999	Jun 12, 2000	Nov 8, 2005
	Option 2	1/1/99	6/12/00	11/8/05
	Option 3	1-1-99	6-12-00	11-8-05
	Option 4	01/01/99	06/12/00	11/08/05
Sequence of information: Day, month, year	Option 5	1 Jan 1999	12 Jun 2000	8 Nov 2005
	Option 6	1 Jan 99	12 Jun 00	8 Nov 05
	Option 7	1.1.1999	12.6.2000	8.11.2005
	Option 8	1.1.99	12.6.00	8.11.05
	Option 9	1/1/99	12/6/00	8/11/05
	Option 10	1-1-99	12-6-00	8-11-05
Sequence of information: Year, month, day	Option 11	1999-01-01	2000-06-12	2005-11-08

Numeric Scale

Sometimes referred to as quantitative, value, amount, or interval scale. A numeric scale consists of numbers organized in an ordered sequence with meaningful and uniform spacings between them. The numbers are used for their quantitative value, as opposed to being used to identify entities. See Scale.

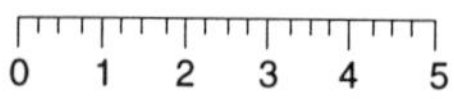

Example of a linear numeric scale. Other types of numeric scales include logarithmic, probability, and power.

Oblique Projection (View)

Most types of charts can be drawn so they appear three-dimensional. One reason for doing this is to provide information, such as displaying a third variable, that cannot be readily done with two-dimensional charts. A second is to improve the appearance of the charts. The two techniques that are most frequently used to give charts a three-dimensional appearance are called oblique projection (view) and axonometric projection (view).

Oblique projection

In the oblique view, surfaces that are parallel to the plane of the paper are shifted slightly with respect to one another. The shapes and dimensions of these surfaces do not change. Since a true two-dimensional object, such as a column on a two-dimensional graph, has only one surface, artificial depth must be added to give the appearance of shifting the back portions. The added depth generally has no significance other than cosmetic. The shifting can occur in any direction and, within reason, can be any amount without affecting the readability of the chart, since the front surfaces remain unchanged. The process is the equivalent of exposing or adding a top or bottom plus a side to the object. Examples are shown below.

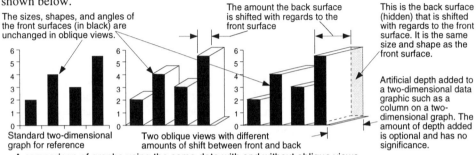

The sizes, shapes, and angles of the front surfaces (in black) are unchanged in oblique views.

The amount the back surface is shifted with regards to the front surface

This is the back surface (hidden) that is shifted with regards to the front surface. It is the same size and shape as the front surface.

Artificial depth added to a two-dimensional data graphic such as a column on a two-dimensional graph. The amount of depth added is optional and has no significance.

Standard two-dimensional graph for reference

Two oblique views with different amounts of shift between front and back

A comparison of graphs using the same data with and without oblique views

In graphs, the direction of the shift sometimes has an effect on how and where the scales, grid lines, and tick marks are positioned. This, in turn, has an effect on how easy or difficult it is to read the values of the individual components. To improve the accuracy of reading a graph with an oblique view, it is sometimes recommended that grid lines and two vertical scales be used, as shown at the right. Even when this is done, some viewers have difficulty determining which plane of the data graphic to read the values from. For example, in the upper two graphs at the right, the left scale applies to the front edges of the data graphics. In the lower two examples, the right scale applies to the front edges.

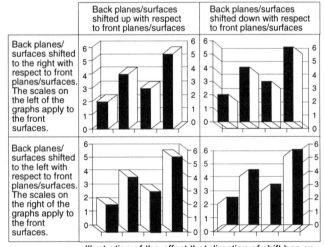

Back planes/surfaces shifted up with respect to front planes/surfaces

Back planes/surfaces shifted down with respect to front planes/surfaces

Back planes/surfaces shifted to the right with respect to front planes/surfaces. The scales on the left of the graphs apply to the front surfaces.

Back planes/surfaces shifted to the left with respect to front planes/surfaces. The scales on the right of the graphs apply to the front surfaces.

Illustration of the effect that direction of shift has on positioning of scales and the legibility of the data

• Oblique views can be used with graphs that have data on three axes, as shown at the left; however, an axonometric view is more often used with this type of graph. • In addition to graphs, oblique views are also used with many other types of charts, including flow charts, diagrams, maps, how-to charts, organization charts, etc. In most cases the primary reason for using oblique views with these types of charts is their aesthetic value. An organization chart using oblique views of the enclosures is shown below. Oblique views of such objects have the advantage over axonometric views in that, with oblique views, the front surfaces retain their true shapes and therefore text or graphics on the surfaces are not distorted.

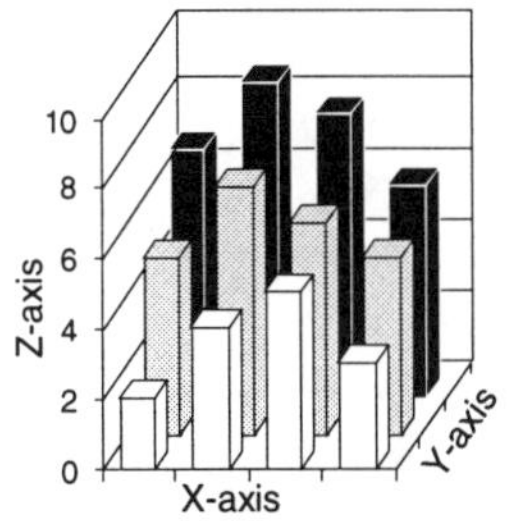

Example of an oblique view of a graph with data on all three axes. An axonometric view is more frequently used with three-axis graphs.

See Projection[2] for a comparison of graphs and maps generated using oblique and axonometric projections.

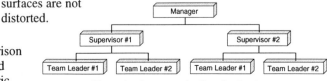

Example of an organization chart using oblique views of the enclosures

Octrant	One of the eight regions formed by the intersection of three perpendicular planes. A well known example is the octrants formed by the intersection of the X-Y, X-Z, and Y-Z planes on a three-dimensional graph.

Ogive Graph

Sometimes referred to as a cumulative frequency graph or summation graph. An ogive graph is the equivalent of a cumulative histogram with the columns replaced by a curve. It is a line graph with values or class intervals typically plotted on the horizontal axis and the cumulative number of times each of those values or groups of values occurs (cumulative frequency) plotted on the vertical axis. The frequency may be plotted in terms of cumulative counts or percentages. See Cumulative Frequency Graph.

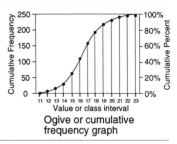

Ogive or cumulative
frequency graph

One-Axis Graph

When the scale and data are displayed along a single axis the graph is sometimes called a one-axis graph. The axis may run vertical or horizontal. For category type information, the data points can often be plotted directly onto the single axis. With quantitative information, the data is sometimes first plotted on a two-axis graph and the key points transferred to a single axis. Examples of one-axis graphs displaying category and quantitative data are shown below, along with two-axis graphs of the same data for comparison. • One-axis graphs offer several potential advantages over two-axis graphs, including:

    – They generally require less space and are easier for some people to read;

    – Additional space is sometimes available for labels and supporting information; and

    – In some situations the information can be decoded more accurately.

One-axis rectangular graph

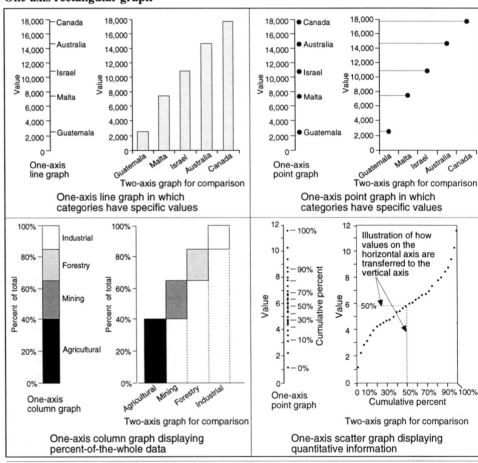

One-axis circle graph

When the same data in the one-axis column graph above is wrapped into a circle, as shown at the right, it is called a circle graph. When the scale is removed from a circle graph, it is called a pie chart. In both the single-axis column graph and the single axis circle graph, either units or percent can be plotted. Generally percent-of-the-whole data is plotted.

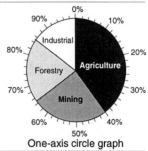

One-axis circle graph

One-Axis Graph
(continued)

Scales on one-axis graphs

One-axis graphs typically have only one scale. Exceptions are when multiple scales present the same information in different forms such as units and percents or miles and kilometers. The variable information is positioned along the axis based on that one scale. When the single scale is a time scale, the display is frequently called a time line chart. See Time Line Chart.

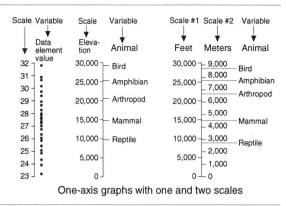

One-axis graphs with one and two scales

One-Dimensional Data Distribution Graph
or
One-Axis Data Distribution Graph

Sometimes referred to as a univariate graph, number line, or number axis. The one-axis data distribution graph is the simplest variation of a data distribution graph. One-axis data distribution graphs show graphically how the data elements of a data set are distributed by using a separate symbol to represent each data element of a data set. The data elements are plotted at their actual values along a single quantitative scale, as shown below. From such graphs the viewer can observe such things as the largest and smallest data elements in a set, whether the points are clustered and if so where, whether the points are skewed towards one end or the other, whether there are many or few data elements in the set, and whether there are any points at the extremes that look like they don't belong with the set.

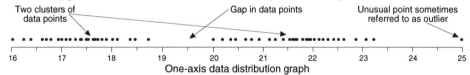

One-axis data distribution graph

One-axis data distribution graphs can also be used to note differences within a data set. For example, if the values are measured times for performing a particular task and half were measured on the day shift and half on the night shift, different symbols might be used for the two shifts to see if there are differences, as shown below. This is called using a dummy variable.

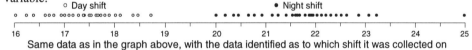

Same data as in the graph above, with the data identified as to which shift it was collected on

Any symbol can be used to designate the data points. Dots and lines/stripes (below) are the most widely used. • Overlapping of symbols can be misleading, since what appears to be a single symbol may actually be multiple symbols representing two or more data elements. Methods for addressing this problem are discussed under the heading Symbol.

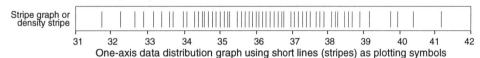

One-axis data distribution graph using short lines (stripes) as plotting symbols

One-axis data distribution graphs are sometimes used to display the distribution of data along the individual axes of scatter graphs. See Marginal Frequency Distribution Graph.

One Hundred Percent (100%) Chart or Graph

Sometimes referred to as a percent-of-the-whole chart or graph. A 100% chart or graph shows what percent each component represents of the whole (total). There are three major applications for this type of chart. They are: to analyze one group of data at a point in time; to analyze multiple groups of data at a point in time; and to analyze one group of data at multiple points in time or multiple situations or conditions. Pie charts, column graphs, bar graphs, and area graphs are most widely used for these applications. Shown below are representative examples. See Percent-of-the-Whole Chart/Graph.

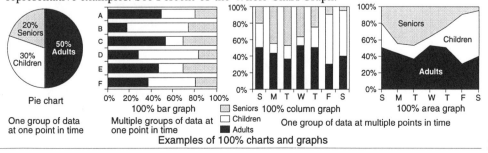

Examples of 100% charts and graphs

One-Way Table	A one-way table is a table displaying two or more variables along either the horizontal or vertical axis. The vertical axis is frequently used, and referred to as the list format of a one-way table. If only one variable is shown in a single column it is called a list. Examples of one-way tables with two, three, and four variables are shown below.

Name	Age
Alice	6
Bill	7
Harry	4
Sue	5

Without grid lines

Name	Age
Alice	6
Bill	7
Harry	4
Sue	5

With grid lines

Two variables along the vertical axis

Two variables along the horizontal axis →

Name	Alice	Bill	Harry	Sue
Age	6	7	4	5

One-way table with two variables

Year	Product	Value
1990	A	2
1990	B	10
1990	C	8
1990	D	22
1991	A	4
1991	B	12
1991	C	7
1991	D	29
1992	A	6
1992	B	17
1992	C	6
1992	D	20

One-way table with three variables

Year	Sex	Age	Value
1980	Male	Under 40	69
1980	Male	Over 40	49
1980	Female	Under 40	21
1980	Female	Over 40	27
1990	Male	Under 40	41
1990	Male	Over 40	66
1990	Female	Under 40	43
1990	Female	Over 40	52

One-way table with four variables

Examples of one-way tables in which all variables are displayed along a single axis

Open-High-Low-Close (OHLC) Graph	Sometimes referred to as a high-low-close-open (HLCO) graph, high-low-close (HLC) graph, high-low graph, bar chart, price chart, or vertical line chart. This type of chart is typically used to record and track the selling prices of securities, commodities, markets, etc. Several prices are typically recorded for each time period with highest, lowest, close, and open being the most common. See Bar Chart.

Operational Charts & Graphs	Operational charts and graphs are those charts and graphs that are used primarily for activities such as analyzing, planning, monitoring, decision making, and communicating in the on-going running of a business, organization, or activity. They are often used to supplement or replace tabulated data and written reports. • Another major classification is presentation charts and graphs, used primarily in formal and semiformal presentations.

• The following are among the reasons for using operational charts and graphs.
  – Data can be organized in such a way that analysis is more rapid and straightforward.
  – Viewers can more rapidly determine and absorb the essence of the information.
  – Large amounts of information can be conveniently and effectively reviewed.
  – Deviations, trends, and relationships stand out more clearly.
  – Comparisons and projections can many times be made more easily and accurately.
  – Key information tends to be remembered longer.
• The following are among the reasons for using presentation charts and graphs.
  – Assists in the rapid orientation of an audience.
  – Aids in the audience's understanding and retention of the material presented.
  – Material presented is frequently looked at more favorably by the audience.
  – The audience perceives the presenter to be more prepared, professional, interesting, etc.
Both types of charts and graphs tend to shorten meetings and expedite group decisions.
• Although many people do not differentiate between operational charts and charts used for presentations at formal and semiformal meetings, the two types generally differ significantly. The following table highlights some of the differences.

Characteristic	Operational Charts and Graphs	Presentation Charts and Graphs
Primary purpose	Gain maximum knowledge and benefits from available information	Communicate a specific, or at most, a limited number of messages
Medium	Primarily hard copy and computer screen	Primarily overhead projection or 35mm slides
Amount of information on one page	No limit as long as it is legible	Many sources recommend limits such as 7 words per line, 8 lines per chart, 30 words per page, etc.
Three-dimensional effects on graphs	Used sparingly because it generally reduces the legibility of data	Widely used for their aesthetic value
Grids and ticks	Frequently used to increase the accuracy of decoding the data	Used sparingly
Pictures, images, and perspective views	Used sparingly	Commonly used

A comparison of key characteristics for operational and presentation charts and graphs

The key difference between the two types of charts is their intended purpose as noted in the first row of the table. With presentation charts and graphs, the objective is generally a very focused communication. With operational charts and graphs the goals are much broader, with primary emphasis on analyzing, absorbing, managing, and communicating a maximum amount of meaningful information in a minimum amount of time.

Opposed Bar Graph	Sometimes called paired, sliding, two-way, or bilateral bar graph. See Bar Graph.

Ordering

Sometimes referred to as sorting. Ordering is the process of arranging information on a chart, graph, or table in some meaningful sequence. Four widely used methods are to alphabetize, to organize in numerical sequence (e.g., by time card number), to rank (e.g., place in ascending or descending order by quantitative value), and to arrange by attribute (e.g., address, product, religion, etc.). An example of each is shown at right. Ordering makes information easier to locate, particularly when material is alphbetized or in numerical sequence, and more meaningful for purposes of analysis, particularly when information is ranked.

• When there are several types of interrelated data, it is sometimes advantageous to order it in several different ways. For example, a company with six different product lines may wish to rank the lines in terms of sales, profit dollars, and profit as a percent of sales. If the data is ordered only one way, whether in tabular or graphical form, the viewer must mentally reorder the information to determine the other rankings. If the data is presented in a matrix form, as shown at the right, the viewer can focus on the analysis without having to do any mental rearranging.

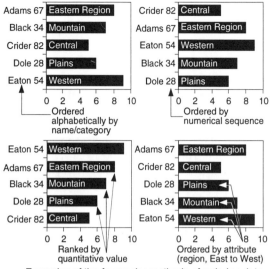

Examples of the four major methods of ordering data. The same information is shown in all four graphs.

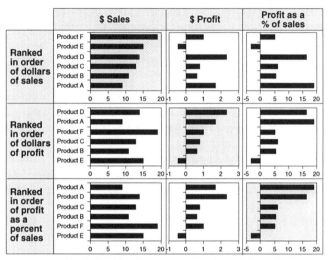

Matrix in which three groups of related information are ordered/ranked with respect to each of the three major characteristics being studied.

Order of Occurrence Scale

An order of occurrence scale is a variation of a sequence scale. It is generally used on graphs that record values for something that is repetitive or for events, activities, samples, etc., in which the sequence in which they occur is important. For example, in monitoring the quality of a process, it is generally important to record test values in the order the tests were conducted since, in addition to being interested in the actual values, one is often looking for trends. These trends may show up only when the data is plotted in the same order the tests occurred or the samples were collected. Order of occurrence scales are almost always on the horizontal axis. They proceed from left to right and generally have equal spacings between entries. Labels may or may not be used. Often, no labels are used since frequently the viewer is looking for trends and changes so the exact sample number is unimportant. Labels often are batch numbers, shop order numbers, patient numbers, serial numbers, etc. See Sequence Graph.

Ordinal Scale

An ordinal scale is a variation of a sequence scale that orders information in nonquantitative terms: first, second, third, fourth; biggest, second biggest, third biggest; small, medium, large; etc. Sometimes, for convenience, numbers are assigned to the entries on an ordinal scale. For instance, the number one might be assigned to small, number two to medium, and three to large. When this is done the numbers are used for identification purposes only and have no quantitative significance. Ordinal scales are most frequently used on the horizontal axis. The entries are generally equally spaced along the axis similar to categorical data. See Sequence Graph.

Ordinate	The distance a data point is located along the ordinate (vertical) axis.
Ordinate Axis	A technical name for the vertical or Y-axis on a two-dimensional graph.

Organization Chart

Organization charts are diagrams that show how people, operations, functions, equipment, activities, etc., are organized, arranged, structured, and/or interrelated. They are applicable with any size of organization. A typical organization chart consists of text enclosed in geometric shapes (sometimes referred to as boxes, enclosures, box enclosures, or symbols) that are connected with lines (sometimes referred to as links) or arrows. Charts of this type generally progress from top to bottom or left to right. Organization charts are sometime considered a variation of flow chart or flow diagram. Examples of several types of information frequently displayed on organization charts are shown below.

Typical types of information included on organization charts

Information on major organizational units

Charts with this type of information show the fundamental structure of an organization. They are applicable to any type of organization, whether a corporation, a government, a nonprofit organization, the military, etc. In some cases the enclosures are not connected with lines, indicating that they are part of a larger organizational structure but have no formal ties.

Information on functions performed

Whereas the above type of chart gives the name of a unit in an organization, this type of chart shows what the various parts of the organization are responsible for. As an example, a unit of a government might be shown as Senate in the above chart, while in this type of chart the corresponding box might read Enact Legislation. In the type below showing titles, the box might read Senator. • This type of chart sometimes gives insight into philosophies (e.g., whether laws are enacted and enforced by the same or separate units).

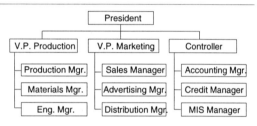

Information by title

This type of chart is often used to convey information about function as well as title and is frequently used in conjunction with the names of individuals holding the titles. When used to compare various organizations, charts with titles sometimes indicate more about personnel philosophies than about the structure of the organization. For example, given the exact same position and responsibilities, one organization might give the title of vice president, while others might give titles such as manager, director, supervisor, etc.

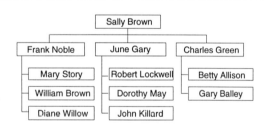

Information by name

One of the most widely used variations of organization charts uses individual's names to indicate who reports to whom. Titles or some description of the individual's position, duties, or role are many times used, since it is generally important or helpful to know such things as whether an individual is a full-time manager or a part-time volunteer. This type of chart is generally more interesting and valuable to persons within an organization. Exceptions to this generality are organizations that deal with the public, where it is important to be able to identify and contact specific individuals.

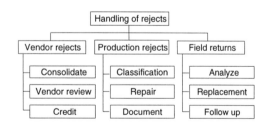

Information by activity

Sometimes organization charts are used to depict how activities are organized. Such charts may or may not make reference to people, departments, etc. For example, the chart at the right shows how rejects are handled, indicating that they are classified into three different types and that the activities associated with each classification are somewhat different. No indication is made with regards to people or organizations involved. Such charts differ from flow charts in that they do not indicate the sequence in which the activities take place.

Special types of information

There are many situations where special information is studied or communicated by means of organization charts. In cases where the organization itself is being studied as opposed to individuals, the chart is sometimes constructed very symbolically, as in the example at the right. This type of chart is used to analyze such things as the number of people with supervisory versus non-supervisory responsibilities, the average number of people reporting to supervisors, the number of layers of managers and supervisors, etc.

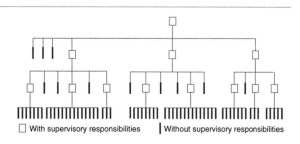

Variations of organization charts

In addition to the simple tree type of organization chart used in the examples on the previous page, there are many others that are either variations of that basic format or that use an entirely different format. Several examples are shown below.

Matrix type

A number of organizations utilize what is called a matrix management. In this type of organization many people report to multiple individuals or positions. For example, in addition to reporting to the manager of engineering, an engineer might also report directly or indirectly to a product manager. This might be true for entire departments or on an individual basis. The example at the right is one way of graphically representing an organization structured on a matrix.

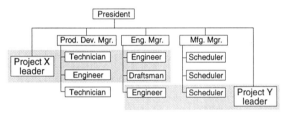

Temporary assignments

It is common in many types of organizations to temporarily assign individuals to special projects that might last from days to years. Typically a project leader is established with full or partial responsibility for the personnel assigned to the project. There are several ways to represent these arrangements on an organization chart. The example at the right leaves the direct reporting relationships intact and indicates the temporary assignment with shading.

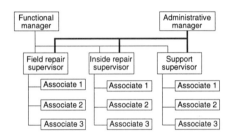

Distribution of responsibilities

In some organizations, particularly highly specialized ones, the responsibilities for managing the organization are distributed between two or more managers. For example, in a research laboratory, one highly skilled in the technical aspects of the work being done in the laboratory might manage the technical activity while another person skilled in the administrative aspects might manage that part of the activity. The example at the right illustrates one way to represent such a structure graphically. This example shows functional and administrative responsibilities merging at the supervisor level. In some cases, these responsibilities don't merge until they get down to the associate level.

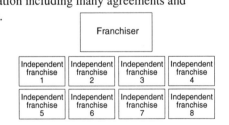

Nonsubordinate relationships

The lines connecting the entities on organizational charts are many times, if not most times, representative of reporting relationships. In some cases, however, the lines denote such things as channels of communication, groups that work closely together, units between which coordination is necessary or where joint approval is required. In charts of this nature it is not uncommon to have multiple solid lines connecting the entities. The chart at the right is an example of a situation where the individual units are not subordinate to one another but, do have the need for close relationships on certain matters.

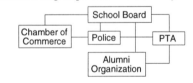

Organization chart without connecting lines

In some situations there are close relationships between various units of an organization including many agreements and transactions; however, on an organizational chart, no lines are drawn between them. For example, independently owned franchises are autonomous businesses that are not subordinate to the franchiser. Thus, an organization chart of a franchise chain with all privately owned franchises may contain no connecting lines at all. Many governmental organizations have a similar situation where groups like a water district, a school district, and a public utility work closely together but have no formal links.

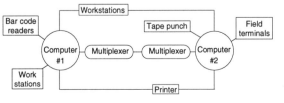

Organization chart of equipment

Organization charts can be drawn of equipment as well as people, organizations, or activities. Such charts are sometimes referred to as network diagrams, schematics, flow charts, or in some cases maps. As with other organization charts, these charts show how many and what type of equipment are involved, how the network is structured, and how the pieces are interrelated. Individual units or pieces of equipment might be referred to by their function (e.g., input orders), their name (e.g., Acme model 972), or their description (e.g., work stations). Organization charts of equipment are seldom hierarchical, use arrows more frequently than other organization charts, and each chart tends to have a unique layout.

Organization Chart (continued)
Variations of organization charts (continued)
──────── **Direct and indirect reporting** ────────

In some cases, individuals report to two or more people for various aspects of their responsibilities. For example, the financial manager in a division might report directly to the general manger of the division and indirectly to the corporate financial manager. Or, a coordinator of new product developments might report directly to the engineering manager and indirectly to the manufacturing manager. A widely accepted method for indicating this on an organization chart is to show the direct reporting relationship with a solid line and the indirect relationship with a dotted or dashed line.

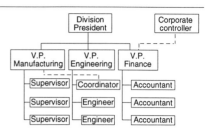

──────── **Using organization chart to place emphasis** ────────

Organization charts are sometimes used to indicate the relative importance of certain functions. The chart at the right is one example of how this might be done. The same information as shown in this chart could be presented in the traditional format; however, it would not be as clear that sales is the major focus of the organization and that all other departments are expected to support sales. Similar effects can be achieved by the use of color, shape, size, etc.

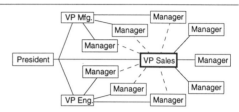

──────── **Staff versus line positions** ────────

Some organizations choose to differentiate between line and staff positions/functions on the organization chart. A typical technique in a large organization is to cluster the staff positions towards one side or the top or bottom of the chart. A separate line and/or a different color might be used to further differentiate them. In smaller organizations or when staff members are scattered throughout the organization, individuals may be shown adjacent to the person they report to. Dashed lines may indicate the positions for which the staff member has an advisory responsibility.

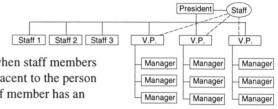

──────── **Partial organization charts** ────────

In some situations it is desirous to know the chain of command for a particular function, position, or individual. A partial organization chart may be adequate and lets the viewer focus directly on the specific area they are interested in.

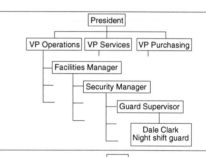

──────── **As a vehicle for recording information** ────────

Every organization has formal and informal channels of communication. Some organizations ignore the informal channels. Others watch them closely and keep a mental picture of who is communicating with whom and how. Still other groups note these channels on an organization chart to look for patterns that might be troublesome or helpful. The example at the right is representative of many specialized ways organization charts can be used to record information about organizations.

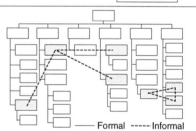

──────── **Analysis of specific characteristics** ────────

When analyses are conducted to study specific characteristics of an organization, unique symbols are sometimes used to represent different categories of the designated characteristics. For example, if educational level versus position are being studied, a different symbol might represent advanced degree, college degree, high school diploma, etc. If technical specialties are being studied, a different symbol might represent scientist, accountant, generalist, marketeer, etc. Names, titles, positions, etc., may or may not be shown.

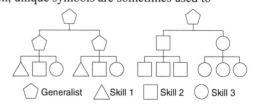

General

Because of the size and complexity of large organizations, multiple charts are generally required. For example, in a large corporation there may be one chart that shows the organization of divisions and subsidiaries into groups or business units. There might be another showing the corporate officials and their positions. In addition, charts are made showing the overall structure of each division, with still another series showing the personnel in each department of each division. • Genealogical charts are a form of organization chart. They are normally drawn in the tree format, either vertical or horizontal, and are sometimes referred to as family trees, descendent charts, or pedigree charts.

Functions of organization charts

Organization charts serve many different purposes including:
- Define lines of authority and responsibility;
- Show who reports to whom;
- Show how people, departments, organizations, or equipment interrelate;
- Make it known where to go to resolve problems or concerns;
- Orient people to a system or organization;
- Obtain insight into organizations such as degree of specialization, hierarchical or matrix type of structure, or degree of centralization;
- Improve understanding of complex organizations, systems, or relationships;
- Assist in the analysis and administration of large organizations;
- Provide a tool for use in the planning process; and
- Provide a tool for comparing the structure of various organizations.

Tree, pyramid, or hierarchical organization chart

One of the most widely used forms of organization chart is the tree, pyramid, or hierarchical type chart. This type of chart is most frequently presented in the vertical position; however, it can be drawn in a number of other configurations. Eight examples are shown below. All, except the stylized pyramid, display the same information.

With such a format the highest ranking position, most important function, primary coordinating unit, etc., is normally shown at the top, with all other entries placed in some meaningful pattern below it.

Traditional tree or pyramid chart

A stylized variation of a pyramid type chart

Sometimes an effort is made to have all positions of comparable rank or importance at the same level to clearly display the number of levels of management in an organization and make it easy to determine how many people are at each level. This method normally makes the chart very wide.

Chart layout used to clearly display the number of levels of management in an organization

Sometimes the shape of an organization chart is selected to deemphasize the hierarchical nature of the organization. The three examples shown here are representative of those types of charts.

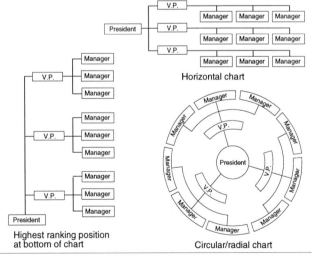

Horizontal chart

Highest ranking position at bottom of chart

Circular/radial chart

To further deemphasize the hierarchical nature of an organization chart, the interconnecting lines are sometimes omitted, as shown in the two examples at the right.

Vertical chart without connecting lines

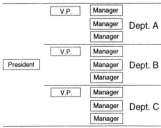

Horizontal chart without connecting lines

Construction options

Shown below are construction variations applicable to many types of organization charts.

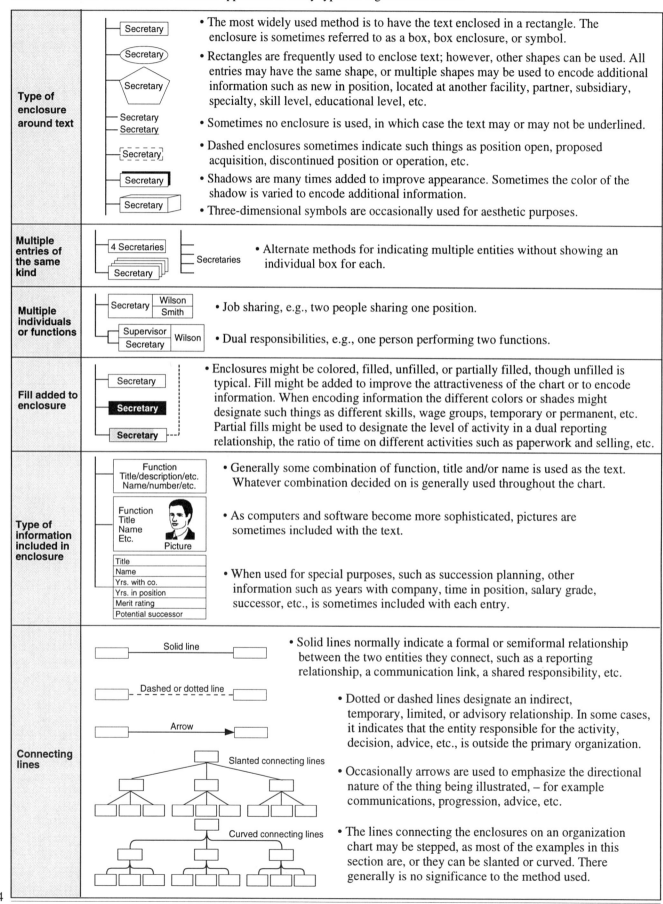

Type of enclosure around text	• The most widely used method is to have the text enclosed in a rectangle. The enclosure is sometimes referred to as a box, box enclosure, or symbol. • Rectangles are frequently used to enclose text; however, other shapes can be used. All entries may have the same shape, or multiple shapes may be used to encode additional information such as new in position, located at another facility, partner, subsidiary, specialty, skill level, educational level, etc. • Sometimes no enclosure is used, in which case the text may or may not be underlined. • Dashed enclosures sometimes indicate such things as position open, proposed acquisition, discontinued position or operation, etc. • Shadows are many times added to improve appearance. Sometimes the color of the shadow is varied to encode additional information. • Three-dimensional symbols are occasionally used for aesthetic purposes.
Multiple entries of the same kind	• Alternate methods for indicating multiple entities without showing an individual box for each.
Multiple individuals or functions	• Job sharing, e.g., two people sharing one position. • Dual responsibilities, e.g., one person performing two functions.
Fill added to enclosure	• Enclosures might be colored, filled, unfilled, or partially filled, though unfilled is typical. Fill might be added to improve the attractiveness of the chart or to encode information. When encoding information the different colors or shades might designate such things as different skills, wage groups, temporary or permanent, etc. Partial fills might be used to designate the level of activity in a dual reporting relationship, the ratio of time on different activities such as paperwork and selling, etc.
Type of information included in enclosure	• Generally some combination of function, title and/or name is used as the text. Whatever combination decided on is generally used throughout the chart. • As computers and software become more sophisticated, pictures are sometimes included with the text. • When used for special purposes, such as succession planning, other information such as years with company, time in position, salary grade, successor, etc., is sometimes included with each entry.
Connecting lines	• Solid lines normally indicate a formal or semiformal relationship between the two entities they connect, such as a reporting relationship, a communication link, a shared responsibility, etc. • Dotted or dashed lines designate an indirect, temporary, limited, or advisory relationship. In some cases, it indicates that the entity responsible for the activity, decision, advice, etc., is outside the primary organization. • Occasionally arrows are used to emphasize the directional nature of the thing being illustrated, – for example communications, progression, advice, etc. • The lines connecting the enclosures on an organization chart may be stepped, as most of the examples in this section are, or they can be slanted or curved. There generally is no significance to the method used.

Origin	In a two-dimensional graph, the origin is the point where the zero base line for the X-axis crosses the zero base line for the Y-axis. On a three-dimensional graph, it is the common crossing point for all three zero base line axes.

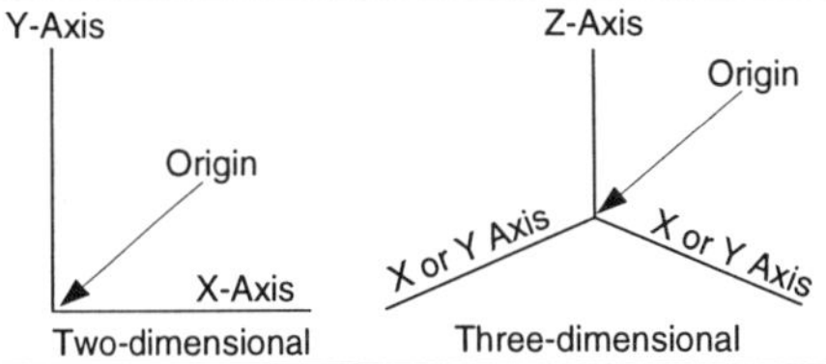

Y-Axis · Z-Axis · Origin · Origin · X or Y Axis · X-Axis · X or Y Axis · Two-dimensional · Three-dimensional

Outlier	An outlier is a data point that is significantly outside of the pattern of the rest of the data points in a given data set or series. For example, if there are 50 numbers in a set of data, 49 of which lie between zero and seven, and one point is 11.5, the 11.5 is considered an outlier since it is distinctly different from all the other data elements in the set. Without looking into each specific situation, there is generally no way to know whether an outlier is caused by an error in collecting or transcribing the data, or whether it is representative of whatever is being studied. When only average or median values are plotted, outliers are not apparent. They generally are easy to spot graphically when individual data points or groups of data points are plotted, as shown at the right.

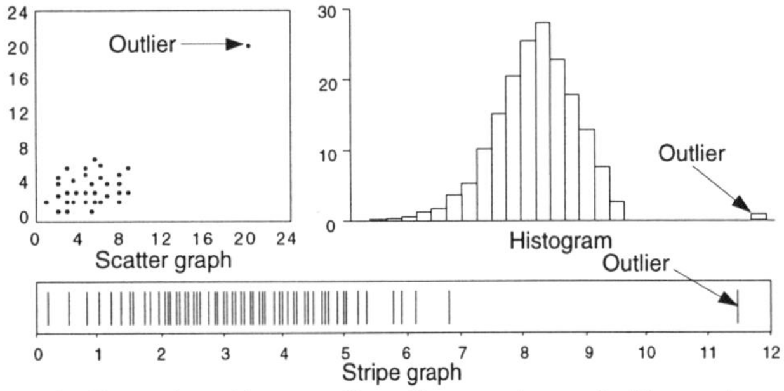

Outlier — Scatter graph — Histogram — Outlier — Stripe graph — Outlier

An illustration of how readily outliers can be spotted in graphs

Overlaid Charts and Graphs	Sometimes referred to as superimposed charts and graphs. An overlaid chart or graph has multiple groups of data superimposed on top of one another. The graph at the right is an example of an overlaid graph with three data series laid on top of one another. A quantitative map is an overlaid map with quantitative information overlaid on a base map.

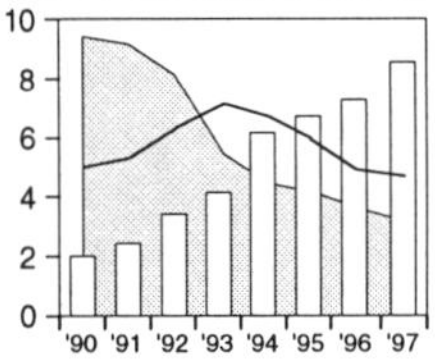

Example of an overlaid graph · '90 '91 '92 '93 '94 '95 '96 '97

Overlapped Area Graph	Sometimes referred to as a multiple or grouped area graph. An overlapped area graph displays multiple data series, all of which are measured or referenced from the same zero base line, usually the horizontal axis. Because all the data series extend down to the zero base line, the values for each data series can be read directly from the graph. This type of area graph should be used with caution because it can easily be confused with the more familiar stacked area graph. See Area Graph.

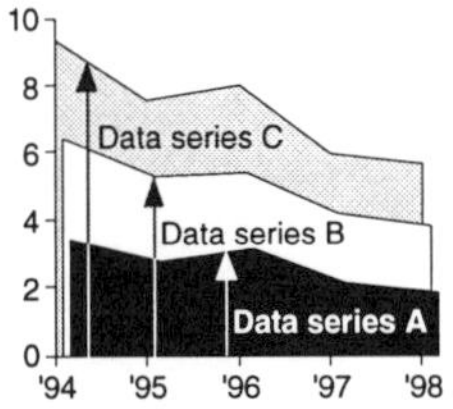

Data series C · Data series B · Data series A · '94 '95 '96 '97 '98 · Overlapped area graph

Overlapped Bar Graph and Overlapped Column Graphs	In order to make grouped bar and column graphs easier to read, the rectangles are sometimes overlapped. In this process the rectangles representing complete data series are shifted such that the rectangles of each successive data series are partially hidden by the rectangles in front of them. They may be shifted by any amount, ranging from barely to completely overlapped (100%). Care should be taken as the limit of 100% overlap is approached, since the graph can easily be mistaken for a stacked bar or column graph. See Bar Graph and Column Graph.

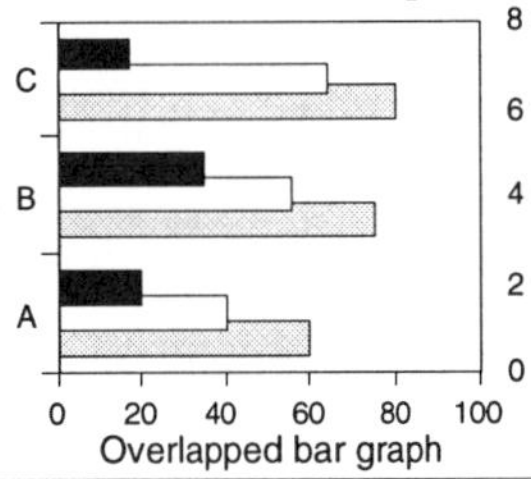

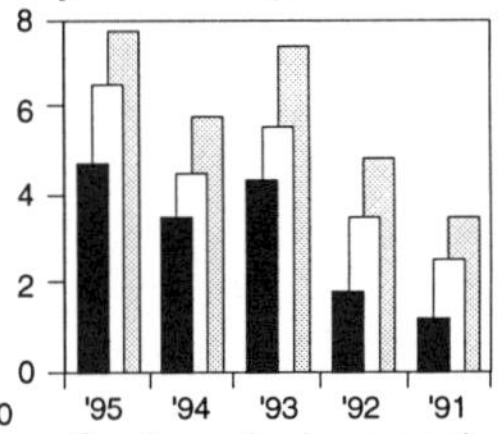

C · B · A · 0 20 40 60 80 100 · Overlapped bar graph · '95 '94 '93 '92 '91 · Overlapped column graph

Overlapped Line Graph	Sometimes referred to as a compound, dual, grouped, or multiple line graph. An overlapped line graph has multiple data series, all of which are measured from the same zero base line axis. See Line Graph.

A · B · C · All curves are plotted from the same zero base line axis. · Overlapped line graph

Overlay Graph	Sometimes referred to as a combination, composite, or mixed graph. An overlay graph displays multiple data series, using two or more types of data graphics to represent them. See Combination Graph.

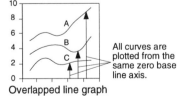
Data series A represented by an area
Data series B represented by a line
Data series C represented by columns

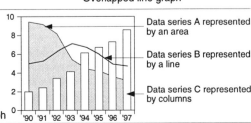
Example of an overlay graph · '90 '91 '92 '93 '94 '95 '96 '97

| **Over-Under Graph** | An over-under graph is a variation of a difference or deviation column graph in which positive and negative values are plotted from a common zero for every interval along the horizontal axis. See Difference Graph. | 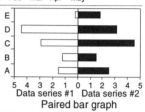 |

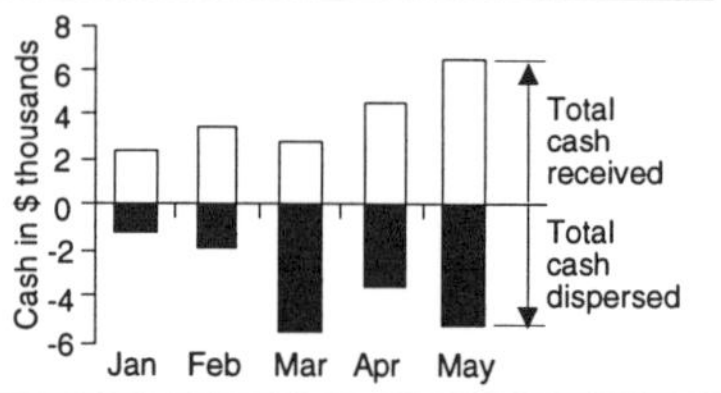

Example of an over-under graph

| **Paired Bar Graph** | Sometimes referred to as a bilateral, opposed, sliding, or two-way bar graph. This type of graph is a variation of a bar graph in which positive values are measured both right and left from a zero on the horizontal axis. The major purpose of such a graph is to compare two or more data series, with particular attention to correlations or other meaningful relationships. See Bar Graph. |

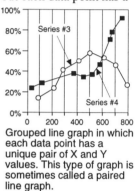

Paired bar graph

Paired Line Graph

When multiple data series are plotted on a line graph the result is frequently referred to as a grouped line graph. There are two major variations of grouped line graphs. One is where all data series on the graph use common values on the horizontal axis, but each data point has a unique value on the vertical axis (left). On this type of graph the data points for all data series align one above the other. • In the second variation, each data point has a separate and unique pair of X and Y values (right). In this variation, the data points for the multiple data series may or may not align above one another. This second variation is sometimes referred to as a paired line graph, since each data point has a unique pair of coordinates.

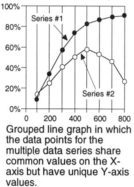

Grouped line graph in which the data points for the multiple data series share common values on the X-axis but have unique Y-axis values.

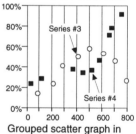

Grouped line graph in which each data point has a unique pair of X and Y values. This type of graph is sometimes called a paired line graph.

Paired Scatter Graph

When multiple data series are plotted on a scatter graph the result is frequently referred to as a grouped scatter graph. There are two major variations of grouped scatter graphs. One is where all data series on the graph use common values on the horizontal axis, but each data point has a unique value on the vertical axis (left). On this type of graph the data points for all data series align one above the other. • In the second variation, each data point has a separate and unique pair of X and Y values (right). In this variation, the data points for the multiple data series may or may not align above one another. This second variation is sometimes referred to as a paired scatter graph, since each data point has a unique pair of coordinates.

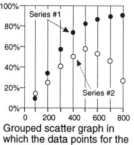

Grouped scatter graph in which the data points for the multiple data series share common values on the X-axis but have unique Y-axis values.

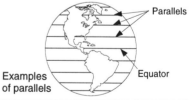

Grouped scatter graph in which each data point has a unique pair of X and Y values. This type of graph is sometimes called a paired scatter graph.

| **Parallel** (as used with maps) | Parallels are grid lines (graticules) running east-west on maps and globes. Since north-south distances (latitudes) are many times referenced from parallels, the east-west grid lines are also sometimes referred to as latitudinal lines. See Map. |

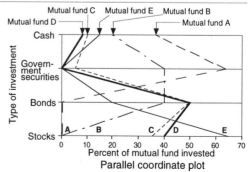

Examples of parallels

Parallel Coordinate Plot

Sometimes referred to as a profile graph. A parallel coordinate graph is a variation of a comparative graph that compares multiple entities and characteristics at a given time. For example, such a graph might be used to compare the investment strategies of five mutual funds in four major types of investments. Those funds having similar strategies will have approximately parallel curves, similar to funds C and D on the graph at the right. See Comparative Graph.

Parallel coordinate plot

Pareto Diagram/Graph

The purpose of the Pareto graph is to highlight the major types, causes, sources, etc., of defects so the primary contributors can be identified and addressed first. In its simplest form, the major defects, causes of defects, locations where defects occur, etc., are plotted in descending order along the category axis of either a column or bar graph. The value axis displays the frequency of occurrence of the defects in terms of units, percents, or both over a prescribed period of time or for a given number of units. From such a chart, it is generally apparent which factors are the biggest contributors to a quality problem.

The use of multiple graphs

Sometimes multiple graphs are generated for the same situation to give further insight into a problem. For example, if there are four parts going through a process and five different types of defects, one graph might plot how many units failed due to each type of defect. Another graph might plot the number of rejects by part number. A third might be a stacked column graph with type of defect on the horizontal axis and number of defective parts on the vertical axis. The segments of the columns would represent individual part numbers. Examples of all three are shown below, with potential observations that might be made from them.

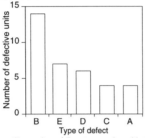 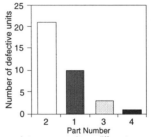 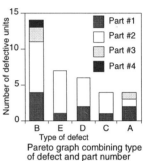

Examples of Pareto graphs with the same data presented two different ways

Pareto graph combining type of defect and part number

Examples of different types of Pareto charts for the same situation. From these graphs it can be seen that part number 2 has the highest number of rejects, and defect B represents the highest percent of the total defects. The graph on the right shows that the type B defect occurred in all four part numbers; defects C, D, and E occurred only in parts 1 and 2, and part #4 had only type B defects.

Combining individual and cumulative data before and after corrective actions

Another variation of a Pareto diagram combines a column graph displaying the quantity and percent of rejects by individual category (e.g., part, defect, location, etc.) with a line graph displaying cumulative data for the same items. Two examples are shown at the right. One incorporates data before corrective actions and another after corrective actions. By using two graphs, improvements can be noted graphically.

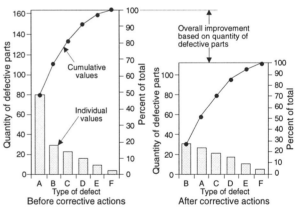

Pareto graphs with cumulative lines. The graph on the right indicates improvements and the new ranking of types of rejects.

General

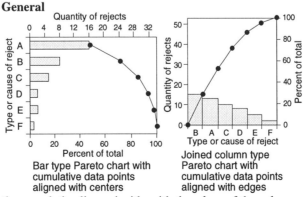

Bar type Pareto chart with cumulative data points aligned with centers

Joined column type Pareto chart with cumulative data points aligned with edges

Column graphs are most frequently used in Pareto charts; however, bar graphs work equally well. • The data points for the cumulative percent line can coincide with the center or edge of the columns. • Standard or joined column graphs can be used for Pareto graphs. The joined variation tends to work better when the data points for the cumulative line coincide with the edges of the columns. • Although Pareto graphs are most frequently used in the field of quality control, they can be employed in any situation where the goal is to reduce something, whether that be rejects, costs, time, parts, etc.

267

| **Partition Plot** | Sometimes called a slice graph, spectral plot, ridge contour, layer graph, or strata graph. A partition plot is a graph in which the data points of a three-dimensional scatter graph are condensed onto a limited number of planes to aid in the interpretation of the data. See Slice Graph. |

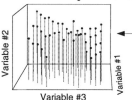

| **Part-of-the-Whole Chart/Graph** | Charts that show what portion each component represents of a whole (total) in terms of actual values. The same data presented as percents is generally referred to as a percent-of-the-whole chart or graph. See Percent-of-the-Whole Chart/Graph. Area, bar, and column graphs as well as proportional charts are most often used in this application. When area graphs are used, it is almost always the stacked variation. With bar graphs, column graphs, and proportional charts, it may be the simple or stacked variation. See Area Graph, Bar Graph, Column Graph, Pie Chart, and Proportional Chart. |

| **Passive Decision Diagram** | In addition to aiding an individual or organization in making their own decisions (active decision diagram), decision charts are sometimes used to estimate outcomes based on decisions made by others (passive decision diagram). For example, to a company that supplies material such as cement to subcontractors, the exact subcontractor awarded a contract is very important. The cement company, however, has no influence on the decision as to which subcontractor will get the contract. Since subcontractors are often closely affiliated with prime contractors, the probability of the company selling cement is further dependent on the decision as to who the prime contractor will be, even though the cement company will not be selling to the prime contractor. If the cement company wants to estimate the probabilities of the various subcontractors getting the order, they might construct a passive decision diagram as shown below. Based on this example, the cement company might focus their attention on subcontractors C and E who have a 25% and 31% probability respectively of receiving a subcontract order and expend very little effort on subcontractor D who has only a 6% probability of receiving a subcontract. |

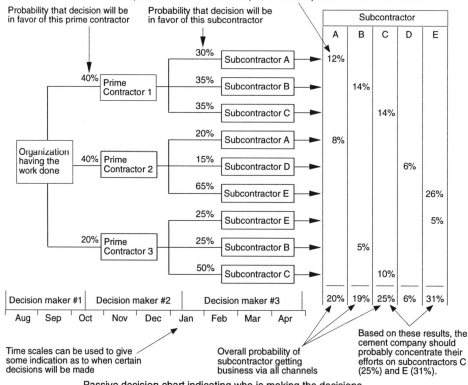

Passive decision chart indicating who is making the decisions, approximately when, and the probability associated with each option

The above decision diagram stops at the subcontractor level. It could be expanded an additional step to include the percent probability of the particular cement company getting the cement order from each of the subcontractors. There are no limitations on the number of levels of decisions that can be included in this type of diagram.

Patch Graph

A patch graph is generated by removing the sides of the columns on a column graph, as shown at the right. Patch graphs basically function the same as column graphs, with the added feature that hidden data is sometimes more visible. Many variations are possible using projections and drop lines. A number of them are shown below.

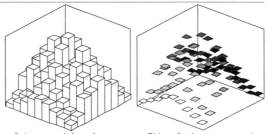

Column graph for reference

Sides of columns removed and shading added to tops

Comparison of a column graph and a patch graph using the same data in both

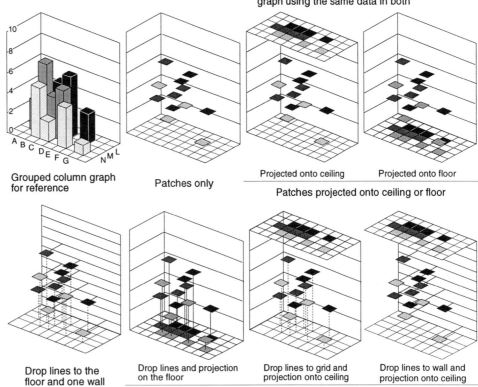

Grouped column graph for reference

Patches only

Projected onto ceiling   Projected onto floor

Patches projected onto ceiling or floor

Drop lines to the floor and one wall

Drop lines and projection on the floor

Drop lines to grid and projection onto ceiling

Drop lines to wall and projection onto ceiling

Combinations of projections and drop lines

Variations of patch graphs

Multiple data series in a single plane

In some cases the multiple data series can be displayed in one plane so all data graphics touch the back surface with the grid lines. In this way it is sometimes easier to determine which data graphics are in the same categories (e.g., A thru G) and to estimate each of their values.

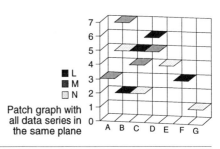

Patch graph with all data series in the same plane

Patch Map

Sometimes referred to as a blot map. Patch maps are a variation of descriptive maps on which such things as natural resources, types of soil, agricultural usage, wetlands, etc., are identified by means of filled areas. The size and shape of the filled areas generally approximate the entities they represent. The meaning of the fills is normally explained in a legend.

Patches or filled areas designate such things as natural resources

Resource A
Resource B
Resource C

Patch map

Percentile

The position of a data element in a data set based on its ranking among all of the data elements in the set. For example, if a student gets a score on a test that is higher than 70% of the other students, that student is said to be in the 70th percentile. See Percentile Graph.

Percentile comparison graphs are used for comparing two sets of data using percentile values. Shown below are two ways of constructing this type of graph.

Values associated with common percentiles

In this method, the value for the 95th percentile of one data series is the X coordinate of one data point on the comparison graph, and the value for the 95th percentile of the data series it is being compared with is the Y coordinate. The value of the 90th percentile of the first data series is the X coordinate of the second data point, and the 90th percentile value of the second data series is the Y coordinate for the second data point. This process continues until enough data points are plotted to define the shape of the curve. An example is shown below, along with individual cumulative frequency graphs for the two data series being compared to show graphically how the percentile comparison graph is constructed. If the distributions of two data series are the same, the data points cluster around a diagonal. • A comparable way to look at the data is to plot the cumulative frequency curves for both data series on the same graph, as shown below. The same basic observations can be made from either type of graph.

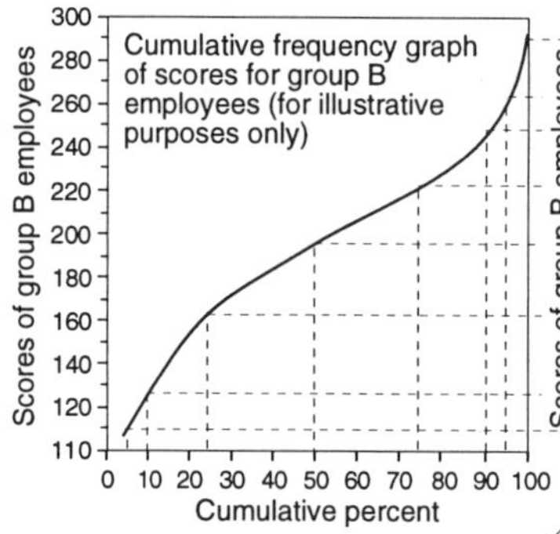

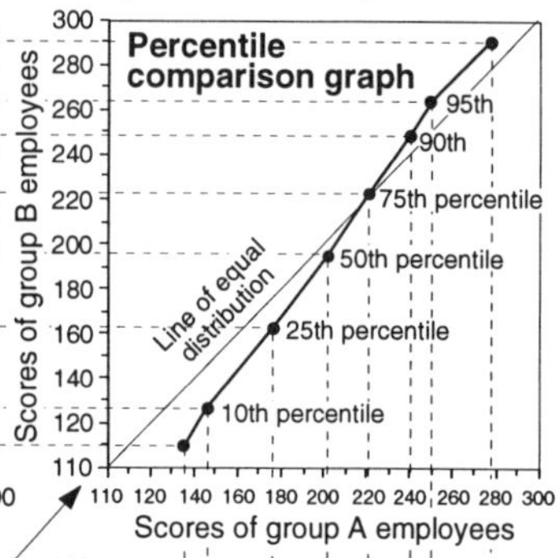

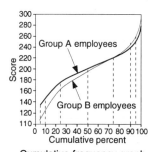

Cumulative frequency graph displaying the same two data series used in the example at the right.

The percentile comparison graph shown here indicates that the lower 75% of group A achieved better scores than the lower 75% of group B, but the 25% of group B with the highest scores had scores better than those of the upper 25% of group A. • The cumulative frequency graphs above and at the right are for illustration purposes only and would not normally accompany the percentile comparison graph.

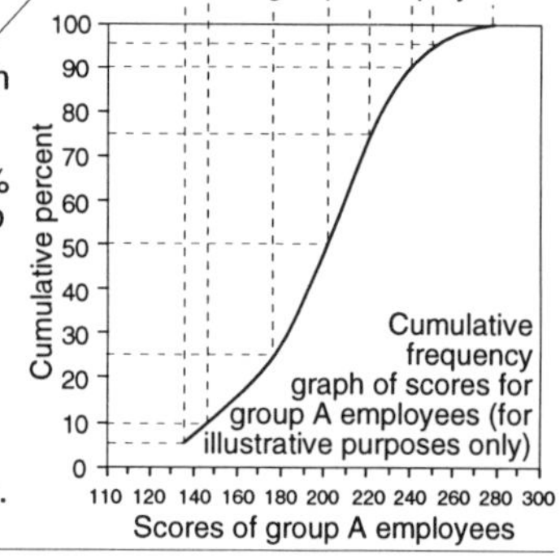

Percentiles associated with common values

This type of percentile comparison graph is constructed similar to the variation above, except common values are selected and the percentiles associated with them plotted. For instance, in the example at the far right, the percentile associated with a $35,000 annual salary for one data set is the X coordinate of one data point on the percentile comparison graph, and the percentile associated with $35,000 for the data set it is being compared with is the Y coordinate. The percentiles associated with a $50,000 annual salary form the coordinates for a second data point, and so on. A percentile comparison graph sometimes

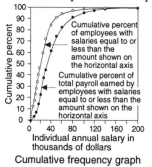

Cumulative frequency graph

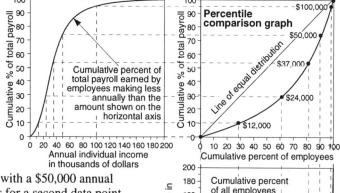

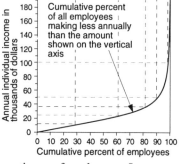

makes the relationship between the two sets of data stand out more clearly than when both sets of data are plotted on a standard cumulative frequency graph, as shown at the left. A graph with percentiles on both axes is sometimes referred to as a Lorenz graph and is widely used for comparing income and population. See Lorenz Graph.

Percentile Graph

Percentile graphs are sometimes thought of as a subcategory of probability graphs. The term percentile graph refers more to how a graph is used than it does to a specific type of graph. When used for percentile purposes, the graph provides a graphical means for determining the percent of data elements in a data set that are above or below a certain value, – for example, the percentage of people in Wyoming with incomes less than $25,000, or the percentage of athletes taller than six feet. Box graphs, cumulative frequency graphs, and quantile graphs are widely used as percentile graphs. Cumulative histograms are used to a much lesser extent. The type of graph chosen for a particular application is subjective and depends to some degree on the nature of the data and the purpose of the graph. In the table below only the name of the graph, a thumbnail sketch, and an example are shown. Detailed information is given under the individual headings.

Cumulative frequency graph

Sometimes referred to as cumulative frequency percent, cumulative relative frequency, or ogive graph. Cumulative frequency graphs are often generated by grouping data into class intervals, as with a histogram, counting the number of data elements in each interval, and cumulating the counts. The counts can be cumulated from smallest to largest or largest to smallest. Among other things, cumulative frequency graphs are used to display percentile rankings and to determine if a data set has a normal distribution.

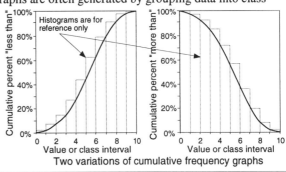

Two variations of cumulative frequency graphs

Quantile graph

In this type of graph individual data elements are plotted as opposed to class intervals. Either quantile values or percentiles (100 times the quantile value) are typically plotted on the horizontal axis. The value of the data element is plotted on the vertical axis. Data is generally cumulated from the smallest to the largest value.

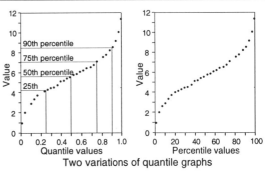

Two variations of quantile graphs

Box graph or box plot

There are several variations of box plots. Those, such as example A (right), are sometimes referred to as percentile plots. In most box plots only the popular percentiles such as 10, 25, 50, 75, and 90 are shown. For special-purpose applications there is no restriction as to how many percent levels might be shown or what values might be used. Compared to the two types noted above, the box plot has the disadvantage of not allowing the viewer to determine percentiles for every value of the data set.

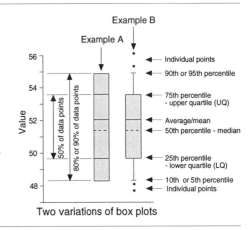

Two variations of box plots

Cumulative histogram

The cumulative histogram is used much less frequently in percentile applications than the other three types shown above. To determine percentiles for values between class interval boundaries, one must interpolate (which, in essence, is what the cumulative frequency graph does).

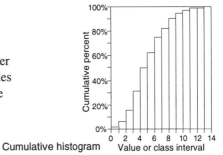

Cumulative histogram

271

| **Percent-of-the-Whole Chart/Graph** | Sometimes referred to as a one hundred percent (100%) chart or graph. If units are used instead of percent, the figure is sometimes referred to as a part-of-the-whole chart or graph. A percent-of-the-whole chart or graph shows what percent each component represents of a whole. For example, if a theater sells a total of 250 tickets (100%), distributed between children, adults, and seniors, a percent-of-the-whole chart might be used to show what percent of the total number of tickets were sold to each of the three age groups. • Stacked graphs and proportional charts are among the most widely used formats for displaying percent-of-the-whole data. With proportional charts, generally the areas of the data graphics represent the percents. With graphs, the data point or the line or edge of the data graphic designates the percents. With both charts and graphs, units may be shown in addition to percents. The examples below are grouped into the three major applications for which percent-of-the-whole charts are often used. See Column Graph, Bar Graph, and Pie Chart. |

One group of data at a point in time

When all that is being presented is a single group of data at a given point in time, circular, column, and bar type charts and graphs are most widely used.

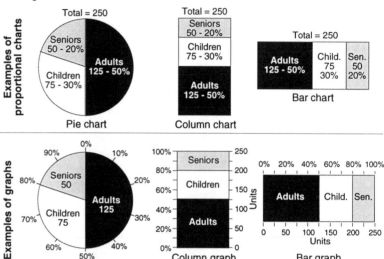

Multiple groups of data at a point in time

When multiple groups of data at a given point in time are displayed, 100% stacked bar or column graphs (right) are commonly used. Multiple pie charts (below) or circle graphs can be used; however, they tend to require more space and comparisons are more difficult to make.

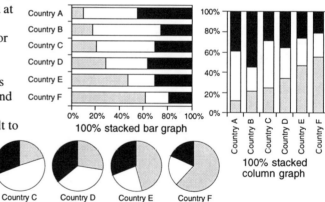

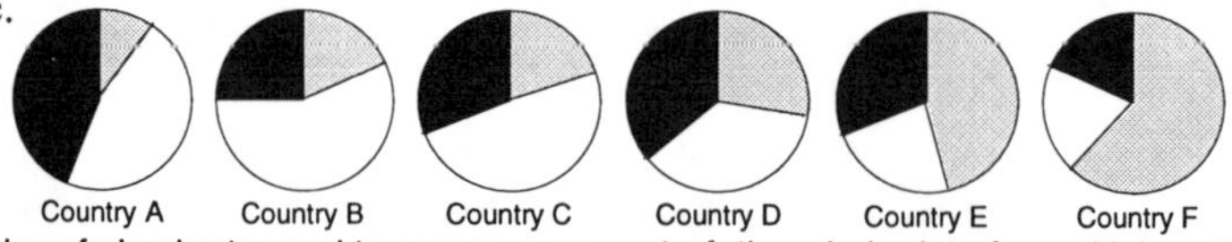

A series of pie charts used to compare percent-of-the-whole data for multiple entities

One group of data at multiple points in time

For tracking changes over time, column or area graphs are frequently used. Multiple pie charts or circle graphs, each representing a different time period, can be used as shown above; however, they tend to require more space and comparisons are more difficult to make.

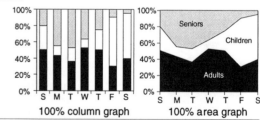

| **Percent Table** | When a table displays how frequently things occur or how many there are of different things, the table is sometimes referred to as a frequency table. When the counts in such a table are replaced by the percents they represent of a whole, the table is sometimes called a percent table. The values may be calculated as a percent of the rows, columns, or the totals for the entire graph. See Table. |

| **Period Data** | Period data is data that applies to some interval of time – for example, the cash received in a month, the average temperature for a week, or the average membership for the year. |

Perspective Projection (View)

When an object is drawn in perspective it tends to simulate one aspect of how the human eye sees objects. That is, the further things are from the viewer, the smaller they appear and the more lines seem to converge into a single point (called vanishing point). The technique is used with some objects to improve the viewer's understanding. In many cases, particularly with graphs, perspective is used primarily to improve appearance. The table below illustrates the more widely used variations of perspective. As the table indicates, perspective can be used with two- or three-dimensional objects. With charts and graphs it is most frequently used with three-dimensional objects. Which view to use in a particular situation is mostly a matter of personal preference.

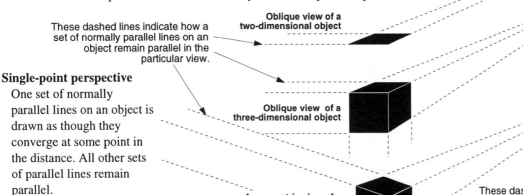

These dashed lines indicate how a set of normally parallel lines on an object remain parallel in the particular view.

Oblique view of a two-dimensional object

Vanishing points

Single-point perspective

One set of normally parallel lines on an object is drawn as though they converge at some point in the distance. All other sets of parallel lines remain parallel.

Oblique view of a three-dimensional object

Axonometric view of a three-dimensional object

These dashed lines indicate how a set of normally parallel lines on an object are drawn as though they converge at some distant point. They are included here for illustrative purposes only and would not be shown in an actual situation.

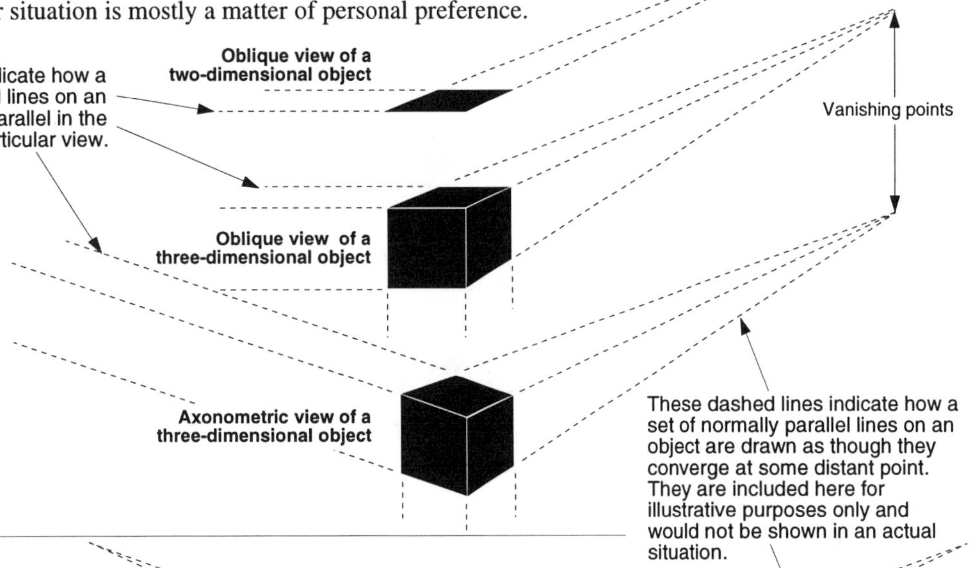

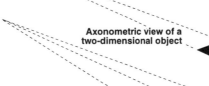

Two-point perspective

Two sets of normally parallel lines on an object are drawn as though they converge at two different points in the distance. Other sets of parallel lines remain parallel. All surfaces are somewhat distorted.

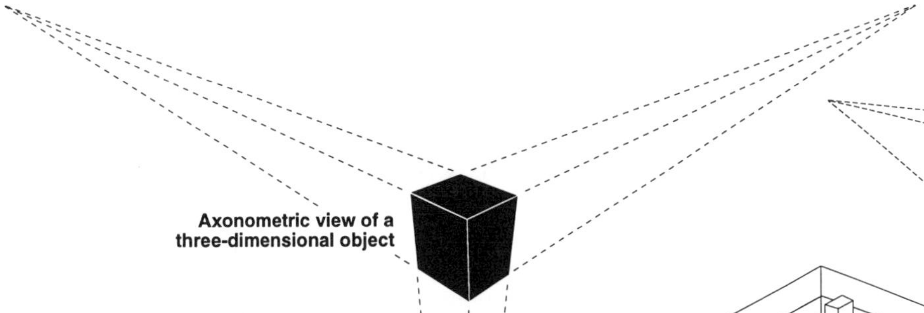

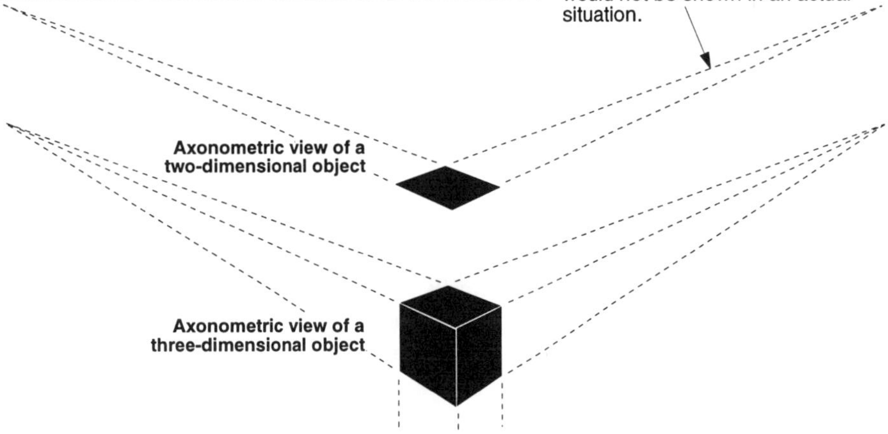

Axonometric view of a two-dimensional object

Axonometric view of a three-dimensional object

Axonometric view of a three-dimensional object

Three-point perspective

All three sets of normally parallel lines on an object are drawn as though they converge at three different points in the distance. All surfaces are somewhat distorted. This view causes the greatest amount of distortion in the surfaces of objects.

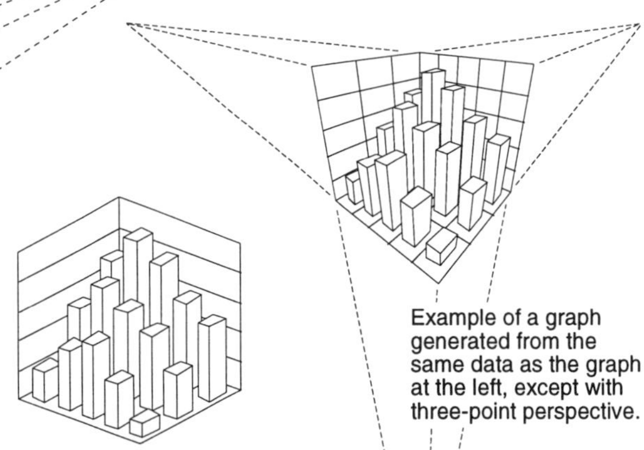

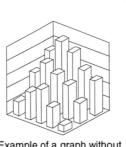

Example of a graph without perspective for reference

Example of a graph generated from the same data as the graph at the left, except with three-point perspective.

For a direct comparison, the graph at the left is not drawn in perspective. The graph at the right contains the same data as the one at the left, except it is drawn in three-point perspective. The dashed lines are included for illustrative purposes only. In an actual situation they would not be shown.

273

Perspective Projection
(continued)

Vanishing points

There is a wide latitude as to where vanishing points might be located on perspective projections in terms of angle and distance from the object. To illustrate, five variations of the same object are shown at the right. In all five examples, the front surfaces (in black) are the same size and shape.

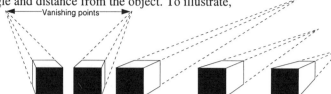

Examples of the same object with the vanishing point at various locations. In each example the front surface is identical in size and shape.

Concerns with perspective projections

The use of perspective views with graphs has been widely discussed and in many cases criticized. The major reason for the criticism is the distortion that perspective views introduce, particularly to graphs. In some cases the distortion is extreme enough that it can mislead the viewer. For instance, in the example at the left, the expenses from 1985 to 1992 increased from $57 to $73. Because the graph is drawn in perspective, the column for expenses in 1992 is actually smaller than the column for 1985. Unless viewer's pay close attention to the grid lines, they might erroneously conclude that expenses went down between 1985 and 1992, while in fact they went up by 28%.

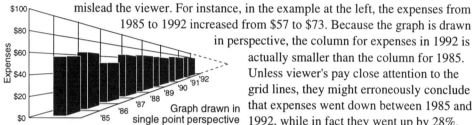

Graph drawn in single point perspective

PERT Chart
or
Program Evaluation and Review Technique

Sometimes referred to as a node diagram, or arrow diagram. PERT charts are time and activity networks that represent the major events and activities of a large program and show their interrelationships. PERT charts are used to plan, analyze, and monitor programs. Among other things, they are helpful in determining how programs can be shortened and identifying which of the many subprograms are the most critical in assuring the overall program is completed on time. PERT charts accomplish basically the same functions as critical path method (CPM) charts. The major difference is that PERT charts focus on events (e.g., complete the foundation) while CPM charts focus on activities (e.g., pour the foundation). A PERT chart uses a separate symbol (circle, square, triangle, etc.) for each event (called node). An event may be a review meeting, a decision point, or simply the time at which one activity is completed and another begins. Lines or arrows (sometimes referred to as links, branches, or arcs) are then used to connect the events in sequential fashion. The main body of a PERT chart generally starts and ends with a single node. The arrows and nodes are arranged in the same order they actually occur. The first event is represented by the first symbol on the left, the last event by the last symbol on the right, and all others in their proper sequence in between. If an activity (B) can not be started until another activity (A) is completed, the tip of arrow A is joined to the base of the arrow B with a node symbol in between. If neither activity is dependent on the completion of the other, the tip of one arrow and the base of the other arrow do not meet. • The bases of multiple arrows often meet at the right side of a node, indicating that the activity(s) on the left side of the node must be completed before any of the activities on the right side can be started. When the heads of multiple arrows meet at the left side of a node, all must be completed before the activity(s) on the right side of the node can proceed. Examples of all three variations are shown at the right. A PERT chart is shown below.

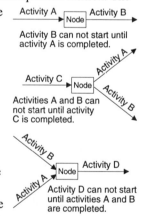

Activity B can not start until activity A is completed.

Activities A and B can not start until activity C is completed.

Activity D can not start until activities A and B are completed.

Examples of how relationships between activities are represented graphically

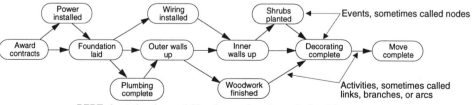

PERT chart showing activities (arrows) and events (ovals).

Methods for identifying activities

In PERT charts it is not uncommon to have dozens or even hundreds of events and activities. A word description of each can be placed on or alongside the arrow or node, as shown on the previous page for the nodes. An alternate method is to use code numbers or letters, as shown in the example below, and include a cross reference table.

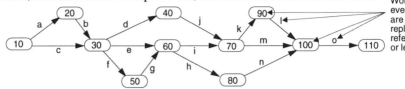

Word descriptions of events and activities are sometimes replaced by reference numbers or letters.

PERT chart with events and activities coded. In an actual chart, a table or legend would match a description with each code number or letter.

Sometimes activities are identified by the numbers at the ends of the arrows (e.g., activity g above might be referred to as activity 50-60, and activity m as activity 70-100).

Incorporating time into a PERT chart
——Times noted on arrows——

Small numbers alongside the arrows sometimes denote information regarding the timing of an activity. Four of the methods currently used are illustrated in the examples below. In one case, a single number is used to represent the estimated time required to accomplish an

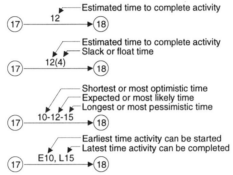

Examples of how times (days, weeks, months, etc.) to complete an activity might be noted

activity. Another method is to show two numbers, the first being the estimated time required to complete the task and the second (many times in parentheses) the amount of additional time (called float or slack) available before that activity or the series of activities in that chain delays the overall program. A third method is to show three times, the shortest (most optimistic), the expected, and the longest (most pessimistic). The fourth alternative is to indicate the earliest time an activity can be started (E) and latest time an activity can be completed (L) without delaying the overall program. In the first three methods, the times noted are incremental times that apply to the length of time the activity will take. In this fourth alternative, the times are referenced from the time the overall program was started.

——Times noted on nodes——

In addition to incremental times for each activity, cumulative times for events are sometimes also shown. Two numbers that are sometimes used are the earliest and latest time each event can occur. For instance, in the example below, event #5 can not occur before the eighteenth week after the start of the overall program, and if it occurs after the twenty-fifth week, the critical events behind it will be delayed.

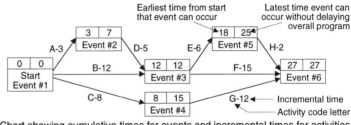

Chart showing cumulative times for events and incremental times for activities

——Time scale for reference——

The length of arrows are normally not proportional to the length of time the activity is estimated to take. Because of this, one normally can not visually estimate how long an activity will take or when an event will occur. In an attempt to give some visual indication of relative times, PERT charts are sometimes arranged along a time scale as shown below.

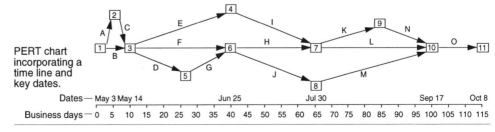

PERT chart incorporating a time line and key dates.

Critical path

Once times are entered, the longest time path can be determined. This is called the critical path and is generally designated on the chart with bold or colored arrows. This is the path along which a delay in any event will cause a delay in the overall program. An example with times designated and the critical path noted is shown at the right.

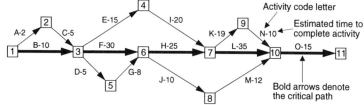

PERT chart showing code letters, activity times, and critical path

In some cases the second most critical path is indicated by arrows drawn with a width of line somewhere between that of a standard path and a critical path.

Dummy arrows

Dummy arrows (indicated with dashes) are sometimes used to overcome some layout and identification problems. For example, if two activities occur between events 7 and 8, as shown below, there is no way to differentiate the two activities if one uses the convention of referring to activities by their start and finish event numbers (i.e., both would be referred to as 7-8). A dummy arrow, as shown at the right, that has no time associated with it solves the problem. After the insertion of a dummy arrow, one activity is still referred to as 7-8 while the other gets the new designation of 7-9. The dummy activity of 8-9 is ignored.

Activity 7-8
Without the use of a dummy arrow, the two activities have the same designation; in this example 7-8.

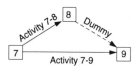

With the use of a dummy arrow, the two activities each have a unique designation and the dummy path/activity is ignored

General

– Arrows generally proceed from left to right, run parallel when possible, and seldom cross.

– Where possible, the number at the base of an arrow is smaller than the number at the head.

– In some cases the charts are updated periodically to reflect new information and/or the status of the program.

– More and more PERT charts are being generated and maintained on computers. When done on computers, in addition to preparing the graphical diagram, the information is frequently also made available in tabular form.

– Several optional construction details are shown below.

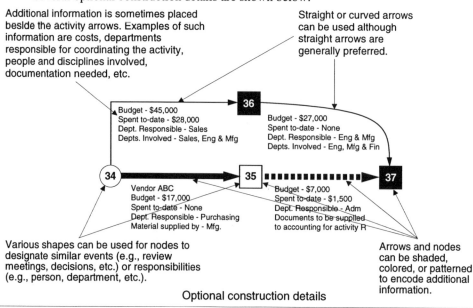

Additional information is sometimes placed beside the activity arrows. Examples of such information are costs, departments responsible for coordinating the activity, people and disciplines involved, documentation needed, etc.

Straight or curved arrows can be used although straight arrows are generally preferred.

Various shapes can be used for nodes to designate similar events (e.g., review meetings, decisions, etc.) or responsibilities (e.g., person, department, etc.).

Arrows and nodes can be shaded, colored, or patterned to encode additional information.

Optional construction details

Sometimes called bank, cycle, deck, or tier. The major interval on a logarithmic scale.

Pictorial Charts and Graphs

Including:
Pictogram
Pictograph
Pictorial Diagram
Pictorial Graph
Pictorial Map
Pictorial Table
Pictorial Unit Chart
Picture Bar
Picture Graph

Pictorial charts and graphs use pictures, sketches, symbols, icons, etc., in place of or in addition to standard data graph elements. For example, bars or columns might be replaced by small pictures or symbols of the thing they represent. Pictorial charts and graphs are used extensively in publications and presentations. Among the reasons for using pictorial charts and graphs are:

- To make the document more interesting and appealing.
- To make the material more understandable to a greater number of people. (The use of pictures sometimes helps overcome differences in language, culture, and education.)
- To improve communication in situations where the appearance of an item is better known than its name or number.
- To facilitate easier reading of a chart or graph by including information to orient the reader that otherwise might have been shown in a legend or note.

There are many different types of pictorial charts. In this section examples of some of the major variations are discussed.

Pictorial graph

Multiple images per column

Sometimes referred to as a pictogram, pictograph, picture bar, or picture graph. In pictorial graphs, pictures, icons, symbols, etc., are used to fill or replace the standard data graphics such as bars or columns. With pictorial graphs, it is the top, end, edge, etc., of the graphic (pictures, icons, etc.) that is used to determine a value on the scale of the graph. Sometimes,

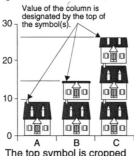

The top symbol is cropped such that the top edge of the graphic is in line with the value the column represents.

as in the example at the left, the size of the symbol is selected to match some even interval on the graph scale such as ten or one hundred. The top symbol is cropped so the upper edge coincides with the value the column or bar is designating. In other cases, as on the right, the outline or frame of the column or bar is the key designation of the value with the symbols simply used as fill. Since the number of units does not determine the

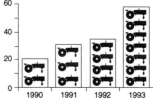

In this example the size and number of symbols bears no relationship to the value the column represents. The symbols are simply used as fill.

value of the column, it is not necessary in either variation to indicate how many units each symbol represents.

Single image per column

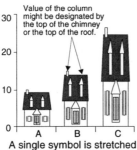

A single symbol is stretched such that the top of the symbol is in line with the value on the scale that the column represents

In some cases a single image is used to represent each column or bar. Each symbol is stretched or shrunk until its top or edge coincides with the value on the graph scale that the column or bar represents. The example on the left illustrates a technique in which the entire symbol is elongated. In the example on the right, the top portion of each symbol is the same in all columns and only the length of the lower portion varies, thus reducing the amount of distortion in the symbols. The example at the left

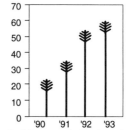

In this example the top of the image remains the same size and the lower portion is elongated.

illustrates one of the potential hazards with pictorial graphs. In this example, it is not clear whether the top of the roof or the top of the chimney designates the value.

The use of a variety of images

In most pictorial graphs, the images used to fill a given bar or column are the same. On occasion a variety of symbols are used, as shown below. Even when multiple symbols are used, there is generally a common attribute or characteristic about all of them.

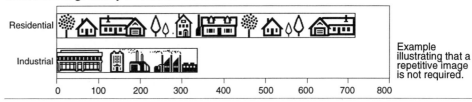

Example illustrating that a repetitive image is not required.

Pictorial graph (continued)

Filled and unfilled columns

In pictorial graphs, the frames of the bars or columns may or may not be shown and the bars or columns may or may not be filled. The examples at the right show a comparison with and without filled columns.

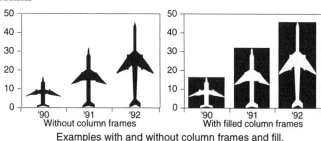

Examples with and without column frames and fill.

Images versus labels

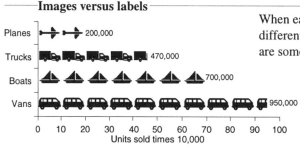

Pictograph with a different symbol for each bar

When each bar or column represents a different type of entity, different symbols are sometimes used for each. When the symbols are obvious, the category labels may or may not be used. As with regular graphs, numeric values can be shown at the ends of the bars to assist the viewer in rapidly interpreting the graph.

Other types of graphs

There are no restrictions on the type of graph that pictures, icons, symbols, etc., can be used with as long as they are legible. An example of a paired bar graph is shown at the right.

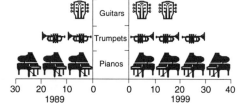

Example of a pictorial graph in the form of a paired bar type

Pictorial proportional chart

Proportional charts are visual devices for graphically communicating differences in size, number, or value of multiple items by means of differences in the sizes of data graphics. The sizes of the data graphics are in the same proportion as the items they represent. For example, if television A is twice as big as television B and both are represented by a symbol of a television, the symbol for television A will be twice as big as the symbol for B.
• Proportional charts with simple geometric shapes as data graphics are used extensively (see Proportional Chart). Pictorial proportional charts with pictures, sketches, icons, etc., are sometimes more difficult to interpret and are used less frequently. The following lists the key elements of pictorial proportional charts.

— In pictorial proportional charts the relative sizes of pictures, sketches, icons, etc., are in the same proportion as the things they represent.
— Most people's ability to accurately estimate sizes of irregular objects such as pictures or sketches is poor, and therefore the graphics are used primarily to orient the reader, with the actual numbers shown alongside to provide the exact values.
— Pictorial proportional charts normally do not have scales, tick marks, or grid lines.
— They are used to compare multiple entities or the same entity at different times or under different conditions
— The relative sizes can be based on length/height, area, or volume. Relative areas are typically used.

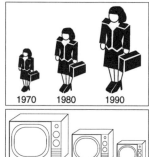

The two charts at the right are examples of pictorial proportional charts, except in actual charts the numeric values are typically shown. The examples illustrate a complication of pictorial proportional charts in that the viewer must decide whether the originator designed the charts with the heights, areas, or volumes proportional to the values they represent. Depending on which one was used, the relative sizes of the data graphics are significantly different.

Examples of pictorial proportional charts. The viewer generally must decide whether the data graphics are based on height, area, or volume.

Pictorial Charts and Graph
(continued)

Pictorial unit chart

A pictorial unit chart is a variation of a unit chart used for communicating numbers of things by making the number of data graphics displayed proportional to the quantity of things being represented. For example, if one graphical unit represents ten cars and five graphical units are shown, the viewer will mentally multiply ten times five and conclude that the graphical grouping represents 50 actual cars.

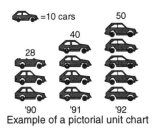

Example of a pictorial unit chart

- General guidelines
  - Each picture, icon, symbol, etc., on this type of chart represents a certain number (e.g., 1, 10, 50, 1000) of actual units of the thing it represents.
  - Actual numeric values may or may not be included, but generally are.
  - Pictorial unit charts generally have no scales, tick marks, or grid lines.
  - In a given unit chart, a graphical unit typically represents the same number of actual units, regardless of its size or shape.
  - Graphical units of the same type are generally all the same size in a given chart.
  - Simple geometric shapes or irregular shapes such as pictures and icons are equally effective in communicating information in this type of chart.
  - Graphical units may be arranged in one, two, or three dimensional groupings.

Barrels of oil to heat an average home in some city

- Unit charts can be used to convey information about a single entity (example at left) or to compare multiple entities (examples at right).

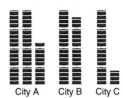

City A   City B   City C

A comparison of how much oil it takes to heat an average home in three different cities

- When each data graphic represents more than one unit, it is sometimes necessary to crop one of the data graphics to have the total number come out to the exact value.

35 Hammers   20 Pliers   29 Wrenches

- A symbol might represent what it looks like or something closely associated with it. For example, the picture of a ship (right) might represent a certain number of ships or a certain number of passengers or tons of cargo transported by ships.

- Additional qualitative information can be encoded into the chart by means of the size, shape, color, or shading of the data graphics.

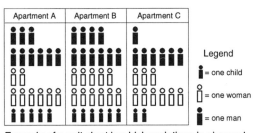

Example of a unit chart in which variations in size and color are used to encode additional information

- Pictorial unit charts are sometimes arranged in the form of a table. This format is sometimes referred to as a pictorial table.

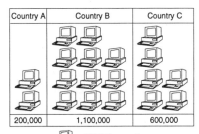

Example of a unit chart in table form

Pictorial Map

Pictures or icons are often selected to resemble the things they represent. Frequently a name or description of the thing represented by the symbol is also included. Sometimes just a few images are used while in other situations the map is almost 100% images, pictures, icons, etc.

Example of a pictorial map

Pictorial Table

Icons and images can be used in the body or headings of tables. In the body of the table they may indicate the presence or absence of something. In other cases their size or quantity may be proportional to the size or number of the thing they represent. Accompanying text is normally required for full understanding.

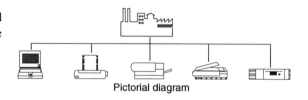

Pictorial diagram

Pictures, icons, images, etc., are used extensively in diagrams. They can be used simply to identify people, things, and/or places or can have their size or quantity proportional to the thing they represent.

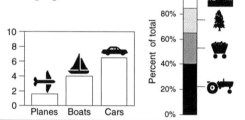

Pictorial diagram

Pictorial images used to identify individual data graphics

Shown at the right are examples in which pictorial images are used to identify individual data graphics. Category labels may or may not be included when such pictorial images are used.

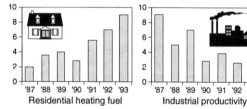

Pictorial images used to supplement chart titles or labels

Pictorial images help orient the viewer but additional information is generally included to specifically identify the data on the chart. For example, the image of the house differentiates the general subject matter from the industrial graph on the right, but the labels or titles specifically identify what the numeric data on the chart applies to.

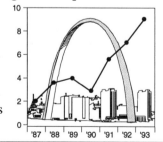

Broad orientation regarding the general scope of data

An image in the background can sometimes rapidly orient the viewer. For instance, in the examples at the right the viewer could safely assume that in one case the data applies to St. Louis while in the other case the data is global in nature.

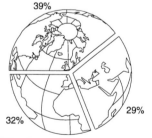

General

Although pictures and images can sometimes be helpful, the full meaning of a chart can generally be determined only by a combination of its contents and its titles, labels, notes, text, etc. For example, if an image of a gun is shown it somewhat orients the viewer; however, additional information is needed to tell the viewer whether the gun stands for crime, violent crime, gun sales, or gun ownership. This additional information is normally provided along with the graph in notes, legend, text, or titles.

Pie Chart

Sometimes referred to as a cake chart, divided circle, circular percentage chart, sector chart, circle diagram, sectogram, circle graph, or segmented chart. Pie charts are members of the proportional area chart family. Their major purpose is to show the relative sizes of components to one another and to the whole. They are used extensively as communication tools in presentations and publications. • A pie chart consists of a circle divided into wedge-shaped segments. The area of each segment (sometimes called slice or wedge) is the same percent of the total circle as the data element it represents is of the sum of all the data elements in its data set. For example, the pie chart at the right shows the distribution of different colors of cars. Since there is a total of 200 cars and 10% of them are red, the segment of the pie representing red cars constitutes 10% of the total pie. The 10% can be determined either of three different ways, all of which yield the same end result. They are:

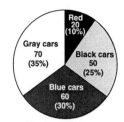

Total cars equals 200

Example of a pie chart in which the area of each segment is the same percent of the total circle as the data element it represents is of the sum of all the data elements in its data set

1. The angle of the wedge (i.e., 10% of 360 degrees)
2. The area of the wedge (i.e., 10% of the total area of the circle)
3. The length of the arc (i.e., 10% of the circumference of the circle)

The same reasoning applies to each different color of car. Twenty-five percent of the cars are black; therefore 25% of the area of the pie is allocated to black cars; 30% is allocated to blue, and 35% to gray. The sum of all of the segments (which obviously adds up to 100%) makes up the complete pie representing all the cars. Pie charts are classified as percent-of-the-whole charts or 100% charts.

Description and terminology

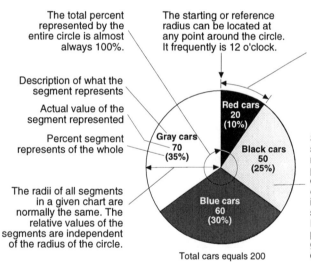

The total percent represented by the entire circle is almost always 100%.

The starting or reference radius can be located at any point around the circle. It frequently is 12 o'clock.

The portion of the circle included between adjacent radii is called a segment, slice, sector, or wedge. It may be measured in terms of angle, arc, or area. The sizes of the pie segments are in the same proportions as the data elements they represent.

Description of what the segment represents

Actual value of the segment represented

Percent segment represents of the whole

The radii of all segments in a given chart are normally the same. The relative values of the segments are independent of the radius of the circle.

Segments can be arranged in any sequence. If the segments have a natural progression, that progression can proceed clockwise or counterclockwise. In this example, they proceed clockwise in terms of the percent that each segment represents of the whole. If there is not a natural progression, the segments are generally arranged so that the chart is most meaningful and easiest to read.

Total cars equals 200

General characteristics of pie charts

- With rare exceptions, negative numbers cannot be displayed on pie charts.
- Percentages over 100% can be used but rarely are, since they go contrary to one of the major advantages of pie charts. That major advantage is that people naturally tend to think of a circle as encompassing 100%.
- From a technical point of view the segments can be in arranged in any order. To make the chart most meaningful and easier to read, segments are typically arranged in some meaningful order such as smallest to largest, natural groupings of the data, alphabetical, etc.
- Information can proceed in a the clockwise or counterclockwise direction. The clockwise direction is normally used.
- The starting or reference radius can be located at any point around the circle. It frequently is located at 12 o'clock.
- With few exceptions the values depicted by the pie segments are independent of the radius of the circle or segment.
- The radii of segments in a given pie chart are generally all the same.

281

Methods for incorporating descriptive and quantitative information into pie charts

Since there is no scale around a pie chart, it is sometimes difficult for the viewer to estimate the sizes of the segments accurately. Because of this, information regarding each segment is generally noted on or adjacent to the segments. Combinations of descriptive information indicating what the segment represents, its value, and its percent of the whole are the most frequently shown types of information. The examples below show some of the more widely used methods for incorporating the key information into the charts.

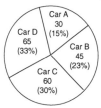
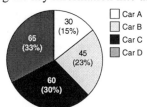
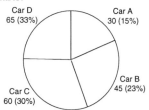

All key data provided inside of the segments. This is generally the recommended method.

Legend used to identify individual segments. Generally this makes it more difficult for the viewer because of the extra eye movement required.

Some or all of the data shown outside of the segments. This is sometimes necessary when the segments are small and the descriptions large.

Reference angle used for plotting pie charts

The point at which one starts measuring off the segments when generating a pie chart is sometimes referred to as the reference angle. This angle is not designated in the final pie chart; however, is sometimes important since it, (along with the sequence of the segments), determines the orientation of all the segments in the final chart. The examples shown below indicate four popular methods for positioning the segments of a pie chart and the resulting location of the reference angle. Some people feel it is advisable to have the most important segment start at 12 o'clock. Others feel the most important segment should be centered at 12 or 3 o'clock, and still others feel the most important segment should be at the 6 o'clock position, particularly when the pie is elongated to give the appearance of depth. • The positioning of the segments is many times accomplished by selecting the proper reference angle for the start. • When a series of pie charts is generated to show such things as changes over time, the same reference angle is generally used for all, and the segments are plotted in the same sequence on each pie chart.

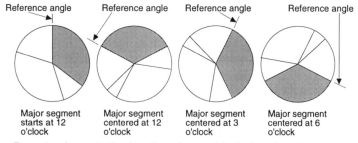

Major segment starts at 12 o'clock

Major segment centered at 12 o'clock

Major segment centered at 3 o'clock

Major segment centered at 6 o'clock

Examples demonstrating how the reference (starting) angle varies depending on where one wants the major segment(s) to be located on the pie chart. This example assumes the major segment (gray) is the first segment plotted. If it is not the first segment plotted, the concept remains the same but the reference angle changes.

Methods for showing changes over time with pie charts

When comparing multiple data series or studying changes in the same series over time or under varying conditions, a series of pie charts is sometimes used. Typically the series of charts is shown side-by-side. Occasionally they are superimposed on top of one another in a stacked pie chart. In both methods the segment representing a given element is kept in the same relative position in each of the multiple pies. More than four or five pie charts side-by-side or more than two stacked pie charts generally becomes confusing. • The relative sizes of the various circles in the stacked variation generally have no significance. Where the pies are placed side-by-side, the relative sizes of the circles may or may not have significance.

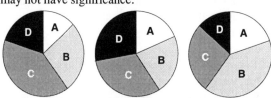
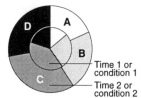

Time 1 or condition 1   Time 2 or condition 2   Time 3 or condition 3
A series of side-by-side pie charts

Stacked pie charts

Examples of how changes with time and/or conditions can be displayed using pie charts

Pie Chart (continued)

Size of circles proportional to overall value of pie chart

The example at the right is a typical application of pie charts. In this example the sales by department are compared at two different points in time. The percents that each departments sales represent of the total sales are shown both numerically and graphically for both years. The overall sales of the company are shown numerically but not graphically.

1995
Total sales $10 million

1996
Total sales $20 million

Two pie charts used to compare sales by department for two different years. Generally there is no graphical indication of differences in the overall values.

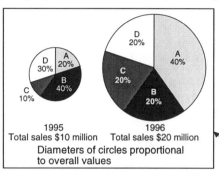

1995
Total sales $10 million

1996
Total sales $20 million

Diameters of circles proportional to overall values

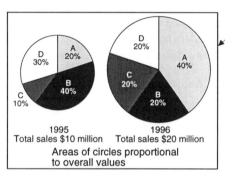

1995
Total sales $10 million

1996
Total sales $20 million

Areas of circles proportional to overall values

• In the two pairs of pie charts at left, in addition to showing the numerical value of the overall sales, the sizes of the circles are drawn proportional to the total sales. In this way the viewer can get a graphical comparison of overall sales, as well as a graphical comparison of individual department sales.

The upper pair of circles are drawn with the diameters of the circles proportional to overall sales.

The lower pair are drawn with the areas of the circles proportional to overall sales. Comparisons based on area are frequently recommended. • Even though the graphical representation gives only a crude indication of the difference, it does alert the viewer to the fact that there is a difference in the overall totals, the direction of the difference, and whether the difference is major or minor.

• Another variation of a pie chart sometimes used to call the viewer's attention to differences in overall values is called a donut chart. In this technique a circle is blanked out of the center of the pie and the overall value written in the blank space. An example is shown at the right. The size, shape, and color of the blanked area are insignificant.

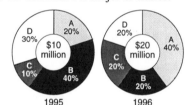

1995

1996

Totals for pie charts are occasionally included in their centers for emphasis. The result is sometimes referred to as a donut chart.

Methods for highlighting selected segments of a pie chart

There are several methods commonly used for emphasizing selected segments of a pie chart:

    – Apply a distinctive color, shade, or fill to the selected segment(s)

    – Separate the segment(s) from the main body of the pie (sometimes referred to as exploded or separated pie chart). Any number of segments can be exploded, ranging from one to all. The more segments that are exploded, the less effective the technique becomes.

    – Graphically tilt (elongate) the pie chart with the selected segment(s) in front

    – Elevate the selected segment(s) (sometimes referred to as multiheight pie chart)

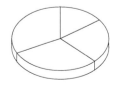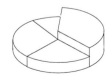

Segment highlighted by adding distinctive color or fill

Segment highlighted by separating it from the main chart (called exploded or separated)

Segment highlighted by tilting and adding depth (sometimes called three-dimensional)

Segment highlighted by changing its height (sometimes called multiheight pie chart)

Examples of methods used to emphasize selected segments. These methods can be used separately or combined.

283

Methods for improving the legibility of small segments

When several of the segments of a pie chart are so small that it is difficult to estimate their relative sizes, they are sometimes displayed in a supplemental chart or graph that may be a column graph or another pie chart. The examples below illustrate the concept.

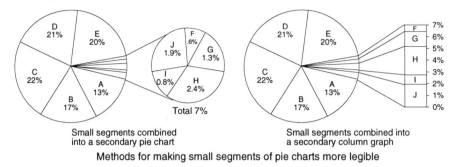

Small segments combined
into a secondary pie chart

Small segments combined into
a secondary column graph

Methods for making small segments of pie charts more legible

Groupings of segments on pie charts

Sometimes it is advantageous to group multiple segments to indicate that they have something in common or are related in some way. For example, one group might represent items that are purchased from domestic sources, another group might represent items purchased from foreign sources, and a third might represent inter-company purchases. Subtotals may or may not be assigned to the subgroupings. Two examples of how this is sometimes done are shown below. One example uses arrows around the outside of the pie chart. The other overlays two pie charts.

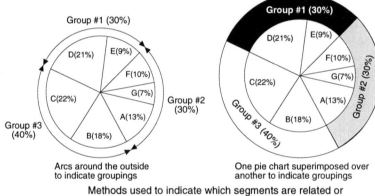

Arcs around the outside
to indicate groupings

One pie chart superimposed over
another to indicate groupings

Methods used to indicate which segments are related or
grouped and what the subtotals are for those groupings

Pie chart used as a histogram

Pie charts are occasionally used as circular histograms. When used in this way, each segment of the chart represents a different class interval, and the size of the segment is proportional to the frequency of occurrence of the values in the class interval. For example, if a data series has 105 data elements and 16 (15%) of those data elements have values that fall in the class interval between 4 and 4.9, the segment representing the class interval 4 to 4.9 would represent 15% of the total area of the circle. The size of each segment is determined by the same procedure. An example is shown on the right along, with its equivalent column type histogram below.

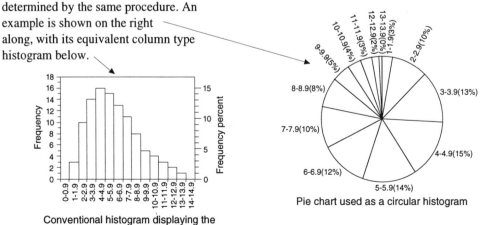

Conventional histogram displaying the
same data as the pie chart at the right

Pie chart used as a circular histogram

Pie Chart (continued)

Encoding an additional quantitative variable

Sometimes additional quantitative variables are encoded into pie charts. One method for doing this is to simply include the numeric values directly onto the pie segments. Graphical methods are sometimes also used, as shown here. In all cases where graphical methods are used, an explanation and/or scale is generally included in the legend so the viewer can decode the information. • With all graphical methods, the accuracy in decoding is generally marginal.

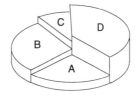

Additional variable encoded into the heights of the individual segments. Sometimes called multiheight.

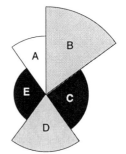

Additional variable encoded into radii of the individual segments

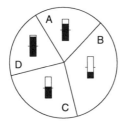

Additional variable encoded with symbols in the individual segments (in this example, framed rectangles)

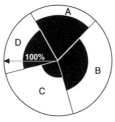

With this variation, each segment of the pie is divided into two pieces. The radii of the two pieces add to 100%.

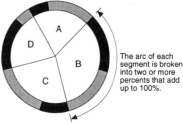

Each segment is divided into two pieces, the sum of which equal 100%. Each segment could have been divided into 3 or 4 pieces.

Methods used to encode an additional quantitative variable into pie charts

Pie charts with the appearance of depth

To improve the appearance of pie charts, they are sometimes given the appearance of depth, as shown at the right. Some people call pie charts with this effect three-dimensional. The technique can be applied to complete pie charts or to individual segments. When the effect is applied, some or all of the segments are distorted, and therefore, observations based only on the graphics can be misleading.

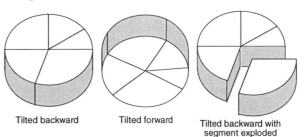

Tilted backward   Tilted forward   Tilted backward with segment exploded

Examples of complete pie charts with the appearance of depth

1991   1992   1993

Examples of individual segments with depth

Overlapping pie charts

In some situations, where the data is just right, space can be conserved and sometimes the appearance improved by overlapping pie charts. This might be done in the vertical or horizontal direction. In cases where one element of the graph is the key focus, only that portion of the pie might be shown or highlighted.

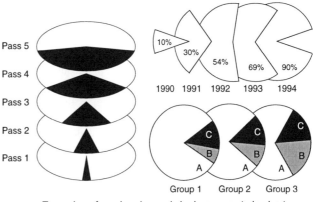

Examples of overlapping and single segment pie charts.

285

Pie Chart (continued)

Decagraph

A seldom-used variation of a pie chart in which the circle around the pie is replaced by a ten-sided polygon. With the circumference made up of ten chords of equal lengths, the points where the chords meet automatically designate every 10% of the complete circle. The objective of the decagraph is to enable the viewer to more easily and rapidly estimate the sizes of the individual segments with the assistance of the chords that indicate 10% increments.

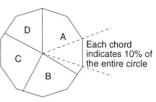

Each chord indicates 10% of the entire circle

Example of a decagraph

Belt Chart

A belt chart is a variation of a stacked pie chart that enables the viewer to concurrently look at the distribution of interrelated data in a number of different ways. By doing this, patterns and relationships sometimes are easier to note. Using the example shown below, the following are representative of the types of observations that can be made from a belt chart.

– 64% of the work force is female and 36% male
– Of the 64% females, 48% are nonexempt and 16% exempt
– Of the 36% males, 15% are nonexempt and 21% male
– 64% of all employees work in the office and 36% in the field
– 63% of the employees are nonexempt and 37% are exempt
– Of the 63% nonexempt, 48% are female and 15% are male
– Of the 37% exempt, 16% are female and 21% male
– Of the 63% nonexempt, 51% work in the office and 12% in the field
– Of the 37% exempt, 13% work in the office and 24% in the field

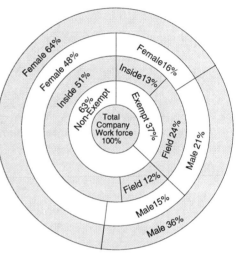

Example of belt chart

Circle graph

When a scale is added to the circumference of a pie chart, as shown at the right, the resulting chart is many times referred to as a circle graph. With the presence of a scale and tick marks, technically the percentages that each segment represents can be read directly from the scale and therefore, the numeric values do not have to be shown on the segments. In practice, it is still easier and more accurate for the viewer to have the actual percents noted in addition to having the scale and tick marks.

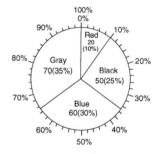

Pin Map

Sometimes referred to as a street map. A pin map is very detailed, typically showing exact locations of structures, events, customers, intersections, etc. In some cases the maps are so detailed that specific addresses, stores, or vacant lots are identified and located as points on the map. Symbols are frequently used to encode information and identify entities. Sometimes pins are used to perform this same function, particularly when the things identified change frequently, such as stops on a delivery route, sales leads, or trouble spots. The symbols used may be simple geometric shapes or icons of the thing being located. The example at the right includes both types of symbols.

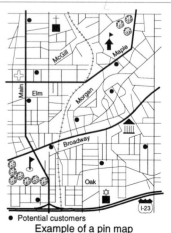

● Potential customers

Example of a pin map

Plane

The term plane is sometimes used with charts and graphs to designate an imaginary surface passing through an object or graph. The planes may or may not stay within the upper and lower limits of the axes and may or may not be parallel to the major axes. When they are parallel, they are sometimes referred to by the axes they parallel. For example, a plane parallel to the X and Y axes is referred to as an XY plane. In a three-dimensional graph, a plane parallel to the X and Z axes is referred to as an XZ plane. Planes that form the sides, top, and bottom of a graph are sometimes called walls, ceiling, and floor, respectively. These planes are many times used to place grid lines on or project data graphics onto for the purpose of analysis and/or communication. When two planes intersect one another, a line is generally drawn to indicate the line of intersection. Several examples illustrating the use of planes are shown below.

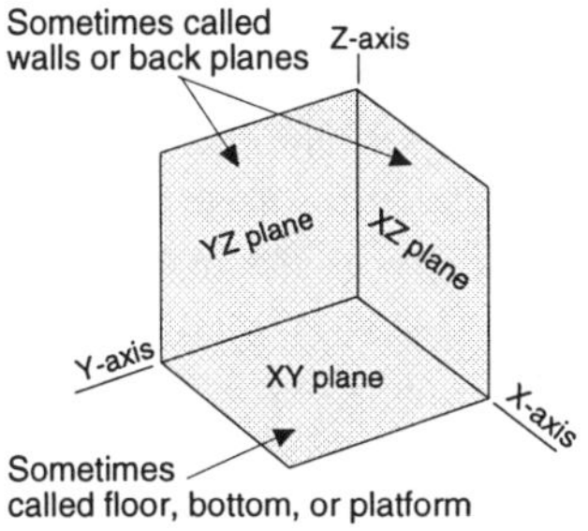

Sometimes called walls or back planes. Z-axis. YZ plane. XZ plane. Y-axis. XY plane. X-axis. Sometimes called floor, bottom, or platform.

Examples of primary planes that coincide with the axes. In this example the planes are bounded by the upper and lower values on the axes.

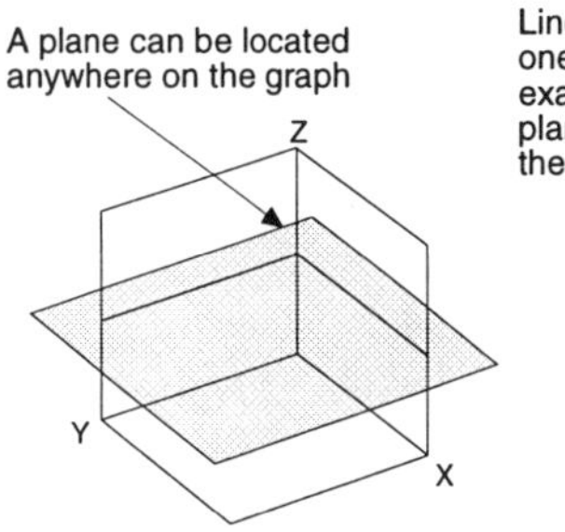

A plane can be located anywhere on the graph. Z. Y. X.

Planes do not have to be coincident with axes and can extend beyond the ends of the drawn axes.

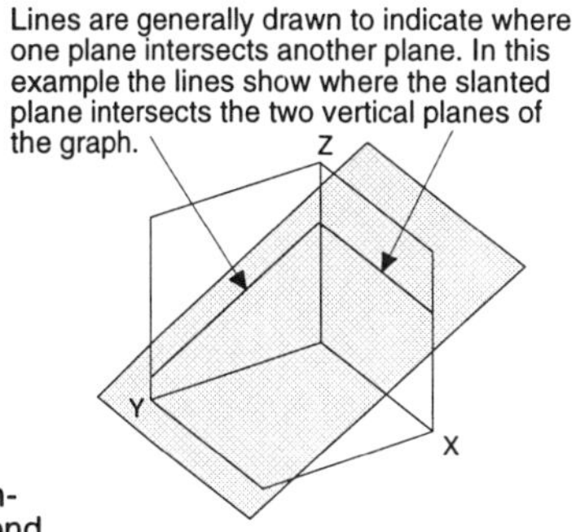

Lines are generally drawn to indicate where one plane intersects another plane. In this example the lines show where the slanted plane intersects the two vertical planes of the graph. Z. Y. X.

Planes do not have to be parallel to axes.

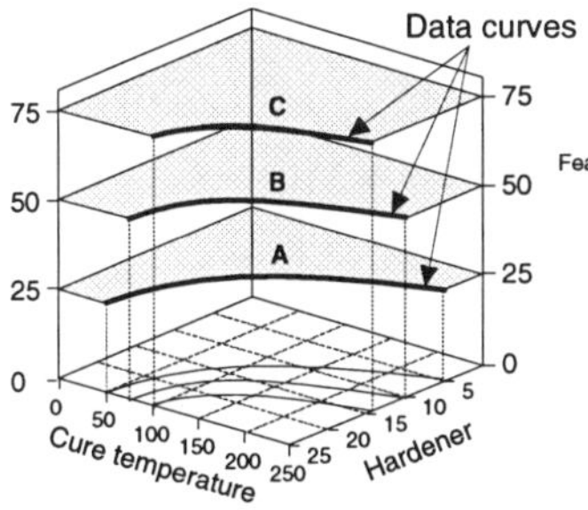

In this example, imaginary planes use data curves to define one boundary.

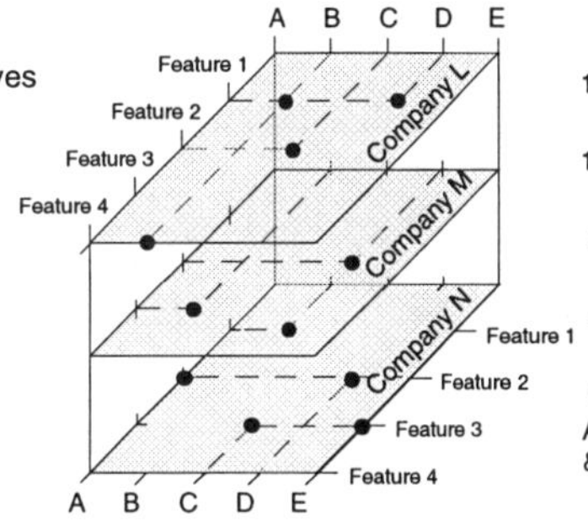

Data curves. C. B. A. Cure temperature. Hardener.

Planes can be used with nonquantitative information and on charts such as tables.

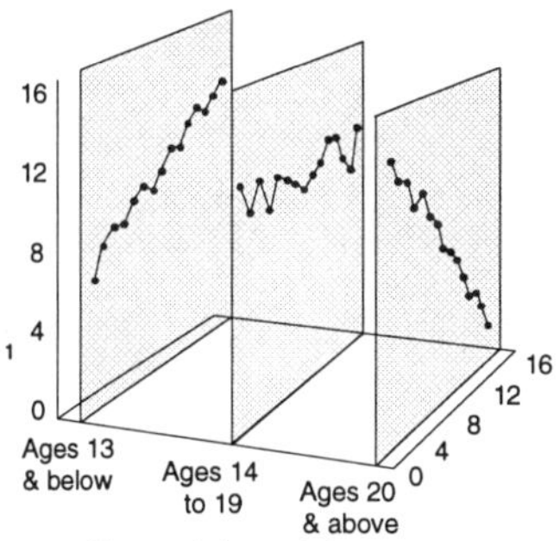

A B C D E. Feature 1. Feature 2. Feature 3. Feature 4. Company L. Company M. Company N. Feature 1. Feature 2. Feature 3. Feature 4. A B C D E.

Here, data points are condensed onto imaginary, vertical planes.

Ages 13 & below. Ages 14 to 19. Ages 20 & above.

Platform

Sometimes referred to as a bottom or floor. The term platform is used to identify the lower plane of a three-dimensional graph.

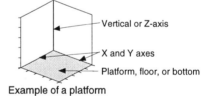

Vertical or Z-axis. X and Y axes. Platform, floor, or bottom. Example of a platform

Plat Map

A variation of a descriptive map that very accurately defines the boundaries of units of land. The maps generally include only minimal information regarding natural or man-made features that exist on the unit of land.

Plot

The terms plot and graph are used interchangeably in this book. See Graph.

Plot Area and Plot Area Border/Frame

Also called data region. The plot area of a graph is the area in which the data is plotted. It is frequently bounded by a border or frame. Grid lines generally end at the edges of the plot area. • The line around the plot area is called the plot area border or frame.

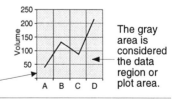

The gray area is considered the data region or plot area.

Border or frame

Plot Symbol or Plotting Symbol

Symbols used to designate data points on a graph. They generally are small geometric shapes such as circles, squares, or triangles. Sometimes their size, shape, or color are used to encode additional information.

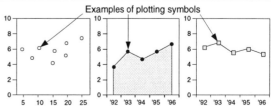

Examples of plotting symbols

Point and Figure Chart
or
P&F Chart

A chart used to record the prices of a stocks, bonds, commodities, exchanges, markets, etc., in such a way that changes in price and the direction of such price changes stand out clearly. Point and figure charts can be used to analyze short-term (e.g., day or week) pricing activity as well as long-term (e.g., months or years). A point and figure chart disregards time completely and focuses on price changes. If there are no changes in the price of the entity being tracked, there are no additional entries on the P&F chart, regardless of how long the period of time or how many sales might be made. • The price of the thing being tracked is indicated on the vertical axis using a linear scale. Actual prices are recorded by means of two different symbols. If the price increases by a certain amount (called box size), one type of symbol, many times an X, is placed in the box representing the new price. As long as the price continues to increase, new Xs are placed one on top of the other in a single column each time the price increase exceeds the specified amount. If the price decreases by the same specified amount, a new column is started and the other symbol, many times an O, is used to record the new prices. For example, if the specified amount or box size is 5 cents for a given commodity, every time the price of the commodity rises an additional 5 cents, another X is added to the top of the stack of Xs. When the price falls by the same increment (5 cents), the second type of symbol is placed in the next column to the right at the new price of the commodity. As the price continues to decrease, additional Os are added in the same column but now each new O is placed below the previous Os. When the price again begins to rise, a new column is started and Xs are used to record the price. The entire process repeats itself endlessly with a new column being started every time the price of the thing being monitored reverses. An example of a point and figure chart is shown below.

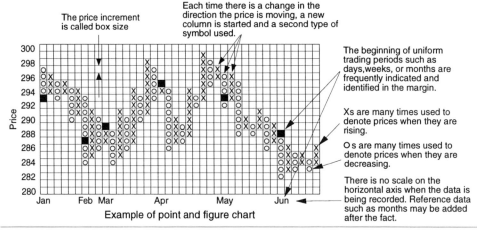

Example of point and figure chart

Box size

Box size is the minimum amount a price must change before an additional X or O is added in the same column. This minimum might be equal to some multiple of the minimum price increment the stock trades at, a fixed dollar or cent amount, or a percent of the selling price. To illustrate, if a stock is normally priced in eighths of a dollar, the box size might be 1/8, 2/8, 3/8, etc. If the box size is set at 1/8 and the price of the stock changes by an 1/8th of a dollar, a new symbol is added. If the box size is set at 2/8ths and the price changes by only 1/8th, no new symbol is added. The size selected for the box is dependent on such things as the price of the thing being tracked (i.e., 1/8 might be fine for a $5 stock but is probably too small for a $200 stock), its volatility, whether one is interested in the short- or long-term nature of the price fluctuations, and how detailed one wants the data to be. In general, the smaller the box size, the more symbols and columns there will be for a given period of time.

Minimum price reversal

This is the minimum price change required to initiate a new column. It is typically equal to a certain number of box sizes (e.g., 1, 2, 3, etc.) or a percent of the closing value. To avoid recording minute fluctuations, analysts many times require the price to increase or decrease by an amount greater than the box size used for adding Xs and Os in the same column. When this method is used it is sometimes referred to as a reversal chart. For example, if the box size is 1/8th dollar and the minimum price reversal is three, a price reversal of 3/8ths is required to initiate a new column. If the box size is one dollar and the minimum price reversal is three boxes, a price reversal of three dollars is required to start a new column. In general, the smaller the price reversal used, the more columns for a given period of time.

Point and Figure Chart
or
P&F Chart (continued)

The examples below illustrate this point. In all three examples, the box size is one dollar. In the graph on the left the minimum price reversal is three boxes. The middle graph has a minimum price reversal of two boxes, and the graph on the right has a minimum of one. The same data is plotted on each of the three graphs.

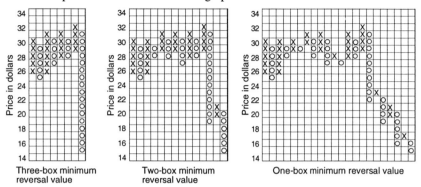

Charts of the same stock price data using three different minimum price reversal values

It may be noted that the example on the right (minimum price reversal of 1 box) is about three times as wide as the one on the left (minimum of 3 boxes). Even with the minimum price reversal equal to the box size, minute price fluctuations can be avoided by making the box size larger. Although the smaller minimum price reversal makes the chart wider, it also yields more information as to what actually happened. For example, the chart on the left might give the impression of an uninterrupted plunge on the right side of the chart, while the expanded chart (right) indicates that several unsuccessful attempts were made to reverse the downward trend. For general reference, the information on these charts might represent about four months of data on a particular stock.

Designating uniform periods of time

Since no time scale is used, there is no way to determine when particular price activities occurred or to know the frequency of price reversals over uniform intervals of time. Some analysts solve this problem by highlighting the first price in each new period and identifying the periods in the lower margin. The highlighting is accomplished in several different ways, including substituting a third symbol such as a black square, bolding the X or O, or replacing the X or O with some identifying letter such as the first letter of the month.

Variations

Although the format of the point and figure chart shown on the previous page is widely used, several other variations accomplish the same objective. Examples of four alternate formats are shown here. All have portions of the same data plotted on them.

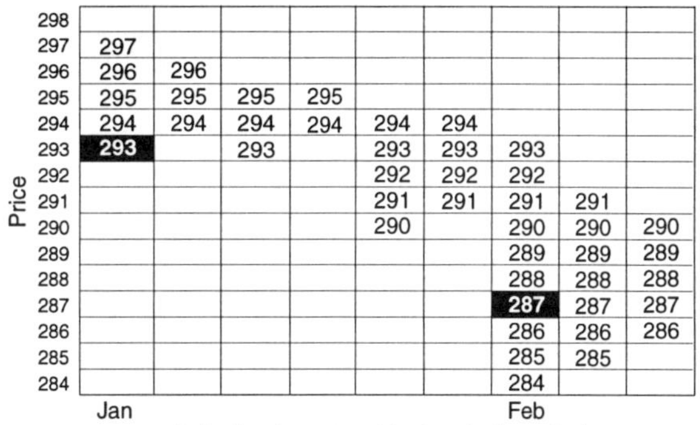

Price	Jan						Feb		
298									
297	297								
296	296	296							
295	295	295	295	295					
294	294	294	294	294	294	294			
293	**293**		293		293	293	293		
292					292	292	292		
291					291	291	291	291	
290					290		290	290	290
289							289	289	289
288							288	288	288
287							**287**	287	287
286							286	286	286
285							285	285	
284							284		

Actual values used instead of symbols

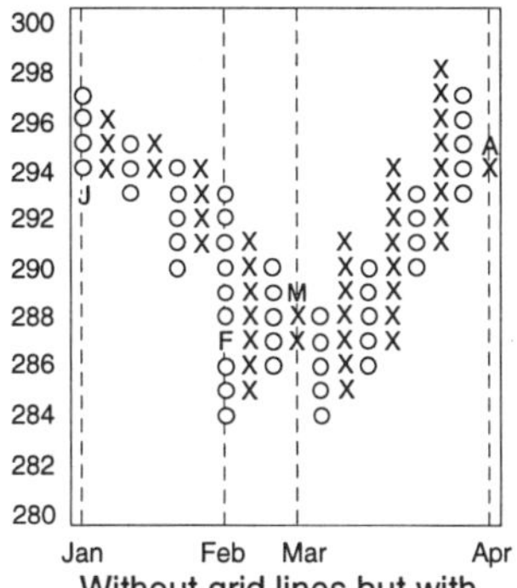

Without grid lines but with reference lines at the start of each time period

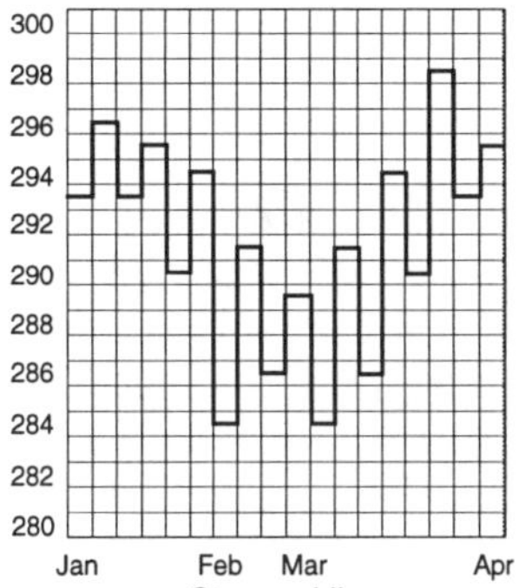

Stepped line (sometimes called swing chart)

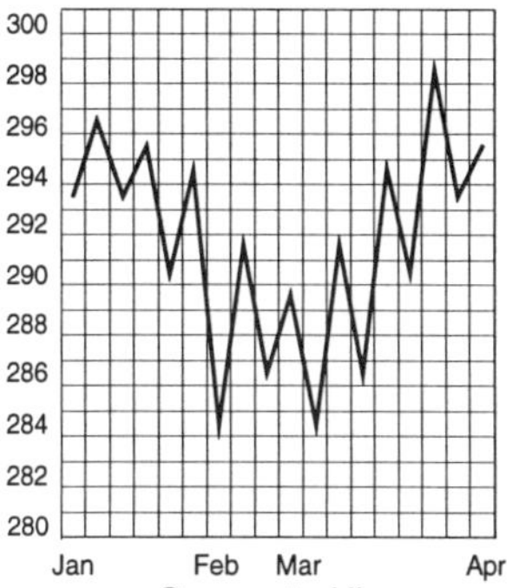

Segmented line

Point Data

Point data is data that applies to a specific point in time – for example, the cash on hand the last day of the month, the temperature at midnight on Sunday, or the membership as of the last day of the year.

**Point Graph
and
Scatter Graph**

Sometimes referred to as dot or symbol graph. Point graphs are a family of graphs that display quantitative information by means of points represented by symbols such as dots, circles, or squares (called plot or plotting symbols). When the points are connected by lines, the graph is generally referred to as a line graph. See Line Graph. • There are several different names used to refer to point graphs. Two major factors sometimes considered in selecting the proper name are the type of scale on the horizontal axis (the scale on the vertical axis is most often quantitative), the number of data series plotted on the graph, and whether the data series use common values on the horizontal axis. The table below summarizes the various names and graph characteristics for two-dimensional point/scatter graphs.

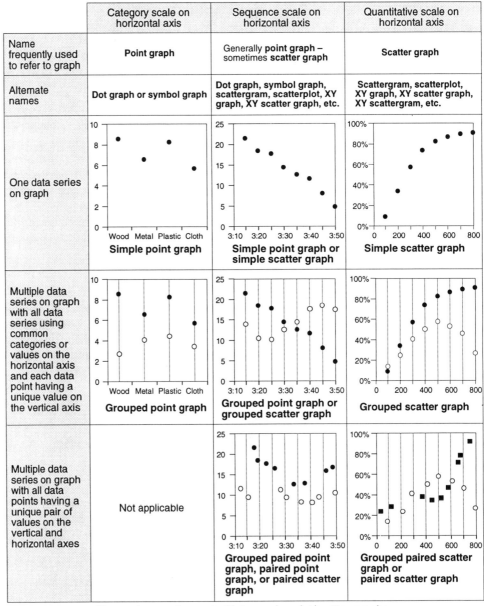

Names frequently used with two-axis point/scatter graphs

Scatter graph

Scatter graphs are probably the most widely used type of point graph. They generally have quantitative scales on both axes and can accommodate many data points (100 in the example at right). Scatter graphs are used extensively for exploring relationships and correlations between two or more sets of data – for example, the relationship between efficiency and speed, dollars spent on food versus dollars of income, rate of chemical reaction versus temperature, etc. After data is plotted on a scatter graph, patterns formed by the data points are used to make observations about the relationships of the data sets graphed. For example, if the data points are tightly clustered around an imaginary slanted line, the two

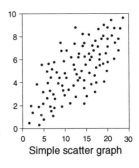

Simple scatter graph

**Point Graph
and
Scatter Graph**
(continued)

Scatter graph (continued)

variables are generally considered to have a strong correlation. If the data points are loosely clustered, the relationship is considered weak. If the imaginary line slopes up, it is considered a direct or positive correlation. If the imaginary line slopes down, it is considered an inverse or negative correlation. The examples below show how some of the more familiar patterns are interpreted.

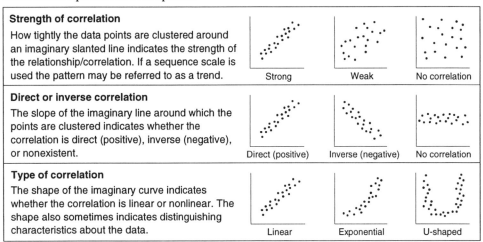

Strength of correlation
How tightly the data points are clustered around an imaginary slanted line indicates the strength of the relationship/correlation. If a sequence scale is used the pattern may be referred to as a trend.

Strong   Weak   No correlation

Direct or inverse correlation
The slope of the imaginary line around which the points are clustered indicates whether the correlation is direct (positive), inverse (negative), or nonexistent.

Direct (positive)   Inverse (negative)   No correlation

Type of correlation
The shape of the imaginary curve indicates whether the correlation is linear or nonlinear. The shape also sometimes indicates distinguishing characteristics about the data.

Linear   Exponential   U-shaped

Basic patterns that are looked for when interpreting the data
point patterns formed on scatter graphs by two data sets

Three-dimensional scatter graph

Three-dimensional scatter graphs generally have quantitative scales on all three axes. The rectangular variation, shown at the right, is sometimes referred to as an XYZ graph or an XYZ scatter graph. Polar variations are referred to as cylindrical or spherical scatter graphs. An example of a cylindrical graph is shown at the left. (See Polar Graph). • When a reference plane passes through the cluster of data points (sometimes referred to as a cloud), the figure is sometimes called an axis scatter graph (right). • Exact values are difficult to read on three-dimensional graphs. This frequently is not a problem since three-dimensional scatter graphs are often, primarily used to observe such things as the general pattern of the data, how the data is distributed, whether there are unusual data points, etc. In an effort to improve the interpretation of three-dimensional scatter graphs, techniques have been developed that enable the viewer to significantly increase the amount of meaningful information that can be derived from such graphs. Two such techniques are the ability to rotate or spin the graph (see Three-Dimensional Graph) and the ability to identify specific data points or groups of data points (See Brushing). • Another technique to assist in the analysis of three-dimensional scatter graphs is to pass imaginary planes through the cloud of data points and observe the patterns on the planes. In some cases only those data points actually at the points where the planes pass through the cloud are recorded. In other cases the data points for a given distance on either side of the imaginary planes are condensed onto the planes and displayed as shown at the left. Such graphs are sometimes called slice graphs. See Slice Graph.

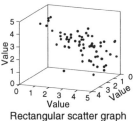

Rectangular scatter graph

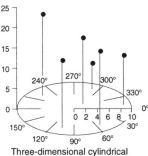

Three-dimensional cylindrical
polar scatter graph with drop lines

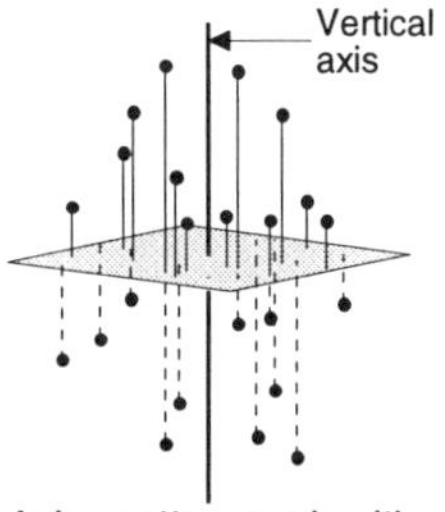

Axis scatter graph with
reference plane through
cloud of data points
Vertical axis

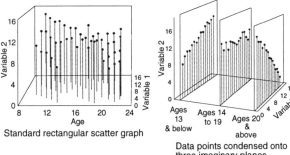

Standard rectangular scatter graph

Data points condensed onto
three imaginary planes

Slice graph resulting from the condensing of data points
of a standard scatter graph onto three imaginary planes

291

Point Graph and Scatter Graph (continued)

Point graphs used to show the distribution of data elements within data series

Point graphs can be effective in giving the viewer a feel for the distribution of data elements within data sets. For example, when studying the test scores of students from different schools, in addition to knowing the average and median scores, plotting each individual score can provide information about how the spread of scores differs from school to school. This can be accomplished by means of a point graph alone or by superimposing a point graph over another graph such as a box or column graph. Three examples are shown below.

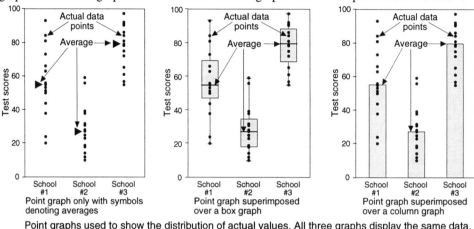

Point graph only with symbols denoting averages — Point graph superimposed over a box graph — Point graph superimposed over a column graph

Point graphs used to show the distribution of actual values. All three graphs display the same data.

One-axis point graph with single scale used to show distribution of data elements

One-axis point graphs can be used to analyze the distribution of data elements within individual data sets. This might be done for an isolated data set (shown below) or in conjunction with a two-axis scatter graph (shown at the right). When doing a distribution analysis, one looks for such things as:

- Clustering of data points (e.g., how many clusters? where? how dense?).
- Whether or not there are gaps in the data
- How big the spread of values is
- Maximum and minimum values
- Unusual data points

When one-axis point graphs are placed in the margins of two-axis scatter graphs, the viewer sees the distribution of the data points along the two axes as well as the relationship between the two sets in the body of the graph. See One-Axis Data Distribution Graph and Marginal Frequency Distribution Graph.

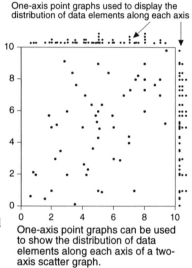

One-axis point graphs can be used to show the distribution of data elements along each axis of a two-axis scatter graph.

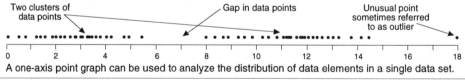

A one-axis point graph can be used to analyze the distribution of data elements in a single data set.

One-axis point graph for displaying data

One-axis graph with category type data

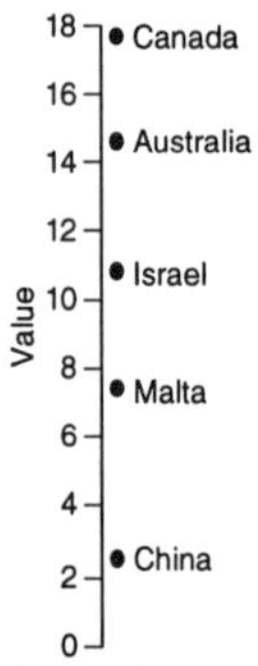

When the scale and data are displayed along the same axis, it is sometimes called a one-axis graph. The axis might run vertical or horizontal. For category type information, the data points can often be plotted directly onto the single axis, as shown at the left. With quantitative information, the data is often first plotted on a two-axis graph and the key points transferred to the other axis, as shown at the right.

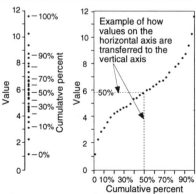

Quantitative data plotted on a two-axis graph (right) and key points transferred to the one-axis graph (left)

Other types of point graphs
Polar graph

There are two- and three-dimensional variations of polar graphs. The two-dimensional variation is by far the most widely used. Each polar graph has one angular axis and one radial axis. Three-dimensional polar graphs have a third axis (frequently called a Z-axis), which is perpendicular to the plane of the other two axes. All scales are quantitative and generally linear. See Polar Graph.

Polar graph

Trilinear graph

Trilinear graphs are point graphs with three axes in a single plane. In most cases they have quantitative scales that show percents from zero to 100%. See Trilinear Graph.

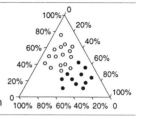

Trilinear graph

Point graphs used for the purpose of monitoring

Simple point graphs utilizing a sequence scale on the horizontal axis are frequently used to monitor processes, machines, production, procedures, etc. Key functions these graphs serve are to give a visible indication whether:

- A given measurement is staying within certain limits, boundaries, etc.
- Measurements are trending towards one of the limits, boundaries, etc.
- Performance is erratic
- The spread of data is becoming more or less consistent

In order to make the graphs as effective as possible, the limits within which the measurements are expected to stay are displayed on the graph by means of reference or control lines, as shown in the example at the right. Typically, data is entered on the graph on a continuous basis as it is generated. In this way corrective actions can be taken as soon as possible. As long as the data points stay within the prescribed limits and show no major trend towards one of the

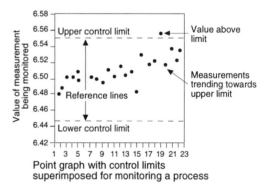

Point graph with control limits
superimposed for monitoring a process

limits, the process is considered to be in control. Sometimes multiple sets of reference lines are shown, with one set serving as a sort of early warning system and the second set as a critical decision point. The reference lines may be straight or stepped. The data points shown might be individual measurements or averages of several measurements. Data points may or may not be connected by lines. See Control Charts.

Scales, grid lines, and tick marks

Details on scales, grid lines, and tick marks are covered under those headings. Since one of the more widely used applications of scatter graphs is for analysis, these graphs tend to incorporate more negative numbers than other types of graphs. Four examples are shown below indicating how axes, scales, grid lines, and tick marks might be arranged to accommodate data that has positive and negative values along both the X and Y axes.

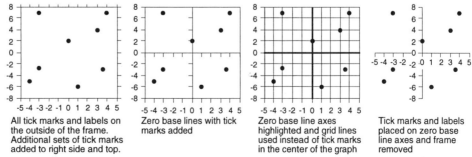

All tick marks and labels on the outside of the frame. Additional sets of tick marks added to right side and top.

Zero base lines with tick marks added

Zero base line axes highlighted and grid lines used instead of tick marks in the center of the graph

Tick marks and labels placed on zero base line axes and frame removed

Methods for displaying scales, tick marks, and grid lines on scatter graphs with positive and negative values on the vertical and horizontal axes. The same data is plotted on all four graphs.

Drop lines used with point and scatter graphs

Drop lines are faint lines extending from data points to an axis, from one data point to another, or in some cases extending completely across the graph. The lines might be solid or dashed and can be used with two- and three-dimensional graph formats. The major function of drop lines is to improve the readability of graphs. Point graphs with drop lines are sometimes substituted for column and bar graphs because they focus the viewer's attention on the data points instead of the data graphics. In some cases, they also tend to reduce problems associated with scale breaks and missing portions of scales. Several variations of drop lines used with point and scatter graphs are shown below.

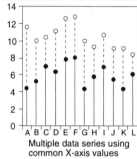

Drop lines from data points
to the horizontal axis

The drop lines in the example on the left extend from the data points to the horizontal axis. In the example on the right, the drop lines run the full length of the graph to deemphasize the distance from the point to the zero axis and instead, focus attention on the location of the data point itself.

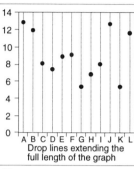

Drop lines extending the
full length of the graph

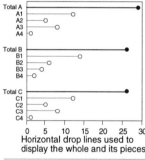

Multiple data series using
common X-axis values

In the example on the left, the data points of the two data series use common values on the horizontal axis. In the example on the right, each data point has a unique pair of X and Y values. Different types of drop lines are used to assist in differentiating the two sets of data points.

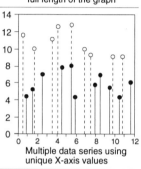

Multiple data series using
unique X-axis values

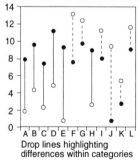

Horizontal drop lines used to
display the whole and its pieces

These two graphs are the equivalent of bar graphs with category scales on the vertical axis. In both cases the components (e.g., A1, A2, A3) and the total (e.g., A) for each group are displayed together. In the example on the right, the symbols for the data points have been eliminated.

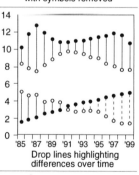

Same graph as at the left
with symbols removed

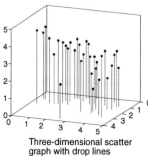

Drop lines highlighting
differences within categories

When the emphasis is on differences, drop lines can be drawn between data points to make what is sometimes called a drop line graph. Differences may be noted at a fixed point in time (left) or over time (right). Two different types of drop lines can be used to indicate reversals in the direction of the difference.

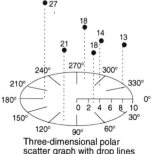

Drop lines highlighting
differences over time

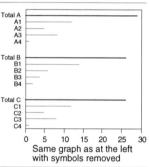

Three-dimensional scatter
graph with drop lines

Drop lines can be of particular value with three-dimensional graphs. They help the viewer more accurately estimate values on two axes; however, the values on the third axis are still a problem (left). When there are just a few data points, the actual values can be noted (right).

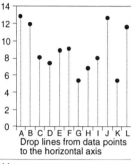

Three-dimensional polar
scatter graph with drop lines

Point Graph and Scatter Graph
(continued)

Curve fitting applied to scatter graphs

Because the data points on scatter graphs many times form irregular patterns, curves are sometimes fitted to the data points to aid in an analysis. Curves can be drawn based on visual estimates; however, usually well-established mathematical procedures are used (see Curve Fitting). The fitted curves are meant to represent what the data points would look like if they were condensed into a single line that best approximated all of the points.

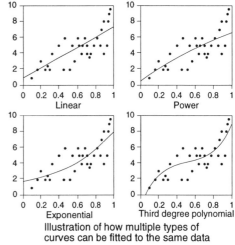

• Computers can generate many different curves for the same data. It is up to the individual to decide which of the curves is right for their purposes. To illustrate the point, four different computer-generated curves based on the same data are shown at the right. The person using the data would have to decide which of these four curves, plus many other possible curves, is most appropriate for a given application.

Illustration of how multiple types of curves can be fitted to the same data

• Fitted curves are used for many different purposes including:
- Establishing the general trend of data
- Determining the type of relationship between two variables such as linear, exponential, or none
- Comparing data series that are intermingled
- Making forecasts or projections
- Determining the degree of variation of individual data points from a theoretical or expected curve or equation
- Determining whether data points vary randomly, symmetrically, uniformly, etc., from a theoretical curve

The examples below illustrate how fitted curves are sometimes used to show differences between data series that are intermingled or to make projections based on sketchy data.

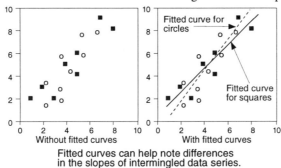

Fitted curves can help note differences in the slopes of intermingled data series.

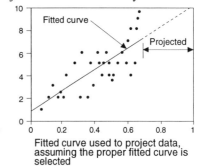

Fitted curve used to project data, assuming the proper fitted curve is selected

Surface fitted to three-dimensional data

Surfaces can be fitted to three-dimensional scatter graphs, just as lines are fitted to two-dimensional scatter graphs. An example is shown at the right. Three-dimensional fitted surfaces are more difficult to interpret than two-dimensional fitted curves.

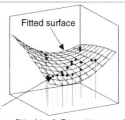

Surface fitted to 3-D scatter graph

Confidence interval displayed with fitted curve

Fitted curves are best estimates based on the data available. Many times the available data is a sampling of a larger family of data. A confidence interval indicates the region in which the fitted curve would probably lie, with a given degree of confidence, such as 90% or 95%, if information for the entire family of data was available. An example of a confidence interval is shown at the right. See Confidence Interval.

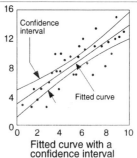

Fitted curve with a confidence interval

295

Symbols

—— **Symbols to identify different groups or series of data** ——

When a single data series is plotted, all of the data points generally use the same type of symbol such as a dot, circle, or square. When multiple data series are plotted on the same graph, each series generally has a different symbol for ease of identification. Different types of symbols are sometimes used within the same data series to determine patterns of subgroups within the series. For example, if automobile fuel efficiencies are being plotted, foreign and domestic manufactures may be given different symbols to see if they are evenly

distributed in the total population. (This technique is referred to as using a dummy variable.) Examples of all three situations are shown at the right.

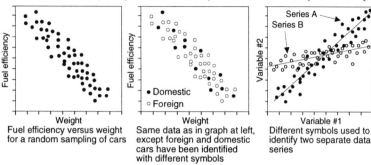

Fuel efficiency versus weight for a random sampling of cars

Same data as in graph at left, except foreign and domestic cars have been identified with different symbols

Different symbols used to identify two separate data series

Examples of symbols used to identify different groups of data

—— **Symbols to encode additional information** ——

In addition to identifying specific groups of data points, symbols are sometimes used to encode additional information into a point graph. Qualitative information such as voltages, new or obsolete product, approved or disapproved, etc., can be encoded by varying the color, shade, size, or shape of symbols. Sometimes the first letter or two of the titles of the data series are used, as shown in the

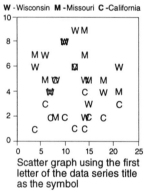

Scatter graph using the first letter of the data series title as the symbol

example at the left. Quantitative information can be encoded by various techniques, such as varying the size of the symbol. When various size circles are used, as shown at the right, the figure is many times referred to as a bubble graph. When any of the encoding methods are used, a legend is generally required so the viewer can decode the information. See Symbol and Bubble Graph.

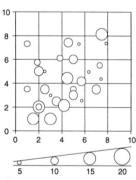

Scatter graph using circles to encode a third quantitative variable. Sometimes called a bubble graph.

—— **Symbols that overlap** ——

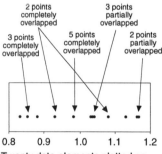

Twenty data elements plotted on a one-axis graph. Approximately 9 plotting symbols are not visible due to overlapping where the data elements have the same coordinates.

When there are many data points on the same graph, overlapping (sometimes referred to as over plotting) can become a problem. If multiple data points have the same, or very nearly the same coordinates, it is difficult and sometimes impossible for the viewer to know how many data points are actually present. To illustrate the problem, the examples on the left and right each have the same 20 data elements plotted on them. In the example at the left, approximately eleven symbols are visible. The

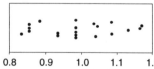

This example contains the same 20 data elements as plotted in the graph at the left. Using a technique called jittering, all 20 plotting symbols are visible.

other nine are not visible because multiple data points have exactly, or almost exactly, the same coordinates and therefore, the symbols are stacked on top of one another. The example at the right uses a technique called jittering to make all 20 symbols visible. In addition to jittering, several other techniques for dealing with overlapping symbols have been developed. Methods for addressing overlapped symbols are discussed under Symbol elsewhere in this book.

Scatter graph matrix

When multiple scatter graphs are used to analyze several variables at one time, they are sometimes organized into what is called a scatter graph matrix, as shown below. This method of organizing the graphs provides a convenient way to analyze large quantities of data, particularly when one is primarily looking for correlations. For example, one might use a scatter graph matrix to check for correlations between twenty different variables from a census report. A separate small scatter graph would be generated for every combination of the twenty variables, and all might then be organized into a single matrix. Matrices do not provide any information that would not be available from the same number of graphs if displayed independently. They do, however, greatly simplify the analysis and sometimes make certain relationships more visible than would be possible with separate graphs. Scales, grid lines, and tick marks are frequently either abbreviated or not used at all. It is common to have a single set of labels apply to an entire row or column. See Matrix Display.

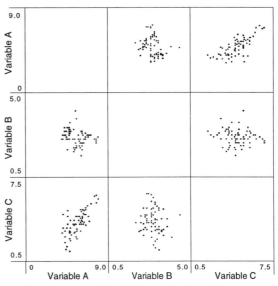

Scatter graph matrix that is used to check for correlations among multiple variables or sets of data

Error bars

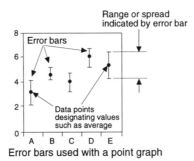

Error bars used with a point graph

The purpose and configuration of error bars varies significantly. In each case, however, the bars are used to indicate the spread or range of something, such as confidence intervals; maximums and minimums; plus and minus standard errors; plus and minus 1, 2, or 3 standard deviations; etc. Error bars can be used with almost any type of point or scatter graph in which the data is appropriate. An example is shown at the left. See Range Symbols and Graphs for more information on error bars, as well as other symbols that serve the same purpose.

Identifying individual data points

If there are just a few data points, they can be individually labeled or coded with a cross-reference. The labels might identify the points, display the values of the points, and/or give descriptive information. When there are many data points it is difficult to identify them on hard copy. With a computer and the proper software, individual data points can easily be identified, even when there are large quantities of them. By selecting one or several data points on the computer screen with the cursor, varying amounts of information about the data elements the symbols represent may be communicated to the viewer in one of several ways, such as displaying it on the screen or highlighting the entries in the spread sheet. This process is many times called brushing. See Brushing.

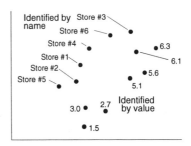

Examples of how individual data points might be identified. The information can be located anywhere as long as it is legible and easily associated with the data point.

Point Size for Type and Lines

Type and line sizes are frequently designated in terms of points, using a system that consists of 72 points to the inch. Thus, if a type size is designated as 12-point, it is 12/72 or 1/6 of an inch tall. Size 36 is 36/72 or 1/2 inch tall, etc. The height of the type is measured from the bottom of the lowest descender on any letter of the typeface to the top of the highest ascender of any letter of the typeface (see example below). • There are several methods for designating line width. One uses the same point system that is used for text, in which there are 72 points to the inch. Thus, an 18-point wide line is 18/72 of an inch or 1/4" thick. See the headings Line and Text.

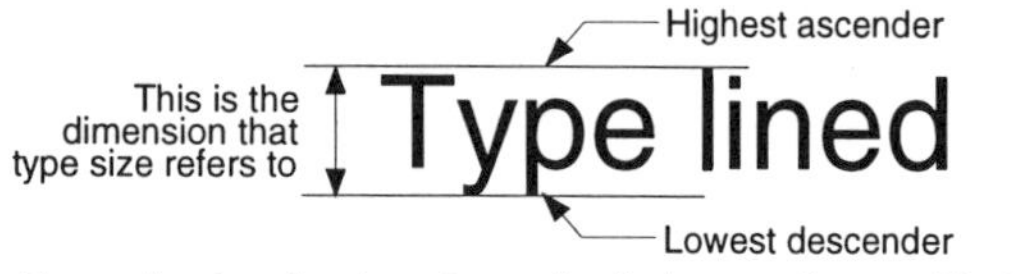

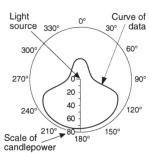

Example showing how type size is frequently specified

Polar Coordinate and Polar Graph and Polar Scale

A polar graph is a circular graph on which data are displayed in terms of values and angles. For example, a polar graph might record the brightness value of light radiating from a lamp at each of the 360 degrees around it. Another graph might record a person's ability to hear (value) the same sound emanating from points at various angles around them. Graphs of these two hypothetical situations are shown below. From the example on hearing ability, one might conclude that the subject can hear sounds on one side better than the other. From the graph on light intensity, it is clear that the lamp was designed to focus the majority of its light in one direction. In both cases, data are shown for a full 360 degrees around the thing being studied. In many cases only selected angles are studied.

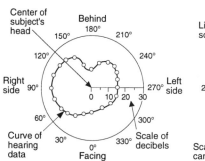

Polar graph used to record a person's hearing ability for sounds emanating from different angles

Polar graph used to show the intensity of light in all directions around a lamp

Scales and terminology

Data points on polar graphs are generally specified by two numbers called polar coordinates. Either or both coordinates can be positive or negative. One coordinate indicates how far a data point is located from the center of a circular grid (origin). This distance from the center is referred to by such terms as radius vector, polar distance, radius, or value. It is sometimes designated with a small letter r and measured along a value scale. Value scales can have any upper and lower values. They typically have their smallest values at the center and largest values at the circumference. The other coordinate designates how many degrees the radius vector (a straight line connecting the data point and the origin) is located from a zero reference angle. This number of degrees or angle is referred to by such terms as polar angle, vectorial angle, or circular angle and is generally designated by the symbol θ. The axis along which the degrees are measured is called a circular or polar axis. The zero reference angle can be located anywhere and the angles can progress in either direction. The circular scale can have any upper and lower values. Zero and 360° are generally used. Shown at the left is a typical polar graph format and associated terminology. In

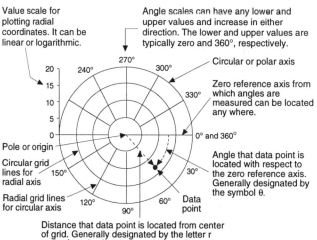

Example of terminology used with polar graphs

some cases conventions and/or an organization's standards govern the choices as to where the zero reference axis should be located and which direction the degrees will progress. In other cases, it is up to the individual and the nature of the data as to which format to use.

**Polar Coordinate
and
Polar Graph
and
Polar Scale**
(continued)

Degrees and radians

In certain fields, angles are measured in radians instead of degrees. The values of the radians are many times expressed in terms of a constant number called pi (designated by the symbol π with an approximate value of 3.14). There are 2π radians in a full 360 degree circle, which makes each degree equal to approximately 0.0175 radians. The example at the right shows the relationship between degrees and radians for each 30 degree increment around the circle. Other than the manner in which the angle is specified, there is no difference between a polar graph plotted in degrees or radians.

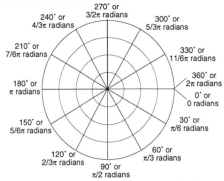

Sometimes the units of measure in polar graphs are radians instead of degrees. This example shows the relationship between degrees and radians.

Angles greater than 360 degrees

In some cases the angles exceed 360 degrees, such as when the force to turn a knob is measured through several revolutions. In such cases, the data can be plotted around the circle multiple times, or the scale on the graph can be expanded to accommodate the additional degrees. When data is plotted around the circle multiple times, the angular scale may or may not reflect the larger angles. The scale in the example below shows two degree

values at each point around the circumference. The upper number in each pair applies to the first time around, and the lower to the second time around. As may be seen, in polar graphs the identical point on a graph can be described by two or more sets of polar coordinates. In this example, points (15, 240°) and (15, 600°) designate the same point and (11, 120°) and (11, 480°) designate another single point. In the example on the left below, the curve goes around the circle four times; however, only one set of labels is shown

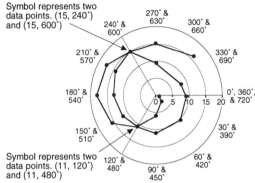

Symbol represents two data points. (15, 240°) and (15, 600°)

Symbol represents two data points. (11, 120°) and (11, 480°)

Example illustrating: 1) how multiple angle scales can be used and 2) how the same plotting symbol can represent two or more sets of polar coordinates

on the circular scale. In the center below, the same data is plotted, except in this case the scale has been expanded to show the full 1440 degrees of the four revolutions. For comparison, the graph at the right below shows the same data plotted on a rectangular graph. Each method has its advantages, depending on the data and the purpose of the graph.

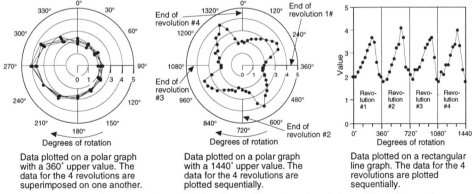

Data plotted on a polar graph with a 360° upper value. The data for the 4 revolutions are superimposed on one another.

Data plotted on a polar graph with a 1440° upper value. The data for the 4 revolutions are plotted sequentially.

Data plotted on a rectangular line graph. The data for the 4 revolutions are plotted sequentially.

Example of the same data plotted on three different graph variations. The data would be typical of something that physically rotates. This example shows four revolutions.

Three-dimensional polar graphs

Three-dimensional polar graphs are occasionally used in technical applications. For example, in addition to measuring the light intensity from a lamp at each of the 360 degrees around it, the intensity of light given off by the lamp might also be measured at different heights. There are two types of three-dimensional polar graphs. One is referred to as cylindrical and the other as spherical. In the cylindrical variation, two of the coordinates are plotted, just as in the case of a two-dimensional polar graph. The third dimension is plotted along the Z axis. In the spherical variation, one coordinate is the straight line

299

Polar Coordinate
and
Polar Graph
and
Polar Scale
(continued)

Three dimensional polar graphs (continued)

distance to the data point from the origin. The second coordinate is an angle with respect to the vertical Z-axis, and the third coordinate is an angle with respect to a zero reference axis perpendicular to the Z-axis. • Because of the difficulty of reading three-dimensional polar graphs, drop lines are sometimes used to assist the viewer. Examples are shown below.

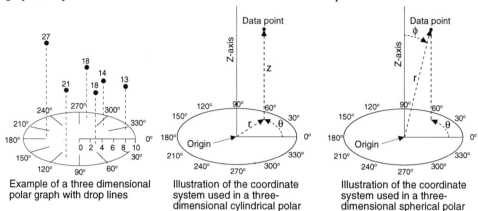

Example of a three dimensional polar graph with drop lines

Illustration of the coordinate system used in a three-dimensional cylindrical polar graph

Illustration of the coordinate system used in a three-dimensional spherical polar graph

Variations of polar graphs

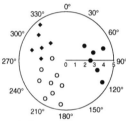

Different symbols can identify multiple data series.

Multiple data series can be plotted on the same polar graph. Different symbols are normally used to differentiate the series. Just the symbols may be shown, as in the example at the left, or the symbols for the same data series can be connected with lines. • Polar graphs can be used as vector graphs by drawing arrows from the center of the graph to the data points, as shown at the right. See Vector Graph.

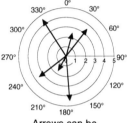

Arrows can be used as vectors.

• Polar data is sometimes plotted on a rectangular grid. When this is done, the center of the rectangular grid corresponds to the center of the circular grid and values increase in all four directions from the center out, just as in the circular grid. The conversion from polar coordinates to rectangular coordinates can be done mathematically or graphically.

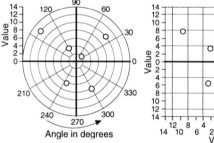

In many cases polar coordinates can be converted to rectangular coordinates and plotted on a rectangular grid.

Polygon

A polygon is a closed figure made up of three or more line segments. In charts and graphs the term polygon is used to describe many different figures. Three representative examples are shown below.

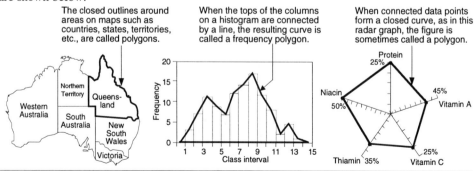

The closed outlines around areas on maps such as countries, states, territories, etc., are called polygons.

When the tops of the columns on a histogram are connected by a line, the resulting curve is called a frequency polygon.

When connected data points form a closed curve, as in this radar graph, the figure is sometimes called a polygon.

Polygon Icon

Miniature graphs, frequently without titles, labels, tick marks, or grid lines, are sometimes called icons. There are many types of icons. The icon at the right has several names, including snowflake, polygon, star, and profile. It is sometimes generated using a radar graph. See Icon and Radar Graph.

Example of a polygon icon

Population Pyramid

Sometimes referred to as an age and sex pyramid. A population pyramid is a pyramid graph or two-way histogram in which age is plotted on the vertical axis and the number of males and females of each age on the horizontal axis. Females are plotted on one side of the vertical axis and males on the other. Because of space considerations, ages are many times grouped into intervals. The most common intervals are two, five, and ten years. The same quantity scale is used for males and females on the horizontal axis, except they progress in different directions from the central vertical axis. Quantities may be plotted in actual units or percents of the total for a given sex. For example, the percent of females in the 40 to 44 age interval can be calculated by dividing the number of females in that age bracket by the total number of females. Generally there are no vertical spaces between the bars. Examples with the vertical scale in two different locations are shown below.

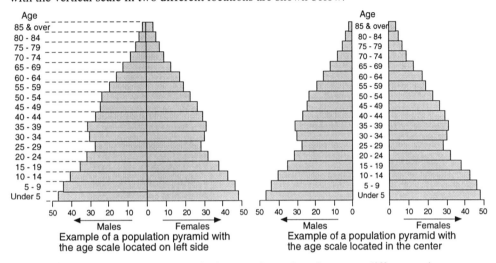

Example of a population pyramid with the age scale located on left side

Example of a population pyramid with the age scale located in the center

Population pyramids provide a graphical means for noting changes or differences in population patterns, either over time within a given population or between different populations. The examples shown below represent distinctly different population patterns for four hypothetical groups of people. Included with each graph is one of several possible interpretations of the pattern. Such graphs are of interest to people in organizations responsible for the analysis and planning for large populations.

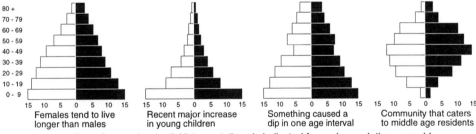

Females tend to live longer than males

Recent major increase in young children

Something caused a dip in one age interval

Community that caters to middle age residents

One of several potential interpretations is indicated for each population pyramid to illustrate how such graphs can be used to analyze various groups

To make comparisons easier, the ends of the bars are sometimes connected by polygons and the polygon for one population superimposed over that of another, as shown at the right. When making comparisons in this fashion, it is generally recommended that the population pyramids be constructed using percents, rather than actual values. • When more detail is desired, instead of grouping ages, every age can be shown. When this is done, lines are sometimes used instead of bars (left). By using lines it is sometimes possible to display the data for each age, which frequently yields a smoother graph.

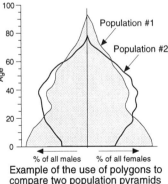

Example of the use of polygons to compare two population pyramids

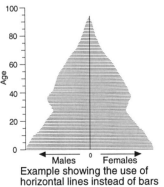

Example showing the use of horizontal lines instead of bars

301

Portrait Chart Layout

The term portrait refers to a chart layout in which the height is greater than the width. Such an orientation is sometimes referred to as vertical. • When the width is greater than the height the orientation is frequently referred to as landscape, horizontal, or sideways.

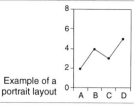

Example of a portrait layout

Power Scale

In a power scale, the values shown are based on some power of the values in a linear scale. For instance, in the example below, the values shown in the linear reference scale are the square roots (0.5 power) of the values shown in the power scale. Looked at another way, the values in the power scale are the squares (power of 2) of the values in the linear scale. Power scales can be constructed using any power. The physical spacings between the values on power scales are all equal.

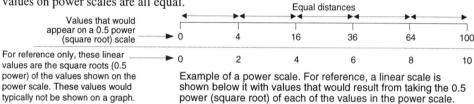

Values that would appear on a 0.5 power (square root) scale

For reference only, these linear values are the square roots (0.5 power) of the values shown on the power scale. These values would typically not be shown on a graph.

Example of a power scale. For reference, a linear scale is shown below it with values that would result from taking the 0.5 power (square root) of each of the values in the power scale.

Presentation Charts & Graphs

Charts and graphs are frequently classified as presentation types when their primary function is for use in formal or semiformal presentations. • Another major classification is operational charts and graphs, used primarily for activities such as analyzing, planning, monitoring, decision making, and communicating in the on-going running of a business, organization, or activity. They are often used to supplement or replace tabulated data and written reports.

• The following are among the reasons for using presentation charts and graphs.
 – Assists in the rapid orientation of an audience.
 – Aids in the audience's understanding and retention of the material presented.
 – Material presented is frequently looked at more favorably by the audience.
 – The audience perceives the presenter to be more prepared, professional, interesting, etc.
• The following are among the reasons for using operational charts and graphs.
 – Data can be organized in such a way that analysis is more rapid and straightforward.
 – Viewers can more rapidly determine and absorb the essence of the information.
 – Large amounts of information can be more conveniently and effectively reviewed.
 – Deviations, trends, and relationships stand out more clearly.
 – Comparisons and projections can many times be made more easily and accurately.
 – Key information tends to be remembered longer.
Both types of charts and graphs tend to shorten meetings and expedite group decisions.
• Although many people do not differentiate between operational charts and charts used for presentations at formal and semiformal meetings, the two types generally differ significantly. The following table highlights some of the differences.

Characteristic	Presentation Charts and Graphs	Operational Charts and Graphs
Primary purpose	Communicate a specific, or at most, a limited number of messages	Gain maximum knowledge and benefits from available information
Medium	Primarily overhead projection or 35mm slides	Primarily hard copy and computer screen
Amount of information on one page	Many sources recommend limits such as 7 words per line, 8 lines per chart, 30 words per page, etc.	No limit as long as it is legible
Three-dimensional effects on graphs	Widely used for their aesthetic value	Used sparingly because it generally reduces the legibility of data
Grids and ticks	Used sparingly	Frequently used to increase the accuracy of decoding the data
Pictures, images, and perspective views	Commonly used	Used sparingly

A comparison of key characteristics for operational and presentation charts and graphs

The key difference between the two types of charts is their intended purpose, as noted in the first row of the table. With presentation charts and graphs, the objective is generally a very focused communication. With operational charts and graphs the goals are much broader, with primary emphasis on analyzing, absorbing, managing, and communicating a maximum amount of meaningful information in a minimum amount of time.

Price Chart
(as used in the field of investment)

Shown below are five types of charts and graphs that are used for monitoring and analyzing the selling price of stocks, bonds, commodities, etc. Some of these are used to track individual securities and commodities as well as exchanges, industries, markets, etc. Others are used primarily to track individual securities or commodities. Each type of chart is discussed under its individual heading.

Point and figure chart

This type of chart displays selling prices on the vertical axis. Xs and Os are used to record actual prices. If prices are increasing, Xs are entered on top of one another and if decreasing, Os are entered beneath one another. A new column is added in sequential order every time there is a price reversal of a certain size. There is no scale on the horizontal axis. See Point and Figure Chart.

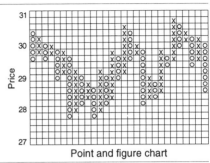
Point and figure chart

Bar chart and candlestick charts

Bar and candlestick charts are the most widely used types of price charts. They record multiple bits of price information (e.g., open, close, high, low) on the vertical axis and have a time scale on the horizontal axis. Bar charts tend to emphasize highs, lows and changes in closing prices over time, while candlestick charts tend to place more emphasis on open, close, and specific groups of symbols and their patterns. See Bar Chart and Candlestick Chart.

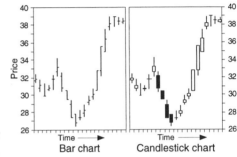
Bar chart   Candlestick chart

Line and column charts

Conventional line and column charts are frequently used for summarizing price information over longer periods of time. For example, line graphs might record closing or midpoint values over a period of months or years. Range column graphs might record the spread of prices for given periods of time, such as the highest and lowest closing price each year for a number of years.

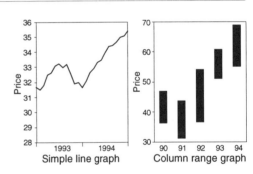
Simple line graph   Column range graph

Primary Axis

Three-axis graphs have three primary axes that are perpendicular to one another. They are generally called the X, Y, and Z axes. Two-axis graphs have two primary axes that are perpendicular to each other, generally called the horizontal and vertical axes or X and Y axes respectively. A one-axis graph has one primary axis, which may run horizontally or vertically. For circular graphs the primary axes are radial and circular.

Prime Meridian

The longitudinal line or meridian passing through Greenwich, England, from which all other meridians are referenced. The prime meridian is considered zero degrees and other meridians are measured east and west from there in degrees, minutes, and seconds. See Map.

Prism Map

Sometimes referred to as a stepped relief, stepped, or block map. In a prism map, areas are elevated in proportion to the values they represent. For example, the height of the various states might be proportional to the grain produced in each of the states. The higher the grain produced per acre, the taller the prism representing that state. See Statistical Map.

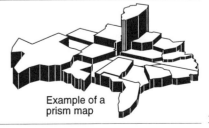
Example of a prism map

Probability Graph

The term probability graph refers more to how a graph is used than it does to a specific type of graph. Cumulative frequency and quantile are two types of graphs that are frequently used for this purpose. Cumulative histograms are used to a much smaller extent. Examples of all three are shown in the table below. There are two main purposes for which probability graphs are used:

- To provide a graphical means for determining the percent probability that a given value, event, measurement, observation, etc., will occur. For example, one might predict the percent probability that a part will pass an inspection, the probability of a student exceeding a particular score on a SAT test, the probability of customers who will be two months late in paying their bills, etc.
- To determine whether a data series has a normal distribution. This is important since many projections and techniques for dealing with a set of data depends on the data having a normal distribution. When used for this purpose, the graph many times is referred to as a normal probability graph.

The type of graph used for a particular application depends on such things as the purpose of the graph, whether a computer or preprinted graph paper is used, and the customs of a particular field. In the table below the name of the graph, a thumbnail description, and an example are shown. Detailed information is given under the individual headings.

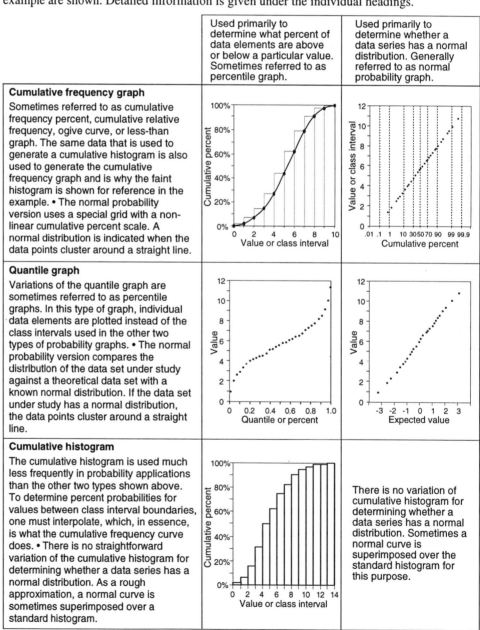

Examples of probability graphs

This book addresses only normal probability graphs. There are other specialized types of probability graphs; however, they are outside of the scope of this book.

Process Chart

Sometimes referred to as a flow chart. Process charts are graphic representations of processes, frequently using symbols to designate the individual steps of the process or procedure. Process charts were originally used most extensively in manufacturing operations; however, they are now used in many other areas including hospitals, grocery stores, and department stores. There are many different types of process charts. The five types outlined below represent some of the more widely used variations and give some indication as to the types of information that may be included. Most process charts are tailored to a specific operation or process; therefore, the number of variations in use is huge. For that reason, the specific variations described in this section are meant only as examples and are not repeated elsewhere in this book. • The symbols at the right are used extensively in process charts as well as in the examples shown in this section. They are shown here for reference purposes. See Symbol for other examples.

○ Operation
⇨ Transportation (move)
☐ Inspection
▽ Storage
◡ Delay

A few common process chart symbols

Operation Process Chart

Sometimes called an operation flow process chart, flow chart, process chart, or outline process chart. An operation process chart is a graphic overview of an entire process or a major portion of a process typically showing only the major operations. Times and locations of the individual operations may or may not be shown. In addition to being applied to processes involving products, parts, equipment, etc., operation process charts can be applied to such things as procedures, forms, information flow, paperwork flow, etc. When used for these purposes they are sometimes referred to by such names as form analysis chart, form process chart, paperwork flow chart, procedure flow chart, or information process analysis.

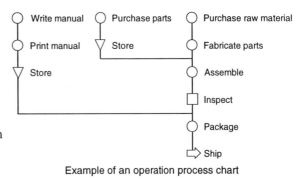

Example of an operation process chart

Flow Process Chart

Sometimes called a flow chart, process chart, product process chart, or product analysis chart. A flow process chart is a graphic, sequential representation of the steps in a relatively uncomplicated process or subprocess. Many times a flow process chart is an expansion on a portion of an operation process chart as shown above. Typical information included are quantity, distance moved, type of work done (generally denoted by symbols) and equipment used. Time may or may not be included. Charts might focus on a specific operator, material, or machine or any combination of these. They can also be applied to paperwork flow, information flow, etc. The major purpose of flow process charts is to analyze, document, and/or communicate information about a process, generally to reduce the time and/or costs associated with the process. The example at the right is representative of a flow process chart. The

Description	Qty.	Distance	Time	Symbol	Remarks
Parts in stockroom				○⇨☐◡▽	
Pick up parts	25		7	●⇨☐◡▽	
Transport to station #3		75ft.	4	○⇨☐◡▽	Forklift
Unload to bench		5ft.	7	○⇨☐◡▽	
Strip parts			37	●⇨☐◡▽	Power drivers
Move to degreaser		18ft.	2	○⇨☐◡▽	Trolley
Load into basket			2	●⇨☐◡▽	
Degrease			6	●⇨☐◡▽	Degreaser
Allow to cool			12	○⇨☐◡▽	
Load onto pallet		7ft.	5	●⇨☐◡▽	
Inspect			16	○⇨☐◡▽	
Move to stockroom		60ft.	4	○⇨☐◡▽	Forklift
Store	▼			○⇨☐◡▽	

Example of a flow process chart for a single part

sequence of the columns and symbols is optional. In some cases the symbols representing the type of activity are shown only at the tops of the columns. When the full set of symbols is shown in each row, the applicable symbol for a given row may or may not be filled, as shown in the example. The example above describes a process involving a single part. When multiple parts are involved, a chart similar to the one shown on the next page is sometimes used so the relationships of the various parts and steps in the overall process can

Process Chart (continued)

Flow process chart (continued)

be seen. In both examples the same types of information are shown, the same symbols are used, and a line is drawn through each of the applicable symbols. • Many times flow process charts focus on an individual's activities, how a part or piece of material is affected by each step of a process, or whether and how a piece of equipment is used. In other cases all three of these are combined into a single chart, as shown at the right. When this is done, some of the other information such as times, distances, etc., are deleted.

Qty. Dist. Time
Activity
200 65 4 B parts to assembly
200 2 Unload
1 1 Inspect part
1 1 Apply adhesive
5 Adhesive sets
1 3 1 Move to fixture
Qty. Dist. Time
Activity
200 65 4 A parts to assembly
200 2 Unload
2 Idle
1 3 1 Move to fixture
1 1 1 Clamp fixture
1 2 Assemble
1 1 Inspect
Store

Example of a flow process chart for a process involving multiple parts

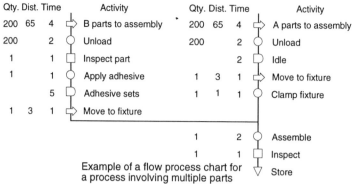

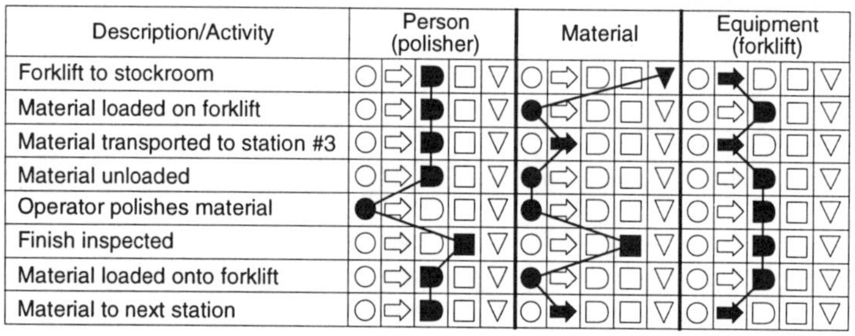

Description/Activity
Person (polisher)
Material
Equipment (forklift)
Forklift to stockroom
Material loaded on forklift
Material transported to station #3
Material unloaded
Operator polishes material
Finish inspected
Material loaded onto forklift
Material to next station

Example of a flow process chart displaying the type of activity for a person, material, and equipment for each step of the process.

Operator Process Chart

Sometimes called a flow process chart, left and right hand process chart, two-handed process chart, simultaneous motion chart (simo chart), film analysis record, micromotion chart, or therblig chart. An operator process chart is a graphic, sequential representation of the motions executed by an individual in the accomplishment of a task or operation. The task or operation might be performed in a single or multiple locations. Such charts are generally used to study tasks that are performed many, many times and where the timing is not controlled by some other outside factor such as availability of supplies or a piece of equipment. These charts serve to analyze, document, and/or communicate the detailed motions and activities in an individual task or operation in order to improve the operation, reduce the time and/or cost to perform the operation, or make it more comfortable for the operator. Sometimes a flow process chart, is used to study the motion and activity of an individual operator, particularly when the task takes place in multiple locations or involves a fair amount of mobility. When the task is very focused, such as an operator standing or sitting in a single location and performing a repetitive operation, a left and right hand operator process chart similar to the one at the right is often used. This type of chart is sometimes called a simultaneous motion or simo chart. The term simo chart is also sometimes used to refer to a left and right hand chart generated from the results of a high speed film of the operator performing a given task. When based on a film of the task, the chart is sometimes referred to as a film analysis record or a micromotion chart. In the micromotion chart the operations are broken down into much more detail than is possible with a flow process chart.

Left Hand
Right Hand
Description
Symbol
Symbol
Description
Hold part
Move to fixture
Insert in fixture
Idle
Hold part
Push part
Rotate 45 degrees
Rotate fixture
Hold part
Pick up tool
Hold tool
Move tool to part
Tap part home
Bend part
Hold tool
Tool to bench
Idle
Bend part

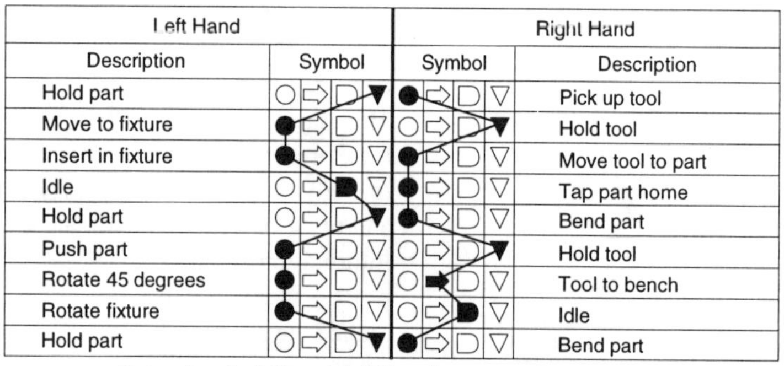

Example of a left and right hand process chart. Sometimes referred to as a two-handed process chart

Multiple-Activity Process Chart

Sometimes called a gang chart, man and machine chart, multi-man process chart, or multiman-machine chart. A multiple-activity process chart is a graphic representation of

Process Chart (continued)

Multiple-Activity Process Chart (continued)

activities performed simultaneously by two or more people, two or more machines, or a combination of people and machines. Their primary purpose is to analyze, document, and/or communicate information about a process involving multiple people and/or machines in order to increase efficiencies, synchronize activities, reduce cost, improve the utilization of resources, etc. An example is shown below.

Operator #1			Operator #2			Helper #1			Machine		
Activity	Time	Sym.	Activity	Time	Sym.	Activity	Time	Sym.	Activity	Time	Sym.
Operate machine	3		Idle	3		Scoop raw material	2		Mold part	3	
						Idle	1				
Idle	2		Label part	3		Unload part	3		Idle	7	
Trim part	3		Adjust settings	3		Load machine	6				
Idle	2		Idle	2							
Operate machine	3		Package part	2					Purge waste	3	
						Idle	1				
■ Idle time	4 - 31%			5 - 38%			2 - 15%			7 - 54%	
▨ Working time	9 - 69%			8 - 62%			11 - 85%			6 - 46%	

Example of a multiple-activity process chart

Process Control Charts

A family of charts and graphs, many times used for the control of processes. They are frequently called control charts. See Control Chart.

Product Life Cycle Graph

The sales of many products follow a typical life cycle over the length of their existence. That cycle generally consists of five major phases:
- The period during which the product is being developed and there are no sales;
- The period during which the product is introduced to the market and sales first begin to be realized;
- During the growth phase the product has been accepted and sales rise at their fastest rate;
- As the product matures and other competitive products are introduced, sales level off; and
- Ultimately need or desire for the product decreases and sales decline.

Example of a typical product line cycle

The graph at the left is many times used to graphically illustrate this cycle. Although the general shape of the sales curves is many times the same from product to product, the length of time required for each phase varies significantly and unlike the idealized graph, the length of the various phases is not always the same. Product life cycle graphs are more frequently used for conceptual and planning purposes than for actual tracking of a product. When used for planning purposes, curves such as profit, cash flow, capital expenditures, etc., are often added to show how these measures are related to the sales cycle (below left). Such graphs are sometimes referred to as cash flow graphs. In other cases, multiple sales curves are drawn to illustrate how the repeated introduction of new products can help to offset the revenue decline of older products and thus assist in maintaining relatively smooth sales (below right).

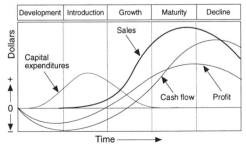

Example of how additional data curves can be superimposed on a product life cycle graph. Each curve may require a different vertical scale.

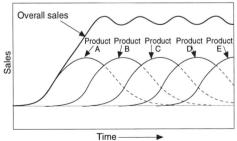

Example of how multiple life cycle curves can be used to illustrate the concept that the continual introduction of additional product lines smooths out overall sales.

307

Profile Graph[1]
or
Profile Map
or
Profile Relief Map

A graph or map consisting of a series of planes or cross-sections stacked side-by-side similar to a three-dimensional area graph. The planes can be spaced irregularly, be nonparallel, and/or be curved. They generally are flat, parallel, perpendicular to the X or Y axis, and regularly spaced as shown in the example at the right. Profile graphs and maps serve several purposes, including easier visualization of certain internal aspects of graphs and the ability to spot changes in subsurface features in the case of maps. For example, when a new building is to be constructed, planes or slices might be shown every 50 feet or so to analyze variations in the soil, clay, and rock stratifications. The planes used in profile graphs and maps may run crosswise or lengthwise. The tops of the planes are the equivalent of isolines on a three-dimensional fishnet display of the data. Shown below

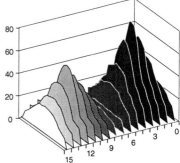

Example of a profile graph or map

are examples of profile planes running in two different directions. A fishnet example is included in the center to show the relationship between profiles and isolines. If a high level of detail is required, gird lines and/or contour lines can be drawn across the surface of each plane as well as being shown on the back planes.

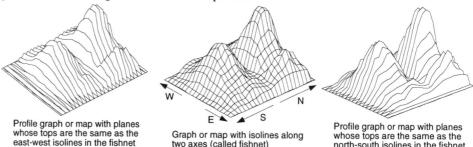

Profile graph or map with planes whose tops are the same as the east-west isolines in the fishnet chart at right

Graph or map with isolines along two axes (called fishnet)

Profile graph or map with planes whose tops are the same as the north-south isolines in the fishnet chart at left

A comparison of a fishnet display and two profile charts for the same data.
In one profile chart the planes run crosswise, and in the other, lengthwise.

Profile Graph[2]

A graph that provides an overview of a particular subject or facet of a situation. It generally summarizes one or a few features or characteristics. For example, the profile graph might summarize the key performance factors of a company, the percent of stocks that went up and down over a given period of time, or how the pricing of various product lines compared during different economic conditions. Almost any basic type of graph can be used as a profile graph. The decision as to whether a particular graph should be called a profile graph is very subjective.

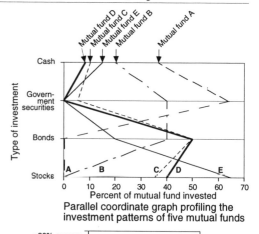

Parallel coordinate graph profiling the investment patterns of five mutual funds

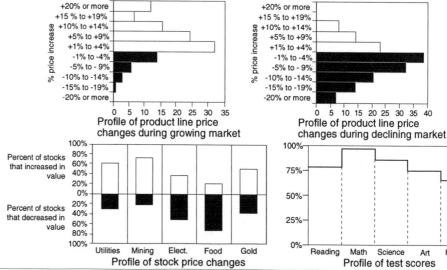

Profile of product line price changes during growing market

Profile of product line price changes during declining market

Profile of stock price changes

Profile of test scores

Profile Icon	Sometimes referred to as polygon, snowflake, or star icon. A profile icon is a miniature graph without titles, labels, tick marks, or grid lines. The length of each spoke or the height of each peak is proportional to some measure or characteristic. The purpose of icons is not to convey specific quantitative information. Instead, they are used as symbols to compare multiple entities with regards to three or more variables. See Icon.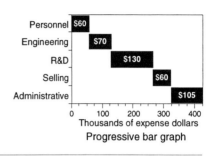

Profile icons |

Progressive Bar Chart

Sometimes referred to as a staggered, stepped, or step-by-step bar graph. A progressive bar graph is the equivalent of a stacked bar graph with only one bar and with each segment of that bar displaced vertically from its adjacent segment(s). The segments can be displaced up or down. The purpose of progressive bar graphs is to add visual emphasis to the individual segments, while still maintaining the concept that the segments all add up to the whole. See Bar Graph.

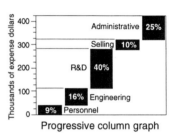

Progressive bar graph

Progressive Column Graph

Sometimes referred to as a staggered, stepped, or step-by-step column graph. A progressive column graph is the equivalent of a stacked column graph with only one column and with each segment of that column displaced horizontally from its adjacent segment(s). The segments, starting from the bottom, are typically displaced to the right. • The purpose of progressive column graphs is to add visual emphasis to the individual segments while still maintaining the concept that the segments all add up to the whole. See Column Graph.

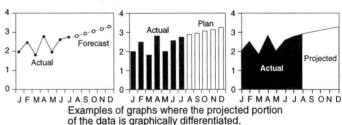

Progressive column graph

Projection[1]

In addition to plotting current and historical information on graphs, it is common to project comparable information into the future. For example, based on the cost of electricity each month for the past twelve months, the monthly cost of electricity for the next six to twelve months might be estimated, planned, or forecast. These predictions for the future are many times called projections and frequently are plotted on the same graph with the current and historical information. When this is done, the projected data is generally clearly identified graphically, as shown at the right. It is not uncommon to also use a word description and/or a change in color, shading, or fill to further call attention to the fact that it is a projection.

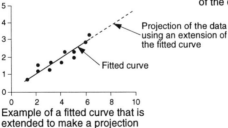

Examples of graphs where the projected portion of the data is graphically differentiated.

Projection of the data using an extension of the fitted curve

Fitted curve

Example of a fitted curve that is extended to make a projection

• Fitted curves are sometimes used to assist in establishing the values for projections. In a number of cases the fitted curve is simply extended and used as the projection, as shown at the left. The projected portion is differentiated graphically from the rest of the curve. • Some projections have a tolerance associated with them, or there might be multiple projections based on such things as possible changes in economic conditions. Several techniques have been developed for tolerances and projection options, some analytical, some intuitive. Shown below are three variations of such projections.

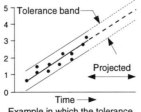

Example in which the tolerance on the projection is the same as on the historical data

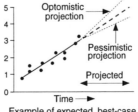

Example of expected, best-case, and worst-case projections

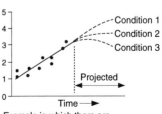

Example in which there are three different projections based on three sets of conditions.

Variations for graphically displaying projections

In charts and graphs, a projection is a method of producing an image or drawning of an object. The process might involve converting information about a three-dimensional object into a two-dimensional image or making a typically two-dimensional object look three-dimensional. The image that results from the process of projection is referred to by several different names. For example, a map of the world generated by a projection technique might be referred to as a cylindriocal projection, a projection of the earth, a view of the earth, a world map, or by a specific name such as Mercator. All are correct. The term view is frequently used in this book. Noted below are four major categories of projections widely used with charts and graphs.

Map projections

With maps, the term projection is used most frequently when discussing the transfer of information from a sphere (earth or globe) to a flat surface (map). The actual transfer process is done mathematically but might be visualized as the process of projecting each point on a globe onto a simple geometric shape such as a cylinder, cone, or plane. When the geometric shape is spread out flat and trimmed, the result is a map as we normally think of one. Shown at the right is an illustration of three of the major methods used for making map projections, plus examples of what a map resulting from each method might look like. See Map.

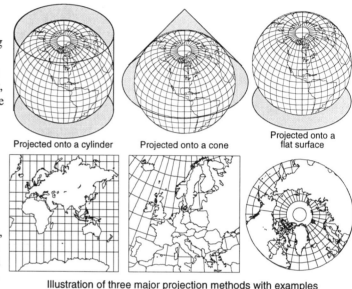

Projected onto a cylinder   Projected onto a cone   Projected onto a flat surface

Illustration of three major projection methods with examples of the types of maps that result from each of them

Projections onto surfaces

Sometimes two-dimensional graphs are incorporated into three-dimensional graphs by projecting certain features of the three-dimensional data graphic onto the floor, ceiling, and/or walls of the three-dimensional graph. In its simplest form, perhaps only a shadow of the three-dimensional graph will be shown. In other cases it might be the contour lines or the representations of surfaces such as those between contour lines. See examples below.

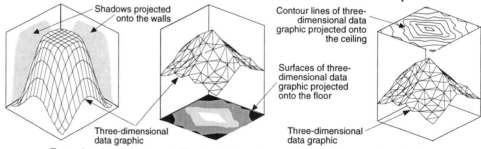

Shadows projected onto the walls

Three-dimensional data graphic

Contour lines of three-dimensional data graphic projected onto the ceiling

Surfaces of three-dimensional data graphic projected onto the floor

Three-dimensional data graphic

Examples showing how two-dimensional graphs are sometimes formed by projecting elements of a three-dimensional data graphic onto the walls, floor, and/or ceiling

Perspective projections

Although perspective is a technique used with other types of projections, it is many times referred to as a type of projection itself. When an object is drawn in perspective, it tends to simulate one aspect of how the human eye sees objects. That is, the further things are from the viewer, the smaller they appear and the more lines seem to converge into a single point. The technique is used with some objects to improve the viewer's understanding. In many cases, such as with graphs, perspective is frequently used to improve their appearance. Two examples are shown at the right. See Perspective Projection (View).

Examples of graphs with perspective projections

Projection[2] (continued)

Oblique and axonometric projections

Almost all charts and graphs can be drawn so they appear three-dimensional. There are two major techniques used for this purpose – oblique and axonometric. This table compares the features of those two methods. See Oblique Projection (View) and Axonometric Projecton (View). Perspective projections are sometimes used in conjunction with these techniques. See Perspective Projection (View).

Oblique view	**Axonometric view**
With this technique the front and back of an object are unchanged, except they are shifted with respect to one another. Since the front and back are basically unaffected, they remain parallel to the plane of the paper and keep their same shape and dimensions as in the original graph.	With this technique, the object is tilted and rotated so the viewer can see the top or bottom and two sides at the same time. None of the planes or surfaces of the object are parallel to the plane of the paper, and each has some distortion when compared against the original object.

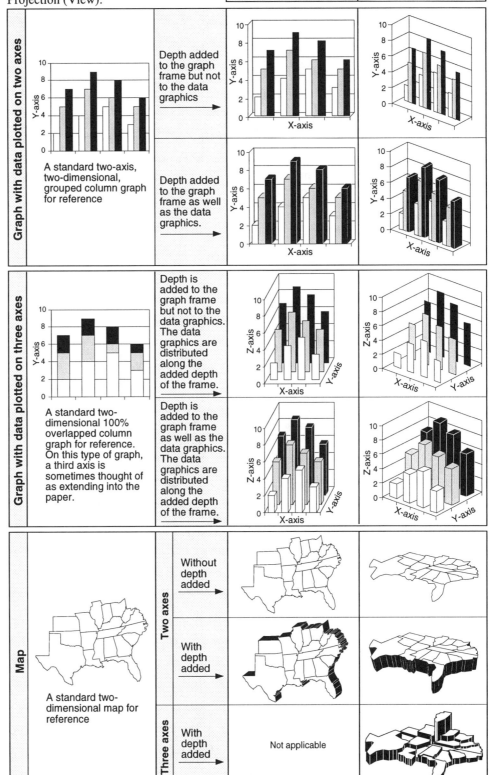

(Table panels, top to bottom, left to right:)

Graph with data plotted on two axes

A standard two-axis, two-dimensional, grouped column graph for reference

Depth added to the graph frame but not to the data graphics

Depth added to the graph frame as well as the data graphics.

Graph with data plotted on three axes

A standard two-dimensional 100% overlapped column graph for reference. On this type of graph, a third axis is sometimes thought of as extending into the paper.

Depth is added to the graph frame but not to the data graphics. The data graphics are distributed along the added depth of the frame.

Depth is added to the graph frame as well as the data graphics. The data graphics are distributed along the added depth of the frame.

Map

A standard two-dimensional map for reference

Two axes — Without depth added

Two axes — With depth added

Three axes — With depth added

Not applicable

311

Proportional Chart

Includes:
Pictorial Proportional Chart
Proportional Area Chart
Proportional Line Chart
Proportional Volume Chart

Proportional charts are visual devices for graphically communicating differences in size, number, or value without the use of scales. This is accomplished by means of a set of data graphics that are in the same proportion to one another as the things they represent. For example, if item A is twice as big as item B and both are represented by squares, the square for item A would be twice as big as the square that represents item B. Proportional charts are used almost exclusively for communication purposes. They are seldom used for analytical purposes. The sizes of the data graphics are not meant to convey exact data but to simply give the viewer some visual indication of the relative size of the items they represent. For instance, the viewer might be able to instantly ascertain such things as the fact that all of the items are about the same size or that there is one huge component and five minor components, etc. If interested in specific data or precise comparisons, viewer's can read the values that are normally included with the data graphics. • The relative size of data graphics can be based on length/height, area, or volume. Examples of each are shown above. Some general observations are noted below.

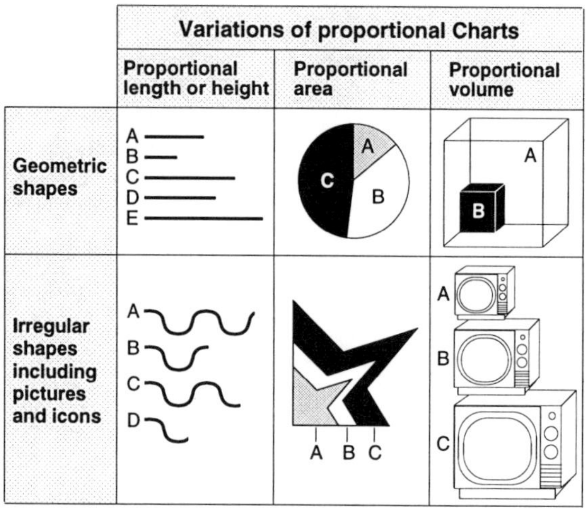

Variations of proportional Charts		
Proportional length or height	**Proportional area**	**Proportional volume**

- Proportional charts are used to compare multiple things at a point in time, or one or more things at various times or under various conditions
- Data graphics based on length and area are most widely used
- Pie charts are one of the most popular types of proportional charts (see Pie Chart)
- For conveying percent-of-the-whole information, a single circular chart is most rapidly understood because the circle tends to more clearly imply 100%
- Data graphics with simple geometric shapes are easier for the viewer to decode than those with complex shapes.
- Data graphics in which the size varies in only one direction are the easiest to accurately decode (e.g., varying the length of a bar versus varying the length and width)

Labels, scales, etc.

Scales, tick marks, and grid lines are almost never used on proportional charts. If a scale is used, the chart would generally be classified as a graph. Numeric values are generally included with the data graphics since in most cases it is difficult to estimate values and differences accurately based solely on the data graphics. The numeric values as well as descriptive information are shown inside the data graphic or immediately adjacent to it. As an alternative, descriptive information is sometimes included in a legend. Several examples of the more widely used methods of displaying numeric information are shown below.

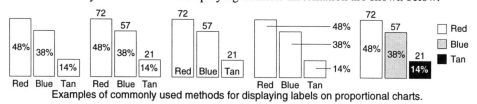

Examples of commonly used methods for displaying labels on proportional charts.

Variations of data graphic arrangements

Most of the variations in data graphic arrangements available with graphs are also available with proportional charts. For example they can be placed side-by-side, overlapped, grouped, stacked, and in some cases arranged in patterns not normally available with graphs.

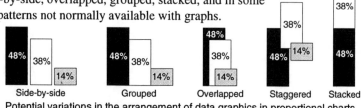

Potential variations in the arrangement of data graphics in proportional charts

Proportional Chart
(continued)

Proportional line chart

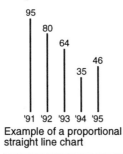

Example of a proportional straight line chart

Charts in which the lengths of straight lines are proportional to the things they represent (example at left) are easy to construct and decode. Adding width to the lines generally improves the readability. Proportional charts based on nonstraight lines (example at right) are difficult to construct and even more difficult to accurately decode.

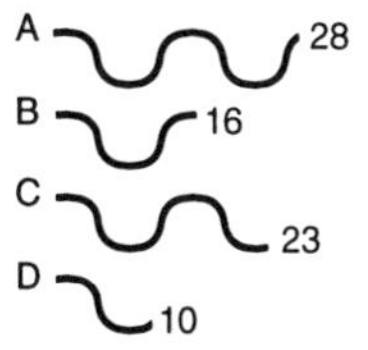

Example of a proportional nonstraight line chart

Proportional area chart

Sometimes referred to an area chart. Shown below are a few of the various geometric shapes and configurations used with proportional area charts. Eye-catching appeal is frequently a major factor in the decision as to which variation to use.

Basic shape	Examples of geometric shapes and configurations sometimes used with proportional area charts			
Column	35 28 10 — Simple	35 25 10 — Overlapped	14% 38% 48% — % of whole (stacked)	21% 50% 29% / 41% 20% 39% — Stacked and linked
Bar	10 28 35 — Simple	10 28 35 — Overlapped	48% 38% 14% — % of whole (stacked)	29% 50% 21% / 39% 20% 41% — Stacked and linked
Square	35 28 10 — Simple	35 28 10 — Overlapped	14% 38% 48% — % of whole (divided rectangle)	35 28 10 — Centered	10 28 35 — Stacked
Circle	35 28 10 — Simple	35 28 10 — Overlapped	14% 48% 38% — % of whole (pie chart)	14% 48% 38% — % of whole (donut)	35 28 10 — Centered
Triangle	35 28 10 — Simple	35 28 10 — Overlapped	48% 38% 14% — % of whole	35 28 10 — Centered
Irregular	35 28 10 — Simple	35 28 10 — Overlapped	48% 38% 10% 14% — % of whole	1991 1992 1993 18% 29% 84% — Comparison over time

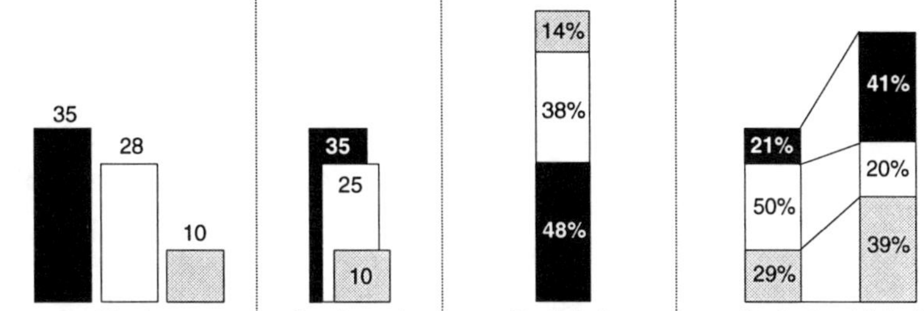

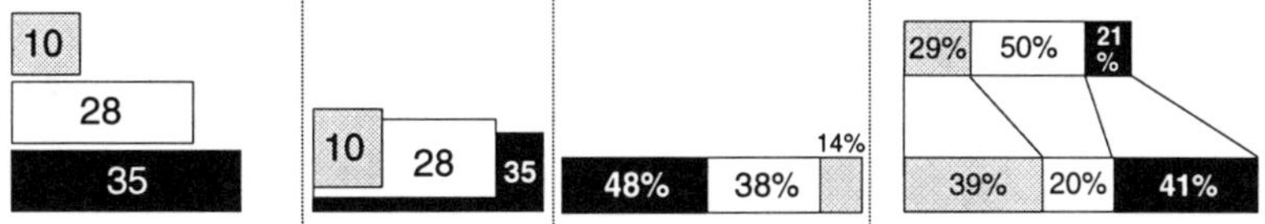

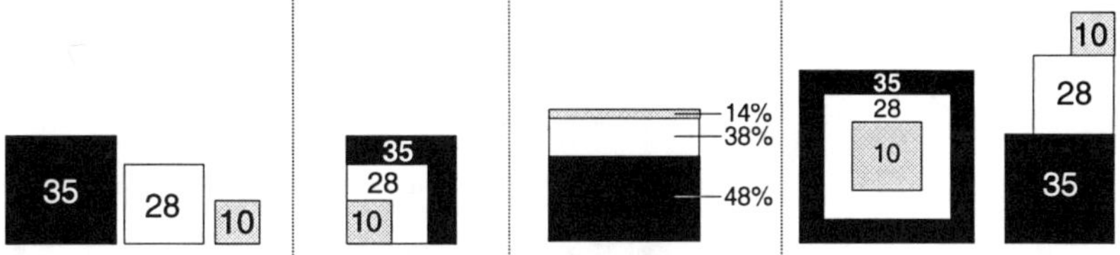

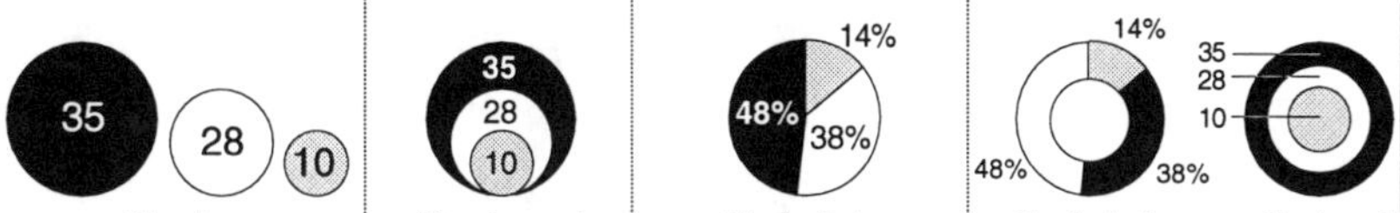

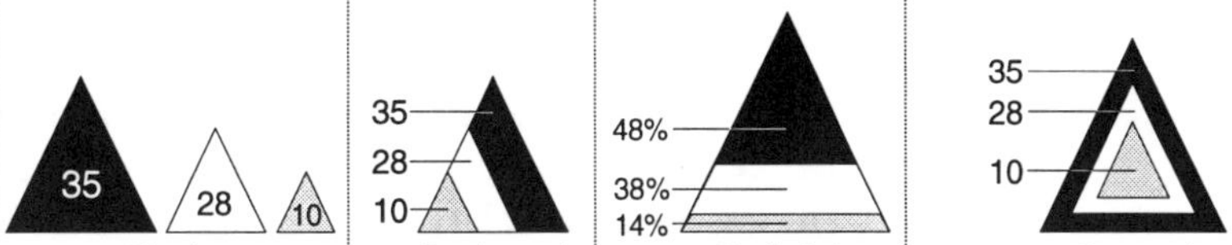

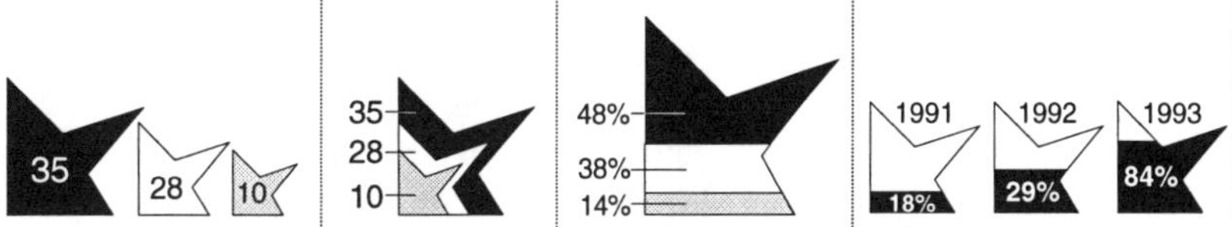

Proportional volume chart

Proportional volume charts use the volume of data graphics to convey information about the relative sizes of the data elements they represent. For example, if data element B is twice as big as data element A, the volume of the data graphic representing element B will be twice as large as the volume of the data graphic for A. Proportional volume charts are difficult to interpret and therefore are generally used only for aesthetic purposes or to give a general impression when values vary significantly. The data graphics can be displayed several different ways, including side-by-side, stacked, and nested. Two examples are shown at the left. Even though each data graphic on a proportional volume chart appears three-dimensional, it

Side-by-side   Nested
Two examples of proportional volume charts.

typically represents a single value. On rare occasions an effort is made to quantify variables on all three sides of the data graphics, as shown at the right. • When "depth" is added to the data graphics on a proportional area chart for aesthetic purposes, it does not reclassify it as a proportional volume chart since the depth is typically uniform on all data graphics and included for cosmetic purposes only.

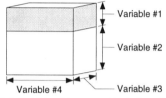

Example of a seldom-used technique for quantifying multiple dimensions on a data graphic of a proportional volume chart

Effect of shape of data graphic on readability

Since viewers tend to estimate linear differences (changes) in lengths better than differences in areas or volumes or nonlinear changes in any dimension, varying the size of data graphic linearly in only one direction can be helpful to the viewer in decoding a proportional chart. In each pair of data graphics shown below, the data graphic on the right has twice the area or volume of the one on its left. Most people find it easier to determine the two-to-one ratio based on the pair of columns at the left because the data graphics change size in only one direction and have a linear relationship.

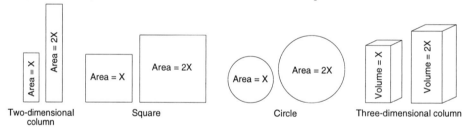

Two-dimensional column   Square   Circle   Three-dimensional column
In each pair of data graphics, the one on the right has twice the area or volume of the one on the left.

Oblique views of proportional area charts

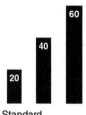

To improve the appearance of a proportional chart it is often shown in an oblique view, which makes it look three-dimensional. Oblique views of area charts do not reclassify them as proportional volume charts. The example at the left is a standard proportional area chart. The one at the right is the same chart, shown at an oblique view. The sizes of the black areas, which are the main portions of the data graphics, are the same in both charts.

Standard proportional chart

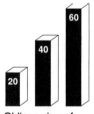

Oblique view of chart shown at the left. The depth has no significance.

Blending features of unit charts with proportional area charts

Stacking certain types of unit charts on top of one another sometimes causes them to function much like proportional area charts. In a unit chart, each unit or segment of the chart equals a certain number of the items the unit represents. For instance, in the example at the right, each square (unit) might represent 10 housing starts in three different price ranges identified by the different shades. The composite represents the total housing starts.

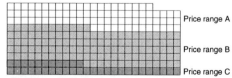

Each square (unit) represents 10 housing starts

Example of multiple unit charts stacked on top of one another to form a variation of a proportional area chart

Proportional Chart (continued)	**The use of icons, pictures, sketches, etc.**

The use of icons, pictures, sketches, etc.

Icons, pictures, or sketches are used extensively in proportional charts (sometimes referred to as pictorial proportional charts) for publications such as newspapers, magazines, advertisements, etc. The icons, pictures, etc., generally indicate only approximate relative values since it is difficult or impossible for the reader to accurately determine the ratios based only on the graphics. Because of this, numeric values are typically included with the

Sized coffee cups indicate the amounts of coffee produced or consumed by three different countries.

data graphics. When the actual values are shown it is less important whether the originator meant for the areas or volumes to be proportional; however, the viewer can be mislead if the graphics are grossly out of proportion.

General

– Since the effectiveness and accuracy of proportional charts rests largely with visual estimates of the sizes data graphics, breaks or discontinuities are seldom used.
– When multiple data series are being compared, similar data elements are generally in the same location and the same color or shade.
– Proportional charts can be used to compare the same types of information for multiple entities (upper example at right), or to display changes over time or in different situations (lower example at right).

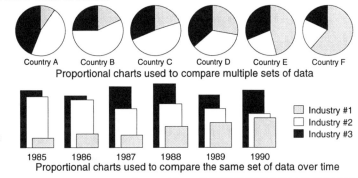

Proportional Map

Sometimes referred to as a distorted or value-by-area map. A proportional map is a variation of a statistical map in which the sizes of the entities on the map are proportional to values other than the land areas of the entities. For example, if a particular resource is being studied by country, the size of each country on the map might be proportional to the availability of the resource in that country. The country with the largest supply of that resource would be drawn with the largest area on the map even if it has the smallest physical area in terms of actual land. Proportional maps are sometimes classified as cartograms, abstract, or thematic maps. See Statistical Map.

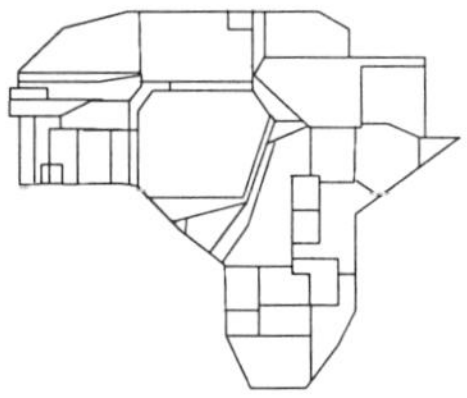

Proportional map

Pyramid Graph

Sometimes referred to as a two-way histogram. A pyramid graph is a variation of a paired bar graph in which the spaces between the bars are eliminated and the elements on the vertical axis are arranged in ascending or descending order. A popular application of this type of graph is called a population pyramid. See Population Pyramid and Bar Graph.

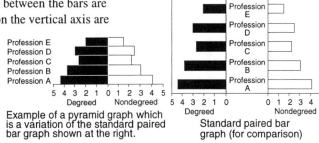

Example of a pyramid graph which is a variation of the standard paired bar graph shown at the right.

Standard paired bar graph (for comparison)

Pyramid Organization Chart

Named for its appearance, a pyramid organization chart is a variation of an hierarchical organization chart. A traditional variation is shown below, and an abbreviated and stylized variation is shown at the right. See Organization Chart.

Conventional pyramid organization chart

Stylized variation of a pyramid organization chart

315

Quadrangle or Quadrilateral or Quad	The area on a map or globe bounded by two lines of latitude (parallels) and two lines of longitude (meridians). The area is frequently designated by the angles included in the enclosure. For example, a seven and one-half minute quadrangle is a quadrangle encompassing 7.5 minutes of latitude and 7.5 minutes of longitude. The U.S. Geological Survey issues a series of maps referred to as 7.5-minute maps or quadrangles.

Quadrant

When a chart or graph is divided into four generally equal sections, it is sometimes described as having four quadrants.

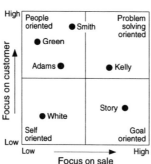

Two widely used applications are the business matrix, shown on the left, and the traditional graph format on the right. In some cases the quadrants are numbered, as in the graph example on the right. In other cases the quadrants are identified by word descriptions, as in the business matrix. Because four quadrants are formed does not mean that all four will be used. The examples below illustrate how various standard graph configurations utilize varying combinations of quadrants.

Example of a chart divided into quadrants with the quadrants identified by word descriptions such as "people oriented" and "goal oriented"

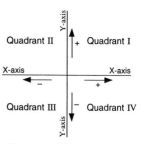

Example of the traditional quadrants formed by the intersection of an X and Y axis. Quadrants in this particular application are historically identified by Roman numerals and are numbered counterclockwise.

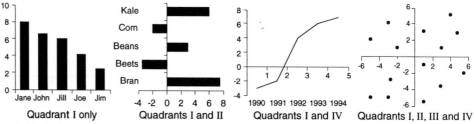

An illustration of how various types of graphs and combinations of data utilize various numbers of quadrants. The numbering of the quadrants is bases on the example shown on the right above.

Qualitative Map	Sometimes referred to as an explanatory, nonquantitative, or descriptive map. This type of map is generally used to show nonquantitative information such as where things are located, how areas are organized and/or subdivided, and where the routes are located that get people and things from one place to another. The types of things included on qualitative maps might be grain fields, buildings, rivers, cities, roads, power lines, sewer lines, sales areas, public facilities, voting precincts, schools, political divisions, resorts, pet stores, sales areas, etc. See Descriptive Map.

Region, territory, area, etc.

Route, road, street, etc.

Examples of qualitative maps

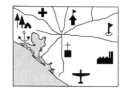

Facilities

Qualitative Scale	Sometimes referred to as a category or nominal scale. A scale consisting of a series of words and/or numbers that name, identify, and/or describe people, places, things, or events. The items on the scale do not have to be in any particular order. When numbers are used on a qualitative scale they are for identification purposes only since these types of scales are not quantitative. Each word or number defines a distinct category that contains one or more entities. In other words, a category might refer to one man or many men. See Scale.

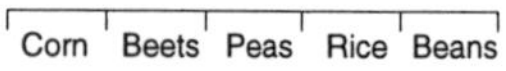

Corn	Beets	Peas	Rice	Beans

Example of a qualitative scale. Sometimes called category or nominal scale.

Quality Control Graphs	A group of charts and graphs often referred to as statistical quality control charts and graphs. See Statistical Quality Control Charts and Graphs.

Quantile Graph

Comparison with cumulative frequency graph

Sometimes referred to as a percentile graph. A quantile graph is very similar to a cumulative percent frequency graph, both in appearance and function. (See Cumulative Frequency Graph). One of the key differences between the two is in their preparation. With the cumulative percent frequency graph, class intervals are formed, data elements are assembled in their appropriate intervals, the total frequencies of occurrence are determined, cumulated values and percents are established, the cumulated percents are plotted at the boundaries of the class intervals, and the points are connected with lines. In the case of quantile graphs, a quantile number is calculated for each data element and the values of the data elements plotted against their respective quantile numbers. Quantile values range from zero to one. They can be multiplied by 100 to arrive at percentiles. Quantile values can be plotted on either the vertical or horizontal axis. For the same set of data, the curves on a cumulative frequency graph and a quantile graph look very much alike. Three examples of quantile graphs are shown below using a data set with a normal distribution.

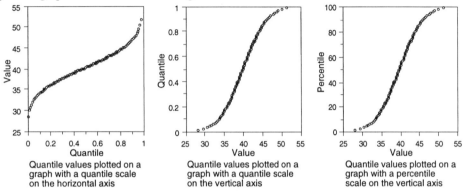

| Quantile values plotted on a graph with a quantile scale on the horizontal axis | Quantile values plotted on a graph with a quantile scale on the vertical axis | Quantile values plotted on a graph with a percentile scale on the vertical axis |

Three graphs with the same quantile data plotted on each. Only the scales are different.

For general information quantile values are determined by arranging the data elements in the data set in ascending order, assigning consecutive numbers to each element starting with one for the smallest, calculating the quantile number by subtracting 0.5 from the assigned consecutive number, and dividing the result by the total number of data elements in the set.

Methods for determining whether a data set has a normal distribution

Using quantiles, there are two ways to determine whether a data series has a normal distribution. One is to plot the data on a special normal probability grid. If the distribution is normal, the data points will cluster around a straight line, as shown at the right. A second way is to compare the data series in question to a theoretical data series that has a normal distribution with an average of

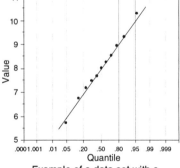

Example of a data set with a normal distribution plotted on a normal probability grid

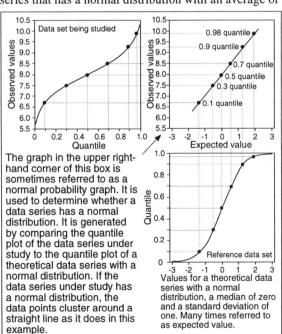

The graph in the upper right-hand corner of this box is sometimes referred to as a normal probability graph. It is used to determine whether a data series has a normal distribution. It is generated by comparing the quantile plot of the data series under study to the quantile plot of a theoretical data series with a normal distribution. If the data series under study has a normal distribution, the data points cluster around a straight line as it does in this example.

Values for a theoretical data series with a normal distribution, a median of zero and a standard deviation of one. Many times referred to as expected value.

zero and a variance and standard deviation of one. If the distribution is normal, the data points will cluster around a straight line as shown in the upper right corner of the box at the left. The graphs alongside and below it are included to graphically illustrate how the observed values relate to the expected values. • When either of these two methods is used to check for a normal distribution, the graph is generally referred to as a normal probability graph.

317

Number of data points plotted

With small to medium-sized data sets, every data element is generally plotted. With very large data series, many times only a fraction of the data points are plotted. In most cases the reduced number of data points detracts little from the effectiveness of the graph. This is particularly true if the distribution is normal and a normal probability type of graph is used. To illustrate the point, two different data series are shown below on quantile and normal probability grids. Series A has 100 data elements and a normal distribution. Series B has 65 data elements and is not distributed normally. In the upper two graphs, all the data elements are plotted. In the lower two graphs every fifth data element is plotted. It can be noted that only a small difference exists between the curves with all of the elements plotted and those with only 20% of the points plotted. For general information, the histograms for the two data sets are shown below.

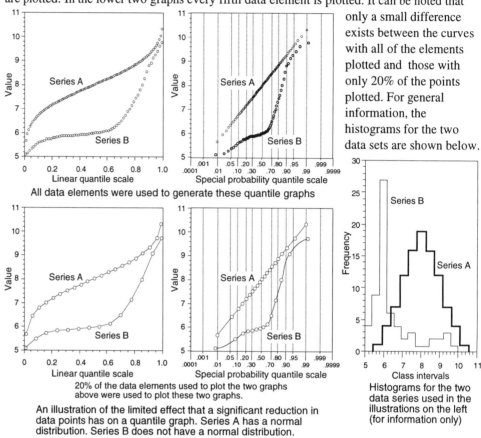

All data elements were used to generate these quantile graphs

20% of the data elements used to plot the two graphs above were used to plot these two graphs.

An illustration of the limited effect that a significant reduction in data points has on a quantile graph. Series A has a normal distribution. Series B does not have a normal distribution.

Histograms for the two data series used in the illustrations on the left (for information only)

Comparing data sets using quantile graphs

Quantile graphs are sometimes used to compare two or more data sets. One of the more widely used methods is to plot the multiple data sets on the same graph, as shown at the right. Another method is to make a combined graph with the values for one data set on the X-axis and the values for the other on the Y-axis, as shown below.

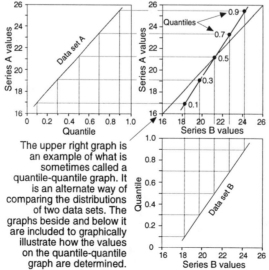

The upper right graph is an example of what is sometimes called a quantile-quantile graph. It is an alternate way of comparing the distributions of two data sets. The graphs beside and below it are included to graphically illustrate how the values on the quantile-quantile graph are determined.

A straightforward method for comparing multiple data sets

This is accomplished by using the values in the two data sets that have common quantile values as the X and Y coordinates on the combined graph. A graphical illustration of how the procedure works is shown at the right. Such graphs are sometimes referred to as quantile-quantile graphs. If the distribution of data elements in the two data sets are basically the same and the X and Y scales are the same, the data points in the quantile-quantile graph will cluster around the diagonal.

Quantile Graph (continued)

Use of quantile graph as a less-than and greater-than graph

In addition to using quantile graphs to analyze the distribution of a data set and to determine whether the set has a normal distribution, these types of graphs are sometimes used to determine the relative positions or rankings of specific data points and to see how the data elements are segmented. The graph at the right has reference lines added to make such observations easier. Based on this example, a few of the potential observations might be:

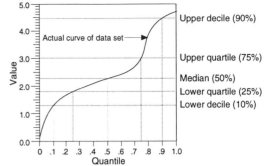
Example of a quantile graph with reference lines

– 50% of the data elements are less than and larger than about 2.3.
– 10% of the data elements are equal to or greater than 4.5 or, looked at in reverse, 90% of the data elements are equal to or less than 4.5.
– 10% of the data elements are equal to or less than 1.3, (90% equal to or greater than 1.3).
– The spread of values between the lower decile and the median is 1.0.
– The spread of values between the median and the upper decile is 2.2.

One-axis quantile graph

When only the major categories such as quartiles and deciles are required, the data can be displayed on a one-axis graph, as shown at the right. In this type of graph the quantile values are transferred to the vertical axis so that the visual task of going from one scale to the curve to the other scale is eliminated. Interpolation of values between quantile labels can potentially yield incorrect results on a one-axis graph.

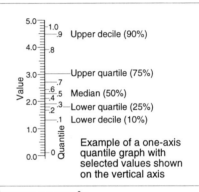
Example of a one-axis quantile graph with selected values shown on the vertical axis

Advantages of quantile graphs versus cumulative frequency graphs

Standard quantile graphs offer certain advantages over cumulative percent frequency graphs. Among these advantages are ease of construction, actual data points are shown as opposed to summaries of class intervals, no decisions are required as to what the best size class interval might be, the same curve functions as a less-than and greater-than curve, and the actual maximum and minimum values are shown on the graph.

Quantitative Axis

An axis with a quantitative scale on it.

Quantitative Map

Sometimes referred to as a statistical or data map. Quantitative maps are used to convey statistical information with respect to areas and locations. Four examples are shown below. See Statistical Map.

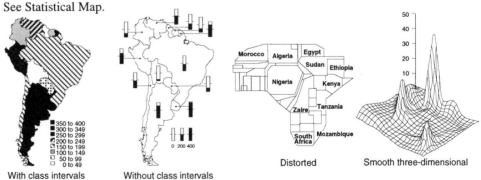

With class intervals   Without class intervals   Distorted   Smooth three-dimensional

Examples of quantitative maps

Quantitative Scale

Sometimes referred to as a numeric, value, amount, or interval scale. A scale consisting of numbers organized in an ordered sequence with meaningful and uniform spacings between them. The numbers are used for their quantitative value, as opposed to being used to identify entities. See Scale.

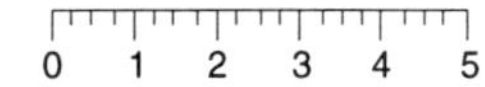

Example of a linear quantitative scale. Other types of quantitative scales include logarithmic, probability, and power.

Querying

Sometimes referred to as brushing. See Brushing.

Radar Graph

Sometimes referred to as a star or spider graph. A radar graph is a circular graph used primarily as a comparative tool. For example, the nutritional content of two different foods might be compared based on the percentage of recommended daily allowances of five different ingredients that each contains. An illustration of such a comparison is shown at the right. In this example, food A would be considered to have a higher nutritional value than food B, based on these five measures.
• Radar graphs can be interpreted by either

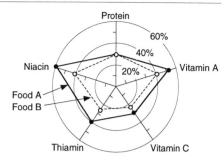

Example of a radar graph comparing the nutritional content of two different foods based on daily recommended allowances

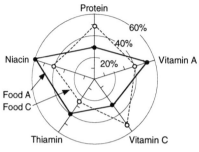

When polygons intersect, as shown here, it is sometimes difficult to determine which has the better overall evaluation.

reading the actual values on the axes or by comparing the areas enclosed by the polygons formed by the data points. In the previous example, the larger the polygon, the "better" the food. In general, whether a larger or smaller polygon is "better" depends on whether the more favorable values are nearer the center or circumference. Which ever convention is selected, it is used consistently throughout the graph. When one polygon is distinctly larger than the others, as in the example above, the interpretation of the graph is straightforward. In other cases, where the curves intersect one another (left), the interpretation is more difficult and many times involves a judgment factor. • In radar graphs, each category axis (radii) represents a different variable. Other than readability, there is no limitation as to the number of variables that can be included in a single graph. Whatever number of variables there are, they are distributed equally around the 360° of the circle. • Each axis typically has a scale along which one characteristic element is plotted for each data series involved in the comparison. • After all of the data elements have been plotted, adjacent data points in the same data series are generally connected by straight lines forming closed polygons.

Multiple data series on individual graphs

In some situations, particularly where many data series are being compared, each data series is placed on a separate graph and the graphs are placed adjacent to one another, as shown at the right. In this type of comparison, labels and titles are many times shown only in the legend. When the major focus is on relative comparisons, tick marks can be removed and axes reduced or eliminated, as shown at the right. In this form, the data graphics are sometimes referred to as icons and the groupings of multiple icons as icon comparison displays. When making

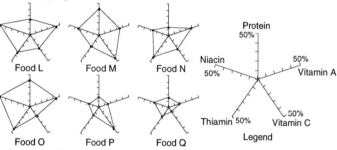

Example of how multiple data series can be compared using individual radar graphs. Axis labels and titles are often shown only in the legend.

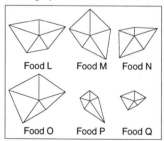

In this variation, tick marks are removed. Radial axes are shortened but remain so the viewer can see the relative size of individual variables. Icons such as these are sometimes referred to as stars, snowflakes, or profiles.

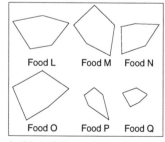

In this variation only the polygon remains. Based on these data graphics, overall evaluations can be made based on area but differences in specific variables is difficult.

overall comparisons of many entities, putting the data in icon form can sometimes make the contrasts between the entities stand out more clearly. See Icon Comparison Display.

Radar Graph (continued)

Scales

Quantitative

There are many variations of scales used with radar graphs. Several variations are shown in these examples. Major, minor, or no tick marks are sometimes used.

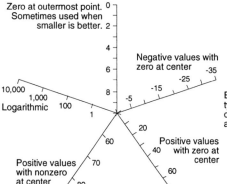

Examples of some of the types and configurations of scales that might be applied to radar graphs

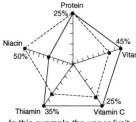

In this example the upper limit on each axis is equal to the largest value in any data series plotted on that axis.

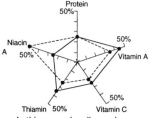

In this example, all axes have the same upper limit and the same size intervals.

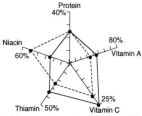

In this example each axis has a different upper limit and intervals appropriate to the upper limit. The relationship of the upper limit and the data plotted on the axis differ from axis to axis.

Examples of variations in scales used on radar graphs

Qualitative

Radar graphs are sometimes used to make comparisons using qualitative as well as quantitative information such as appearance, ability to get along with people, user friendliness, etc. This is many times accomplished by establishing an arbitrary scale, such as 0 to 10, and then ranking each applicant, product, group, etc., along the scale. These rankings can then be plotted on the radar graph. Since each axis has its own scale, qualitative information can be included in the same graph with quantitative information, as shown in the example at the right.

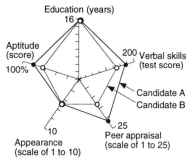

Example of a radar graph incorporating quantitative and qualitative information

Labels

The examples shown here illustrate several methods for displaying labels on radar graphs. To avoid labels interfering with the information being presented, occasionally only the upper limit of a scale is shown. When the sole purpose of the graph is to compare two or more entities, it is not uncommon to leave the labels off completely. Categories may progress clockwise or counter-clockwise.

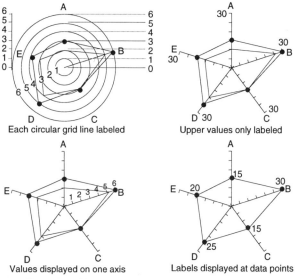

Examples of how labels are displayed on radar graphs

Grid lines

The grid lines that radiate from the center of the graph are called radial axes, category axes, or category grid lines. They generally appear only where categories are shown; however, they are sometimes included between categories for appearance purposes. The circles (sometimes multisided polygons) are called circular grid lines, radius grid lines, or value grid lines. As with rectangular graphs, there can be major and minor value grid lines. Any combination of circular and radial grid lines can be used. The general tendency is to keep the number of grid lines to a minimum; frequently circular grid lines are not used at all.

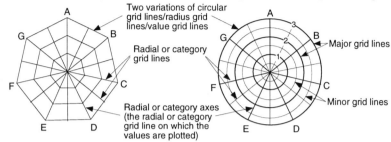

Examples of grid lines used with radar graphs.

General

Symbols and lines

There are a number of combinations of symbols and lines used to identify individual data series. Three of the major variations are shown at the right.

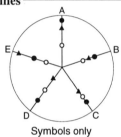
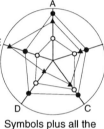
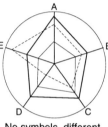

Symbols only — Symbols plus all the same type lines — No symbols, different types of lines

Shading

The areas within the polygons may or may not be colored or shaded. Most of these choices affect the appearance or ease of reading the graph more than its functioning. • The shading in the example at the upper right tends to make the differences between the two entities stand out more clearly than the lower example, without the shading.

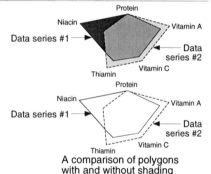

A comparison of polygons with and without shading

Grouping

Arcs can be drawn around groups of categories to indicate they have something in common. For example, when comparing personnel for a new position, one grouping might be associated with technical skills, another with people skills, etc.

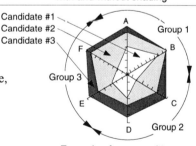

Example of arcs used to indicate how various categories might be grouped

Comparing data series with many variables

Radar graphs are well-suited for comparing data series that have many variables. Two different series are compared with regards to 26 variables in the graph at the right. When more than two data series with large numbers of variables are to be compared, each series is generally plotted on a separate graph.

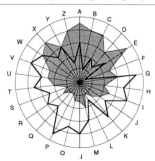

Radial Column Graph and Radial Line Graph	Sometimes referred to as a star graph or a circular column graph. This type of graph is the equivalent of a rectangular column graph wrapped into a circle. The horizontal axis of the rectangular graph becomes the circular axis of the circular graph, and the vertical axis of the rectangular graph becomes the radius or value axis of the circular graph. When the width of the columns approach that of a line, it is sometimes called a radial line graph. See Circular Column Graph.

Radial column graph

Radial Organization Chart	A circular hierarchical organization chart with the highest ranking person or entity in the center and lower ranking persons or entities arranged in descending order as their names, positions, etc., appear further from the center on the chart. The names or descriptions of the lowest ranking persons or entities form the circumference of the circular chart. See Organization Chart.

Range Symbols and Graphs	A family of graphs and symbols that graphically designate upper and lower boundaries/values of groups of data. The boundaries might be measured or observed values, calculated values, tolerance values, theoretical values, projected values, specification values, etc. Two additional values or sets of values, referred to here as intermediate and central values, are sometimes also designated at or between the upper and lower boundaries. Examples of all three types of values are shown below using four of the most widely used types of symbols.

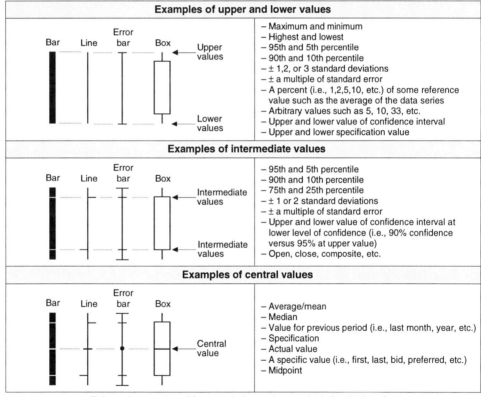

Enlarged examples of four symbols used extensively for designating ranges of data and the types of data encoded into them

In many cases symbols incorporate a combination of two or more of the three types of information shown above. In a given graph, the elements of a symbol always designate the same value. From graph to graph, the elements of the same symbol can designate different values. For example, in one graph a symbol might designate the minimum and maximum values, the tenth and ninetieth percentile values, and the average value of a data set. In another graph, that same configuration of symbol might designate plus and minus two and three standard deviations and the median value of the data. Since the same symbol can have many different meanings, it is generally recommended that a legend be included with the graph clearly identifying what each element of a symbol designates. In each of the examples above, the elements of the symbols designating the three different types of information are all symmetrical. In actual applications this may or may not be the case and frequently is not the case. The symbol used and the type of information conveyed by each element of the symbol depends on the purpose of the graph and the nature of the data.

Graphs that indicate ranges are very diverse and include most of the basic graph types. They are referred to by a number of different names including range bar, error bar, high-low, box symbol, floating column, and sliding bar. The examples shown below give some idea of the many ways that the ranges of values can be displayed.

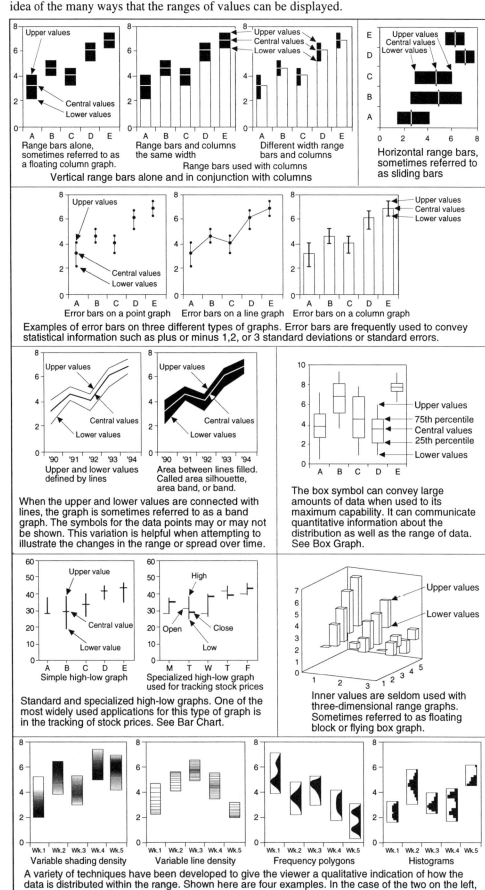

Range bars alone, sometimes referred to as a floating column graph.

Range bars and columns the same width

Different width range bars and columns

Range bars used with columns

Vertical range bars alone and in conjunction with columns

Horizontal range bars, sometimes referred to as sliding bars

Error bars on a point graph

Error bars on a line graph

Error bars on a column graph

Examples of error bars on three different types of graphs. Error bars are frequently used to convey statistical information such as plus or minus 1,2, or 3 standard deviations or standard errors.

Upper and lower values defined by lines

Area between lines filled. Called area silhouette, area band, or band.

When the upper and lower values are connected with lines, the graph is sometimes referred to as a band graph. The symbols for the data points may or may not be shown. This variation is helpful when attempting to illustrate the changes in the range or spread over time.

The box symbol can convey large amounts of data when used to its maximum capability. It can communicate quantitative information about the distribution as well as the range of data. See Box Graph.

Simple high-low graph

Specialized high-low graph used for tracking stock prices

Standard and specialized high-low graphs. One of the most widely used applications for this type of graph is in the tracking of stock prices. See Bar Chart.

Inner values are seldom used with three-dimensional range graphs. Sometimes referred to as floating block or flying box graph.

Variable shading density

Variable line density

Frequency polygons

Histograms

A variety of techniques have been developed to give the viewer a qualitative indication of how the data is distributed within the range. Shown here are four examples. In the case of the two on the left, the darker the shading or denser the lines, the more concentrated the data. The two on the right are conventional frequency polygons and histograms of the data included in the range bars.

Error bars

Error bars are frequently used to indicate some statistical characteristic of a data set or of a specific data point. For example, they might be used to indicate standard deviations, standard errors, confidence intervals, etc. If the bars run horizontal, they are referred to as X error bars. If they run vertical, they are referred to as Y error bars. Both can be used with the same data point if, for example, the data is variable in both directions. When X and Y error bars are used, an oval is sometimes drawn around them and on occasion only the oval is shown. Error bars generally have a symbol (dash, circle, dot, etc.) at their ends which is sometimes referred to as a cap or, if it is a line, a serif. Sometimes the cap itself is referred to as the error bar. On occasion the size of the cap encodes additional information. In some situations the bar is eliminated and only the cap is shown, or the bar and cap are both eliminated and the value the bar would have represented is shown. The bars can be of equal length on both sides of the data point or, if for example the data is skewed, they might be of different lengths on the two sides. Error bars might be shown on one or both sides of the data point, depending on the nature of the data. The lengths of the error bars are equal to the values they represent and are drawn to the same scale as the corresponding data series. Variations of error bars are shown below.

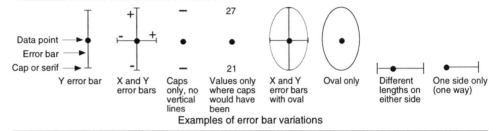

Examples of error bar variations

Multiple values designated by range symbols

Range symbols are not limited to just one of each type of value (e.g., upper/lower, intermediate, and central). For example, if the upper and lower values represent a confidence interval with a 95% level of confidence, two sets of intermediate values might represent the confidence intervals with 75% and 50% levels of confidence. If the upper and lower values represent plus and minus three standard deviations, two sets of intermediate values might represent plus and minus one and two standard deviations. With regards to central values, it is common to display both the average and median values.

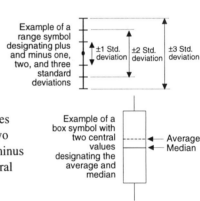

Range graphs displaying multiple data series

Range symbols for multiple data series can be plotted on the same graph. Only the range symbols might be displayed to compare the spread of data in two or more sets of data, as

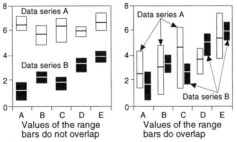

Two examples of range symbols used to convey primary information about multiple data series

shown at the left, or the range symbols may provide supplemental information and be displayed in conjunction with other data graphics such as lines, columns, bars, etc., as shown at the right.

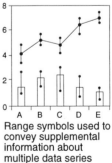

Range symbols used to convey supplemental information about multiple data series

Examples of data that lends itself to having ranges designated:

– Maximum, minimum, and average salaries by profession
– Average test values and the 10th and 90th percentile values by test series
– High, low, open, and close stock prices by day
– Average, high, and low temperatures by month
– Highest, lowest, average, and last year's sales by month
– Projected performance and associated confidence limits

Ranking	Ranking is a form of ordering based on quantitative values. Data may be ranked in ascending order, which means placing the smallest value first, the second smallest value second, the third smallest value third, and so forth. The other way that data is frequently ranked is in descending order, which is the reverse of ascending order and starts with the largest value, followed by the second largest value, etc. In graphs, when data is plotted in ascending order with the quantitative scale on the vertical axis, it typically means that the

data element with the smallest value is the first data graphic on the left, and the data element with the largest value is the last data graphic on the right of the graph. When the quantitative scale is on the horizontal axis, the custom is not as clear-cut as to which should be called ascending and descending. Examples are shown at the right.

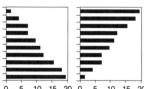

Ascending order   Descending order
When the quantitative scale is on the vertical axis, the orders shown here are generally agreed to as being ascending and descending.

When the quantitative scale is on the horizontal axis, there is no consensus as to which is called ascending and descending.

Ranking Chart

A ranking chart is a variation of a comparative graph in which multiple rankings of several entities are compared. For example, in 1990, five different models of cars might have been ranked based on fuel efficiency. In 1995, those same five models of cars might have been ranked again on the same feature. A ranking chart displays those two sets of rankings on two vertical axes with lines connecting the pairs of rankings for each model. If there are no changes in rankings, all the connecting lines are horizontal. When the ranking for a given model changes from one period to the other, the slope of the connecting line makes it clear whether the ranking increased or decreased. See Comparative Graph.

Example of a ranking chart in which the rankings on fuel efficiency for five models of cars are compared for two different time periods.

**Rate-of-Change Graph
or
Ratio Graph**

Sometimes referred to as a semilogarithmic graph. This type of graph has a logarithmic scale on one axis and a linear quantitative or sequential scale on the other. One of the major uses of such graphs is to determine how rapidly and consistently one variable changes with respect to another or the rate of change of some variable with respect to time. See Logarithmic Graph.

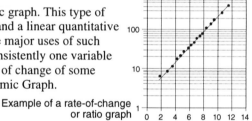

Example of a rate-of-change or ratio graph

Ray Map

A map that shows multiple locations and uses straight lines to show how the locations are related. For example, a newspaper might have two printing locations and fifty cities and communities the paper is delivered to. A ray map might show all the locations and use straight lines to indicate which locations are serviced by which printing facility.

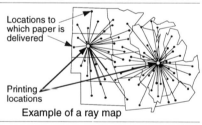

Locations to which paper is delivered

Printing locations

Example of a ray map

Rectangular Coordinates

Sometimes referred to as Cartesian coordinates. Rectangular coordinates are the values by which data points are located on rectangular graphs. On two-dimensional graphs, two coordinates or numbers define the location of each data point. The two values are written one behind the other with a comma between (2,4 ; 5,1; 4,-3). The first number indicates how far the point is located from zero on the horizontal or X-axis. The second number indicates how far the point is located from zero on the vertical or Y-axis. Both positive and negative values can be used. With three-dimensional graphs, the same procedure is used except there are three coordinates to designate the location along all three axes.

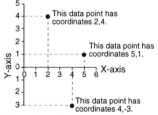

This data point has coordinates 2,4.

This data point has coordinates 5,1.

This data point has coordinates 4,-3.

Examples of points located by means of rectangular coordinates

**Rectangular Graph
or
Rectilinear Coordinate Graph
or
Rectilinear Graph**

Each of these three terms refers to a graph in which the axes are straight and perpendicular to one another. Although perpendicular, when drawn in certain projections or views the axes sometimes appear not to be perpendicular.

Examples of rectangular/rectilinear graphs

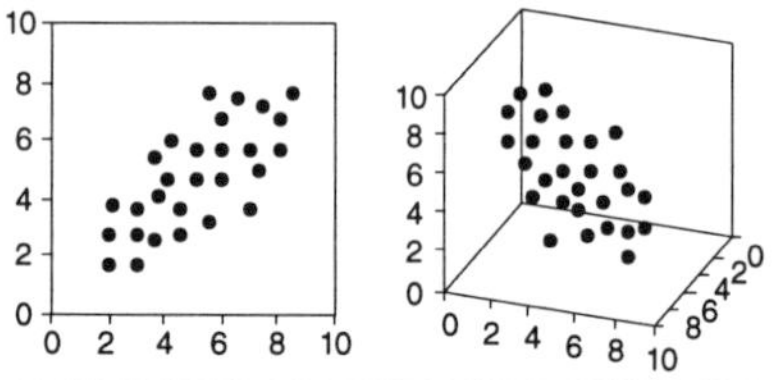

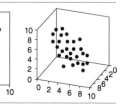

| **Reference** | Sometimes referred to as a datum or base line. A reference is a line, point, surface, etc., from which values or information are compared, referenced, or measured. In many cases the reference is zero. In other cases it might be things such as dividing lines, industry averages, the edge of a property, an arbitrary elevation, etc. | 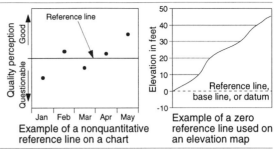 |

Example of a nonquantitative reference line on a chart

Example of a zero reference line used on an elevation map

Reference Axis
or
Reference Line[1]

Sometimes referred to as a shifted reference axis. Reference axes are typically located at some value other than zero and are included on graphs to denote differences between actual values and some reference value(s), while still showing the actual values. For example, if one wanted to show actual profit values but at the same time clearly designate deviations from budget, the budget values might be used as the reference axis. The data points, measured from the zero base line axis, would designate the actual profit values, while the data graphics between the data points and the reference axis would indicate the deviations from budget. A reference axis might indicate the same value for each interval or it might vary from interval to interval. Examples of these two variations are shown below. The data graphics between the data point and the reference axis might be columns, lines, bars, or areas. Coloring or shading may or may not be used to differentiate favorable and unfavorable values.

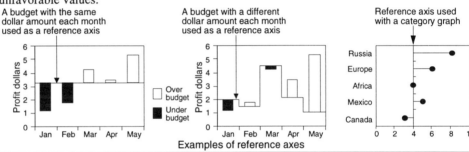

A budget with the same dollar amount each month used as a reference axis

A budget with a different dollar amount each month used as a reference axis

Reference axis used with a category graph

Examples of reference axes

Reference Line[2]

A reference line is a faint or dashed line that is normally drawn completely across the graph to designate or call attention to a value, event, item, etc., that is significant. Two widely used applications of reference lines are to indicate values such as budget, specification, control limits, etc., along quantitative scales and to highlight specific events along time scales. The words or numbers that identify the reference lines are sometimes referred to as markers.

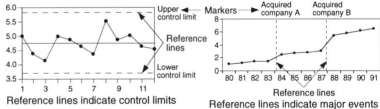

Reference lines indicate control limits

Reference lines indicate major events

Reference Table

Sometimes referred to as a source table. The primary purpose of a reference table is to make data easy to retrieve. Information is generally arranged in alphabetical order, numerical order, chronologically, by geographic area, etc., so specific bits of information can be located rapidly. Examples would be interest tables, census tables, tax tables, comfort index tables, etc.

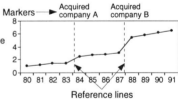

Air Temperature (°F)	10	20	30	40	50	60	70	80	90	100
110	105	112	123	137	150					
105	100	105	113	123	135	149				
100	95	99	104	110	120	132	144			
95	90	93	96	101	107	114	124	136		
90	85	87	90	93	96	100	106	113	122	
85	80	82	84	86	88	90	93	97	102	108
80	75	77	78	79	81	82	85	86	88	91
75	70	72	73	74	75	76	77	78	79	80
70	65	66	67	68	69	70	70	71	71	72

Comfort Index
Percent Relative Humidity
Example of a reference table

Regression Line

Sometimes referred to as linear regression line. See Linear Regression Line.

Relational Graph

Technically all graphs are relational graphs, since they depict the relationships between the things plotted along the axes. In some instances the term relational graph is used more narrowly by excluding graphs with certain types of scales from the definition. For example, graphs with category scales are sometimes excluded. Other times graphs with time scales as well as category scales are excluded. Scatter graphs, with quantitative scales on both axes, are prime examples of what most people would agree fit the definition of relational graphs. Relational graphs with quantitative scales on multiple axes are sometimes more difficult to interpret than graphs with time or category scales. • Relational graphs, such as scatter graphs, are used extensively for analysis purposes when attempting to determine what, if any, relationship exists between two quantitative variables.

Relationship Diagram

A relationship diagram is a graphical tool for organizing and displaying things in such a way that their interdependencies can be established, analyzed, recorded, and communicated. A wide variety of things can be studied using relationship diagrams, including events, concepts, actions, problems, social preferences, etc. • Relationship diagrams are constructed by first generating a symbol for each thing that bears on the issue under investigation. For example, if a program is falling behind, one would make symbols for all the things that might be causing delays. Next, arrows are drawn between all things that have an interdependency, such as the castings being late because the raw material was delivered late, or the material being delivered late because the purchase order was placed late. The arrows can point either direction as long as the same procedure is used throughout. In one case the arrow might point from A to B, indicating that B was caused by A. Or the arrow might point from B to A, indicating that A was the cause of B. Both methods yield the same result when used consistently. In the example below, the thing at the base of the arrow causes, affects, influences, etc., the thing at the head. • After all the arrows are drawn, it many times is clear which things are having the greatest effect, which are isolated items, where attention should be focused, etc. In the example below, for instance, item C is affecting four other items (A,B,E, and H), while none are affecting C. C would generally be considered a major or root event, activity, or cause which might need particular attention and which when addressed may resolve some of the items that are dependent on it. On the other hand, item J is an isolated issue that may have to be addressed independently. The size, shape, and/or fill of the symbols are sometimes used to encode additional information such as the relative importance of the item the symbol represents, what organization is responsible for or affected by the item, etc.

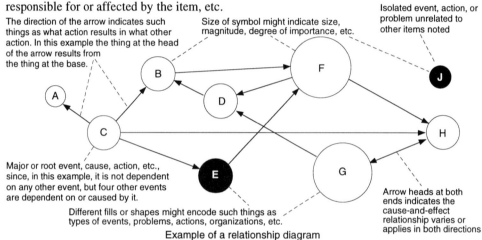

The direction of the arrow indicates such things as what action results in what other action. In this example the thing at the head of the arrow results from the thing at the base.

Size of symbol might indicate size, magnitude, degree of importance, etc.

Isolated event, action, or problem unrelated to other items noted

Major or root event, cause, action, etc., since, in this example, it is not dependent on any other event, but four other events are dependent on or caused by it.

Different fills or shapes might encode such things as types of events, problems, actions, organizations, etc.

Arrow heads at both ends indicates the cause-and-effect relationship varies or applies in both directions

Example of a relationship diagram

Relief Map

Sometimes called a hypsometric or contour map. A relief map shows the surface features of the terrain such as locations, configurations, and elevations with particular emphasis on elevations. This is accomplished largely through the use of contour lines which connect points of equal elevation. Relief maps might have the appearance of being two- or three-dimensional. See Topographic Map.

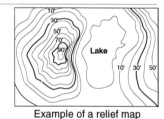

Example of a relief map

Repeated Time Scale Graph

In a repeated time scale graph, a single data series is broken into pieces and each of those pieces represented by a different curve on the same graph. For example, a three-year data series might be broken into three one-year periods, as shown at the right. The same quantitative and time scales apply to all of the curves. This type of graph makes it easier to compare a given point in time (i.e., March, June, etc.) from time period to time period (i.e., year to year) and sometimes makes cyclical patterns stand out more clearly. On the other hand, these graphs tend to make it more difficult to see longer term trends. See Time Series Graph.

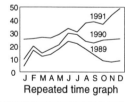

Repeated time graph

Representative Fraction
or
RF

A type of scale used on maps to relate a length on the map to the distance it represents on the ground or earth. For example, the fraction 1/1,000,000 or ratio 1:1,000,000 means that 1 unit on the map (e.g., one centimeter, one inch, etc.) represents 1,000,000 of the same units on the ground. The representative fraction shown at right means that one centimeter on the map represents 1,000,000 centimeters (10,000 meters or 10 kilometers) on the ground.

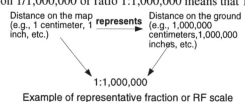

Distance on the map (e.g., 1 centimeter, 1 inch, etc.) **represents** Distance on the ground (e.g., 1,000,000 centimeters, 1,000,000 inches, etc.)

1:1,000,000

Example of representative fraction or RF scale

Residual

A residual is the vertical distance between a data point and a fitted curve or regression line. Residuals apply to individual data points as well as to a continuous string of data points in the form of a line. The fitted curve or trend line to which the residuals are measured might be straight or curved. When a data point is above the fitted curve, it is considered positive, and when it is below the curve, it is considered negative. Residuals are sometimes displayed on the same graph as the data series they apply to by means of vertical lines between the data points and the appropriate fitted curve. (The reference to vertical lines assumes the dependent variable is on the vertical axis.) • For a more detailed analysis, just the residuals are sometimes plotted on a separate graph, as shown below. On such graphs, only the data points might be plotted, drop lines might be added to the data points, or in some cases only the drop lines might be plotted. Generally speaking, more meaningful observations can be made when the residuals are plotted by themselves.

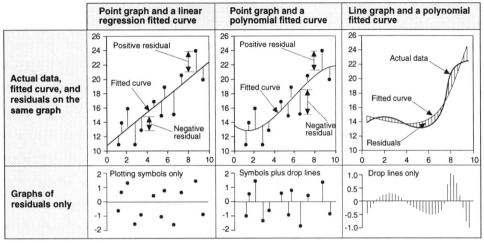

The upper graph in each case displays actual data, a fitted curve, and the differences between the two (residuals). The lower graph plots only the residuals. Positive values on the residual-only graphs indicates that the actual data point is above the fitted curve.

Residuals are analyzed for several reasons including:
- To determine how well the fitted curve or trend line represents the data series. (The smaller and more homogeneous the residuals, the better the fit)
- To look for cyclical or seasonal aspects of the data
- To see if deviations from the fitted curve follow some pattern, such as getting larger or smaller as time progresses or as the variable on the horizontal axis gets larger
- To check for unusual data points (outliers)

Examples are shown below.

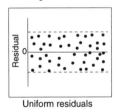 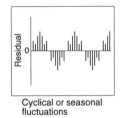 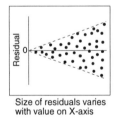 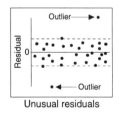

Examples of patterns sometimes formed when residuals are plotted on a separate graph

In some cases it is advantageous to convert the residuals to a percent of the value on the fitted curve where the residual touches the curve. If the average values of the residuals remain relatively constant but the actual values go up or down significantly, the percent that the residuals represent of the actual values will change from one end of the data series to the other. This will not show up on a plot of actual values of residuals, but will be noticeable on a percentage graph.

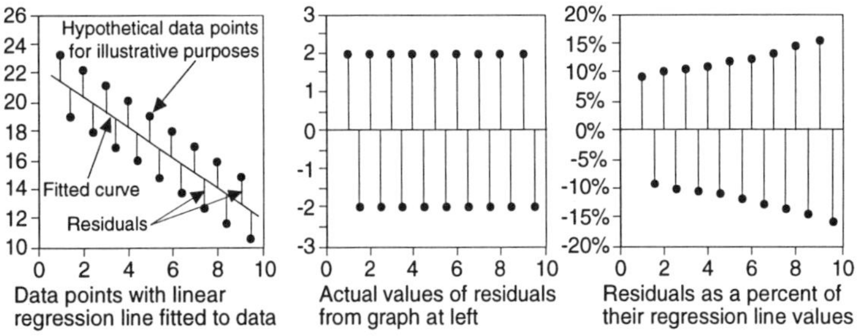

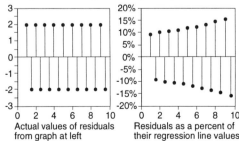

Comparison of residuals plotted as their actual values and as a percent of the corresponding value on the fitted curve.

329

Reveal Chart

Sometimes referred to as a build chart. One of a series of charts used to develop or present an overall message, idea, or concept. For example, if used in a presentation where only text is used to discuss a series of points, the first chart of the series will have only the first point on it. The next chart will have points one and two on it, and the next, points one, two, and three. This continues until the subject is fully explained and all points have appeared. A similar technique can be used with graphs, maps, diagrams, etc.

Examples of reveal charts in which each chart includes the material from the previous chart, plus one or more additional points

Reversal Chart

Sometimes referred to as a Point and Figure chart. See Point and Figure Chart.

Ribbon Graph

A three-dimensional line graph in which the lines appear to have width and depth. Data series are typically distributed along a third axis. The data series can be stacked but generally are not. Ribbon graphs are most commonly used for presentation purposes.

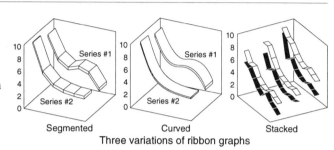

Three variations of ribbon graphs

Ridge Contour Graph

Sometimes called a slice graph, spectral plot, layer graph, strata graph, or partition plot. A ridge contour graph is a graph in which the data points of a three-dimensional scatter graph are condensed onto a limited number of planes to aid in the interpretation of the data. See Slice Graph.

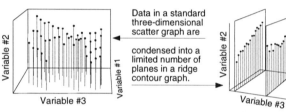

Data in a standard three-dimensional scatter graph are condensed into a limited number of planes in a ridge contour graph.

Ring Buffer

A variation of a buffer used with maps. See Buffer Map.

Ring Graph

A graph in which the widths (not radii, diameters, or areas) of concentric rings are proportional to the values they represent. For instance, in the example below, A and B are each equal to two ounces, C equals three ounces and D and E each equal one and one-half ounces. The total all of the data elements is ten, as designated by the outer circumference of the graph. When a scale is not used, the data graphic becomes a variation of a proportional chart. Ring graphs can be very misleading because it is easy for the viewer to assume values are proportional to areas instead of widths. For example, although A and B in the example are equal, it would be easy for the viewer to assume B is larger than A because it is represented by a larger area.

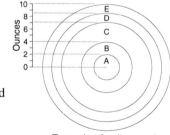

Example of a ring graph

Robinson Projection Map

A well-known variation of a world map. Two of its key features are an appearance that many people feel comfortable with and the fact that it reduces the large distortions in land masses at the higher latitudes that exist with some of the other types of world maps such as Mercator. The Robinson map has varying degrees of distortion in size, shape, distance, and direction.

Robinson projection map

Rolling Average

Sometimes referred to as a moving average or trend line. A rolling average smooths the curve of a data series and makes general trends more visible. To generate a rolling average curve, each point is calculated by averaging the value for the current period plus a fixed number of prior periods. Rolling average curves are used with sequential data and are generally superimposed over a graph of the actual data. See Moving Average.

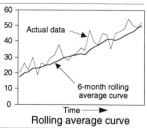

Rolling average curve

| **Rolling Total Curve** | Sometimes referred to as a moving total curve. On a rolling total curve, the value plotted for each interval is equal to the sum of the incremental value for that interval, plus the values for a fixed number of preceding intervals. For instance, if one is preparing an annual rolling total curve, the value plotted in May, for example, would be equal to the actual |

value for May, plus the values for the preceding eleven months, making it a 12-month or annual total. In June, the plotted value would be equal to the actual value for June plus the values for the preceding eleven months. Any number of intervals can be used in the total. Whatever number of intervals are selected, that same number is used throughout the entire graph. Incremental values are many times plotted on the same graph with the rolling total, as shown at the right. The purpose of rolling total curves is to smooth out the interval-

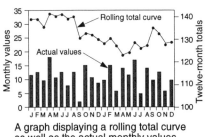

A graph displaying a rolling total curve as well as the actual monthly values

to-interval fluctuations somewhat and provide a graphic indication of the overall trend of the data. As an illustration, in the example, the rolling total indicates a general decline in the annualized totals until about the first quarter of the second year at which time it tends to increase. Such a trend is not as obvious in the monthly values.

| **Rootogram** | A seldom used variation of a histogram with the square root of frequency instead of the actual frequency, plotted on the vertical axis. |

| **Rose Chart/Graph** | The term rose encompasses several types of charts and graphs. See Sector Graph, Wind Rose Graph, and Four-Fold Chart/Graph. |

| **Rotate** | Sometimes referred to as spin. One of the major limitations of three-dimensional charts and graphs is the fact that it is sometimes difficult or impossible to see or interpret all of the data graphic when viewed from a single angle. One technique developed to overcome this problem when viewing the chart on a computer screen is to rotate the entire chart so it can be seen from almost any angle, e.g., front, back, side, top, etc. In this way patterns, configurations, trends, relationships, distributions, outliers, anomalies, etc., can many times be readily observed. Various software programs allow the graph to be rotated in 1, 2, or 3 directions (axes). In most programs, the rotation can be done in increments at the command of the viewer. In some programs the graph is rotated continuously until the viewer gives the command to stop. The example below shows nine views (including the top view) of the same three-dimensional graph. In this case it is a floating column graph; however, the same technique is applicable to almost any type of three-dimensional chart. In an actual situation many more than nine views would be possible. Since the identification and value of specific points in such an analysis is sometimes extremely difficult to determine, some programs enable the viewer to point to (sometimes referred to as brush) a particular data point or group of points and have the information for those points appear on the screen or be highlighted in the data table. |

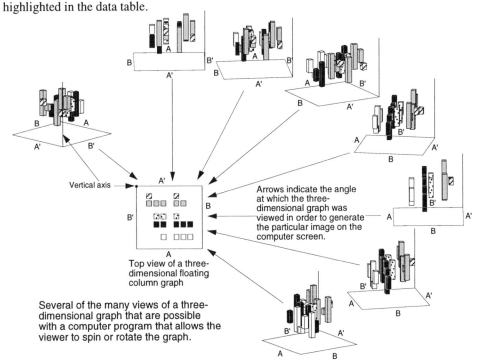

Arrows indicate the angle at which the three-dimensional graph was viewed in order to generate the particular image on the computer screen.

Top view of a three-dimensional floating column graph

Several of the many views of a three-dimensional graph that are possible with a computer program that allows the viewer to spin or rotate the graph.

Rotated Bar Graph	Sometimes referred to as a bar graph, vertical bar graph, or column graph. See Column Graph.	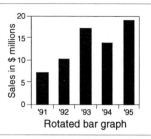

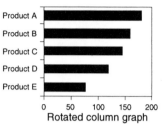

Rotated Column Graph	Sometimes referred to as a horizontal bar graph, horizontal column graph, or bar graph. See Bar Graph.	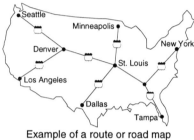

Route Map or **Road Map**	A route or road map is a variation of a descriptive map. It shows the paths things take to get from one place to another. The paths or routes might be walkways, corridors, roads, railroad tracks, streets, shipping lanes, overhead cables, buried pipes, etc. Extensive amounts of information can be encoded into the lines used for route maps, such as type pavement, number of lanes, diameter of pipe, voltage of cable, etc. When route maps connect multiple locations, they are sometimes called networks.	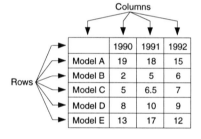

Row — Rows are the horizontal arrangements of words or numbers used in tables and spreadsheets, as shown at the right. Rows may or may not be separated by lines that are frequently called rules.

Rule — Sometimes referred to as a grid line. Rules are the lines used in tables to separate rows and columns of information and to identify, highlight, and organize titles, headings, subheadings, etc. See Table.

Ruling — Sometimes called a grid line, grid rule, coordinate line, or scale line. Rulings are thin lines used on graphs as visual aids for the purpose of encoding and decoding information. See Grid Line.

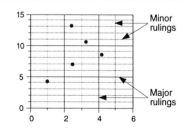

Run Chart or **Run Diagram** — Sometimes referred to as a trend chart. A run chart is a point or line graph used to record sequential data regarding a repetitive process or task. Examples include the diameter of a machined part that is mass produced, the quantity of milk dispensed into milk cartons at a dairy, how long patients have to wait to see a doctor in a busy hospital, etc. The purpose of run charts is to provide the viewer with a simple graphical tool for observing, analyzing,

**Run Chart
or
Run Diagram**
(continued)

monitoring, and/or controlling processes, with particular attention to such things as establishing base lines, noting trends or sudden changes, studying erratic or cyclical performance, etc. The horizontal axis of a run chart is always sequential. The scale might be in terms of time, lot or sample number, sequence number, or there may be no scale at all and the data points simply plotted one after the other in sequential order. The vertical scale is almost always quantitative and linear. For reference and/or control purposes, horizontal lines are frequently included on the chart. One of the horizontal lines that is many times used represents the average or median. It might be the average or median of just the points plotted on that graph or one that was previously established for the process. Other horizontal lines might represent specification midpoints, maximums, and minimums; targets or goals; customer preferences; etc. An example of a run chart is shown below with several examples of the types of things a viewer might look for in such a chart.

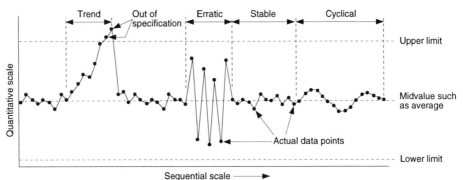

Typical patterns the viewer might look for in a run chart

- Run charts are similar to control charts but differ in several key ways including:
  - Run charts frequently plot individual data points, as opposed to control charts that often plot averages.
  - Run charts are often used as an inspection tool to weed out unacceptable performance, while control charts are more frequently used to assure that a process continues to perform as it has in the past.
  - Run charts typically use specific limits such as specified by a manufacturer, customer, or government. Control charts use limits established statistically based on past performance.
  - Control charts are sometimes based on the results of run charts.

Sans Serif Typeface

When the letters of a typeface have small lines (serifs) projecting from the ends of each of their main strokes, as shown at the right, the typeface is described as a serif typeface.

Serif typeface

Sans serif typeface

When the letters of a typeface do not have the small projecting lines, it is said to be a sans (without) serif typeface, as shown at the left.

Sawtooth Curve

A sawtooth curve is so named because it resembles the teeth on the cutting edge of a saw. With this type of curve, a quantitative variable is plotted on the vertical axis. Generally time is plotted on the horizontal axis. The distinguishing characteristic of a sawtooth curve is that its value increases or decreases for a given period of time, at which point the value abruptly changes direction and returns to some base value. That base value may be a fixed value (e.g., a horizontal line) or a value that increases or decreases with time (an inclined line). Examples of both types are shown below. An application where the values jump abruptly,

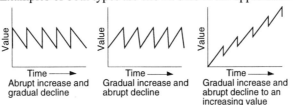
Graphs with sawtooth curves

for example, might be inventory levels that gradually decline as parts are used or product is sold. Once the stock gets to a predetermined point, the parts or products are replenished; the level of inventory jumps abruptly and again starts a gradual decline. An application in which the values drop abruptly might be a hopper whose weight gradually increases as it is filled and then abruptly drops once it is emptied. The downward portion of each cycle on a sawtooth curve is sometimes referred to as decay.

Classification of scales

Scales are classified several different ways. In this book the three major classifications used are category, quantitative, and sequence. The table below describes and differentiates the three classifications and indicates other terms that are sometimes used when referring to these scales.

	Summary of Major Classifications of Scales		
	Category	**Quantitative**	**Sequence**
Other terms the scales are sometimes called by	– Qualitative – Nominal	– Value – Interval – Numeric – Amount	
Description	A scale consisting of an ordered or unordered series of words and/or numbers that name, identify, and/or describe people, places, things events, etc. Each word or number defines a distinct category that contains one or more entities.	A scale consisting of numbers organized in an ordered sequence with meaningful and uniform spacings between them. The numbers are used for their quantitative value, as opposed to being used for identification. A ratio scale is a variation of a quantitative scale.	A scale consisting of words and/or numbers in an ordered sequence with uniform spacings between them. When numbers are used, they are for identification and/or ordering purposes only and have no quantitative significance. Time series and order-of-occurrence scales are the most widely used sequence scales.
Type of labels	Words or numbers. When numbers are used, they are for identification purposes only and have no quantitative significance. Sometimes referred to as nominal information.	Always numbers	Sequential numbers or words (e.g., 1,2,3,4,5; Jan, Feb, Mar; etc.). In some cases, such as the sequential collection of samples, no labels are used and the values are simply placed in successive order along the sequential axis. Frequently terms such as sales order numbers, lot numbers, etc., are used as labels for identification purposes.
Can labels be reordered without degrading the information in the graph?	Yes. Labels are frequently reordered. Alphabetizing is one of the most common methods. When data is ranked by its quantitative values, category labels are automatically reordered.	No	No
Are scales linear or nonlinear?	Not applicable	Can be linear or nonlinear	Generally linear. Sometimes time scales covering large periods of time are nonlinear.
Do the scales represent independent or dependent variables?	Always independent	Can be independent or dependent	Always independent
Can information plotted along the axis with this scale be mathematically manipulated in a meaningful way? (e.g., calculating averages, multiplying by ten, etc.)	No	Yes	No
Example	Corn Beets Peas Rice Beans	0 1 2 3 4 5	Jan Feb Mar Apr May

Scale (continued)

Examples of quantitative scales

Linear and logarithmic are the two most widely used quantitative scales for rectangular graphs. There are also a number of lesser-used quantitative scales. Shown below are examples and descriptions of linear and logarithmic scales, as well as two of the lesser-used types, probability and power. Also shown is an angular scale of the type used as a quantitative scale on many circular graphs.

Linear scale

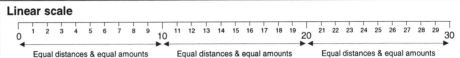

A linear scale is sometimes referred to as an arithmetic scale. A linear scale is arranged such that the distance and value between any two major tick marks (major intervals), is always the same in a given scale. For example, a major interval might be one inch and represent ten units of measure. On the same scale, the distance between any two minor tick marks, called minor interval, might be one-tenth of an inch and represent one unit of measure. On a linear scale, these two criteria would apply at any point along the entire length of the scale.

Logarithmic scale

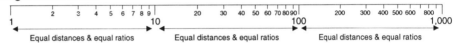

A logarithmic scale is arranged such that the distances between major tick marks (major intervals) are of equal length and equal ratios. For example, the distance between any two major tick marks or major values might be one inch and a ratio of ten. Since, in this example, the ratio between any two major values is ten, that means that each successive major value is ten times larger than the previous major value (i.e., 100 is ten times larger than 10; 1,000 is ten times larger that 100, etc.). This pattern applies at any point along the entire length of the scale. The spacing of minor tick marks follows a similar pattern. See Logarithmic Graph.

Probability scale

One of the lesser-used scales is labeled in percent and is used on a probability graph to determine whether a data series has a normal distribution. If the cumulative percent frequencies of a data series form a straight line when plotted on a graph with this type of scale, it indicates that the data series has a normal distribution. No distances between any two major tick marks on the same side of the scale are the same. The spacings on the right side are; however, mirror images of those on the left side.

Power scale

In a power scale, the values shown are based on some power of the values in a linear scale. For instance, in the example above, the values are the squares (power of two) of the the linear values of 0, 2, 4, 6, 8, and 10. Looked at another way, a power scale is a scale with values such that taking each of the values to a particular power yields a linear scale. In this example, that power is 0.5 (square root) which identifies the scale. Power scales can be constructed using any power. The physical spacings between major scale values are all equal.

Angular (polar) scale

Angular scales are sometimes used with circular graphs. They typically are arranged so the labels are uniformly distributed around the circumference of the graph. The labels generally designate even fractions of 360 degrees such as 1/24 (15°), 1/12 (30°),1/6 (60°), 1/4 (90°), etc. The labels might be shown as degrees or radians (1 degree equals about 0.0175 radian). Portions of degrees might be in terms of minutes and seconds or decimals of a degree.

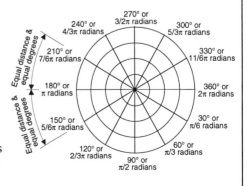

Scale (continued)

Example of category scale

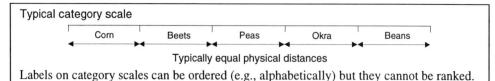

Typical category scale

| Corn | Beets | Peas | Okra | Beans |

Typically equal physical distances

Labels on category scales can be ordered (e.g., alphabetically) but they cannot be ranked.

Examples of sequence scales (see sequence scale)

Time series scale (sometimes called chronological scale)

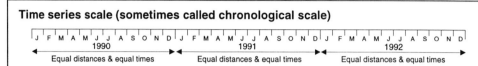

J F M A M J J A S O N D J F M A M J J A S O N D J F M A M J J A S O N D
1990   1991   1992
Equal distances & equal times   Equal distances & equal times   Equal distances & equal times

A time series scale has intervals that are of equal physical size and represent equal increments of time. The time might be shown in terms of dates, specific times, or successive increments of time. Unlike some sequence scales, time scales often have minor intervals as well as major intervals. For example, years can be broken into months, weeks into days, etc. See Time Series Graph.

Order of occurrence scale

1   2   3   4   5   6   7   8   9   10   11   12

Equal physical distances but no indication of how the times, distances, frequencies, etc., between the actual data elements compare

This type of scale plots data in the same order or progression as it occurred or is planned to occur. An order of occurrence scale indicates which came before and which came after, but not necessarily how long the gap was between the two entries. The numbers or words used as the labels are primarily for identification, location, and ordering purposes and generally have no quantitative significance.

Ordinal scale

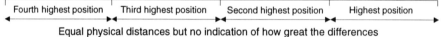

| Fourth highest position | Third highest position | Second highest position | Highest position |

Equal physical distances but no indication of how great the differences are between entries or whether or not the differences are equal

Ordinal designations rank information in nonquantitative terms such as small, medium, large; first, second, third; lowest position, second lowest position; etc. Ordinal labels appear on the scale in the same sequence as they occur in the ranking process.

Scales for circular symbols

Sometimes symbols such as circles are used to convey quantitative information, as in the bubble graph at the right. When this is done a scale or legend is normally required so the viewer can relate the symbol size to a value. The examples below use circles to illustrate some of the more widely used methods of constructing such scales. • Studies on circles have shown that viewers tend to underestimate the relative sizes of larger circles. In order to compensate for this problem, techniques have been developed to systematically enlarge the circles in proportion to their diameters, i.e., the larger the circle, the greater the enlargement.

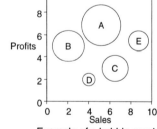

Example of a bubble graph using circles to convey quantitative information

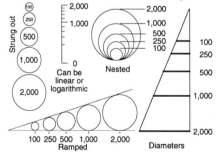

Examples of techniques for relating circle size to the values represented.

The example at the right compares modified and unmodified circles. It can be seen how the difference between the modified and unmodified circles becomes greater as the circles get larger. Sometimes the unadjusted circles are referred to as having an absolute scale and the modified circles as having an apparent-magnitude scale. Modified circles are used primarily on maps.

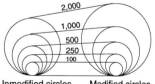

Unmodified circles   Modified circles

Comparison of circles based on traditional (absolute) and modified (apparent-magnitude) methods

Scale (continued)

Scales for maps (also see Map)

Surface distance scale

This type of scale is used primarily for determining the size or relative location of places and things, for example, to determine the width of Kansas or the distance from Paris to Rome.

Simple fraction or ratio - Sometimes referred to as representative fraction or RF. This type of scale relates a unit distance on a map to the distance it actually represents on the ground or earth. For example, the fraction or ratio of 1/1,000,000 or 1:1,000,000 means that 1 unit on the map represents 1,000,000 of the same units on the ground.

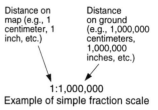

1:1,000,000
Example of simple fraction scale

Written statement or verbal - A notation on a map such as "one inch equals one mile" means that one inch on the map is the equivalent of one mile on the ground.

Graphic representation -

Bar scales are frequently used for determining distances.

Distorted distance scales are sometimes used when distances on the map are distorted. The example at the right addresses a situation where the further one gets from the equator, the shorter the actual distance represented by a unit length on the map.

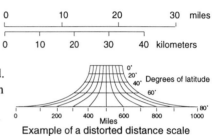

Example of a distorted distance scale

Elevation or vertical axis scale

When color or shading are used to indicate elevation, a scale similar to those at the right is frequently used. The color/shading concept can be applied to surfaces above or below water.

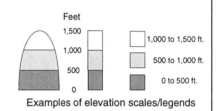

Examples of elevation scales/legends

Latitude and longitude

For determining the absolute location of places and things on the surface of the earth, a system of latitudes and longitudes is used. See Map.

Supplementary scale

Sometimes referred to as a supplementary amount scale. The grid lines for a supplementary scale are sometimes a series of curves that are extended as data becomes available. For instance, in the example at the right, the supplementary grid lines represent the profit dollar values associated with various percents of sales. Consequently, the values of the supplementary grid lines cannot be calculated until the sales and profit figures are available. A supplementary scale provides much of the same basic information as a second value scale (example below) but, in addition, lets the

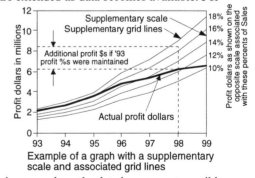

Example of a graph with a supplementary scale and associated grid lines

viewer analyze the data in ways not possible with a pair of conventional scales. For instance, in the example above it can be seen that the profit was 17% in 1993. By means of the supplementary scale, one can see that if the 1993 profit level of 17% had been maintained instead of dropping to 12.5%, the company would have made about two million dollars more profit in 1998. • Supplementary scales are typically ratio or percent scales.

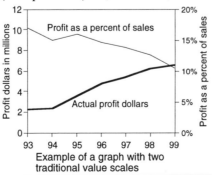

Example of a graph with two traditional value scales

Upper and lower values

The smallest value on a quantitative scale is normally equal to or less than the smallest coordinate plotted on that axis. The largest value on the scale is normally equal to or greater than the largest coordinate plotted on that axis. The smallest and largest values on a scale are referred to by several different terms. The examples at the right are some of the more widely used terms.

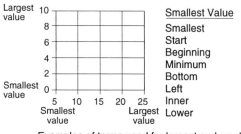

Examples of terms used for largest and smallest scale values

Smallest Value	Largest Value
Smallest	Largest
Start	End
Beginning	End
Minimum	Maximum
Bottom	Top (vertical axes)
Left	Right (horizontal axes)
Inner	Outer
Lower	Upper

Scale Breaks

Sometimes referred to as split grid. Scale breaks are intentional discontinuities in the scales of graphs. They are one of the more controversial aspects of graphs and are a significant factor in misinterpretations. The major reasons scale breaks are used are to reduce the size of the graph when scales are expanded, and to call changes in scales to the viewer's attention. Three of the major situations in which scale breaks are used are, 1. large differences exist between the high and low values of the data being plotted, 2. the situation where differences between data points are small compared to the actual values of the data points, and 3. there are changes in scale intervals or units of measure (e.g., years and quarters on the same axis). Shown below are some of the more generally accepted methods for incorporating scale breaks into graphs.

———**Full versus partial scale break**———

A full scale break is one that runs from side to side or from the top to the bottom of a graph. A partial scale break typically appears at the two sides or the top and bottom of the graph. Full scale breaks are better than partial breaks at bringing the scale break to the viewer's attention.

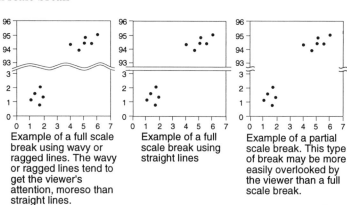

Example of a full scale break using wavy or ragged lines. The wavy or ragged lines tend to get the viewer's attention, moreso than straight lines.

Example of a full scale break using straight lines

Example of a partial scale break. This type of break may be more easily overlooked by the viewer than a full scale break.

———**Data graphics extending across a scale break**———

Even though it is the tops of columns and the ends of bars that denote the values the data graphics represent, viewers many times equate the area of the data graphics to the values they represent. As a result of this, when a scale break exists, but is not extended across the data graphic to clearly bring it to the viewer's attention, viewers are apt to draw erroneous conclusions. For this reason, it is generally recommended that scale breaks be prominently shown in the data graphics as well as in the scales and background.

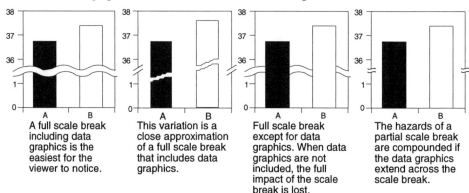

A full scale break including data graphics is the easiest for the viewer to notice.

This variation is a close approximation of a full scale break that includes data graphics.

Full scale break except for data graphics. When data graphics are not included, the full impact of the scale break is lost.

The hazards of a partial scale break are compounded if the data graphics extend across the scale break.

Scale breaks (continued)

Scale breaks used to expand a scale

Sometimes a scale break is introduced so the physical size of the scale intervals in the larger values can be enlarged to improve the ease and accuracy of reading the data. The lower portion of the scale is occasionally retained to help viewer's orient themselves as to how the scale intervals are laid out. (Such scale breaks are sometimes called interior scale breaks.) Even though there are no data points below the break, it is advantageous to have a full scale break to alert the viewer that the distances of the data points from the zero axis are not proportional to the values of the data points shown.

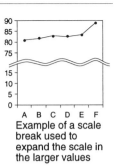
Example of a scale break used to expand the scale in the larger values

Scale breaks in time scales

Scale breaks are seldom used in time scales. One of the major reasons for this is that breaks in time scales many times distort the data, particularly when the data is being used to determine trends. When a scale break becomes necessary, it is generally advisable to use a full break that converts the graph into the equivalent of two separate graphs. This is also true when units of measure change, as from years to quarters (right). As an additional precaution, the data graphic fill is sometimes also changed. • If data is missing on a time series graph, the intervals are included and the data is left blank or bridged over.

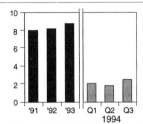
Example of a full scale break and change in fill used to highlight a change in time intervals on the horizontal scale.

Scale breaks for unusual data points

When one or a few data points are far above or below all the others, the top or bottom of the graph is sometimes cropped and the value of the point(s) noted. The shape of the remaining data graphics is generally unchanged. This particular procedure is sometimes referred to as breaking off the top or bottom.

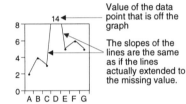
Value of the data point that is off the graph

The slopes of the lines are the same as if the lines actually extended to the missing value.

Scale breaks with graphs whose data graphics are based on areas

Breaks generally are not used with any graph in which the areas of the data graphics are proportional to the values they represent, such as histograms, area bar graphs, and area column graphs. The introduction of scale breaks in such graphs distorts the data and makes it difficult for the viewer to make meaningful observations.

Horizontal versus vertical

Even though the examples used here to illustrate scales breaks all have the quantitative scale on the vertical axis, the same principles apply to graphs having a quantitative scale on the horizontal axis. • Scale breaks have little significance when used with category scales on either axis.

Tick marks, labels, and intervals

Scales are a series of lines (generally tick marks in conjunction with a label axis) and labels (generally numbers or words) that are used systematically to encode and decode information onto charts, graphs, maps, and diagrams. There are three types of tick marks: major, minor, and intermediate. The distances between major tick marks are generally referred to as major intervals. The distances between minor tick marks are referred to as minor intervals.

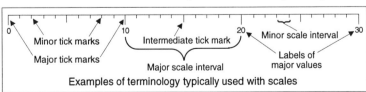
Examples of terminology typically used with scales

Intervals are generally of a numerical size that will make the chart easy to read such as 1,2,5,10, etc. Unusual intervals such as 3,7,9, etc., are normally avoided. Minor intervals are generally a logical division of the major intervals. For example, weeks might be broken into seven minor intervals to represent the days of the week. Feet might be subdivided into 12 inches, years into four quarters, standard measures into minor intervals of 5, 10, 20, 100, etc. There are many options as to the labels used. See Numbers, Times, and Dates.

Expanded scales

When a scale is expanded, the physical distance is increased between the labels on the scale. For example, if a scale has one-fourth inch between the values of 57 and 58 before expanding, after expanding the two values might be separated by one-half inch. When a scale is expanded, the graph obviously becomes larger, unless a portion of the scale is eliminated. If one wants to keep the graph the same overall size, a section is generally removed from the center (scale break) or bottom. Although offering advantages, the use of scales that do not start at zero have generated much criticism. When graphs are to be viewed by persons not familiar with them, scales frequently start at zero. For persons more familiar with graphs, expanded scales many times offer significant advantages, even at the risk of the viewer overlooking the fact that the scale does not start at zero. The examples below illustrate how using expanded scales in the area under study increases the viewer's ability to see trends and determine values more accurately. The examples also illustrate how changing the scale can have a dramatic effect on the appearance of the graph and therefore potentially mislead the casual observer.

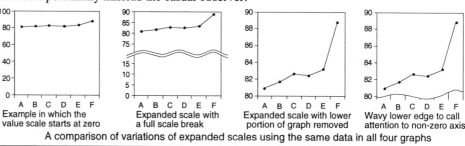

Example in which the value scale starts at zero

Expanded scale with a full scale break

Expanded scale with lower portion of graph removed

Wavy lower edge to call attention to non-zero axis

A comparison of variations of expanded scales using the same data in all four graphs

Shifted frame and scale labels

Most graphs have the scales and axes coincident with the frame of the graph. This sometimes makes it difficult to read data points on or close to the axes. One way to overcome this problem is to shift the frame and scale labels slightly, as shown in the example at the right. The dashed lines in the example on the far right may or may not be used in an actual application.

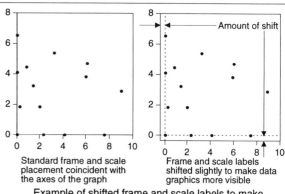

Standard frame and scale placement coincident with the axes of the graph

Frame and scale labels shifted slightly to make data graphics more visible

Example of shifted frame and scale labels to make data points on and near the axes more legible

Scale bars

Sometimes when studying one or more data series, a graph of the data using a zero axis is used to get an overview of the information. For more detailed analyses, the same data is plotted on graphs with expanded scales. When this is done it is sometimes difficult to rapidly relate the scales on the various graphs without carefully comparing the labels. Scale bars help to overcome this problem by giving a graphical representation of the same unit of measure on each of the graphs. For instance, in the example below, the black vertical bar represents the same unit of measure on each graph. Thus, without reading any numbers, one can observe that about one vertical inch on the graph on the right is equivalent to about one-sixteenth of a vertical inch of the graph on the left. Scale bars can be applied to horizontal as well as vertical scales. The bars can be displayed inside the frame of the graph, as they are here, or adjacent to the graph. Lines, arrows, or other symbols can be substituted for the bars. The technique works equally well whether comparing related or unrelated data.

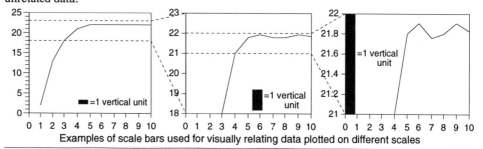

Examples of scale bars used for visually relating data plotted on different scales

Scale (continued)

Number and location of scales

The major factors regarding the number and placement of scales include the degree of accuracy required in reading the graph, whether or not grid lines are used, size of the graph, and the purpose of the graph. There is significant flexibility and variety as to where scales are located on graphs. Examples of some of the major variations are shown below.

Rectangular

On rectangular graphs a single scale is most frequently used on the horizontal axis at the bottom of the graph and on the left vertical axis. Values typically increase from left to right and bottom to top. An example is shown at the right.

Sometimes the data is such that the scales are more helpful to the viewer when located on the top, right side, or within the graph. Examples are shown below.

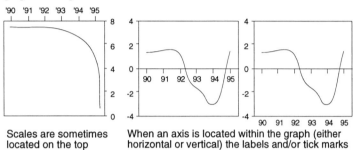

Scales are sometimes located on the top and/or right side.

When an axis is located within the graph (either horizontal or vertical) the labels and/or tick marks are sometimes shown with the relocated axis.

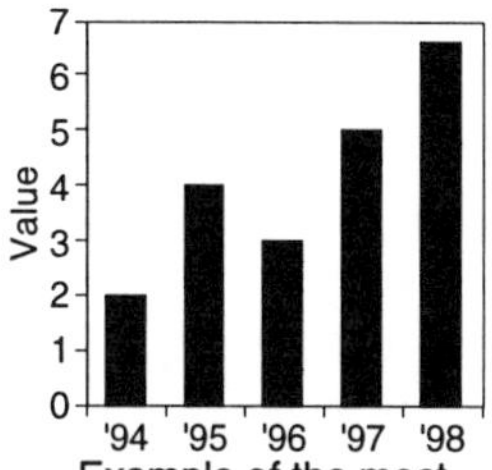

Example of the most frequently used combination of scales

There are several situations where multiple scales are used. One situation is when multiple data series are plotted on the same graph and either the units of measure are different for two or more of the data series, or the values of two data series are different enough that it would be difficult to read the values for one of the data series if a single scale were used. Examples are shown below.

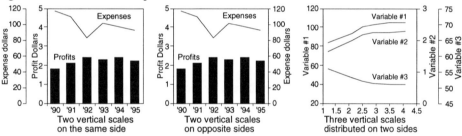

Two vertical scales on the same side

Two vertical scales on opposite sides

Three vertical scales distributed on two sides

Examples of graphs with multiple value scales

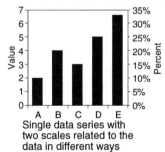

Single data series with two scales related to the data in different ways

In a good number of cases, multiple scales are used to enable the viewer to estimate the values of the data elements in two different units of measure, as shown at the left. For example, data may be read as actual values and percents-of-the-whole, Fahrenheit and centigrade, feet and meters, etc. Multiple scales can be used on the horizontal axis as well as the vertical, although it is done much less frequently. Two examples of multiple scales on the horizontal axis are shown below.

There is no limit as to how many scales can be used on a given graph; however, more that two scales per side, top, or bottom is unusual. In some cases a scale is simply duplicated on the opposite side or the top because the graph is large, grid lines are not used, or a higher degree of accuracy in estimating values is desired/required. When two or more vertical scales are used, the

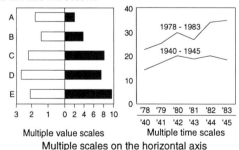

Multiple value scales

Multiple time scales

Multiple scales on the horizontal axis

graph is sometimes referred to as having multiple value scales, dual vertical scales, dual Y-axis scales, double Y-axis scales, a second Y-axis scale, or binumerical scales. When two or more horizontal scales are used, the graph is sometimes referred to as having dual horizontal scales, dual or double X-axis scales, a second X-axis scale, or dual/multiple category scales. • When there are multiple scales, each generally has its own set of tick marks; however, typically only one set of grid lines is used for each axis.

Number and location of scales (continued)

Circular

With circular graphs the angular or circular scale is almost always placed just outside the largest diameter. The scale might start at any angle and progress in either direction. They are normally linear. There are three major ways in which radial scales are displayed. In almost all cases the lowest value on the scale is at the center and the largest at the circumference. The radial scales might be linear or logarithmic.

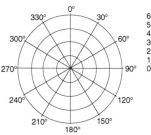

Typical location for angular/ circumferential scale

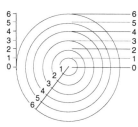

Examples of typical locations for radial scales

Trilinear

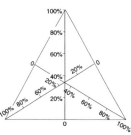

Trilinear graph with the scales in the plot area

The examples shown here illustrate the two different ways scales are generally applied to trilinear graphs. The example at the left is the easiest to understand, but many times the labels tend to interfere with the plotted data. Scales on the outside of the triangle, as shown at the right, clear up the plot area but are more difficult to interpret.

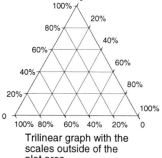

Trilinear graph with the scales outside of the plot area

• Scales on trilinear graphs always have their zero at the baseline, are linear, and most times have 100% as their upper values. See Trilinear Graph.

Three-dimensional

Placement of scales on three-dimensional graphs becomes more difficult because of the additional axis, the fact that the scales are more apt to interfere with the data graphics, and the inclined angle at which the graph is drawn. The examples shown here illustrate some of the scale placement variations used with three-dimensional graphs. Although technically the zero base line axes are at the back of the graph, placing the scales back there, as shown on the left, would cause significant interference with the data graphics. Consequently, the scales are generally placed in the front, as shown at the right.

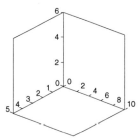

Basic graph with scales on the zero base line axes. With this layout the labels tend to interfere with the data graphics.

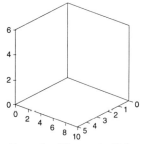

Example of the most widely used layout with scales and labels in the front and outside of the graph frame

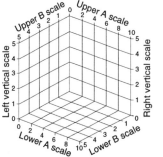

The use of duplicate scales is more common in three-dimensional graphs than in two-dimensional. Grid lines are included in the example for reference only.

• The use of duplicate scales, as illustrated on the left, is much more common on three-dimensional graphs than on two-dimensional graphs. The use of multiple scales with different values on a given axis is much rarer on three dimensional graphs than on two. • When a zero base line axis or reference axis is placed within the graph, as shown in the example at the right, the scales generally remain on the outside of the frame.

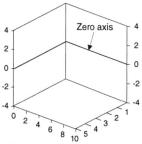

Even though zero base line or reference axes might be drawn on the walls, the scales generally remain outside the frame.

| **Scale Bars** | Sometimes the same data is plotted on two or more graphs. One with the quantitative scale starting at zero might be used for an overview. Another might have an expanded scale for a more detailed analysis. When this is done it is often difficult to rapidly relate the scales on the various graphs without carefully comparing the labels. Scale bars help to overcome this problem by giving a graphical representation of the same unit of measure on each of the graphs, as shown below. |

Scale bars used for visually relating data plotted on different scales
In this example, each bar represents the same unit of one.

Scale Break	Scale breaks are intentional discontinuities in the scale of a graph. See Scale.
Scale Line[1]	Sometimes called a grid line, grid rule, coordinate line, or ruling. Scale lines are thin lines used on graphs as visual aids for the purpose of encoding and decoding information. See Grid Line.
Scale Line[2]	Sometimes referred to as a label axis. Scale lines are the lines or axes on which the scales and labels are located. They may or may not have tick marks and titles associated with them. They may or may not be coincident with the zero base line axis.
Scale Point	Sometimes called a tick mark, tick, tic, stub mark or stub. Scale points are short lines that intersect or abut the axes or scale line to mark off intervals for encoding and decoding information and/or to serve as markers for labels. There are three sizes of scale points. The largest is called major and frequently has a label associated with it. The other two sizes are called intermediate and minor, with minor being the smallest. Intermediate and minor scales points often do not have labels associated with them. See Tick Mark.
Scatter Graph **or** **Scattergram** **or** **Scatterplot**	Sometimes referred to as point, dot, or symbol graph. Two-dimensional scatter graphs are sometimes referred to as XY graphs or XY scatter graphs. Three-dimensional scatter graphs are sometimes referred to as XYZ graphs or XYZ scatter graphs. Scatter graphs are a variation of point graphs with all quantitative scales or a combination of quantitative and sequence scales. They are used extensively for exploring relationships, particularly correlations between two or more data sets. See Point Graph.
Scatter Graph Matrix	Sometimes referred to as a stacked scatter graph. A scatter graph matrix is a variation of a matrix display that utilizes scatter graphs to make up the matrix. Their major purpose is to simplify the analysis of large quantities of data, particularly when one is looking for correlations or other meaningful relationships. The cells where the same variable appears on both axes are left blank, since a given variable plotted against itself always produces a straight line that yields no useful information. Sometimes distribution graphs for each of the variables are shown in the blank cells. See Matrix Display.

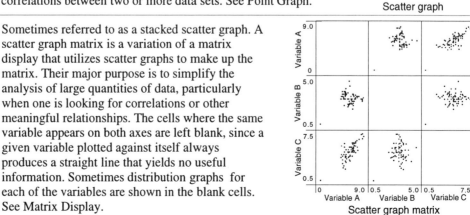

Scatter Line Graph	A line graph with quantitative scales on both axes. If there are multiple data series, it is called a grouped scatter line graph. On line graphs with category or sequence scales, the data points are connected sequentially from left to right. The data points on scatter line graphs are sometimes connected this way also. In other cases the data points are connected in the order in which the data were collected or plotted. This many times correlates with the order in which they appear on the spread sheet from which the graph is generated. As shown at the right, the shape of the curve can be significantly different, depending on which method is used. Only a person familiar with the data can decide which is the proper sequence for a given set of data.

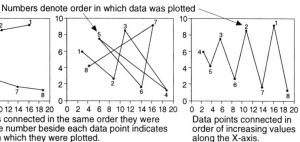

Numbers denote order in which data was plotted

Data points connected in the same order they were plotted. The number beside each data point indicates the order in which they were plotted.

Data points connected in order of increasing values along the X-axis.

Line graphs on which the same data points are connected two different ways

Scheduling Chart	There are many different types of scheduling charts. Most have several things in common; 1. A tabular format using rows and columns; 2. The things being scheduled are shown on the vertical axis; 3. Time is generally displayed on the horizontal axis; and 4. When something is scheduled, an entry is made in the cell(s) corresponding to the item scheduled and the appropriate time. An example is shown at the right.

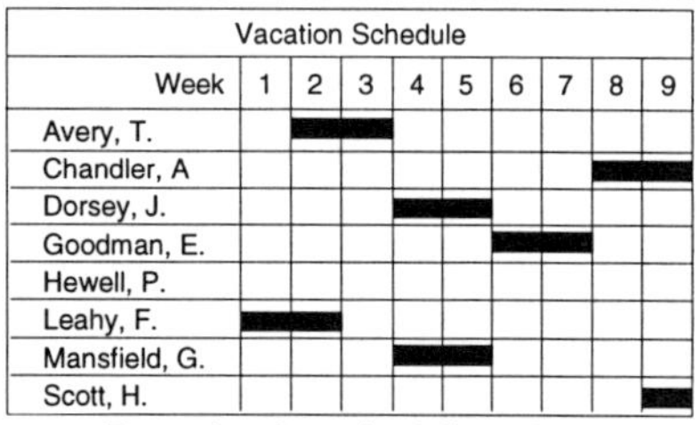

Vacation Schedule									
Week	1	2	3	4	5	6	7	8	9
Avery, T.			▓	▓					
Chandler, A								▓	▓
Dorsey, J.					▓	▓			
Goodman, E.						▓	▓		
Hewell, P.									
Leahy, F.	▓	▓							
Mansfield, G.				▓	▓				
Scott, H.									▓

Example of a scheduling chart

Schematic or Schematic Diagram	The term schematic is applied to certain types of diagrams. Exactly which types is not well defined. In some cases the terms schematic and diagram are used interchangeably. Both are made up of symbols, images, words, etc., frequently connected by lines. Schematics generally show how the individual pieces of something are arranged, organized, interconnected, interrelated, etc., or how things work, evolve, function, proceed, etc. In a few cases schematics show how to do something. They seldom show how individual pieces are arranged physically. Schematics often deal with the broad overview of something and are sometimes classified as conceptual diagrams. On the other hand, schematics can be very detailed diagrams such as flow charts, electronic circuits, PERT charts, etc. See Chart

Scientific Graph	There are a large number of specialized charts and graphs used in the fields of science and engineering. Although most are outside the scope of this book, most are variations or specific applications of the basic types of charts and graphs described in this book.

Seasonal Graph	Some organizations experience major seasonal fluctuations in certain aspects of their operations. For example, with some companies, most of their sales occur over the year-end holiday season. With others, fluctuations in sales, expenses, manpower, etc., might occur during the summer or be associated with some annual occurrence such as schools opening or closing. When data are plotted for these organizations, there sometimes are such significant fluctuations that it is difficult to spot trends or note unusual deviations (below).

To address this problem, specialized index graphs have been developed that display mathematically adjusted data that smooth out seasonal fluctuations. In this way trends and variations in the data can more easily be observed. This is accomplished by using historical data to develop a series of index

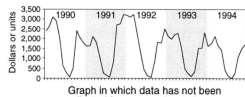

Graph in which data has not been adjusted for seasonal variations

numbers, sometimes called seasonal indexes. Index numbers are normally calculated for each month of the year; however, other periods such as weeks or quarters might be used. On the assumption that seasonal fluctuations are the same from year to year, the seasonal indexes are then applied to past data or future projections and the resulting values plotted.

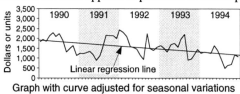

Graph with curve adjusted for seasonal variations

The graph at the left shows the same data as used in the example above, except it has been seasonally adjusted (sometimes referred to as seasonalized) and a regression line added. After the data has been season-ally adjusted, the downward trend of the

Seasonal Graph (continued)

data in this example becomes clearer. In addition, certain unusual or unexplained variations stand out, – for example, the greater than normal swing of values in 1991 and the fact that the values in the second half of 1991 were higher than the first half which is contrary to the general pattern in other years.

Seasonal index numbers used to forecast monthly performance

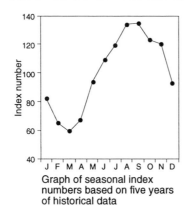

Graph of seasonal index numbers based on five years of historical data

The index numbers used to smooth the curve in the preceding graph can be used to quantify the seasonal cycle by plotting only the index values, as shown at the left. The index numbers can also be used to graph future monthly values when only the overall annual value is known. For example, if a company projects next year's sales at $20,000 (an average of $1,667 per month) and the seasonal indexes as shown in the graph at the left are applied, the projected sales for each month can be calculated by multiplying the index number (divided by 100) times $1,667. The resulting graph of projected sales is shown at the right.

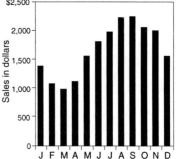

Projected monthly sales based on estimated annual sales of $20,000 and the index numbers shown in the graph at the left.

Graph of seasonal curve plus monthly trends

The above technique gives insight as to how a data series tends to vary from month to month throughout the year. Another way to look at the data is to see how it varies from year to year for each month. In this type of graph, values for the same month throughout the entire period under study, say five years, are plotted on small graphs called icons. For example, if there are five years being studied, the values for the five Januarys are plotted on one icon. The values for the five Februarys are plotted on another icon, etc. The values plotted might be actual values (used in the example below), deviations from the mean of the monthly values, variation from a standard, etc. These small graphs or icons are then positioned on a larger graph along a curve of the standard seasonal cycle. The seasonal cycle on which the icons are placed might be plotted using average values, index numbers, or some other measure representative of the seasonal cycle.

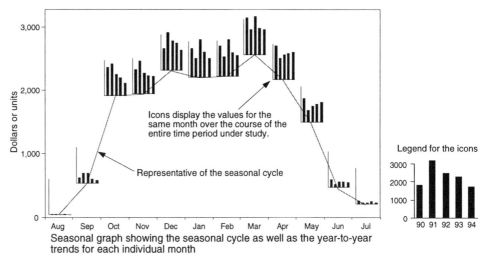

Seasonal graph showing the seasonal cycle as well as the year-to-year trends for each individual month

With such a graph, year-to-year trends for a given month can be related to the relative position of the month on the seasonal cycle. For instance, in the example above, in months during the up part of the seasonal cycle (e.g., October, November and December), the year-to-year values decreased for the last four years. At the same time, the year-to-year values for April and May, in the down part of the annual cycle, have increased for the past four years. The time scale on the horizontal axis might start with any month. For example, it might start with the first month of the year or the first period of increase in the cycle, as in the example above.

Sectogram or **Sector Chart**	Sometimes referred to as a pie chart, cake chart, divided circle, circular percentage chart, circle diagram, circle graph, or segmented chart. Sectograms or sector charts, which are variations of proportional area charts, consist of a circle divided into wedge-shaped segments. Each segment is the same percent of the total circle as the data element it represents is of the sum of all the data elements in its data set. The purpose of this type of chart is to show the relative sizes of components to one another and to the whole. The sizes may be shown in terms of actual values, percentages, or both. See Pie Chart.

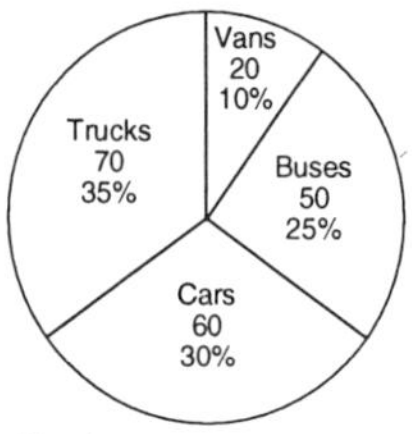

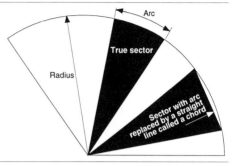

Sectogram or sector chart

Sector

A sector is a portion of a circle bounded by two radii and the included arc. In practice the arc is sometimes represented by a straight line which is called a chord. Examples of both are shown at the right. • When used in conjunction with pie charts and circle graphs, a sector is sometimes referred to as a segment, slice, or wedge.

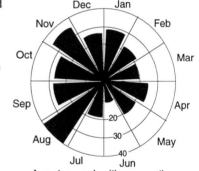

Sector Graph

Sometimes referred to as a rose or Nightingale rose chart. A sector graph is a circular graph that is divided circumferentially into equal size segments (i.e., segments with equal angles). The number of segments is equal to the number of data elements in the data series. Each segment of the circle has a sector plotted in it, the relative size of which is proportional to the data element it represents. Categories are distributed equally around the circumference of the circle, and values are plotted along the radii with zero at the center of the circle. An example of a sector graph is shown at the right. • Sector graphs differ significantly from circle graphs and pie charts. The wedges of a circle graph or pie chart generally all have the same radii and the sizes of the segments are varied by changing the angle of each. In a sector graph, all sectors have the same angle and the relative size of the sectors are varied by changing the radii. Sector graphs can be constructed so that either the areas or radii of the sectors are proportional to the values they represent.

A sector graph with arcs on the ends of the sectors and spaces between the sectors

There may or may not be space between the sectors. The sectors are sometimes filled with color or patterns for emphasis, appearance, or to encode additional information. Sector graphs are sometimes used to plot the distribution of elements in a data set, in which case the class intervals are shown as categories and the graph is called a circular histogram (left). Other than readability, there is no limitation on the number of categories that can be plotted. When the same categories are compared in multiple sector graphs, the location of each category remains the same so the viewer can easily compare the same item from graph to graph. Linear, power, or logarithmic scales can be used on the radii. A sector graph with only four segments is sometimes called a two-by-two or four-fold chart. See Four-Fold Chart/Graph.

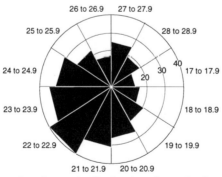

A sector graph with chords on the ends of the sectors and no space between sectors. Sometimes used as a circular histogram.

Segment

Sometimes referred to as a wedge, sector, or slice. The term segment is sometimes used to refer to one portion of a pie chart or circle graph. Each segment typically represents one data element of a data set. The example at the right shows a circle divided into four segments.

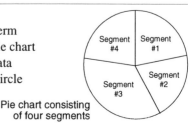

Pie chart consisting of four segments

Segmented Bar Graph	Sometimes referred to as a divided, composite, extended, stacked, or subdivided bar graph. A segmented bar graph has multiple data series stacked end-to-end. This results in the far right end of each bar representing the total of all of the components contained in that bar. Segmented bar graphs are used to show how a larger entity is divided into its various components and the relative effect that each component has on the whole. See Bar Graph.	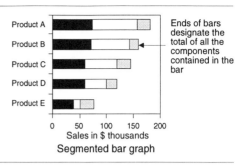

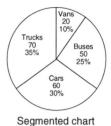

Segmented Chart	Sometimes referred to as a pie chart, cake chart, divided circle, circular percentage chart, sector chart, circle diagram, sectogram, or circle graph. A segmented chart (a variation of a proportional area chart) consists of a circle divided into wedge-shaped segments. Each segment is the same percent of the total circle as the data element it represents is of the sum of all the data elements in its data set. The purpose of this type of chart is to show the relative sizes of components to one another and to the whole. The sizes may be shown in terms of actual values, percentages, or both. See Pie Chart.	

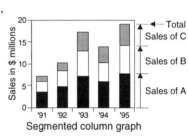

Segmented Column Graph	Sometimes referred to as a divided, composite, extended, stacked, or subdivided column graph. A segmented column graph has multiple data series stacked on top of one another. This results in the top of each column representing the total of all the components shown in that column. Such graphs are generally used to show how a larger entity is divided into its various components, the relative effect that each component has on the whole, and how the sizes of the components and the total change over time. See Column Graph.	

Segmented Line Graph	Sometimes referred to as a fever, broken line, thermometer, or zigzag graph. A segmented line graph is a variation of a line graph in which the data points are connected by straight lines, as shown at the right.	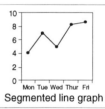

Semilogarithmic Graph	Sometime referred to as a rate-of-change or ratio graph. The semilogarithmic graph is the most widely used variation of a logarithmic graph. It has a logarithmic scale on one axis and a linear scale on the other. One of the major purposes of such a graph is to determine how rapidly and consistently one variable is changing with respect to a second variable. That second variable frequently is time. See Logarithmic Graph.	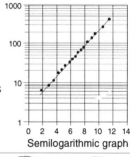

Separated Pie Chart	Sometimes referred to as an exploded pie chart. A separated pie chart is a variation of a pie chart in which one or more wedges of the pie are moved radially away from the center of the circle for visual emphasis.	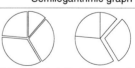

Sequence Axis	An axis that has a sequence scale on it.

Sequence Chart	Sometimes referred to as a time line chart. A sequence chart is generally a one-axis chart used to display past and/or future events, activities, requirements, etc., in chronological order. The major function of sequence charts is to consolidate and graphically display time-related information for purposes of analysis and communication. See Time Line Chart.

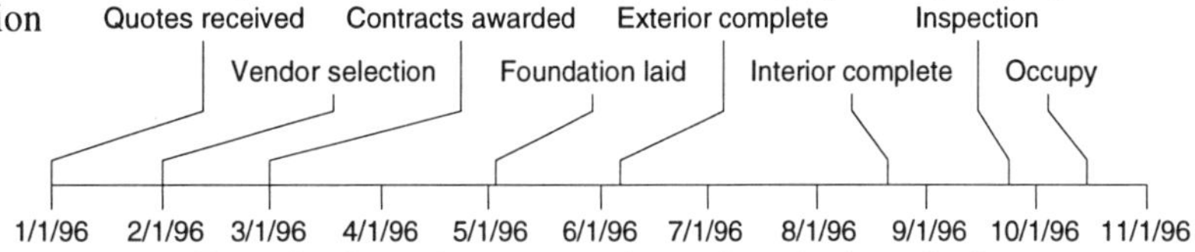

Quotes received   Contracts awarded   Exterior complete   Inspection
Vendor selection   Foundation laid   Interior complete   Occupy

1/1/96 2/1/96 3/1/96 4/1/96 5/1/96 6/1/96 7/1/96 8/1/96 9/1/96 10/1/96 11/1/96

Example of a basic sequence chart using a horizontal axis

Sequence Graph

A sequence graph is a graph with a sequence scale on one axis and a value scale on the other. In two-dimensional graphs the sequence scale is generally on the horizontal axis. In three-dimensional graphs the sequence scale is on either the X or Y axis, and the value scale is on the vertical axis. A sequence scale consists of words or numbers in an ordered or progressive sequence. When numbers are used, they are only for locating, identifying, or ordering purposes and have no quantitative significance. The three examples below illustrate three of the major types of sequence scales and graphs.

Time series

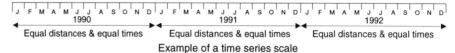

Example of a time series scale

Sometimes called a chronological scale. A time series scale typically has intervals that are of equal physical size and that represent equal increments of time. The time might be shown in terms of dates (e.g., June 23, June 24, June 25), specific times (e.g., 9:18, 9:19, 9:20), or successive increments of time (e.g., 42nd week, 43rd week, 44th week). Unlike most other sequence scales, time scales many times have the major intervals subdivided into minor intervals. For example, years can be broken into months, weeks into days, minutes into seconds, etc. See Time Series Graph.

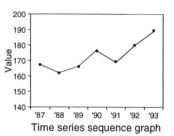

Time series sequence graph

Order of occurrence

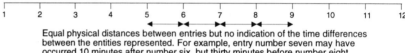

Example of order of occurrence scale

This type of scale plots data in the same order or progression it occurred or the data was collected. The time between successive data points may or may not be equal. For example, if the data points represent samples tested, there is no indication on the graph as to how long the time intervals were between samples. One interval may have been 10 minutes and the next one 30 minutes. All that an order of occurrence scale indicates is which came before and which came after. In some cases, sequence numbers are not assigned to data points and the values on the vertical axis are simply plotted successively along the horizontal axis. This method is especially likely when the data is plotted as it is collected; in these cases a trend is often the key thing looked for and the actual number of the sample is unimportant. In other cases, numbers such as lot number, sample number, patient number, etc., might be the labels along the horizontal axis.

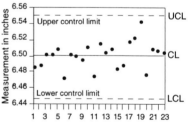

Order of occurrence type control chart. In this example, the numbers along the horizontal axis represent the sequence/order in which the samples were taken.

Ordinal information

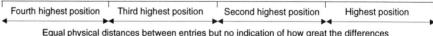

Example of an ordinal scale

Ordinal headings rank information in nonquantitative terms such as small, medium, large; first, second, third; or lowest position, second lowest position, and third lowest position. Such scales will many times indicate that one thing is larger than another, but not how much larger. Sequence scales use the same terms and order of information along the axis as established in the ranking process.

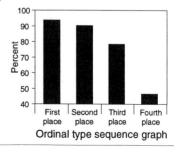

Ordinal type sequence graph

Sequence Graph (continued)

Factors that differentiate sequence scales from quantitative and category scales

- A sequence scale is the only type of scale that incorporates the concept of nonquantitative progression. The progression might be in terms of time, the order in which things occur irrespective of time, or the progression or succession of ordinal information such as small, medium, large or first place, second place, third place.
- A sequence scale is differentiated from a category scale in that the labels on a sequence scale cannot be reordered without degrading the information plotted on the graph. (reversing the scale so it progresses from right to left instead of left to right is excluded.) For example, if the years on a time scale were reordered such that 1990 was the first year, 1996 the second year, 1992 the third, 1994 the fourth, etc., the data plotted on the graph would be almost meaningless, particularly if one were looking for trends.
- Sequence scales are differentiated from quantitative scales in that sequence information cannot be mathematically manipulated in any meaningful way. For example, even if it could be done, taking the logarithm of dates or lot numbers would yield meaningless numbers.

Applications for sequence graphs

Sequence graphs are used extensively for many different purposes including:
- Monitoring and controlling processes, activities, performance, etc.
- Analyzing for trends
- Looking for repetitive fluctuations such as seasonal and cyclical
- Forecasting and projecting
- Comparing the performance of multiple entities
- Studying repetitive phenomenon
- Communicating sequential type information

Missing data

In most cases, when data is missing an empty space is provided so the pattern formed by the data is not distorted and the viewer is alerted to the missing information. For example, if the data for 1989 is missing, a space is normally provided for it between 1988 and 1990 even though there is no data point. If tests are being conducted on every fifth part, but the data for the 25th part is missing, a space is frequently left between the 20th and 30th sample. If the data is plotted on a line graph, the line may or may not be broken, or a dashed line may bridge the gap where the missing data point would have been. Examples of both options are shown at the right.

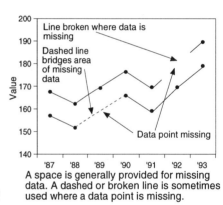

A space is generally provided for missing data. A dashed or broken line is sometimes used where a data point is missing.

General

- Point, line, area, and column type graphs are most frequently used for displaying sequential information.
- Multiple data series can be plotted on the same graph

Sequence Scale

A sequence scale consists of words or numbers in an ordered sequence. When numbers are used they are only for identification and/or ordering purposes and have no quantitative significance. A time series is one of the most widely used types of sequence scales. Other types of sequence scales include order of occurrence and ordinal scales. Examples of these three major types of sequence scales are shown below. See Scale and Sequence Graph.

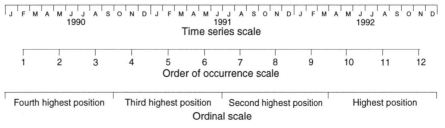

Examples of sequence scales

349

Series Axis

A series axis is a category type axis where the categories are the names of the data series being plotted. It is normally used on a three-dimensional graph.

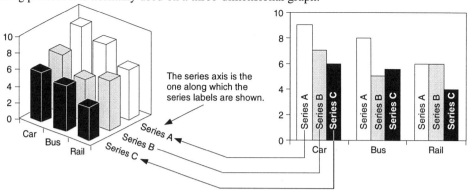

The series axis is the one along which the series labels are shown.

An illustration of how the information plotted on the series axis of a three-dimensional graph relates to the same data plotted on a two-dimensional graph

Serif Typeface

When the letters of a typeface have small lines (serifs) projecting from the ends of each of their main strokes, as shown at the right, it is some-

times called a serif typeface. When the type face does not have the small lines, it is generally called a sans (without) serif typeface, as shown at the left.

Set Diagram

Also referred to as a Venn diagram or Ballantine diagram. Set diagrams are typically used to describe a relationship between two or more sets of information. They accomplish this by the relative positioning of geometric shapes (normally circles) representing the sets of information. For example, the two overlapping circles shown below represent two sets of buyers. Circle A represents the set of buyers who bought brand X cars, and circle B the set of buyers who bought red cars. The area where the circles overlap indicates the buyers who fell into both categories, i.e., buyers who purchased red, brand X cars. Set diagrams might be quantitative or qualitative. See Venn Diagram.

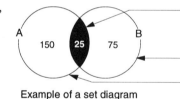

The area in which the two circles overlap represents the 25 buyers who bought brand X cars that were red.

Circle B represents 100 buyers of red cars

Circle A represents 175 buyers of brand X cars

Example of a set diagram

Seven and One-Half Minute Map

An area bounded by two lines of latitude and two lines of longitude is sometimes called a quadrangle, quadrilateral, or quad and is frequently designated by the angles included in the enclosure. For example, a seven and one-half minute map or quadrangle is a map encompassing 7.5 minutes of latitude and 7.5 minutes of longitude. The U.S. Geological Survey issues a series of maps referred to as 7.5-minute maps or quadrangles.

Shaded Map

Sometimes referred to as a choropleth, crosshatched, or textured map. A shaded map is a variation of a statistical map that displays data with regards to areas by means of shading, color, or patterns. The areas (sometimes called areal units) might be countries, states, territories, counties, zip codes, trading areas, etc. The data is generally in terms of ratios, percents, rates, etc., as opposed to absolute units. For example, incomes are typically given in terms of dollars per capita, dollars per household, etc., as opposed to total dollars for the area. Data is often organized into class intervals, as shown in the example at the right. See Statistical Map.

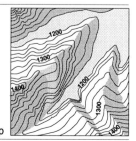

350 to 400
300 to 349
250 to 299
200 to 249
150 to 199
100 to 149
50 to 99
0 to 49

Example of a shaded map

Shaded Relief Map

A map on which elevations and variations in surface features are indicated by means of shading. The shading is meant to simulate shadows cast by an imaginary light source representing the sun, thus giving the map a three-dimensional appearance. The shadows are typically positioned as though the imaginary light source was in the Northwest. The technique is most frequently used with topographic maps. Shaded relief maps generally incorporate contour lines as shown at the right.

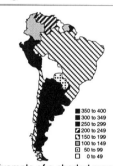

Shaded relief map

Shadow or Shading

Shadows are generally used to enhance the appearance of charts and graphs. In a small percentage of cases they are used to identify and/or highlight selected data graphics very much as color and fill might be used. For example, on an organization chart the symbols with shadows may indicate senior employees or employees with a certain type of training. Shadows can be used with almost every type of chart element, ranging from the text in a heading to an entire chart. Three of the major types of shadows are shown at the right. Examples of specific applications are shown below.

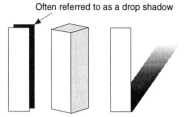

Often referred to as a drop shadow

Examples of three of the major types of shadows and shading used in charts and graphs

Graphs

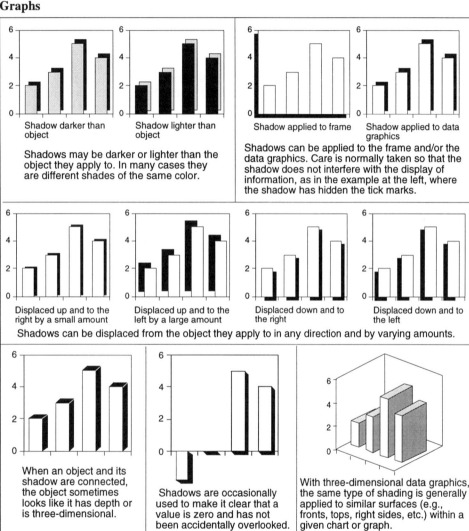

Shadow darker than object

Shadow lighter than object

Shadows may be darker or lighter than the object they apply to. In many cases they are different shades of the same color.

Shadow applied to frame

Shadow applied to data graphics

Shadows can be applied to the frame and/or the data graphics. Care is normally taken so that the shadow does not interfere with the display of information, as in the example at the left, where the shadow has hidden the tick marks.

Displaced up and to the right by a small amount

Displaced up and to the left by a large amount

Displaced down and to the right

Displaced down and to the left

Shadows can be displaced from the object they apply to in any direction and by varying amounts.

When an object and its shadow are connected, the object sometimes looks like it has depth or is three-dimensional.

Shadows are occasionally used to make it clear that a value is zero and has not been accidentally overlooked.

With three-dimensional data graphics, the same type of shading is generally applied to similar surfaces (e.g., fronts, tops, right sides, etc.) within a given chart or graph.

Maps

Shadows are sometimes used with maps to add emphasis and to improve their appearance. In some cases the shadows used with maps are more elaborate than those used with other chart types. The technical difference between an outline and a shadow sometimes becomes blurred as shown at the right. Shadows and

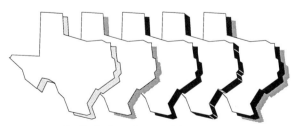

An illustration of some of the many variations of shadows used with maps

shading are widely used with topographic maps to indicate changes in elevations. These maps are frequently called shaded relief maps.

| Shifted Frame and Scale | On most graphs the frame and scale lines coincide with the axes. This sometimes makes it difficult to read data points on or close to the axes. One way to overcome this problem is to shift the frame and scale labels slightly away from the axes, as shown in the example at the far right. This technique can be applied to most types of axes. The dashed lines in the example may or may not be used in an actual application. | 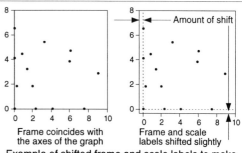 |

Example of shifted frame and scale labels to make data points on and near the axes more legible

| Shifted Reference Axis | Sometimes referred to as a reference axis or reference line. Shifted reference axes are typically located at some value other than zero and are included on graphs to denote differences between actual values and some reference value(s) while still showing the actual values as measured from zero. For example, if one wanted to show actual profit values while clearly designating deviations from budget, the budget values might be used as the reference axis. The data points, measured from the zero base line axis, would designate the actual profit values, while the data graphics between the data points and the reference axis would indicate deviations from budget. See Reference Axis. | 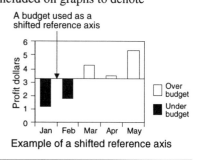 |

Example of a shifted reference axis

| Side-by-Side Bar Graph | Sometimes called a clustered bar, grouped bar, or multiple bar graph. A bar graph with two or more data series plotted side-by-side for comparison purposes. The bars for a given data series are always in the same position in each group throughout a given graph. Each data series typically is a different color, shade, or pattern. See Bar Graph. | 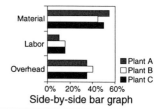 |

Side-by-side bar graph

| Side-by-Side Column Graph | Sometimes called a clustered, grouped, or multiple column graph. A column graph with two or more data series plotted side-by-side for comparison purposes. The columns for a given data series are always in the same position in each group throughout a given graph. Each data series typically is a different color, shade or pattern. See Column Graph. | 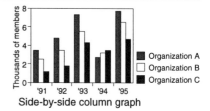 |

Side-by-side column graph

Silhouette Graph[1]

Silhouette graphs are formed by using color, shading, or patterns to fill the areas between lines representing multiple data series or to fill the areas between a data series and a reference line. Examples of the major variations and their alternate names are shown below.

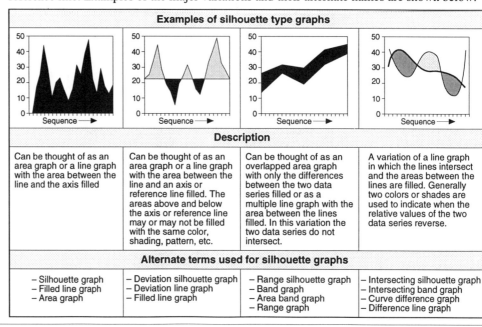

Silhouette Graph[2]

A series of small, related graphs stacked on top of one another to provide an overview of a subject, project, program, financial situation, etc. The graphs are generally of the times series type. All use the same horizontal scale located on the lower axis, but each has its own vertical axis. Units of measure on the vertical axes of the various graphs may or may not be the same. When the units of measure are the same, the intervals on the scales are frequently the same also so direct comparisons can be made. The graphs might display values for individual periods or cumulative values. An example of a series of graphs summarizing costs is shown at left.

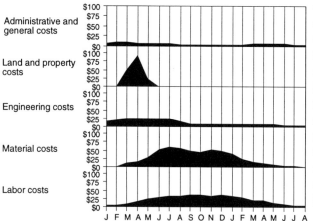

Silhouette graph used as an overview of costs on a major project. Similar graphs can be used to show other aspects such as manpower, equipment, and cash requirements.

Similar graphs can be generated for manpower, cash, equipment usage, requirements from vendors, etc. For repetitive information, a circular silhouette graph might be used, as shown at the right. • Silhouette graphs are used primarily for overview and planning purposes rather than for tracking purposes.

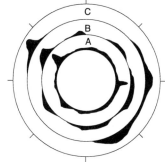

Circular silhouette graph

Simple Graph

A simple graph is typically the most basic format of a particular type of graph with only one data series displayed on it. Four examples are shown at the right. See individual graph headings.

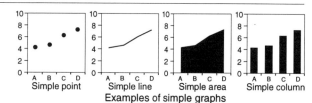

Examples of simple graphs

Single Column Chart

A variation of a proportional chart. A single column chart consists of a single stacked column with no grid lines, tick marks, or scales. In a single column chart, only one data series is displayed, and each segment of the column represents a different data element of the data series. The individual segments typically abut one another. When a small space is placed between them the chart is sometimes said to be exploded. Multiple single column charts can be used to compare multiple data series or to show changes or differences with time or conditions.

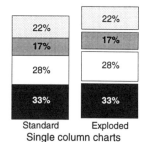

Single column charts

Skewhart Charts

Sometimes called control charts. Skewhart charts are a series of charts and graphs used primarily in the field of quality control. See Control Charts.

Skirt

Sometimes referred to as a side. Skirts are the surfaces that enclose the sides of certain three-dimensional graphs. In some cases plain skirts are added to emphasize the shape of the data along the two outer planes. In other cases grid lines are projected onto the sides resulting in the equivalent of two, two-dimensional graphs. In still other cases the skirts are added for aesthetic purposes and to minimize the distraction of the grid lines under the surface of the data.

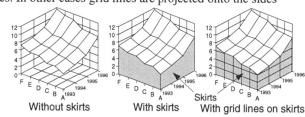

Slice[1]

Sometimes referred to as a wedge, sector, or segment. The term slice is often used to refer to one portion of a pie chart or circle graph. Each slice typically represents one data element of a data set.

353

Slice Graph and Slice[2]

Sometimes referred to as a spectral plot, ridge contour, layer graph, strata graph, or partition plot. A slice graph aids in the analysis of three-dimensional scatter graphs by condensing and displaying the data points on several planes or slices passed through the cloud of data points. With a slice graph, it is sometimes possible to spot trends and/or relationships, clusters of data, unusual data points, etc., that would be difficult or impossible to spot with the standard three-dimensional scatter graph. The axis along which the data points are condensed is largely dependent on the nature of the data and the type of analysis being done. Usually the X or Y-axis chosen. The number and location of planes and the size of the interval of data that is to be condensed onto a single plane are optional. The planes may or may not be equally spaced. They generally are equally spaced. The planes do not necessarily have to be displayed at the midpoints of the data being condensed although it is frequently done this way. An example of a slice graph in which the data points have been condensed onto three planes is shown below.

Variable #2
Variable #1
Age
Ages 13 & below
Ages 14 to 19
Ages 20 & above

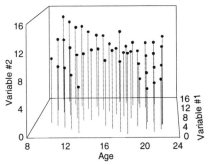

Three-dimensional scatter graph with drop lines

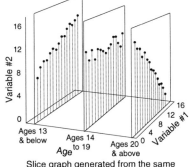

Slice graph generated from the same data used in the graph at the left

A slice graph and a three-dimensional scatter graph generated from the same data

The outlines or frames of the selected planes may or may not be shown. Depending on the number and nature of the data points, drop lines, connecting lines, and/or fill might be used to make the graph easier to decode. The examples below illustrate each of these potential variations.

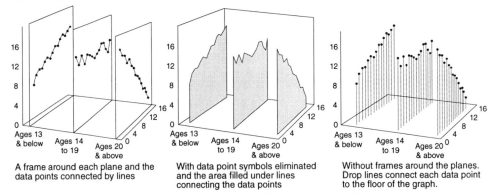

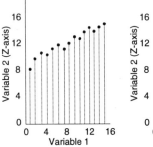

Ages 13 & below
Ages 14 to 19
Ages 20 & above

A frame around each plane and the data points connected by lines

With data point symbols eliminated and the area filled under lines connecting the data points

Without frames around the planes. Drop lines connect each data point to the floor of the graph.

Examples of potential variations available for use with slice graphs

Slices displayed side-by-side

It is sometimes easier to analyze the data if the planes on which the data are condensed are displayed side-by-side as multiple two-dimensional graphs. This is particularly true if there are many planes. An example is shown below. Such a group of graphs is sometimes called a casement display. Scales, grid lines, and tick marks on the slices are optional depending on the purpose of the graph and the type of data.

Variable 2 (Z-axis)
Variable 1
Ages 13 & below
Ages 14 to 19
Ages 20 & above

Example of the slices shown in the examples above arranged side-by-side (Sometimes referred to as a casement display)

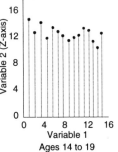

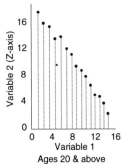

Sliding Bar Graph

Sometimes referred to as a bilateral, opposed, paired, or two-way bar graph. This type of graph is a variation of a bar graph in which positive values are measured both right and left from a zero on the horizontal axis. The major purpose of such a graph is to compare two or more data series with particular attention to correlations or other meaningful relationships. See Bar Graph.

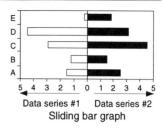

Sliding bar graph

Smoothing

Sometimes referred to as curve fitting or line-of-best-fit. Smoothing is a process in which a curve that most closely approximates a data series is superimposed over a plot of the data points of that data series. There are many different types of curves that can be fitted to a given set of data points. Two examples are shown at the right. See Curve Fitting.

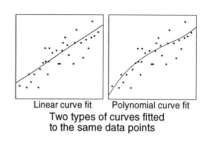

Linear curve fit   Polynomial curve fit
Two types of curves fitted
to the same data points

Smooth Statistical Map

A smooth statistical map is one in which the height (sometimes referred to as the Z-axis value) of the surface is proportional to the value of a variable being studied. For example, to plot pollution levels, one might plot latitude and longitude along the X and Y axes and pollution levels along the vertical or Z-axis, as illustrated at the right. The higher the level of pollutants, the higher the level of the surface of the map. Normally only data that is continuous with smooth transitions from one point to another or one area to another are displayed on this type of map. See Statistical Map.

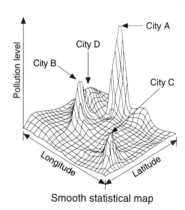

Smooth statistical map

Snowflake Icon

Miniature (1" square or less) graphs, frequently without titles, labels, tick marks, or grid lines are sometimes called icons. The purpose of these types of icons is not to convey specific quantitative information. Instead, they are many times used as symbols to show such things as relative sizes, overall comparisons, trends, etc. There are many types of icons. The icon at the right has several names including snowflake, polygon, star, and profile. It is sometimes generated using a radar graph. See Icon and Radar Graph.

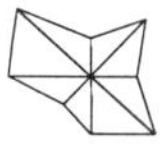

Example of an
icon sometimes
called snowflake,
polygon, star, or
profile

Sociogram

A sociogram is a variation of a relationship diagram that displays sociometric data for the purpose of studying the social interaction of groups. Sociograms are used most extensively in the field of sociometry, a branch of social psychology. One of the key functions of this type of chart is to provide a tool for determining the relative roles of individuals in group settings. Sociograms are helpful in identifying potential leaders, informal groupings, individuals who might require attention, changes in group dynamics over time, etc. They are also helpful in such undertakings as organizing harmonious work groups and assuring compatibility of individuals who must live together in confined environments. • Information for constructing sociograms is frequently obtained via confidential questionnaires completed by the members of the group under study. Each member is asked to answer one or more questions such as: "Who would you like to be on a task force with? Who would you like to see as chairperson of that task force? Who would you like to share a dormitory room with?" The same or similar questions may also be asked in reverse: "Is there anyone you would not like on a task force with you? Is there anyone you would prefer not to share a dormitory room with?" The information from the questionnaires is then displayed on a chart containing symbols representing each member of the group. An example is shown on the next page. Individual names, code numbers, or letters might be used to identify the members. If person A answers that she would like to be on a task force with person B, an arrow is drawn from A to B. If A answers that she does not want to be on a task force with B, a line is drawn from A to B but a short, perpendicular line replaces the arrow head. The same procedure is followed for each response.

Sociogram (continued)

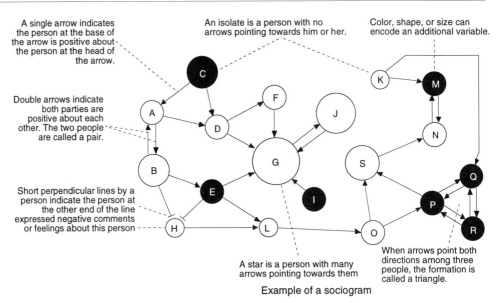

A single arrow indicates the person at the base of the arrow is positive about the person at the head of the arrow.

An isolate is a person with no arrows pointing towards him or her.

Color, shape, or size can encode an additional variable.

Double arrows indicate both parties are positive about each other. The two people are called a pair.

Short perpendicular lines by a person indicate the person at the other end of the line expressed negative comments or feelings about this person

A star is a person with many arrows pointing towards them

When arrows point both directions among three people, the formation is called a triangle.

Example of a sociogram

People who have a higher than normal number of arrows pointing at them are generally referred to as stars (symbol G in example). Stars may be social leaders or power figures, depending on who the arrows are emanating from. If no arrows go towards an individual, he or she is referred to as an isolate (symbols C and K). If arrows go both directions between two individuals, they are referred to as a pair (symbols A & B, M & N). If arrows go both directions among three people, they are referred to as a triangle (symbols P, Q, & R). Groups of individuals who have chosen one another are referred to as cliques. Additional information such a male or female can be coded into the chart by varying the shape or color of symbols. The size of the symbol can be used to encode such things as length of time in the group, ranking in the organization, etc. There are three methods commonly used to indicate that two people have selected one another as shown at the right. • Sometimes this type of chart is prepared at various intervals of time to note changes that have taken place as a result of actions initiated, or to note the effects of time alone. Different questions asked of the same group of people many times yield entirely different sociograms. For example, the sociogram for the question, "Who would you like study with?" may be quite different from the chart for the question, "Who would you like to go to the prom with?"

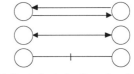

Methods for indicating when two members have selected each other

Sorting

Sometimes referred to as ordering. Sorting is the process of arranging information on a chart, graph, or table in some meaningful sequence. There are several ways information can be ordered. Four widely used methods are to alphabetize, to organize in numerical sequence (e.g., by time card number), to rank (e.g., place in ascending or descending order by quantitative value or ordinal position), and to arrange by attribute (e.g., address, product, religion, etc.). Examples of all four are shown here. Two major reasons for sorting are to make information easier to locate (particularly when material is alphabetized or in numerical sequence) and to make the information more meaningful for analysis (particularly when the information is ranked).

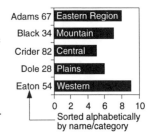

Sorted alphabetically by name/category

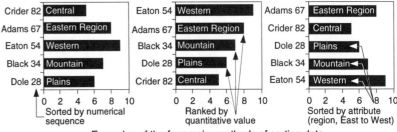

Examples of the four major methods of sorting data. The same information is shown in all four graphs.

| Source Note | A notation on a chart or graph indicating where the information used in constructing the chart or graph came from. In some instances it indicates the source of the chart or graph itself. |

Source Note   A notation on a chart or graph indicating where the information used in constructing the chart or graph came from. In some instances it indicates the source of the chart or graph itself.

Source Table   Sometimes referred to as a reference table. A source table's primary purpose is to make data easy to retrieve. Information is generally arranged in alphabetical order, by geographic area, in numerically ascending or descending order, chronologically, etc., so specific bits of information can be located rapidly.

Air Temperature (°F)									Comfort Index
110	105	112	123	137	150				
105	100	105	113	123	135	149			
100	95	99	104	110	120	132	144		
95	90	93	96	101	107	114	124	136	
90	85	87	90	93	96	100	106	113	122
85	80	82	84	86	88	90	93	97	102	108
80	75	77	78	79	81	82	85	86	88	91
75	70	72	73	74	75	76	77	78	79	80
70	65	66	67	68	69	70	70	71	71	72
	10	20	30	40	50	60	70	80	90	100

Percent Relative Humidity

A source table arranged for easy access/reference

Spatial Grid   A phrase sometimes used to refer to the grid of a three-dimensional graph.

Spectral Plot   Sometimes called a slice graph, ridge contour, layer graph, strata graph, or partition plot. A spectral plot is a graph in which the data points of a three-dimensional scatter graph are condensed onto a limited number of planes to aid in the interpretation of the data. See Slice Graph.

Data in a standard three-dimensional scatter graph are condensed into a limited number of planes in a spectral plot.

Spider Graph   Sometimes referred to as a radar or star graph. A spider graph is a circular graph with three or more radial axes. Each axis may have its own unique scale, or all axes might have similar scales. Data elements are plotted on the axes and all the data points for a given data series are connected by lines forming polygons. When used for comparative purposes (which they often are), the larger the polygon the higher the overall evaluation of the entity being evaluated. See Radar Graph.

Spider graph

Spike Graph   Sometimes referred to as a needle or vertical line graph. On this type of graph a separate vertical line is used to designate each individual data point. The tops of the vertical lines designate the actual data points. Such graphs are used when many data points are to be plotted at uniform intervals, as in many sequence type graphs. Spikes can be used with either two- or three-dimensional graphs. See Line Graph.

Spike graph

Spin   One of the major limitations of three-dimensional graphs is the fact that it is sometimes difficult to see or interpret all of the data graphics when viewed from a single angle. A technique developed to overcome this problem when viewing a three-dimensional graph on a computer screen is to spin or continuously rotate the chart so the viewer can look at it from almost any angle, e.g., front, back, side, top, etc. In this way, patterns, configurations, relationships, distributions, outliers, and anomalies can better be observed. See Rotate.

Spline Graph   A variation of a line graph in which all data points are connected by gently curving lines as opposed to straight lines. The exact shape of the curve might be based on visual estimates or established mathematical techniques.

Spline graph

Split Bar Graph   A split bar graph is a variation of a difference or deviation bar graph in which positive and negative values are plotted from a common zero for every interval along the vertical axis. See Difference Graph and Bar Graph.

Split bar graph

Split Grid   Sometimes referred to as scale break. Split grids are intentional discontinuities in the scale and grid of a graph. See Scale.

Spreadsheet	A type of table containing predominately numeric values. Spreadsheets are frequently used for the systematic recording, organizing, and manipulation of large quantities of data. It is common for many of the values on a spreadsheet to be mathematically interrelated such that a change in one value or series of values affects all or many of the other values in the table. • Spreadsheets are frequently used with computers to conduct what-if analyses. In a what-if analysis, hypothetical, predicted, alternative, worst case, etc., type values are entered into a matrix of information and their effect on the other values noted. For example, sales values that are 25% below forecast might be entered into a spreadsheet of the key financial information for a company to see what would happen to profit, cash flow, return on assets, employment levels, etc., if there was a major down-turn in sales.

Stacked Area Graph

Sometimes called a layer area, multiple-strata, strata, stratum, divided, subdivided area, or subdivided surface graph. A stacked area graph has multiple data series stacked on top of one another, as shown at the right. This is the area graph format most frequently used. See Area Graph.

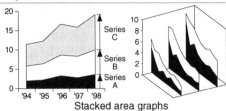
Stacked area graphs

Stacked Bar Graph

Sometimes referred to as a divided, composite, extended, segmented, or subdivided bar graph. A stacked bar graph has multiple data series stacked end-to-end. The far right end of each bar represents the total of all components contained in that bar. Stacked bar graphs are used to show how a larger entity is divided into its various components and the relative effect that each component has on the whole. See Bar Graph.

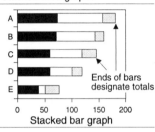
Stacked bar graph

Stacked Column Graph

Sometimes referred to as a divided, composite, extended, segmented, or subdivided column graph. A stacked column graph has multiple data series stacked on top of one another. The top of each column represents the total of all components shown in that column. Such graphs are generally used to show how a larger entity is divided into its various components, the relative effect that each component has on the whole, and how the sizes of the components and the total change over time. See Column Graph.

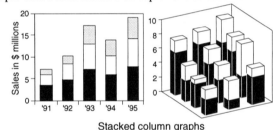
Stacked column graphs

Stacked Histogram

Stacked histograms are sometimes used to show how subcategories of a data series affect the overall frequency distribution pattern. For example, a histogram similar to the one at the right might be generated to study the distribution of orders based on dollar value. The black portions might represent the number of export orders and the white portions the number of domestic orders. The tops of the columns would indicate the total number of orders. More than two data series can be plotted on this type of graph; however, interpretation of the subportions of the graph sometimes becomes difficult.

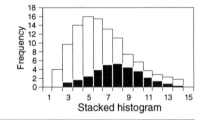
Stacked histogram

Stacked Line Graph

Sometimes called a layer line graph. A stacked line graph has multiple data series stacked on top of one another. In this type of graph, the top curve represents the total of all the components or data series below it. This format of line graph is used cautiously because of the potential confusion with the more familiar grouped line graph format, in which each curve is referenced from the same zero axis. See Line Graph.

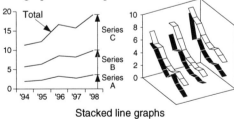
Stacked line graphs

Stacked Pie Chart

When comparing multiple data series or studying changes in the same series over time, a series of pie charts is sometimes used. Typically the series of charts is shown side-by-side. Occasionally they are superimposed on top of one another, as shown at the right, in an arrangement called a stacked pie chart. The segment representing a given data element is kept in the same relative position in each of the multiple pies. See Pie Chart.

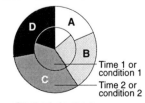
Stacked pie chart

Stacked Scatter Graph

Sometimes referred to as a scatter graph matrix. A stacked scatter graph is a variation of a matrix display that utilizes scatter graphs to make up the matrix. Their major purpose is to simplify the analysis of large quantities of data, particularly when one is looking for correlations or other meaningful relationships. The cells where the same variable appears on both axes are left blank, since a given variable plotted against itself always produces a straight line which yields no useful information. Sometimes distribution graphs for each of the variables are shown in the blank cells. See Matrix Display.

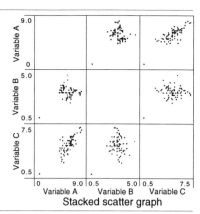
Stacked scatter graph

Stacked Surface Graph

A graph in which multiple quantitative surfaces are located directly above one another. The values of the lowest surface are plotted from the base of the graph. The values for all other surfaces are measured from the surfaces directly below them. This results in the values for the upper surface being the cumulative total for all of the surfaces shown in the graph.

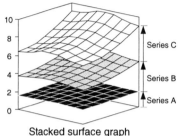
Stacked surface graph

Staggered Bar Graph

Sometimes referred to as a stepped, step-by-step, or progressive bar graph. A staggered bar graph is the equivalent of a stacked bar graph with only one bar and with each segment of that bar displaced vertically from its adjacent segment(s). The segments can be displaced up or down. The purpose of staggered bar graphs is to add visual emphasis to the individual segments while still maintaining the concept that the segments all add up to the whole. See Bar Graph.

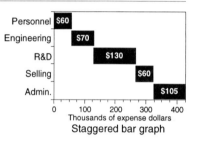
Staggered bar graph

Staggered Column Graph

Sometimes referred to as a stepped, step-by-step, or progressive column graph. A staggered column graph is a stacked column graph with only one column and with each segment of that column displaced horizontally from its adjacent segment(s). The segments, starting from the bottom, are typically displaced to the right. The purpose of staggered column graphs is to add visual emphasis to the individual segments while still maintaining the concept that the segments all add up to the whole. See Column Graph.

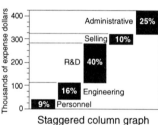
Staggered column graph

Stair Chart

A chart designed to compare a current five-year plan with previous five-year plans. This is accomplished by vertically aligning each of the plans estimates for a given year (e.g., all plan estimates for 1997 are aligned). Each chart generally compares only one aspect of the plan, such as projected sales, projected capital expenditures, projected contributions, etc. By comparing the current projection for the coming year(s) with those made in previous years, one can sometimes determine such things as whether the plans have become more or less optimistic and whether the forecasts are consistent or erratic. By comparing the projections of multiple years, basic changes in strategic thinking can sometimes be learned.

3 Years ago e.g., 1993	2 Years ago e.g., 1994	Last year e.g.,1995	Current year e.g., 1996	Coming year e.g., 1997	2 years out e.g., 1998	3 years out e.g., 1999	4 years out e.g., 2000	5 years out e.g., 2001
4 year ago ('92) plan for 1993 sales	4 year ago ('92) plan for 1994 sales	4 year ago ('92) plan for 1995 sales	4 year ago ('92) plan for 1996 sales	4 year ago ('92) plan for 1997 sales				
	3 year ago ('93) plan for 1994 sales	3 year ago ('93) plan for 1995 sales	3 year ago ('93) plan for 1996 sales	3 year ago ('93) plan for 1997 sales	3 year ago ('93) plan for 1998 sales			
		2 year ago ('94) plan for 1995 sales	2 year ago ('94) plan for 1996 sales	2 year ago ('94) plan for 1997 sales	2 year ago ('94) plan for 1998 sales	2 year ago ('94) plan for 1999 sales		
			Last year ('95) plan for 1996 sales	Last year ('95) plan for 1997 sales	Last year ('95) plan for 1998 sales	Last year ('95) plan for 1999 sales	Last year ('95) plan for 2000 sales	
				Current year ('96) plan for 1997 sales	Current year ('96) plan for 1998 sales	Current year ('96) plan for 1999 sales	Current year ('96) plan for 2000 sales	Current year ('96) plan for 2001 sales

Example of stair chart used for comparing current five-year plan with previous year plans.

359

Standard Deviation	A measure of how spread out (dispersed) the data elements are in a set of data. As a general rule, the larger the standard deviation, the more dispersed the data elements are. A standard deviation can be calculated for each data set. The assumption is generally made that the distribution of the data elements is normal, which may or may not be the case. See Normal Distribution Curve.
Standard Deviation Graph	A graph on which symbols, bars, lines, etc., are used to designate plus and minus one, two, or three standard deviations with regards to a set of average or median values. The basic graph designating the average or median values might be point, line, area, bar, or column type. There are a number of techniques for designating the standard deviations, – four are shown below, see Range Symbols and Graphs for others.

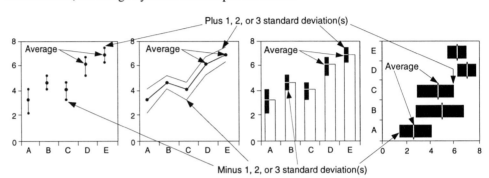

Examples of standard deviation graphs on which plus and minus standard deviations are displayed with reference to the average values for the same data.

Star Graph[1]	Sometimes referred to as a radar or spider graph. A star graph is a circular graph with three or more radial axes. Each axis may have its own unique scale, or all axes might have similar scales. Data elements are plotted on the axes, and all of the data points for a given data series are generally connected by lines forming polygons. When used for comparative purposes, as they often are, the larger the polygon, typically the higher the overall evaluation of the entity being studied. See Radar Graph.

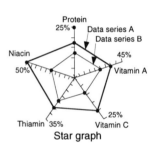

Star Graph[2]	Sometimes referred to as a circular column, or radial column graph. A star graph is the equivalent of a rectangular column graph wrapped into a circle. The horizontal axis of the rectangular graph becomes the circular axis of the star graph, and the vertical axis of the rectangular graph becomes the radius or value axis of the star graph. When the widths of the columns are very thin, the graph is sometimes called a radial line graph. See Circular Column Graph.

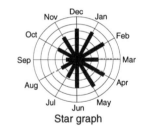

Star Icon	Miniature (1" square or less) graphs, frequently without titles, labels, tick marks, or grid lines, are sometimes called icons. The purpose of such icons is not to convey specific quantitative information. Instead, they are many times used as symbols to show such things as relative sizes, overall comparisons, trends, etc. There are many types of icons. The icon at the right has several names including snowflake, polygon, star, and profile. It is sometimes generated using a radar graph. See Icon and Radar Graph.

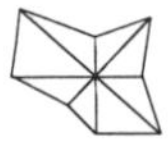

Example of an icon sometimes called snowflake, polygon, star, or profile

Statistical Graph	There are a large number of specialized charts and graphs used in the field of statistics. Although many of them are outside of the scope of this book, most are variations or specific applications of the basic types of charts and graphs described in this book. There is no clear-cut differentiation between business charts and graphs and those used for statistical purposes. Some types of graphs that at one time tended to be used exclusively by statistical personnel are now becoming commonplace in the day-to-day operations of many organizations.

Statistical Map

Sometimes referred to as a quantitative or data map. A statistical map presents quantitative information regarding areas, locations, distances, etc. Data might be shown as actual values (e.g., the total number of people in each country in Europe) or derived values such as percents, ratios, etc., (e.g., the density of people per square mile in each country in Europe). The quantitative data that is used is sometimes referred to as attribute data. In most cases, the base map containing geographic information onto which the attribute data is superimposed is deemphasized with only enough detail included to orient the viewer. If city names, highways, mountains, etc., are not germane to the theme of the map or required to orient the viewer, they normally are left off. Statistical information on maps generally references areas (countries, states, counties), points (city, store, intersection), distances, or lines/bands along which all values are the same or in the same range.

Information referenced to areas

There are two widely used methods for encoding statistical information by area. One uses class intervals; the other uses symbols. A third method relies on shades of color to encode values; however, this method is used to a much lesser extent.

─────**Class intervals used to encode information** ─────

When this system is used, the total range of values to be displayed is broken into five to fifteen groups called class intervals. Each class interval is assigned a shade, pattern, and/or color which is applied to all areas having values within the range of the class interval. For example, if black is assigned to the class interval covering the range of 350 to 400, all areas with values between 350 and 400 would be filled black. If 80% gray is assigned to the class interval covering the range of 300 to 350, all areas with values between 300 and 350 would be filled with 80% gray. Generally, the higher value of the area, the darker the fill used. An example of a class interval type map is shown at the right. This type of map is also referred to as shaded, crosshatched, textured, or choropleth map.

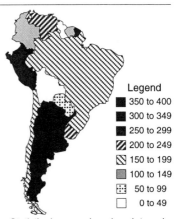

Legend
■ 350 to 400
■ 300 to 349
■ 250 to 299
▨ 200 to 249
▨ 150 to 199
▨ 100 to 149
⊞ 50 to 99
☐ 0 to 49

Statistical map using class intervals to encode quantitative information

─────**Symbols used to encode information** ─────

In this system of encoding, variations in the size, shape, number, etc., of symbols are proportional to the values associated with the areas. For example, a framed rectangle symbol might be used with a full rectangle (all black) representing 400 and an empty rectangle (all white) representing zero. The degree to which each of the individual rectangles is filled indicates the value of the area (example at right). To illustrate, if an area has a value of 310, the rectangle for that area would be slightly over three-fourths full (the lower three-fourths of the rectangle black). Overlapping symbols can become a problem with this type of map. Coding systems that do not use class intervals are sometimes referred to as unclassed, no class, or classless methods.

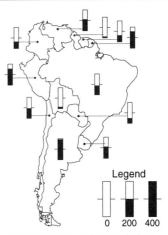

Legend

0   200   400

Statistical map using symbols to encode quantitative information

─────**Concerns with class interval system** ─────

Some of the methods that do not use class intervals were developed in response to concerns about the class interval method of encoding statistical information. The following specific concerns are sometimes expressed about the use of class interval with maps.

- The selection of class intervals and associated fill patterns, shades, and colors is somewhat arbitrary. (The methods for selecting class intervals are described at the end of this section.)
- The size of the geographic area can have a greater effect on the viewer's perception than the intensity of the fill used.
- In many cases the fill must be within the area it refers to, which can be a problem with very small areas.
- Sometimes a good deal of eye movement and distraction results from looking back and forth between the map and the legend.

Statistical Map (continued)

Information referenced to points

Many of the techniques for graphically encoding area type information can also be applied to statistical information related to points. The points may be such things as cities, factories, schools, stores, etc. The problem of data graphics overlapping and interfering with one another is sometimes even greater with points than with areas. Locating the data graphic away from the point and connecting it to the point with a line can lessen the problem.

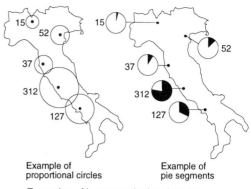

Example of proportional circles   Example of pie segments

Examples of how quantitative data that applies to specific points might be displayed

Information referenced to distances

Distance maps differ from nonquantitative route maps in that route maps show how to get from point A to point B, while distance maps quantify how far it is from A to B. Since the distances are many times designated on the map with straight or gently curving lines, a notation is sometimes included stating whether the distances are via a straight line, by road, by train, etc. The thicknesses of the lines generally do not vary in proportion to the distances represented. Arrow heads may or may not be used. Many times the actual values are shown on or alongside of the line. In some cases a scale of miles and/or kilometers is also shown. When a scale is used it generally requires that the entire map be laid out to the same scale and the paths be accurately drawn.

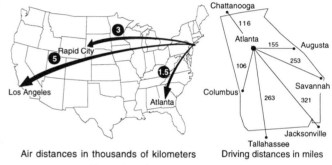

Air distances in thousands of kilometers   Driving distances in miles

Examples of maps showing distances

Information referenced to lines or bands

Sometimes referred to as an isogram map. With line or band maps, the data graphics are used to designate lines or areas along which the values are all the same. When points of equal value are connected the resulting map is sometimes referred to as a contour or isoline map. When it is bands of equal values, it is sometimes referred to as an isopleth map. Common examples are contour maps showing equal elevations, weather maps showing bands or lines of equal temperature or pressure, zone maps indicating where shipping charges or message rates are the same, sociological maps showing equal levels of education, equal crime rates, equal costs of living, etc. Generally the intervals between lines of equal value are the same, such as five between each line or ten between each line. In some cases the values vary geometrically between lines – for example, two between the first and second lines, four between the second and third lines, eight between the third and fourth lines, etc. In still other cases, the lines are chosen to fit the data available irrespective of interval size. The lines of equal value may or may not vary in boldness as the values they represent increase. The areas between the lines of equal value may be colored or shaded to visually aid in reading the map.

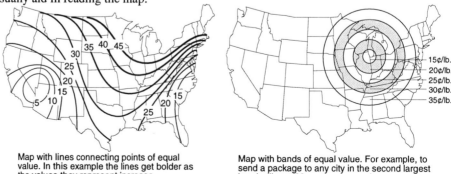

Map with lines connecting points of equal value. In this example the lines get bolder as the values they represent increase.

Map with bands of equal value. For example, to send a package to any city in the second largest band will cost 30 cents per pound.

Maps that display points and areas of equal value

Statistical Map (continued)

Distorted map

Sometimes referred to as a proportional or value-by-area map. A distorted map is a variation of a statistical map in which the size of the entities on the map are proportional to values other than the land surface areas of the entities. For example, if a particular resource is being studied by country, the size of each country on the map is proportional to the availability of the resource in that country. The country with the largest supply of that resource is drawn with the largest area on the map even if it has the smallest physical area in terms of actual land. An example is shown at the right.

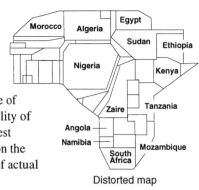
Distorted map

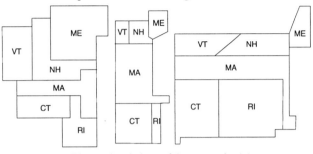
Distorted variations of the same six states. Different data are used in each map.

• There are no well established guidelines as to how to construct distorted maps, therefore, there are an unlimited number of possible variations. Shown at the left are three examples of a cluster of Northeastern states to give some idea of how diverse designs can be. Each example displays a different set of data.

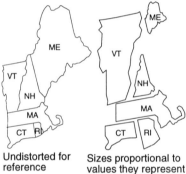
Undistorted for reference   Sizes proportional to values they represent

• Another type of distorted map keeps the shape of the areas unchanged and simply varies the size of the entities in proportion to the values they represent. This generally requires that there be no shared boundaries between the areas (i.e., a noncontiguous map), as shown at the left. Even though the entities are not contiguous, their relative locations are maintained as near as possible. The retaining of their original shapes and relative locations generally makes it easier for the viewer to recognize the entities.

• The distorting of the entities allows one variable to be encoded into a map. By means of coloring, shading, stepping, or a number of other techniques, additional variables can be encoded. For example, the size of the states on the distorted map at the right might be proportional to the total value of the available market. The degree of shading might be proportional to average household income for each state.

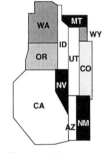
One variable encoded by distorting and a second by coloring or shading

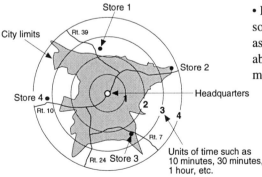
Area distorted to indicate points of equal driving time from a central location taking into consideration types of roads, terrain, traffic, congestion, etc.

• Distorted maps are sometimes classified as cartograms, abstract, or thematic maps.

• Although distorted maps generally have data encoded by making the size of the areas shown on the map proportional to the values represented, other variations are possible. One example is a map in which the distance between various locations on the map are proportional to the time it takes to travel from one point to the other. For example, two stores might be the same number of miles from their headquarters but because one has a major highway running to it while the other has only two-lane mountain roads, the times to get to the two of them might be significantly different. A distorted map similar to the one above might take into consideration such things as the type of roads, terrain, traffic, congestion, etc., so the viewer can read directly from the map how much time it will take to get to the various locations.

Stepped three-dimensional map

Sometimes referred to as stepped relief, prism, or block map. In a stepped statistical map, areas are elevated in proportion to the values they represent. For example, the height of the various states may be proportional to the average grain produced per acre in each of the states. The higher the grain produced per acre, the taller the prism representing that state. Because of the difficulty of accurately estimating actual values, this type of map is generally more qualitative than quantitative.

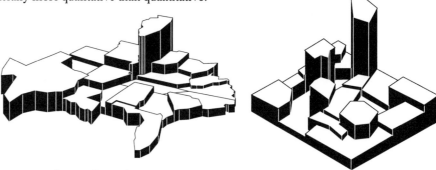

Areas on the map have the same shapes as the areas they represent

Areas on the map are distorted and do not have the same shapes as the areas they represent

Stepped statistical maps with actual and distorted area shapes

Each area (prism) of a stepped three-dimensional map might have its unique height (class-less) based on its actual value, or class intervals might be established and discrete heights assigned to each class interval. • The actual shape of the area being represented is generally shown; however, the areas can be distorted to encode a second variable. Examples of distorted and undistorted stepped maps are shown above. A legend is normally required to aid the viewer in decoding the map. Grid lines, tick marks, and scales are seldom used.

Smooth three-dimensional map

A smooth statistical map is one in which the height (sometimes referred to as the Z-axis value) of the surface is proportional to the value of the variable being represented. For example, if latitude and longitude are shown on the X and Y axes, the level of pollutants associated with each combination of latitude and longitude might be shown by the height of the surface as shown at the right. The higher the level of pollutants, the higher the level of the surface. Normally only data that is continuous, with smooth transitions from one point to another or one area to another, are displayed on this type of map. Atmospheric pressure or temperature are examples. When data or information changes abruptly from one area to another, such as the tax rates in individual states, a stepped statistical map is often used when a three-dimensional effect is desired.

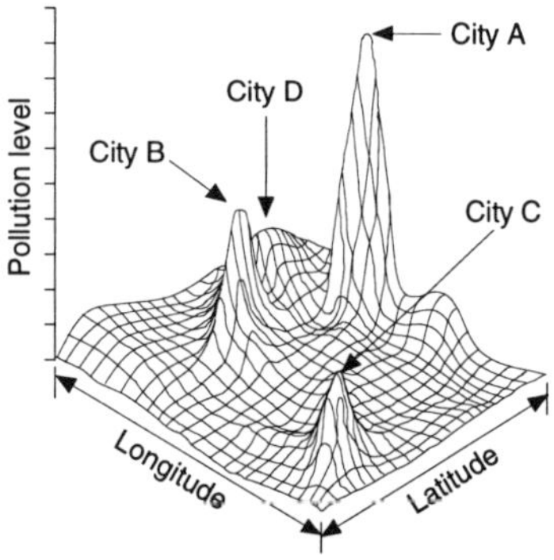

Smooth statistical map using smooth lines to indicate levels of pollution at various locations

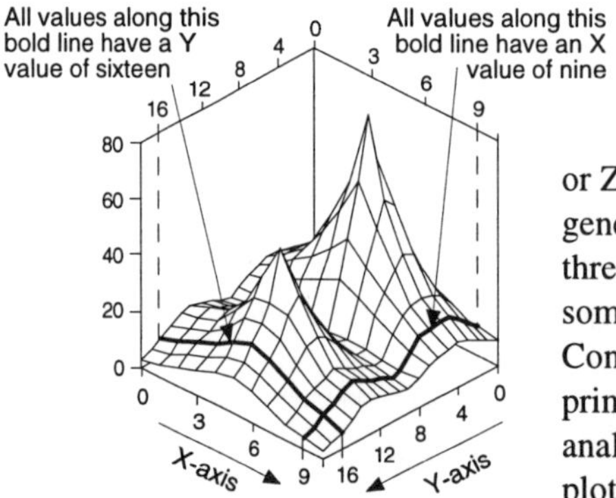

All values along this bold line have a Y value of sixteen

All values along this bold line have an X value of nine

Straight lines have been used to connect equal X and Y values.

• The individual line segments connecting data points can be straight or curved. The lines may represent points of equal value along the X, Y, or Z axes or may simply be drawn to illustrate general trends in the values. In order to display the three axes, the maps are shown tilted, which sometimes makes quantitative analysis difficult. Consequently this type of map is many times used primarily for qualitative analysis. For example, to analyze the acidity of the soil in a region, one might plot distances on the X and Y axes and the acidity values on the Z-axis. The areas where the acidity is the highest and lowest will be immediately apparent. Precise levels can then be obtained from the tabulated data or a series of two-dimensional profile maps. The angles of rotation and tilt of the map can be varied to analyze or present the information more clearly. See Three-Dimensional Map.

Statistical Map (continued)

Methods for encoding a single variable

The following table summarizes methods for encoding a single variable into statistical maps by area. Many can also be used to encode information for specific points.

Description	Information shown in legend	Example of High Value	Example of Mid Value	Example of Low Value
Numeric value only - Used alone or with additional data graphics.	Notation indicating units of measure	857	421	98
Pattern - Sometimes referred to as hatched, crosshatched, or textured. Can include variations in color and shades. Each pattern represents a different class interval.	Table matching pattern with class interval			
Pattern in circle - Same as above, except pattern is enclosed in a circle so the viewer is not visually biased by the relative sizes of the areas to which the data applies.	Table matching pattern with class interval			
Shaded or colored - Sometimes referred to as graduated color or shade. Can be used with a class interval type map or an unclassed map with the gradation of color or shading proportional to the value represented.	Table matching color or shade with class interval or scale relating color or shade and values.			
Dot or dot density - Each dot might represent an absolute number such as 100 people or a ratio such as 100 people per square mile. Spacing between dots is arbitrary. Dots can be enclosed in a square or circle. The more dots there are, the higher the value represented.	Notation indicating how many units each dot represents (i.e., 1 dot = 1,000 cars)			
Distributed dots - Same as above, except distribution of the dots throughout the area gives the viewer some idea as to where the items represented are located.	Notation indicating how many units each dot represents (i.e., 1 dot = 1,000 cars)			
Circles, filled or unfilled - Sometimes referred to as proportional symbols. The diameter or area of the circle is proportional to the value it represents. Shading, color or pattern, if used, can be the same or vary to convey additional information. Map can have class intervals or be unclassed. Other geometric shapes are sometimes used.	Scale relating diameter or area of circle to value represented, plus note explaining use of shading			
Column - Sometimes referred to as block. The height of the column is proportional to value represented. The column might be two- or three-dimensional.	Scale relating height of column to value represented			
Framed rectangle - Degree to which the rectangle is filled is proportional to the value represented. For example if an empty rectangle represents 0 and a full rectangle 400, a one-fourth full rectangle represents 100, one-half full 200, three-fourths full 300, etc.	Examples indicating how much a full and empty rectangle represent			
Pie/Sector - The size of the filled segment is proportional to the value represented. For example, if an empty circle represents 0 and a full circle 400, a one-fourth segment represents 100. Can also be used to designate percent from 0 to 100%.	Examples indicating how much a full and empty circle represent			
Sphere/volume - Volume of data graphic is proportional to value represented. Other shapes can be used, and the same principle applies. Sometimes used when there are large differences in values.	Scale relating volume of object to value represented			

365

Methods for encoding multiple variables

Occasionally two or more variables are displayed on the same map. Four methods sometimes used for this purpose are shown below. The problem of overlapping symbols sometimes becomes severe in these types of maps. Any of these methods can be used to encode data be area (e.g., country, state, etc.) or specific point (e.g., city, intersection, etc.).

Description	Area A	Area B	Area C
Pie chart - In a map application, pie charts are used in the same way as in any other application. All the same size pies can be used, or the size can vary from area to area in response to some variable.	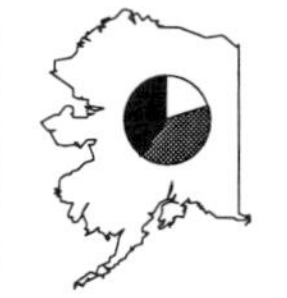		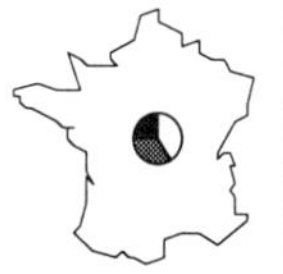
Icons - Stars, as shown here, are one of several icons used to encode multiple quantitative variables. A different star is assigned to each location on the map. With star icons, the length of each spoke is proportional to a different variable. The location of a given variable is the same on every star on a particular map, so one can easily compare a particular variable in several different locations. See Icon.			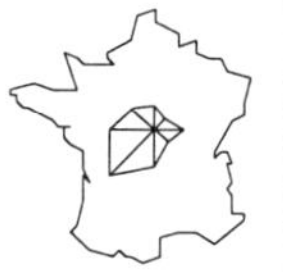
Bar/Column graphs - Bars or columns might represent different variables or the same variable over time. If there is enough space, scales can be included with each graph. If not, a legend is required. Other types of graphs such as line and area can be used in place of the bar or column graph.			
Multi point - There are several variations of this type of data graphic. The horizontal line might represent the average or median. The length of the bar might represent such things as quartiles or standard deviations, and the length of the lines might represent the total spread of values. Values are generally shown beside the symbol.	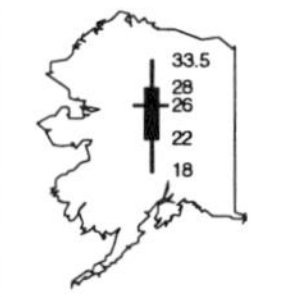		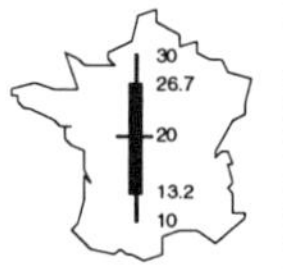

Graphic methods for displaying multiple quantitative variables on maps

Methods for displaying data that changes over time

Techniques for displaying changes in data over time are shown below. To make these graphics most meaningful, a fair amount of space is required. Because of this, they frequently are placed in the area outside of the map. In some cases icons are used instead of individual graphs. See Icon.

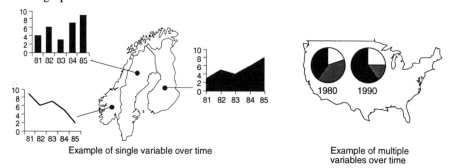

Example of single variable over time

Example of multiple variables over time

Examples of how changes over time can be incorporated into statistical maps

Inclusion of numeric values

Data graphics used on maps tend to give a rough approximation of the values they represent. When a higher degree of preciseness is desired, the actual numeric values are sometimes included with the data graphics as shown below.

Examples of how numeric values can be included with the graphics, in addition to explanatory information shown in the legend.

Statistical Map (continued)

Methods of establishing class intervals for statistical maps

When class intervals are used with statistical maps, there are several methods for choosing the sizes of the intervals, depending on the nature of the data and the purpose of the map. For example, if the data is equally distributed and the map is meant for general reference, intervals of equal value (steps) might be selected. On the other hand, if the data has a normal distribution and those entities that deviate from the norm are to be highlighted, intervals based on standard deviations might be employed. The ranges for class intervals are generally shown adjacent to the map. The method used to determine the intervals may or may not be noted. Examples of some of the most widely used techniques for determining class intervals are shown below.

Equal size intervals

The sizes of the intervals are equal to the total range of the data, divided by however many intervals are desired. For example, if the range from the smallest value in the set to the largest value is 100 units and five intervals is decided on, each interval will be 20 units.

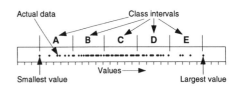

Quantiles

This process provides that each interval contains an equal number of observations. Any quantile can be used (e.g., 0.1, 0.2, 0.25). Whichever quantile is selected is then used for all intervals on the map. 0.2 (20%) is used in the example at the right.

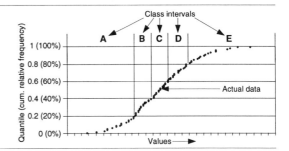

Natural breaks

In some instances a histogram of the data indicates clear-cut natural dips or breaks where the data can logically be separated. In other cases, a dip in the histogram has no significance and, therefore, dips alone cannot be used as the only determining factor.

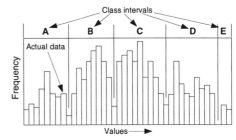

Standard deviations

This method generally is applicable only when the data approximates a normal distribution. It is most useful when one of the major purposes of the map is to note areas that deviate from the norm. Fractional standard deviations can be used for greater detail. There are several options for grouping the standard deviations into intervals.

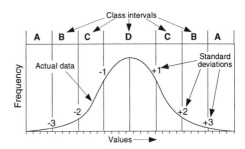

Arithmetic or geometric progression

The nature of some data is such that if a reasonable number of equal intervals were used, the majority of the observations would fall into just one or two of the intervals. Systematic progressions can help alleviate this problem.

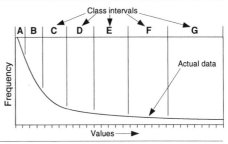

The selection of class intervals is an important and difficult step in the preparation of a statistical map. Because of the complexities and differences of opinion associated with the selection process, this becomes a key element in the debate of a class interval map versus an unclassed or classless map where class intervals are not used.

367

Statistical Quality Control Charts and Graphs

Shown below is a list of the charts and graphs most widely used for statistically analyzing, monitoring, and controlling quality. Although most of the charts and graphs were developed for use in high-volume manufacturing processes, they are now also being used extensively in many other areas. See the individual headings for more detail.

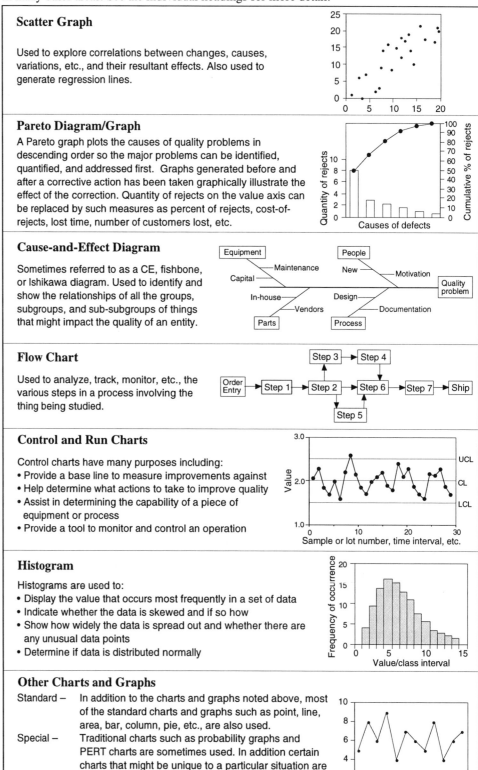

Scatter Graph

Used to explore correlations between changes, causes, variations, etc., and their resultant effects. Also used to generate regression lines.

Pareto Diagram/Graph

A Pareto graph plots the causes of quality problems in descending order so the major problems can be identified, quantified, and addressed first. Graphs generated before and after a corrective action has been taken graphically illustrate the effect of the correction. Quantity of rejects on the value axis can be replaced by such measures as percent of rejects, cost-of-rejects, lost time, number of customers lost, etc.

Cause-and-Effect Diagram

Sometimes referred to as a CE, fishbone, or Ishikawa diagram. Used to identify and show the relationships of all the groups, subgroups, and sub-subgroups of things that might impact the quality of an entity.

Flow Chart

Used to analyze, track, monitor, etc., the various steps in a process involving the thing being studied.

Control and Run Charts

Control charts have many purposes including:
• Provide a base line to measure improvements against
• Help determine what actions to take to improve quality
• Assist in determining the capability of a piece of equipment or process
• Provide a tool to monitor and control an operation

Histogram

Histograms are used to:
• Display the value that occurs most frequently in a set of data
• Indicate whether the data is skewed and if so how
• Show how widely the data is spread out and whether there are any unusual data points
• Determine if data is distributed normally

Other Charts and Graphs

Standard – In addition to the charts and graphs noted above, most of the standard charts and graphs such as point, line, area, bar, column, pie, etc., are also used.

Special – Traditional charts such as probability graphs and PERT charts are sometimes used. In addition certain charts that might be unique to a particular situation are occasionally used such as decision matrixes, relationship diagrams, and tree diagrams.

Major charts and graphs used in statistical quality control

Statistical Table

Sometimes referred to as a summary or analytical table. This type of table is used primarily for analyzing information as opposed to referencing, scheduling, etc. There are always two or more variables and the data in the body is normally numeric. Information is generally arranged, ranked, sorted, etc.; to make such things as relationships, trends, comparisons, distributions, and anomalies stand out. See Table.

Stem and Leaf Chart

A chart that displays the distribution of data elements in a set of data similar to a histogram or dot array. By displaying a data set in a stem and leaf arrangement, one can observe such things as the spread of the data, the location of the mode, whether the data is skewed and if so, in what direction, whether there are gaps in the data and whether there are any unusual data points. (example at right). A stem and leaf chart is constructed on two sides of a vertical line.

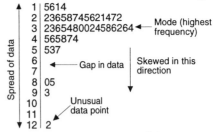

```
Spread of data
        1 | 5614
        2 | 23658745621472
        3 | 2365480024586264  ← Mode (highest frequency)
        4 | 565874
        5 | 537
        6 |     ← Gap in data      Skewed in this direction
        7 |
        8 | 05
        9 | 3
       10 |     Unusual data point
       11 |
       12 | 2
```
Example of a stem and leaf chart

The common first or leading digit (referred to as stem) of a group of numbers is placed on the left side of the line. The other digits (referred to as leaves or trailing digits) are placed

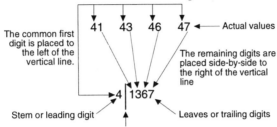

The common first digit is placed to the left of the vertical line.

41 43 46 47 ← Actual values

The remaining digits are placed side-by-side to the right of the vertical line

4 | 1367

Stem or leading digit

Vertical line separating the stems and leaves

Leaves or trailing digits

An illustration of how stem and leaf charts are constructed

on the right of the vertical line. For example, with the numbers 41, 43, 46, and 47, the common first or leading digit, four, is placed on the left of the line and the remaining digits 1, 3, 6 and 7 are placed side-by-side on the right. An illustration as to how the digits are placed is shown at the left.

Comparison with other data distribution methods

An advantage of the stem and leaf chart over simply aligning the data elements themselves is the large reduction in the number of digits that have to be written. An advantage of the stem and leaf chart when compared to histograms, tally charts, and dot arrays is that the identity of each number is retained in the stem and leaf chart. These advantages are illustrated in the table on the next page which shows a number of variations possible with stem and leaf charts.

Ordering of leaves

```
3 | 232
4 | 5687459862
5 | 452536587458565
6 | 02136018405
7 | 4520202
8 | 4502
9 | 98
```
Example of a stem and leaf chart with the leaves unordered

To the left is an example of a typical stem and leaf chart. The digits to the right of the vertical line are in a random order, typical of the way they appear when unordered raw data is used to make the chart. On the right is a stem and leaf chart of the same data, except that the leaves have been ordered by ascending values.

```
3 | 223
4 | 2455667889
5 | 234455555566788
6 | 00011234568
7 | 0022245
8 | 0245
9 | 89
```
Example of a stem and leaf chart with the leaves ordered

By ordering the information, observations can sometimes be made that would otherwise be difficult. For instance, in this example it might be noted that the leaves in the second row tend to be five or larger, while in the fourth row the reverse is true. Also, in the third row, there is an abundance of fives.

Back-to-back stem and leaf charts

When making comparisons between two data sets, stem and leaf charts can be placed back-to-back similar to a paired bar graph. An example is shown at the right.

```
          8 | 1 |
       5682 | 2 | 21
     546235 | 3 | 5874
 3654128901 | 4 | 15483252
       3264 | 5 | 57845689213
   12032014 | 6 | 6541225
        512 | 7 | 356
         38 | 8 | 84
          7 | 9 | 7
```
Example of a back-to-back stem and leaf chart comparing two data sets

The use of a grid when generating stem and leaf charts

When there are many values and they are hand-written, an accumulation of varying widths of numbers can give misleading indications, since a row with lots of narrow numbers (such as ones) might appear shorter than a row of wider numbers, even though they have the same number of leaves. One way this is addressed is to use a preprinted grid similar to the one shown at the right. By this means, the same width is allocated to each number, and the person generating the chart has an on-going count in each row.

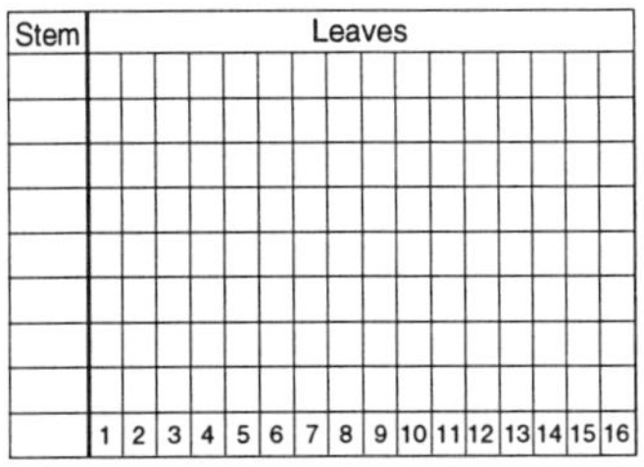

Stem | Leaves
1 2 3 4 5 6 7 8 9 10 11 12 13 14 15 16

Example of a grid sometimes used for generating stem and leaf charts

Stem and Leaf Chart (continued)

Description	Actual values arranged in rows (arrangement is for reference only)	Stem and leaf chart	Dot array chart for reference
Basic stem and leaf chart This example shows one of the most straightforward variations of the stem and leaf chart. In order to keep the number of digits to the right of the vertical line the same in every case, the lower two groups of numbers have two digits to the left of the vertical line.	37,38 40,41,41,43 50,51,52,52,53,56,58, 60,60,62,63,63,64,66,67,67,68, 71,71,73,74,74,76,78,78,79 82,82,83,85,86,88,89 93,95,96,96,98 101,106 115	3 | 78 4 | 0113 5 | 0122368 6 | 0023346778 7 | 113446889 8 | 2235689 9 | 35668 10 | 16 11 | 5	30 - 39　•• 40 - 49　•••• 50 - 59　••••••• 60 - 69　•••••••••• 70 - 79　••••••••• 80 - 89　••••••• 90 - 99　••••• 100 - 109　•• 110 - 119　•
Values with few first digits When the data elements have just a few different first digits (three, in this example, 6, 7, and 8), multiple stems are sometimes generated for the same first digit. In this example, one stem includes the last digits from 0 to 4 (e.g., 60 to 64) and the second stem encompasses the last digits from 5 to 9 (e.g., 65 to 69).	60,62,63 65,66,67,67,68, 70,71,71,73,74,74 75,75,76,77,77,78,78,79 81,82,82,83,84,84 86,88,89	6 (0-4) | 023 6 (5-9) | 56778 7 (0-4) | 011344 7 (5-9) | 556778889 8 (0-4) | 122344 8 (5-9) | 689	60 - 64　••• 65 - 69　••••• 70 - 74　•••••• 75 - 79　••••••••• 80 - 84　•••••• 85 - 89　•••
Values with a single first digit In the case where all data elements have the same first digit, five or ten stems are sometimes used. In this example ten stems are used, one for each second digit. Note that, just as in a histogram, even though an interval has no data elements in it (in this example, numbers 77 and 78), spaces are provided in the chart.	70 71,71,71 72,72,72,72,72,72,72 73,73,73,73 74,74,74 75 76 79	7 | 0 7 | 111 7 | 2222222 7 | 3333 7 | 444 7 | 5 7 | 6 7 | 7 | 7 | 9	70 | • 71 | ••• 72 | ••••••• 73 | •••• 74 | •••• 75 | • 76 | • 77 | 78 | 79 | •
Values with three digits With three digit values the last two digits are sometimes entered to the right of the vertical line, instead of the single digit as above. When this is done a comma is generally placed between the pairs of digits.	489 515,537,572 620,641,654,672,679,691 708,714,768,782 845,867	4 | 89 5 | 15,37,72 6 | 20,41,54,72,79,91 7 | 08,14,68,82 8 | 45,67	400 - 499 | • 500 - 599 | ••• 600 - 699 | •••••• 700 - 799 | •••• 800 - 899 | ••
Values with four or more digits When the data elements are large numbers, one or more of the trailing digits are sometimes dropped. In this example the last two are dropped. In most cases this technique has little or no effect of the shape of the chart, but it does detract from the value of such a chart since the viewer can no longer see the actual values.	48,903 49,565;49,734;49,287 50,019;50,130;50,472;50,251;50,984;50,119 51,832;51,428;51,804;51,264 52,517;52,709	48 | 9 49 | 5,7,2 50 | 0,1,4,2,9,1 51 | 8,4,8,2 52 | 5,7	48,000 - 48,999 | • 49,000 - 49,999 | ••• 50,000 - 50,999 | •••••• 51,000 - 51,999 | •••• 52,000 - 52,999 | ••
Values in the form of decimals When the data elements are in the form of decimals, they are sometimes multiplied by 10, 100, 1000, etc., to get the data elements into a form that is usable in a stem and leaf chart. In this example the data elements were multiplied by 10,000.	.0013,.0015 .0020,.0021,.0021,.0024 .0032,.0033,.0033,.0036,.0037,.0039 .0040,.0042,.0042,.0046 .0052,.0053,.0057 .0063,.0068 .0075	1 | 35 2 | 0114 3 | 233679 4 | 0226 5 | 237 6 | 38 7 | 5	.0010 - .0019 | •• .0020 - .0029 | •••• .0030 - .0039 | •••••• .0040 - .0049 | •••• .0050 - .0059 | ••• .0060 - .0069 | •• .0070 - .0079 | •
Few values for each first digit or a very wide spread Sometimes the sparsity or spread of data elements necessitates the use of class intervals. In this example two first digits are combined into a single stem at each level. Because of this, the entire number must be shown on the right of the vertical line.	04,13 22,26 44,46,51 62,64,65,70,71,76 81,88,90	0 & 1 | 04,13 2 & 3 | 22,26 4 & 5 | 44,46,51 6 & 7 | 62,64,65,70,71,76 8 & 9 | 81,88,90	0 - 19 | •• 20 - 39 | •• 40 - 59 | ••• 60 - 79 | •••••• 80 - 99 | •••

Step-by-Step Chart	There are several time and activity bar charts that are sometimes referred to as step-by-step charts. Gantt and milestone charts are two examples. In step-by-step charts, a major project or program that has a distinct beginning and end is subdivided into a series of subprojects. Each of these subprojects is represented by a bar on a chart that generally has time on the horizontal axis and the names of the subprojects on the vertical axis, as shown at the right. Step-by-step charts are often used for planning, monitoring, and controlling programs. See Time and Activity Bar Chart.

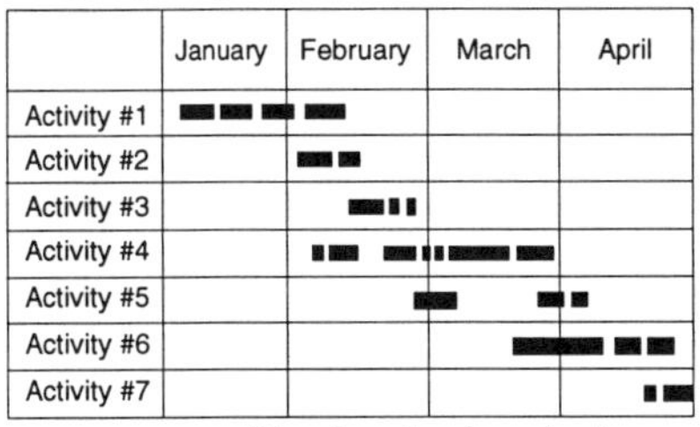

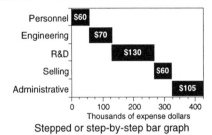
Step-by-step bar chart

Stepped Bar Graph **or** **Step-by-Step Bar Graph** (Also see Stepped Graph)	Sometimes referred to as a staggered or progressive bar graph. A stepped bar graph is the equivalent of a stacked bar graph with only one bar and with each individual segment of that bar displaced vertically from its adjacent segment(s). The segments can be displaced up or down. The purpose of stepped bar graphs is to add visual emphasis to the individual segments while still maintaining the concept that the segments all add up to the whole. See Bar Graph.

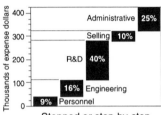
Stepped or step-by-step bar graph

Stepped Column Graph **or** **Step-by-Step Column Graph** (Also see Stepped Graph)	Sometimes referred to as a staggered or progressive column graph. A stepped column graph is the equivalent of a stacked column graph with only one column and with each individual segment of that column displaced horizontally from its adjacent segment(s). The segments, starting from the bottom, are typically displaced to the right. The purpose of stepped column graphs is to add visual emphasis to the individual segments while still maintaining the concept that the segments all add up to the whole. See Column Graph.

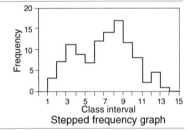
Stepped or step-by-step column graph

Stepped Frequency Graph	Typically a histogram is made up of a series columns. Occasionally the lines between the columns are eliminated in which case the figure is sometimes referred to as a stepped frequency graph, as shown at the right. Functionally a stepped frequency graph is the same as a histogram.

Stepped frequency graph

Stepped Graph	Graphs in which the data graphic forms a stair-step appearance are sometimes called stepped graphs. Examples of the four major graph types are shown below. See Area Graph, Bar Graph, Column Graph, and Line Graph.

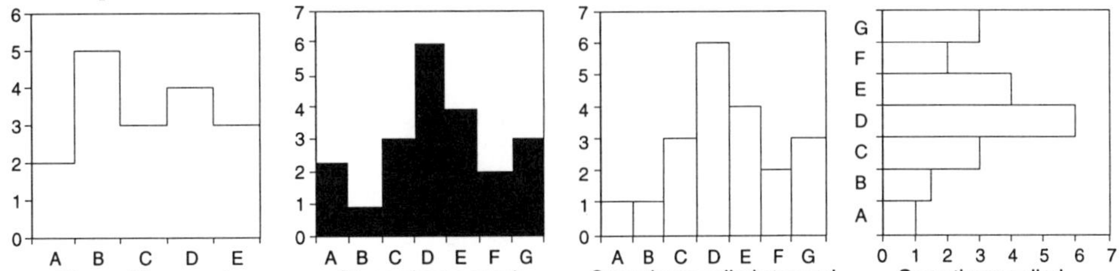

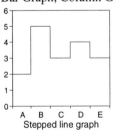
Stepped line graph

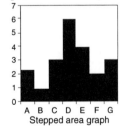
Stepped area graph

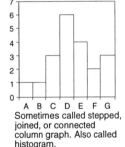
Sometimes called stepped, joined, or connected column graph. Also called histogram.

Sometimes called stepped, joined, or connected bar graph

Examples of stepped graphs

Stepped Map **or** **Stepped Relief Map**	Sometimes referred to as a prism or block map. In this type of statistical map, areas are elevated in proportion to the values they represent. For example, the height of the various states might be proportional to the grain produced per acre in each of the states. The higher the grain produced per acre, the taller the prism representing that state. See Statistical Map.

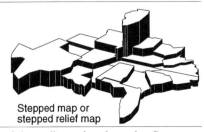
Stepped map or stepped relief map

Stereogram	A term occasionally used to refer to certain variations of three-dimensional graphs. See Three-Dimensional Graph.

Strata Graph[1] or Stratum Graph	Sometimes called a layer area, multiple-strata, stacked area, divided, subdivided area, or subdivided surface graph. A stratum graph is an area graph in which multiple data series are stacked on top of one another. This is the area graph format that is most frequently used. See Area Graph.	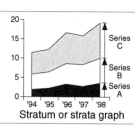 Stratum or strata graph

Strata Graph[2]	Sometimes called a slice graph, spectral plot, ridge contour, layer graph, or partition plot. See Slice Graph.

Strip Chart	A long narrow chart generally used to record data automatically over time. The paper (sometimes film) used for strip charts typically is supplied on rolls ranging from a few inches to a foot or more in width. Time is measured along the length of the chart.

Stripe Graph

Sometimes referred to as a density stripe graph. A stripe graph is a variation of a one-axis distribution graph. The major purpose of this type of graph is to show graphically how the data elements of a data set are distributed. This is accomplished by having a separate symbol represent each data element of a data set, as shown below. From density stripe

Example of a stripe graph displaying the distribution of data points in a single data set.

graphs one might observe such things as the largest and smallest data elements in a set, whether the points are clustered and if so where, whether the points are skewed towards one end or the other, whether there are many or few data elements in the set, and whether there are any points at the extremes that look like they don't belong with the set. The widths of the stripes in a given graph are uniform and typically have no significance. Overlapping of stripes can be misleading since what appears to be a single line may actually be multiple lines representing two or more data elements. Additional information such as average, median, etc., are sometimes superimposed over stripes in the form of symbols, columns, bars, or boxes. Stripes can be used to display the distribution of a single set of data or to compare the distribution of multiple data sets, as shown at the right.

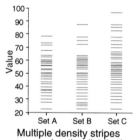

Multiple density stripes comparing the distribution of data elements in multiple data sets

Strip Map

Strip maps are sometimes classified as diagrammatic maps, cartograms, or abstract maps. They are frequently drawn in a long narrow format. They might run vertically or horizontally and are sometimes as much as several feet long. Exceptionally long strip maps are often segmented and bound. They are generally not to scale, although distances many times are noted. The main focus is generally on the sequential relationship of the items shown, as opposed to directional or geographic relationships. For example, the purpose might be to relate highway exit numbers with route numbers and their relative locations with regards to cities. Selected information that might orient the viewer is frequently included. Shown below is an example of a highway strip map showing exit markers.

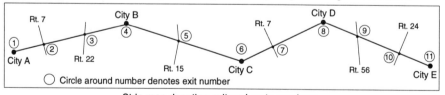

Strip map denoting exit and route numbers

Stub[1] or Stub Mark	Sometimes called tick mark, tick, tic, or scale point. See Tick Mark.

Stub[2] or Stub Item	Sometimes referred to as a row heading, row label, category, line category, or caption. Stubs are the descriptive words in the first column of a table. Occasionally the entire first column is referred to as a stub and the words at the top of the column as the stub heading.	The text at the top of the first column is sometimes called the stub heading. These descriptive words are sometimes called stubs or stub items. The entire first column with the descriptive text is occasionally referred to as a stub.	Type car	1990	1991	1992 | Model A	19	18	15 | Model B	2	5	6 | Model C	5	6.5	7 | Model D	8	10	9 | Model E	13	17	12	

| Subdivided Area Graph or Subdivided Surface Graph | Sometimes called a layer area, multiple-strata, stacked area, strata, stratum, or divided area graph. A subdivided area graph has multiple data series stacked on top of one another. This is the area graph format that is most frequently used. See Area Graph. |

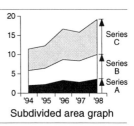

Subdivided area graph

Subdivided Bar Graph

Sometimes referred to as a divided, composite, extended, segmented, or stacked bar graph. A subdivided bar graph has multiple data series stacked end-to-end. This results in the far right end of each bar representing the sum or total of all of the components contained in that bar. Subdivided bar graphs are used to show how a larger entity is divided into its various components and the relative effect that each component has on the whole. See Bar Graph.

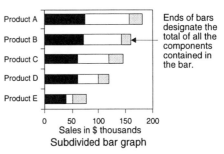

Ends of bars designate the total of all the components contained in the bar.

Subdivided bar graph

Subdivided Column Graph

Sometimes referred to as a divided, composite, extended, segmented, or stacked column graph. A subdivided column graph has multiple data series stacked on top of one another. This results in the top of each column representing the sum or total of all of the components shown in that column. Such graphs are generally used to show how a larger entity is divided into its various components, the relative effect that each component has on the total entity, and how the sizes of the components and the total change over time. See Column Graph.

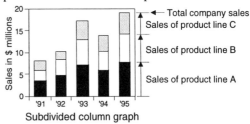

Subdivided column graph

Summary Diagram

Sometimes referred to as a conceptual or explanatory diagram. Summary diagrams are typically made up of geometric shapes, lines, arrows, sketches, and text to aid in the visualization and understanding of relationships. The key purpose of summary diagrams is to provide a graphical overview of how people, things, ideas, influences, actions, etc., interrelate. These diagrams can effectively condense large amounts of information into a single chart; therefore, they are sometimes used as the starting point for the study or discussion of both simple and complex subjects and ideas. See Conceptual Diagram.

An example of a simple summary diagram

Summary Table

Sometimes referred to as a statistical or analytical table. This type of table is used primarily for analyzing information, as opposed to referencing, scheduling, etc. There are always two or more variables, and the data in the body is normally numeric. Information is generally arranged, ranked, or sorted to make such things as relationships, trends, comparisons, distributions, and anomalies stand out.

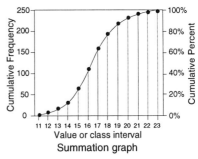

Summary table for analyzing information

Summation Graph

Sometimes referred to as a cumulative frequency graph or ogive graph. A summation graph is the equivalent of a cumulative histogram with the columns replaced by a curve. It is a line graph with values or class intervals plotted on the horizontal axis and the cumulative number of times each of those values or groups of values occurs (cumulative frequency) plotted on the vertical axis. The frequency might be plotted in terms of cumulative counts, cumulative percentages, or both. Summation graphs are used to graphically display the probability of a given value or group of values occurring, to determine whether a data set has a normal distribution, and to compare the distributions of multiple data sets. See Cumulative Frequency Graph.

Summation graph

Sunflower Symbol	When multiple data points with the same coordinates are plotted on a scatter graph, it is impossible for the viewer to know how many data points are present unless some additional method is used to encode that information into the data graphic. One such method is to use what is called a sunflower symbol. A sunflower symbol is a dot with short, radiating lines called petals. The number of radiating lines corresponds to the number of data points the symbol represents. Two radiating lines means there are two data points represented. Three radiating lines means three data points, etc. If only one point is represented there are no radiating lines.

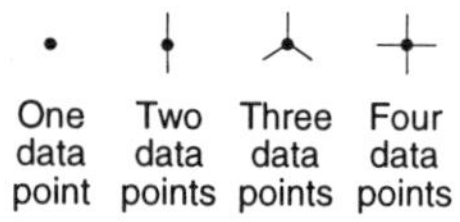

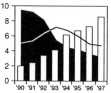

One data point / Two data points / Three data points / Four data points

Examples of sunflower symbols for representing multiple data points plotted on identical or almost identical coordinates.

Superimposed Charts and Graphs

Sometimes referred to as overlaid charts or graphs. A chart that has multiple groups of data laid on top of one another. The groups of data may or may not use the same types of data graphics. The graph at the right has three data series superimposed on top of one another. • A statistical map is a superimposed chart with quantitative information superimposed on a base map. • The opposite of

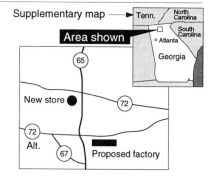

Three data series superimposed over one another

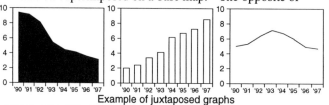

Example of juxtaposed graphs

superimposed is juxtaposed where the various groups of information are placed adjacent to one another, as shown at the left.

Supplementary Map

Sometimes referred to as an inset or ancillary map. If a viewer is not familiar with the area shown on a detailed map, a small supplementary map may be used to help with orientation. For example, the larger map at the right shows a hypothetical area in northern Georgia. If shown just the detailed map, most people would not know what state it was located in, whether it was in the north, south, east, or west portion of the state, and what was close to it that they might recognize. A supplementary map, as shown in the upper right corner of the example, supplies answers to most of these questions.

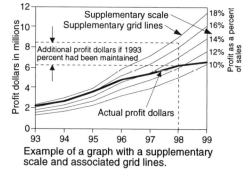

Supplementary Scale/Grid

The grid lines for a supplementary scale are sometimes a series of curves that are extended as data becomes available. For instance, in the example below, the scale on the supple-mentary grid lines is in terms of profit as a percent of sales. Consequently, the locations of the supplementary grid lines can not be determined until the sales and profit figures are available. • A supplementary scale provides much of the same information that would be available from a conventional graph (example below) but, in addition, lets

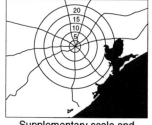

Example of a graph with a supplementary scale and associated grid lines.

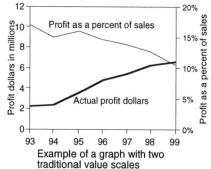

Example of a graph with two traditional value scales

the viewer analyze the data in ways not possible with a pair of conventional scales. For instance, in the example above it can be seen that the profit was 17% in 1993. By means of the supplementary scale, one can see that if the 1993 profit level of 17% had been maintained in 1998, instead of dropping to 12.5%, the company would have made about two million dollars more profit in 1998. Supplementary scales on graphs are frequently in terms of ratios or percents. • On maps, supplementary scales are sometimes included to enable the viewer to estimate the distances from a specific point to any other location within the scope of the supplementary scale. An example is shown at the right.

Supplementary scale and grid lines on a map

Supply and Demand Graphs

There are many different supply and demand graphs used to describe how various aspects of a market economy function and interrelate. One of the most basic graphs (shown at right) has both a supply curve and a demand curve plotted on it. The demand curve illustrates the principle that, with all other things being equal, the number of people willing to buy a product or service (demand) is inversely proportional to its price. For example, as the price goes up, the quantity sold goes down. The supply curve illustrates the principle that, with all other things being equal, the number of companies willing to supply a product or service is directly proportional to its selling price. For example, if the price customers are willing to pay for a product or service goes up, more companies will want to supply more of it. The point at which the two curves intersect is called the equilibrium point which defines the equilibrium price and quantity. For example, at any price above the equilibrium price there is more product supplied than there are people willing to buy it and the price drops to the equilibrium point.

Basic supply and demand graph

Supply and demand graphs used to depict changing conditions

As conditions change, the basic graph is modified to reflect those changes. One example at the right illustrates a change in demand while supply stays the same, and the other example illustrates a change in supply while demand stays the same. In reality, a change in either aspect generally results in a change in the other; consequently, a new curve is generated for both, resulting in a new equilibrium point.

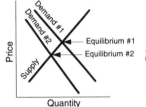

A market change such as new fashions or government regulations sometimes generates a new demand curve and a new equilibrium point/selling price.

A producer change such as a major new supplier or a plant destroyed by fire sometimes generates a new supply curve and a new equilibrium point/ selling price.

Supply and demand graphs used to depict outside influences

Supply and demand graphs are also used to illustrate the effects of actions or conditions that sometimes impact a market economy system. Examples are price ceilings and price support as shown at the left.

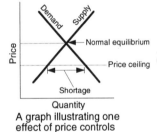

A graph illustrating one effect of price controls

A graph illustrating one effect of price support

Supply and demand graphs used to depict market sensitivity to price

Based on the slope and location of the supply and demand curves, certain observations or predictions can be made. For example, the more vertical the curve is, the less elastic it is said to be and the less effect changes in price have on volume. Conversely, the more horizontal the curve is (more elastic), the greater the impact that price changes have on volume.

The slope of the curve indicates how elastic it is. The steeper it is, the less effect changes in price have on volume.

Supply and demand graphs used to depict things other than price and quantity

In addition to being used to analyze relationships of price and quantity, supply and demand graphs are used to study and describe other relationships. Two of these are shown at the right.

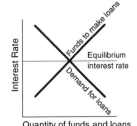

Interest rates and their effect on loans and availability of funds

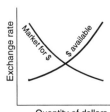

Supply & demand for dollars in foreign exchange

375

Surface Graph[1]

Surface graphs are three-dimensional wireframe graphs in which the areas between the wires (lines) are opaque. When color or shading is added to the opaque areas, the charts are sometimes called filled surface graphs. In a number of situations, wireframes are generated with the sole purpose of making a surface graph. For example, instead of looking at a large quantity of three-dimensional data in the form of a column graph, a wireframe might be generated by connecting the tops of adjacent columns and the areas between the lines made opaque to form a surface graph. In applications such as this, a surface graph sometimes makes the high points and patterns of data stand out more clearly than does a three-dimensional column graph. In the case of three-dimensional equations, wireframes and surface graphs are sometimes the only way to get a graphical representation. Examples showing the relationship between line and column graphs, wireframes, and surface graphs are shown below. See Three-Dimensional Graph.

	Line graph changed to surface graph	Column graph changed to surface graph	Surface graph generated from equation
Conventional graph for reference With many three-dimensional equations there is no conventional counterpart.	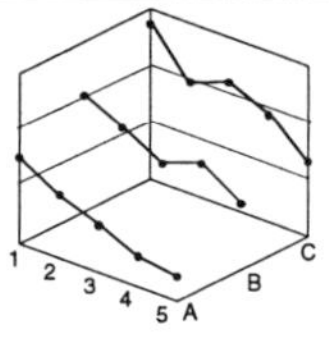 Line graph	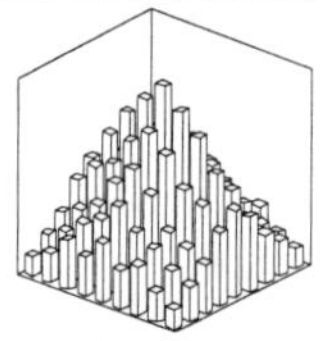 Column graph	In the case of three-dimensional equations, there often are no conventional counterparts to wire frame or surface graphs.
Wireframe graph When a finite number of data elements are plotted, all adjacent data points are connected with lines. With equations and certain continuous data, uniformly spaced isolines are frequently drawn.	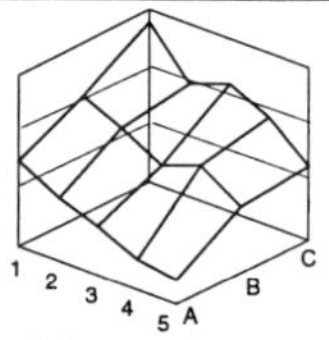 Adjacent data points connected	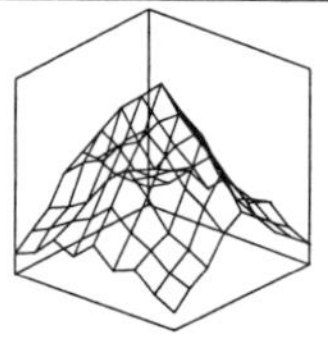 Tops of all adjacent columns connected	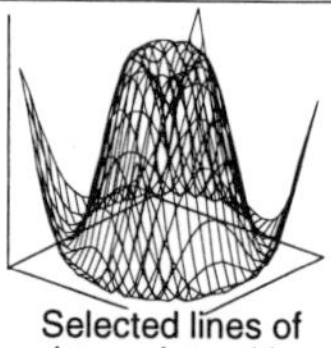 Selected lines of data points with equal values
Surface graph In the simplest form of surface graph, the areas between the lines are made opaque, forming a continuous surface. Sometimes called fishnet or lined surface graph.	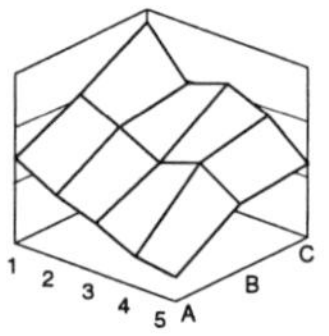	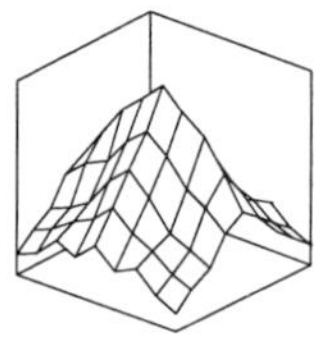	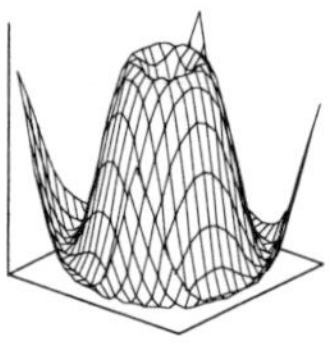
Filled surface graph When shading or color is added to the surfaces, the graph is many times called a filled surface graph. The fill might be uniform or varied to encode quantitative information.	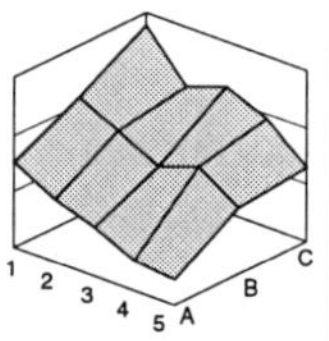	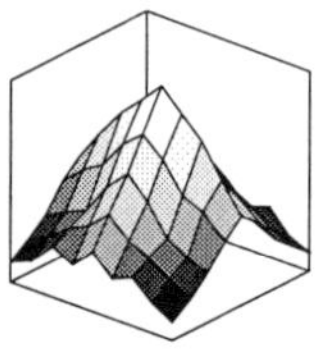	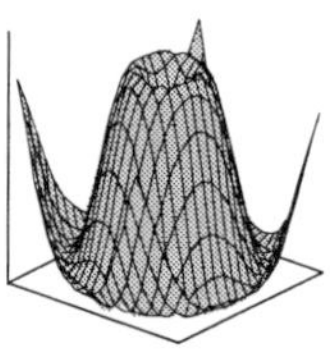

Equally spaced lines connecting data points of equal value

When specific data points are available for plotting, as in the line and column graph examples above, the construction of a wireframe is straight forward. In certain situations where continuous data is graphed, if all the possible data points were plotted and connected the data graphic would appear as a solid silhouette and thus yield little useful information. The way this problem is generally solved is to plot only data points along equally spaced lines of equal values, sometimes called isolines. The two major options are lines of equal X and Y values (called fishnet) or lines of equal Z values (called contour). The spacing between lines is optional. Determining the optimum distances between the lines sometimes requires some trial and error. Examples with various distances between isolines are shown at the right.

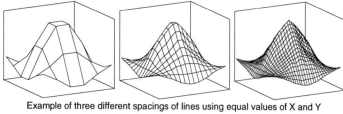

Example of three different spacings of lines using equal values of X and Y

Example of three different spacings of lines using equal values of Z (contour lines)

Examples of three different spacings. Generally speaking, the greater the spacing the more distortion of the data graphic.

Surface Graph [1] (continued)

Equally spaced lines connecting data points of equal value (continued)

Sometimes it is difficult to envision the equal value lines, as shown in the examples on the previous page. The following two draftsman's displays provide top and side views of two additional graphs as further illustrations of the concept.

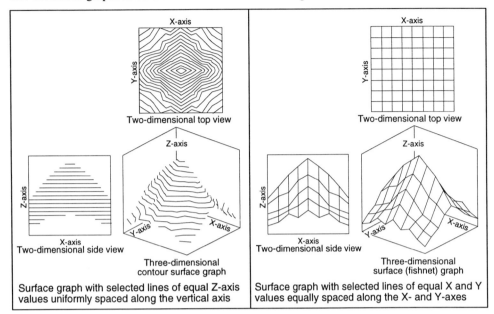

Diagonal lines added to form a mesh pattern

In addition to the lines of equal value along the X and Y axes, diagonal lines are occasionally added to give a denser pattern, sometimes call a mesh. An example is shown at the right. The same process can be applied to wire frame graphs and surface graphs with contour lines. The number and location of the additional diagonal lines are optional.

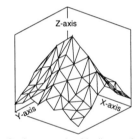

Surface graph with diagonals added between isolines

Filled surface graph

When color or shading is added to the areas between the isolines, the figure is frequently referred to as a filled surface graph. Examples are shown below with top and side views for illustrative purposes. The boundaries between the filled areas represent the lines of equal value. Actual lines may or may not be included between the shaded areas. Legends are many times used to enable the viewer to decode the values (fill) on the graph.

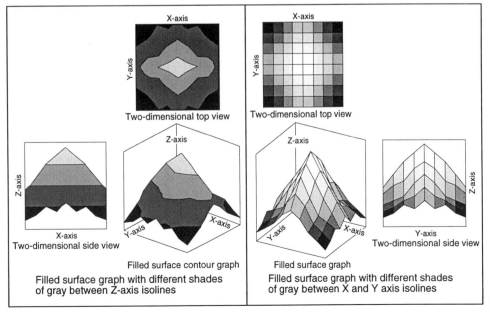

Filled surface graph (continued)

Legend

With some filled surface graphs, particularly contour types, a legend is frequently used to indicate the values associated with the various colors or shades. In cases where the isolines are clearly marked, a legend may not be necessary. Shown at the right are two ways legends are constructed for filled contour graphs. When color or shading is used for nonquantitative purposes, a legend may not be required.

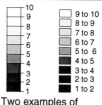

Two examples of legends relating color or shading and values

Fill without isolines

On some surface graphs, where the configuration is not complex and all that is required is a general feel for the shape, isolines are eliminated and shading is used to show the general outline as shown below.

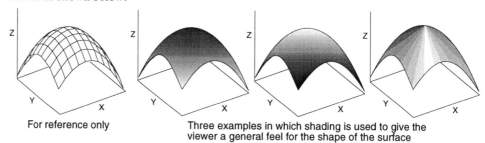

For reference only

Three examples in which shading is used to give the viewer a general feel for the shape of the surface

Combination of filled surface and wire frame

On complex surfaces it is sometimes advantageous to use a combination of fill and wireframe in order to obtain additional information about inner surfaces of the data graphic (center below). In other cases an entire section of the filled surface is removed (right below) to enable the viewer to see hidden surfaces.

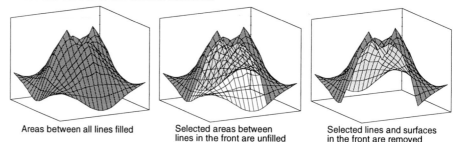

Areas between all lines filled

Selected areas between lines in the front are unfilled

Selected lines and surfaces in the front are removed

Illustration of how selected portions of the surface can be removed or left unfilled to provide better visibility of hidden lines and surfaces

Multiple surfaces on the same graph

Most surface graphs have a single surface displayed. On occasion, multiple surfaces are plotted on the same graph. The example on the right shows multiple surfaces in a layered format. On such a graph, all of the surfaces can be generated using the XY plane as a reference from which each of the surfaces is measured (grouped surface graph) or they might be arranged in such a way that the lower surface is measured from the XY plane and the other surfaces referenced off of the surfaces immediately below them (stacked surface graph). In this second variation, the top surface represents the total of all the surfaces shown on the graph. • On the left is an example of multiple planes that intersect. This type of graph is sometimes used to determine the line along which the two planes have points in common. On such graphs there might be two or more surfaces. The surfaces might be flat or curved and might run any direction.

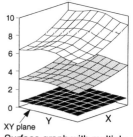

Surface graph with multiple nonintersecting surfaces

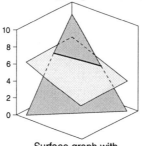

Surface graph with intersecting surfaces

Scales

The scale on the vertical axis is almost always quantitative. The scales on the other two axes might be quantitative, categorical, or sequential. When equations are plotted or the data is entered electronically, all three axes normally have quantitative scales.

Surface Graph²

The term surface graph is sometimes used to refer to an area graph. See Area Graph.

Symbol

Symbols are typically small graphic representations of things, either tangible or intangible. The same symbol might serve two or more functions at the same time in the same chart, e.g., location and type of facility. In other cases, the same symbol might be used for one function on one chart and for a completely different function on another chart. Symbols serve a variety of functions. Seven of the major functions are illustrated below.

Major functions of symbols

Convey quantitative information

In graphs, the sizes, shapes, and colors of the plot symbols are sometimes varied in proportion to the values they represent. In maps and tables, the size of the symbols are sometimes varied to denote different values, or in some cases, the number of symbols shown is proportional to the value represented. In still other situations, different symbols are assigned different values or the width of a line is proportional to the value represented.

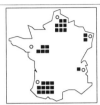

Number of symbols proportional to quantity

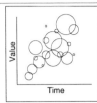

Size of symbols proportional to value

Convey descriptive information

Descriptive symbols are used extensively on maps. Such symbols not only indicate where something is located but frequently also convey considerable additional information such as what it is, what it might be used for, how big it is, etc. In tables, symbols are sometimes used to minimize text, overcome language differences, denote availability of various services, etc.

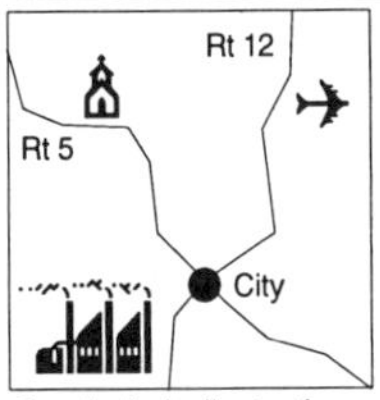

Symbols indicate the type of facility

	55	61	28
	42	27	72
Total	97	88	100

Symbols used as row and column headings

Designate location

Almost any type of symbol can be used for this purpose. Small geometric symbols called plotting symbols are most frequently used to locate data points on graphs. On maps, small geometric shapes are frequently used to denote specific locations. Lines are used to designate the locations of highways, paths, railroads, etc., and polygons are most widely used to designate the boundaries of areas. On diagrams enclosures are used to locate things and events with regards to one another.

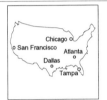

Map with symbols locating cities

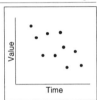

Graph with symbols locating data points

Differentiate, identify, etc.

When multiple data series are plotted on the same graph, it is important to identify all the data points that belong to the same series and to differentiate them from other data series. Different symbols are typically used for this purpose. In diagrams, different shapes of enclosures are sometimes selected so that various types of events or activities such as decisions, government approvals, etc., can be easily identified.

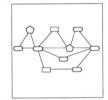

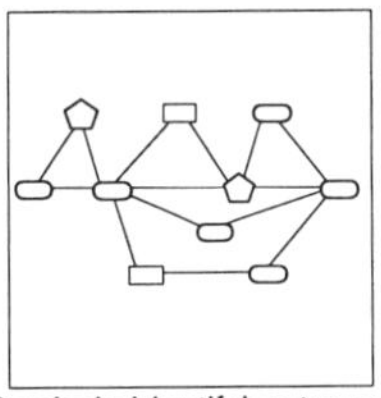

Symbols identifying types of events in a diagram

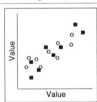

Graph with symbols identifying data series

Serve as an enclosure

Enclosures around text on organization charts, flow charts, diagrams, etc., tend to organize the information, make the chart more legible and easier to understand, and improve the appearance in general. The use of enclosures also makes the termination of connecting lines crisper, allowing patterns in the material being diagrammed to stand out better. The values derived from the use of enclosures are realized whether one or multiple shapes are used.

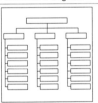

Organization chart with symbols as enclosures

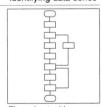

Flow chart with symbols as enclosures

Highlight specific information

One of the simplest uses of symbols is to highlight information on charts (sometimes called bullets). When used on text charts, symbols indicate to the viewer that a new idea, thought, bit of data, etc., is beginning. When different types of symbols are used they might help to differentiate between major points and subordinate points. Symbols such as arrows are occasionally used on all types of charts to call attention to a particular location, a specific value, a change in direction, etc.

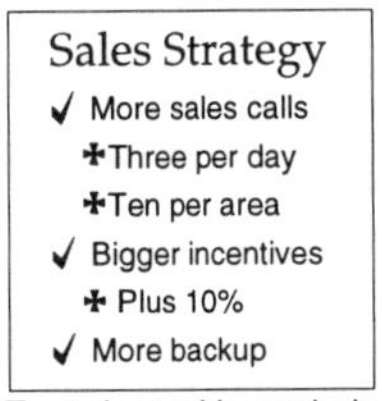

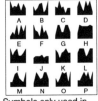

Text chart with symbols highlighting key points

	Experiment		
	Case 1	Case 2	Case 3
Plan A	125	100	75
Plan B	150	120	90
Plan C	75	60	45
Plan D	50	40	30
Plan E	100	80	60
Plan F	0	0	0

Table with symbols highlighting key data

Form meaningful displays by themselves

In some cases the symbols by themselves are used to convey meaningful information. In the example of the icon comparison display shown at the right, many bits of information are encoded into each symbol (icon). In the example of the three-dimensional array at the far right, a single bit of information is communicated using many symbols (cubes).

Symbols only used in icon comparison display

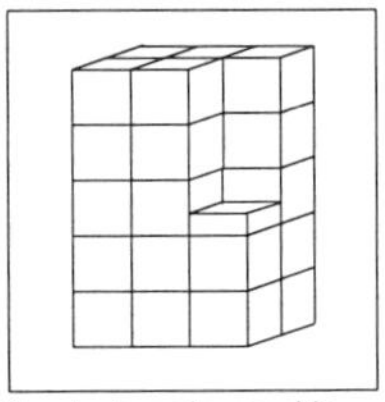

Symbols only used in three-dimensional array

Symbols for encoding descriptive and quantitative information

The table below illustrates how relatively standard forms of symbols can be used to represent quantitative and descriptive information. In most of the examples, the symbol represents a single bit of information. With each basic form (i.e., literal, point, line, area, and volume) a row of examples is included to show how multiple bits of information might be incorporated into that type of symbol. Some symbols are applicable to only one type of information. In most cases the same symbol can be used with several or all types of data.

Type of symbol / Classification of information		Descriptive information		Quantitative information	
		Nominal (qualitative)	**Ordinal** (Ranked nonquantitatively e.g., first, second, third, etc.)	**Interval** (quantity, value)	**Proportional** (e.g., ratio, relative, etc.)
Literal (e.g., words, numbers, etc.)	Standard	library, coal mine, plant	L, M, S — large, medium, small	10 ton / 20 passengers / 50 births	2.7 tons per acre / 2,080 passenger-miles / 1.2 births per family
	Multiple bits of information encoded into symbol	**Portland** = In the west / *New York* = In the east / <u>Tulsa</u> = In the south	**NEW YORK** = Largest / Portland = Medium / TULSA = Smallest	Portland = 100k to 500k / New York = 50k to 100k / Tulsa = .5k to 1k	**2.7** = current year / 2.1 = previous year / 1.2 = 5yr. average
Point (geometric or pictorial)	Size	• unincorporated city / ● incorporated city	• = minor / ● = average / ● = major	• = $100 Sales / ● = $200 Sales / ● = $300 Sales	• = X$ purchases per household / ● = Y$ purchases per household / ● = Z$ purchases per household
	Shape	✈ airport / ⛪ church / △ has feature / ☐ tin deposit	▲ = third place / ● = second place / ■ = first place	▲ = 10 to 99 employees / ● = 100 to 499 employees / ■ = 500 to 999 employees	▲ = 1/2 depleted / ● = 3/4 depleted / ■ = fully depleted
	Multiple	NA - if used in multiples they become quantitative	• = least / •••• = mid-value / •••••••••• = most	• = 100 / •• = 200 / ••••• = 500	size of A / B twice size of A / C four times size of A
	Multiple bits of information encoded into symbol	Bridge / Bridge closed / ● Existing O Proposed	✈ minor airport / ✈ average airport / ✈ major airport	62% / 38% / • = 100 kilograms / • = 200 kilograms / O = 300 kilograms	size of A / B twice size of A / C three times size of A / O seasoned ● not seasoned
Line	Thickness (weight)	original contours / natural changes / man-made changes	unimproved road / light duty road / primary highway	10 messages / 100 messages / 1000 messages	2 messages per hour / 4 messages per hour / 8 messages per hour
	Solid or dashed	property boundary / township boundary / county boundary	highest elevation / middle elevation / lowest elevation	300 ft. elevation / 200 ft. elevation / 100 ft. elevation	10 X elevation / 5 X elevation / 1 X elevation
	Patterned	wall / stream / power line	most curves / medium / least curves	100 hertz / 200 hertz / 300 hertz	100% / 200% / 300%
	Multiple	walking trail / bike tail / walking & bike trail	narrowest road / average width road / widest road	10 ton capacity / 20 ton capacity / 30 ton capacity	1 ton per axle / 2 tons per axle / 3 tons per axle
	Symbols included	—•——•— fence line / truck route	densest / medium / sparsest	|| or ▮ = width of road in meters	3 X base / 2 X base / base
	Multiple bits of information encoded into symbol	railroad / abandoned railroad	steepest incline / intermediate incline / no incline	50 to West by truck / 60 to East by air	speed of A / 2 X speed of A / 3 X speed of A
Area and volume (includes geometric and nongeometric shapes)	Size	children's park / adult's park	smallest / medium / largest	500 1,000 2,000	all products / product A / product B
	Shape	foot print / plot / oil reserves	#1 rank #2 rank #3 rank	10 10² (10^2) 10³ (10^3)	50% complete / 75% complete / 100% complete
	Multiple bits of information encoded into symbol	patterns designate features	by class, category, feature, etc.	500 1,000 2,000	by type, category, characteristic, etc.

Examples of symbols used to represent quantitative and qualitative information

Symbol (continued)

Variations possible with a single type of symbol

The previous chart gives an indication of the breadth of relatively standard symbols that are available. The chart below illustrates the depth that is possible within a single type of symbol. The circular symbol is used since it is the most versatile and widely used of all the symbols. The variations depicted here can also be used with other types of symbols.

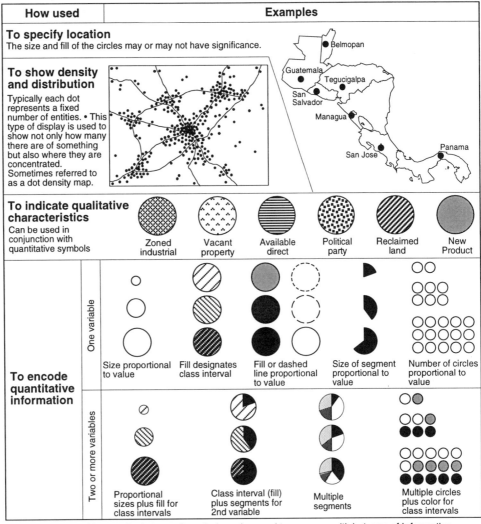

Examples of how a single type of symbol can be used to convey multiple types of information

Legends used with circles that encode quantitative information

When symbols are used to encode quantitative information, a scale or legend is normally required. The examples at the right illustrate some of the more widely used methods of constructing such a scale or legend. • Studies on circles have shown that viewers tend to underestimate the relative sizes of larger circles. In order to compensate for this problem, techniques have been developed to systematically enlarge the circles in proportion to their diameters, i.e., the larger the circle, the greater the enlargement. Sometimes the unadjusted circles are referred to as having an absolute scale and the modified circles as having an apparent-magnitude scale. There is not universal agreement as to which scale should be used. The series of circles on the right side of the adjacent illustration were drawn using the modified technique. It can be seen how the difference between the modified and unmodified circles becomes greater as the circles get larger. Modified circles are used primarily on maps.

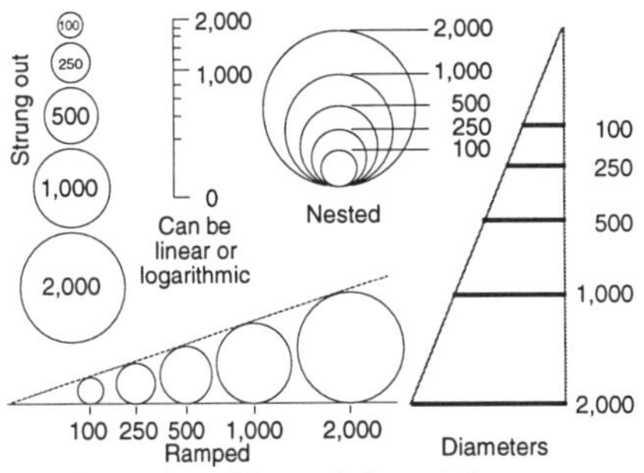

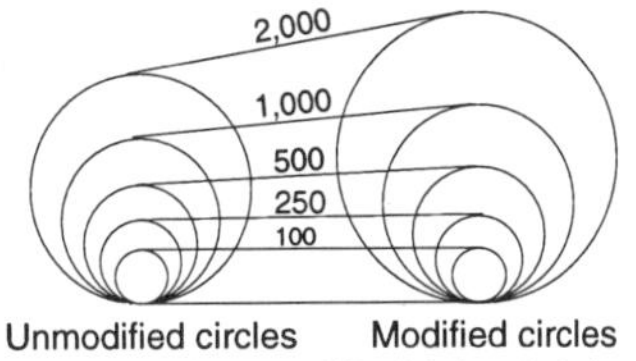

Examples of legends for relating circle size to the value it represents

Unmodified circles   Modified circles
Comparison of unmodified (absolute) and modified (apparent-magnitude) circles

General variations

The following options are available with a wide variety of symbols. The various techniques might be used to encode additional information, improve the legibility of the chart, or to improve the appearance of a chart.

Option	Examples
Opacity Solids stand out. Transparency sometimes improves readability when data points overlap.	Solid / Transparent (St. Louis)
Orientation Symbols might be oriented in any of 360° either for appearance or to encode specific information.	Above, upward, etc. / Below, downward, etc. / Increasing or positive / Decreasing or negative
View The symbol might appear as though one is looking down on the top of an object or at the side of the object. For improved appearance, an oblique or perspective view is occasionally used.	Factory / Tree — Plan or top view / Factory / Tree — Profile or side view
Two- or three-dimensional Either two- or three-dimensional symbols might be used. Two-dimensional symbols are the most common.	Two-dimensional / Three-dimensional
Color, shading, fill, etc.	Used extensively for encoding many types of information on both a quantitative and qualitative basis (e.g., size, magnitude, similarities, condition, status, proposed, etc.). Color, shading, and/or patterns can be used with most types of symbols.

Framed rectangle symbol

Even though circles are widely used to encode quantitative information, there is a general concern about the accuracy of decoding such information. The framed rectangle symbol (described below) offers an alternative that improves the accuracy of decoding.

– The symbol consists of a rectangle that is taller than it is wide.
– There are tick marks on both sides that generally designate a distance half way up the rectangle.
– The symbol can be any size. Typically all rectangles on the same chart are the same size. Exceptions are when the width of the rectangle is used to encode an additional variable.
– When the rectangle is empty (all white), a lower value is indicated. When the rectangle is full (all black), an upper value is indicated. When it is half full (the lower half black), a value midway between the lower and upper is indicated. The same reasoning applies to any fractional fill of the rectangle.
– The upper and lower values can be any numbers, actual or percent. They generally are explained in a legend.
– A framed rectangle is used primarily to designate the values of a single variable. It is particularly useful for encoding quantitative information onto maps.

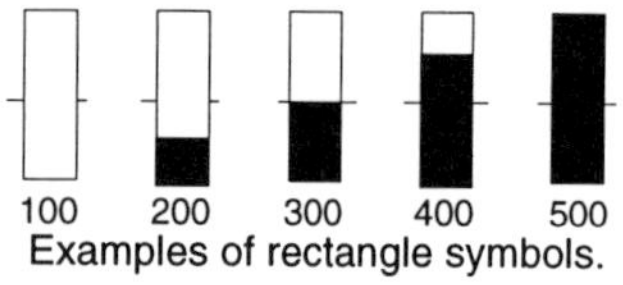

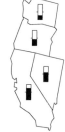

100 200 300 400 500
Examples of rectangle symbols.

Map using framed rectangles

Dingbat symbols

Small symbols, typically entered by means of a keystroke, are many times called dingbats. In specific applications they are sometimes called bullets, plot symbols, markers, etc. Typical applications of this type of symbol are designating data points on graphs, differentiating one data series from another, emphasizing points on a text chart, locating cities on maps, organizing information, or simply improving the appearance of a chart.

A group of conventional dingbats that are frequently available on computers is shown at the right. Examples of some

Examples of conventional dingbats

of the less conventional dingbats that are more and more available with computer software are shown at the left. The symbols that incorporate two geometric shapes (e.g., a circle or an X in a square) or combinations of a single shape (e.g., filled and unfilled, bold and fine line) enable one to convey two or more bits of information with a single symbol.

Examples of less conventional dingbats

Symbols used to encode descriptive information

In addition to other functions that symbols perform, they can also encode descriptive/nonquantitative information. For example, symbols may denote what a facility is, what is done at a location, what type of equipment is involved, whether it is day shift or night shift, etc. The descriptive aspect of symbols is closely related to the function of differentiating and identifying. In addition to literal symbols (e.g., words and numbers), there are three other widely used types of descriptive symbols as shown below.

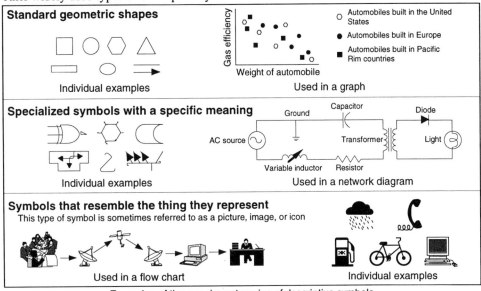

Examples of three major categories of descriptive symbols

Plotting symbols

Plotting symbols (sometimes called plot symbols or markers) are the symbols used to designate the location of data points on graphs. They generally are simple geometric shapes; however, they can be any shape, size, configuration, or color. Two problems that are occasionally encountered when using plotting symbols are, 1. the difficulty of spotting patterns, relationships, and trends when multiple data series are plotted on the same graph and 2. the problem of data points overlapping one another and thus making it difficult or impossible for the viewer to know how many data elements are present at a particular location (set of coordinates).

———**Multiple data series on one graph**———

When multiple data series are plotted on the same graph, a different type of symbol is used for each series. In one graph, a particular combination of symbols might make each of the data series perfectly visible. In another graph, due to the number and location of the data points, a different set of symbols might make it easier to analyze the data. Examples of several combinations of symbols are shown at the right using the same set of data in each example. These examples illustrate how one combination of symbols might make the symbols stand out somewhat better, another might make the overlapping data points most visible, and still another may make it easiest to identify the data series without repeatedly referring to a legend. Sometimes trial and error is required to determine the combination of symbols that best satisfies the objectives of the graph.

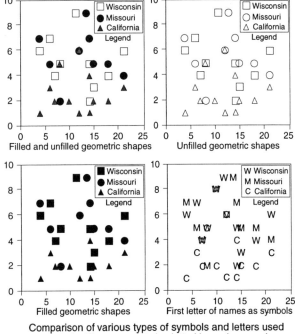

Comparison of various types of symbols and letters used as symbols. The same data is plotted in each example.

Plotting symbols (continued)
——**Overlapping plotting symbols**——

Hidden data points due to overlapping symbols can present a major problem when analyzing point and scatter graphs. Shown below are methods for addressing this problem. The top graph is included for reference to show how the data points appear without any modification.

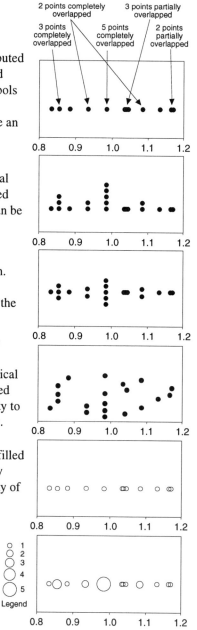

- **Reference** – A point graph with 20 data points distributed along a single axis is used to illustrate the problem and potential solutions. The completely over-lapping symbols make it impossible to know how many data points are represented. The partially overlapping symbols require an estimate as to how many points are represented.

- **Stacking** – In this technique overlapping symbols are stacked on top of one another on one side of their actual location. This method makes the completely overlapped symbols visible. A variation of the stacking process can be used with the partially overlapped symbols.

- **Distributing** – In this technique the symbols are distributed equally on two sides of their actual location. This technique is similar to the stacking method. However, in this method the data points at the ends of the stacks are closer to their original location.

- **Jittering** – With jittering, all data points are randomly shifted by a slight amount in the vertical direction for horizontal plots and in the horizontal direction for vertical plots. The distance points are shifted may be determined visually or mathematically. This method gives visibility to both the completely and partially overlapping symbols.

- **Unfilled circles** – When solid dots are replaced by unfilled circles it gives some increased visibility to the partially overlapping symbols but does not improve the visibility of the completely overlapped symbols.

- **Graduated unfilled circles** provides a partial solution to both the total and partially overlapping problems. With this technique a legend is generally required.

- **Sunflower symbols** – In this method the number of lines radiating from a data point corresponds to the number of overlapping data points, thus solving the completely overlapping problem, but not the partially overlapping situation. A modification of the technique is to group partially overlapping symbols and represent them by a single symbol.

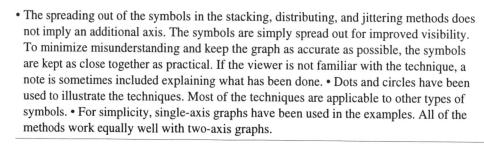

- The spreading out of the symbols in the stacking, distributing, and jittering methods does not imply an additional axis. The symbols are simply spread out for improved visibility. To minimize misunderstanding and keep the graph as accurate as possible, the symbols are kept as close together as practical. If the viewer is not familiar with the technique, a note is sometimes included explaining what has been done. • Dots and circles have been used to illustrate the techniques. Most of the techniques are applicable to other types of symbols. • For simplicity, single-axis graphs have been used in the examples. All of the methods work equally well with two-axis graphs.

Enclosures as symbols

Enclosures around text on diagrams like organization charts, flow charts, tree diagrams, PERT charts, etc., help organize the information, sometimes convey additional information, make it easier easier to read and understand, and improve the appearance. In many charts the enclosures are all the same and have no significance. In other charts, features of the enclosures/symbols are used to encode additional information such as position, skill, type of equipment, function, timing, etc. Shown below are examples.

Methods for encoding additional information into enclosure symbols

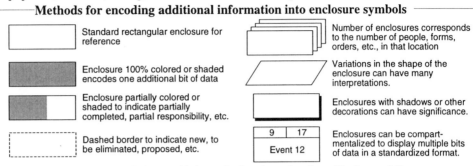

Standard rectangular enclosure for reference
Enclosure 100% colored or shaded encodes one additional bit of data
Enclosure partially colored or shaded to indicate partially completed, partial responsibility, etc.
Dashed border to indicate new, to be eliminated, proposed, etc.
Number of enclosures corresponds to the number of people, forms, orders, etc., in that location
Variations in the shape of the enclosure can have many interpretations.
Enclosures with shadows or other decorations can have significance.
9
17
Event 12
Enclosures can be compart-mentalized to display multiple bits of data in a standardized format.

Each method might be used independently or in conjunction with other variations.

Examples

Less than one year
Position vacant
Green
Pink
White
Invoice
To file
Gold
To customer
To accounts receivable
Pink
To accounting
3
7
Event #2
18
25
Event #5
0
0
Start Event #1
12
12
Event #3
27
27
Event #6
8
15
Event #4
3
5
6
2
12
15
8
12

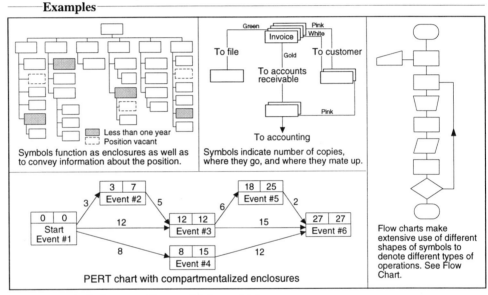

Symbols function as enclosures as well as to convey information about the position.

Symbols indicate number of copies, where they go, and where they mate up.

PERT chart with compartmentalized enclosures

Flow charts make extensive use of different shapes of symbols to denote different types of operations. See Flow Chart.

Unique symbols for encoding multiple variables

Most symbols encode only one or two pieces of information. Some specialized but widely used symbols have been developed that routinely encode three or more bits of data. Two examples of these specialized symbols are shown below.

Wind speed and direction symbol

This symbol is used on meteorological maps to encode multiple bits of information about weather conditions.

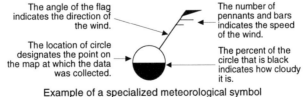

The angle of the flag indicates the direction of the wind.
The location of circle designates the point on the map at which the data was collected.
The number of pennants and bars indicates the speed of the wind.
The percent of the circle that is black indicates how cloudy it is.

Example of a specialized meteorological symbol

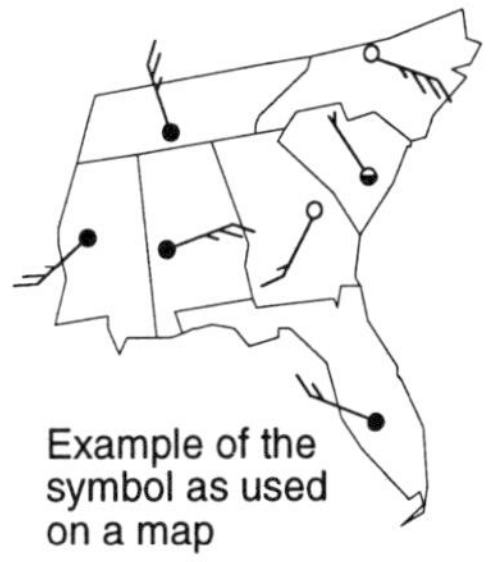

Example of the symbol as used on a map

Stock price symbol

This symbol is used on stock price charts to record various combinations of stock prices for given periods of time.

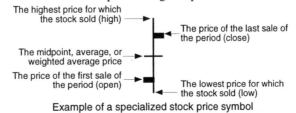

The highest price for which the stock sold (high)
The price of the last sale of the period (close)
The midpoint, average, or weighted average price
The price of the first sale of the period (open)
The lowest price for which the stock sold (low)

Example of a specialized stock price symbol

36
35
34
33
32
31
30
29
28
Stock price
Time

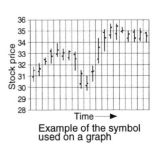

Example of the symbol used on a graph

Symbol (continued)

Icons as symbols

Miniature graphs without titles, labels, tick marks, or grid lines are sometimes used as symbols. In this form they are sometimes referred to as icons (See Icons). As many as 10 to 20 variables can be encoded on this type of symbol. The variables might be quantitative or qualitative. Four examples are shown below.

Icon /Symbol	Names sometimes used to describe	Description
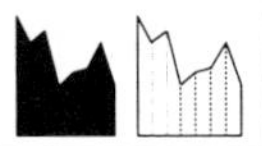	Area or profile	Each point on the curve represents a different variable. The higher the point, the more favorable that characteristic. The greater the area of a symbol, the higher the overall rating of the thing represented.
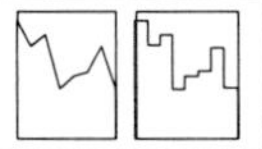	Line	Similar to an area icon except a border is added as a frame of reference for estimating relative values. When a scale of zero to one is used, the bottom of the frame is zero and the top is one.
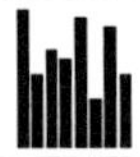	Column or histogram	Each column represents a different variable. The height of the columns are proportional to the values they represent.
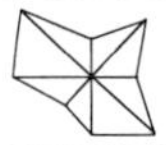	Polygon, star, snowflake, or profile	Each point on the polygon represents a different variable. The length of the spoke leading to the point is proportional to the value it represents.

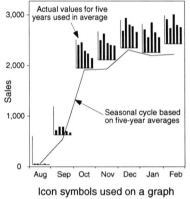

Icon symbols used on a graph

Icon symbols used on a map

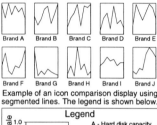

Example of an icon comparison display using segmented lines. The legend is shown below.

Symbols used as a meaningful display by themselves

Legends

Legends are generally used when there are multiple types of symbols, a standard symbol is used in some unusual way, or the viewer is unlikely to be familiar with the symbols.

Symbol Display or Symbolic Display

Sometimes called an icon comparison display. A symbol or symbolic display is a group of icons or symbols assembled for the purpose of simultaneously comparing and/or screening a sizable number of entities with regards to three or more variables. In this type of display a separate icon/symbol is generated for each entity under review. The function of the icons is not to convey specific quantitative information. Instead, they are used to show such things as relative sizes, overall comparisons, trends, patterns, etc. The icon used in the example below is only one of many that are used for this purpose. See Icon Comparison Display.

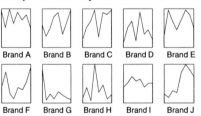

Example of a symbol or symbolic display using a separate icon for each entity being compared. Since there are no scales on the icons, a legend always accompanies such a display.

Symbol Graph

Sometimes referred to as point, scatter, or dot graph. See Point Graph.

Symbol Table

Some tables use symbols to convey information instead of or in addition to words and numbers. Such tables are sometimes called symbol tables. In the example below, symbols are used to indicate whether certain features are available in two different products produced under five different brand labels. See Table.

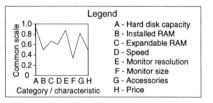

Example of a symbol table

Table

Sometimes referred to as a matrix. Tables are charts with information arranged in rows and columns in some meaningful way. The following list summarizes some of the major reasons for using tables.

- They are one of the best ways to convey exact numerical values.
- They present data more compactly than in sentence form.
- They assist the viewer in making comparisons, determining how things are organized, noting relationships between various sets of data, etc.
- They are one of the most convenient ways of storing of data for rapid reference.
- They are an excellent vehicle for recording and communicating repetitive information (forms).
- They organize information for which graphing would be inappropriate.

Terminology and location of key elements

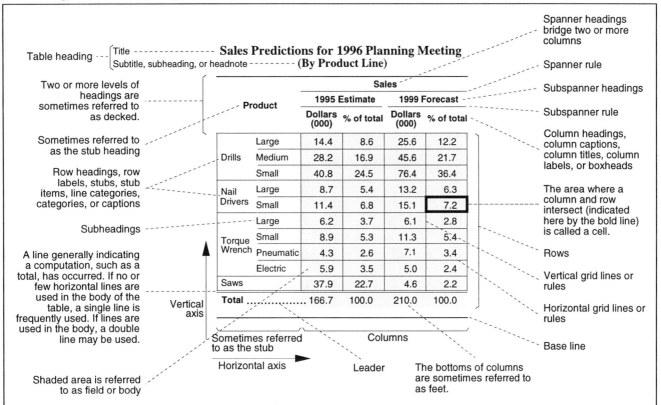

If multiple terms are used to describe the same thing, each of the terms are shown above. • When tables are used in specific applications, parts of the tables are sometimes given additional names related specifically to the application. For example, when used in a database, a column might be called a field and a row referred to as a record.

General

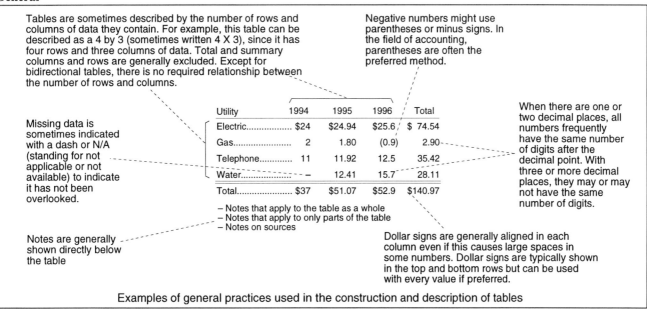

Examples of general practices used in the construction and description of tables

Applications in which tables are used

In many cases the application or purpose of a table influences how the table is designed and how the data in it are organized. For example, when used for reference, material is many times alphabetized or sorted in numerical sequence. When used for analysis, the information is many times ranked in ascending or descending order. The examples below illustrate some of the major ways tables are categorized by application or purpose.

Analysis

Tables used for analysis are sometimes called statistical, summary, or analytical tables. Their major function is to assist the viewer in analyzing the information included in the table. There are always two or more variables and the data in the body is normally numeric. Information is generally arranged, ranked, sorted, etc., to make such things as relationships, trends, comparisons, distributions, and anomalies stand out.

Number of students with given combination of scores

Score on test A	0%	10%	20%	30%	40%	50%	60%	70%	80%	90%	100%	Total
Total	3	8	13	22	21	20	14	10	4	3	1	119
100%										1		1
90%									1	1	1	3
80%						2	4	3	2	1		12
70%				1	3	4	3	4	1			16
60%			1	4	5	6	4	2				22
50%			1	6	6	4	2	1				20
40%	1	2	4	6	3	2	1					19
30%	1	2	4	2	2	1						12
20%	1	3	2	3	2	1						12
10%		1	1									2
0%												

Score on test B

Example of a table for analyzing sizable quantities of data

Reference

This type of table is frequently called a reference or source table. The major emphasis of such tables is on easy retrieval of data. Information is generally arranged in alphabetical order, by geographic area, numerically in ascending or descending order, chronologically, etc., so specific bits of information can be located rapidly.

Comfort Index

Air Temperature (°F)	10	20	30	40	50	60	70	80	90	100
110	105	112	123	137	150					
105	100	105	113	123	135	149				
100	95	99	104	110	120	132	144			
95	90	93	96	101	107	114	124	136		
90	85	87	90	93	96	100	106	113	122	
85	80	82	84	86	88	90	93	97	102	108
80	75	77	78	79	81	82	85	86	88	91
75	70	72	73	74	75	76	77	78	79	80
70	65	66	67	68	69	70	70	71	71	72

Percent Relative Humidity

A table with data arranged for easy access/reference

Scheduling

Tables used for scheduling purposes are many times referred to as charts, such as machine loading charts, Gantt charts, milestone charts, etc. The things being scheduled are typically listed on the vertical axis and the time intervals or increments on the horizontal. Names, numbers, symbols, color, etc., might be used in the body of the table.

Vacation Schedule

Week	1	2	3	4	5	6	7	8	9
Avery, T.		■							
Chandler, A								■	■
Dorsey, J.					■				
Goodman, E.							■	■	
Hewell, P.									
Leahy, F.	■	■							
Mansfield, G.					■	■			
Scott, H.									■

Example of a scheduling chart

Calendar

Calendars represent one of the simplest forms of tables. They have a unique characteristic in that the information shown on them flows from one line to the next the same as text does.

Sun	Mon	Tue	Wed	Thur	Fri	Sat
			1	2	3	4
5	6	7	8	9	10	11
12	13	14	15	16	17	18
19	20	21	22	23	24	25
26	27	28	29	30	31	

Example of a calendar table for a month

Forms

Endless varieties of forms utilize a tabular format.

Ordered by: _____   Ship to: _____

Line Item	Quantity Ordered	Description	Unit Price	Amount

Example of a purchase order form

General

There is a huge number of tables, such as the one at the right, that cannot be easily categorized. They probably constitute the majority of the tables in use today.

	Over-all Rating	Health Factors Value			Housing Value	Social Value
		Method A	Method B	Method C		
Hospital A	High	125	100	75	50	0
Hospital B	High	150	120	90	0	1
Hospital C	Med.	75	60	45	30	6
Hospital D	Med.	50	40	30	25	25
Hospital E	Low	100	80	60	0	0
Hospital F	Low	0	0	0	10	20

Example of one of many types of tables that cannot be easily categorized

Table format

How a table is formatted depends on a number of things including the type of data, the amount of data, the number of variables, the purpose of the table, and how it will be used. Shown below are examples of three commonly used formats.

One-way table

A one-way table is a table displaying two or more variables (sets of data) along the same axis which might be either the horizontal or vertical. The vertical axis is most frequently used and the resulting table is sometimes referred to as the list or vertical format. If only one variable is shown in a single column, it is called a list. Examples of one-way tables with two, three, and four variables are shown on this page.

Name	Age
Alice	6
Bill	7
Harry	4
Sue	5

Two variables along the vertical axis without rules

Name	Age
Alice	6
Bill	7
Harry	4
Sue	5

Two variables along vertical axis with rules

Two variables along horizontal axis with rules

Name	Alice	Bill	Harry	Sue
Age	6	7	4	5

Three variations of one-way tables

Two-way table

Year	Product	Value
1990	A	2
1990	B	10
1990	C	8
1990	D	22
1991	A	4
1991	B	12
1991	C	7
1991	D	29
1992	A	6
1992	B	17
1992	C	6
1992	D	20

A one-way table of the same data shown in the two-way tables at the right

A table displaying three variables, one along each of the two axes and the third in the body of the table, as shown at the right. The source of the data for the two tables at the right is the one-way table at the left. A comparison of the two formats indicates how much easier it is to analyze data in the two-way format than in the one-way.

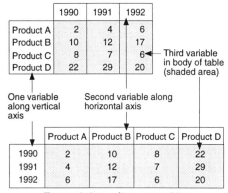

	1990	1991	1992
Product A	2	4	6
Product B	10	12	17
Product C	8	7	6
Product D	22	29	20

Third variable in body of table (shaded area)

One variable along vertical axis / Second variable along horizontal axis

	Product A	Product B	Product C	Product D
1990	2	10	8	22
1991	4	12	7	29
1992	6	17	6	20

Two variations of two-way tables using the same data

Multiway table

A table displaying four or more variables distributed between the two axes and the body of the table. Two examples of three-way tables with four variables are shown at the right. In one example, two variables are displayed along the horizontal axis, one on the vertical axis and the fourth in the body of the table. In the other, two variables are displayed on the vertical axis and one on the horizontal. The data for both three-way tables was taken from the one-way table below. In this type of multi-way table, an additional set of columns or rows is introduced for every additional variable over three.

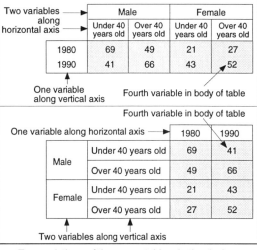

Two variables along horizontal axis

	Male		Female	
	Under 40 years old	Over 40 years old	Under 40 years old	Over 40 years old
1980	69	49	21	27
1990	41	66	43	52

One variable along vertical axis / Fourth variable in body of table

Fourth variable in body of table

One variable along horizontal axis →

		1980	1990
Male	Under 40 years old	69	41
	Over 40 years old	49	66
Female	Under 40 years old	21	43
	Over 40 years old	27	52

Two variables along vertical axis

Two variations of three-way tables. In the design above, an additional set of columns or rows is introduced for every additional variable over three.

Year	Sex	Age	Value
1980	Male	Under 40	69
1980	Male	Over 40	49
1980	Female	Under 40	21
1980	Female	Over 40	27
1990	Male	Under 40	41
1990	Male	Over 40	66
1990	Female	Under 40	43
1990	Female	Over 40	52

A one-way table of the same data displayed in the multiway tables shown at the right

• An alternate multiway format uses the equivalent of a series of two-way tables, as shown at the right. In this design, an additional set of two-way tables is generated for each additional variable over three. In this way, large numbers of variables can be displayed. As the number of variables increases, however, it becomes more and more difficult to spot trends and relationships within the data.

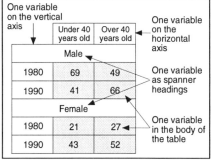

One variable on the vertical axis / One variable on the horizontal axis

	Under 40 years old	Over 40 years old
Male		
1980	69	49
1990	41	66
Female		
1980	21	27
1990	43	52

One variable as spanner headings / One variable in the body of the table

Example of a multiway table design in which an additional two-way table is generated for each variable over three.

Table format (continued)

Full and one-half (bidirectional) matrix

When the same headings are used for both the rows and columns, the table is referred to as matrix or bidirectional table. Such tables are frequently used to indicate distances between locations. In a full bidirectional table, shown on the left below, the data is repeated in the upper and lower triangles. In some cases different units of measure such as miles and kilometers are used in the two halves to enhance the value of the table. In other cases the duplicate values are eliminated, in which case the table is referred to as a one-half bidirectional table (shown on right below).

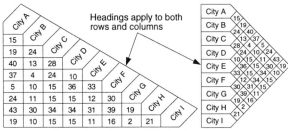

	City A	City B	City C	City D	City E	City F	City G	City H	City I
City A		15	19	40	37	5	24	43	19
City B	15		24	13	4	10	11	30	10
City C	19	24		28	24	15	15	34	15
City D	40	13	28		10	36	15	34	15
City E	37	4	24	10		33	12	31	11
City F	5	10	15	36	33		30	39	16
City G	24	11	15	15	12	30		19	2
City H	43	30	34	34	31	39	19		21
City I	19	10	15	15	11	16	2	21	

Example of full bidirectional table

City A	City B	City C	City D	City E	City F	City G	City H	City I
15								
19	24							
40	13	28						
37	4	24	10					
5	10	15	36	33				
24	11	15	15	12	30			
43	30	34	34	31	39	19		
19	10	15	15	11	16	2	21	

Headings apply to both rows and columns

Examples of one-half bidirectional tables

Irregularly shaped tables

There are no restrictions as to the shape of a table, although most are rectangular. Irregularly shaped tables can sometimes be helpful in highlighting relationships that might otherwise be difficult to spot. This example shows how the estimated sales for a given year vary from yearly plan to yearly plan.

Year	1992	1993	1994	1995	**1996**	1997	1998	1999	2000
1992 Plan	1,942	2,224	2,318	2,467	**2,681**				
1993 Plan		2,375	2,411	2,500	**2,602**	2,657			
1994 Plan			2,309	2,388	**2,411**	2,482	2,512		
1995 Plan				2,014	**2,137**	2,258	2,390	2,471	
1996 Plan					**1,986**	2,054	2,169	2,354	2,466

All sales values in thousands of dollars

Frequency and percent tables

When a table displays how frequently various things occur or how many there are of different things, the table is sometimes referred to as a frequency table. An example is shown at the right. When the counts are replaced by the percents they represent of a whole, the table is sometimes called a percent table. Three different types of percents are sometimes calculated depending on the nature of the data. The values might be calculated as a percent of the totals for the rows, the totals for the columns, or the totals for the entire table. Examples of all three are shown at the left.

	Reject type A	Reject type B	Reject type C	Total rejects	
Product A	2	4	6	12	8%
Product B	10	12	17	39	27%
Product C	8	7	6	21	15%
Product D	22	29	20	71	50%
Total	42	52	49	143	100%

Example of a frequency table noting the number of times various things occurred

Sometimes called marginal values or marginal distribution because the values are shown in the margin of the table.

	Total rejects (N)	Reject type A	Reject type B	Reject type C	Total
Product A	12	17%	33%	50%	100%
Product B	39	26%	31%	43%	100%
Product C	21	38%	33%	29%	100%
Product D	71	31%	41%	28%	100%
Total	143	30%	36%	34%	100%

Numbers indicate what percent each value is of the total for the row

	Total rejects (N)	Reject type A	Reject type B	Reject type C	Total
Product A	12	5%	8%	12%	8%
Product B	39	24%	23%	35%	27%
Product C	21	19%	13%	12%	15%
Product D	71	52%	56%	41%	50%
Total	143	100%	100%	100%	100%

Numbers indicate what percent each value is of the total for the column

When percentages, totals, etc., are shown in the margin just to the right and/or bottom of the table, they are sometimes referred to as marginal values or marginal distributions.

	Total rejects (N)	Reject type A	Reject type B	Reject type C	Total
Product A	12	1%	3%	4%	8%
Product B	39	7%	8%	12%	27%
Product C	21	6%	5%	4%	15%
Product D	71	16%	20%	14%	50%
Total	143	30%	36%	34%	100%

Numbers indicate what percent each value is of the total for the entire table

In percentage tables, an additional column is many times added to designate the total number of items from which the percentages were calculated (many times denoted by the letter N). This column might be added in the body of the table (left) or placed in the margin.

Table format (continued)
————**Three-dimensional tables**————

Although three-dimensional tables can be constructed, they are not widely used, primarily because they are difficult to construct and interpret. Three-dimension tables are sometimes used for their aesthetic value.

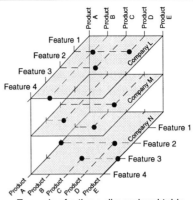

Example of a three-dimensional table

————**Doubling up long tables**————

	1988	1989	1990	1991	1992	1993	1994	1995	1996	1997	1998	1999
Store #1	8	6	5	8	4	7	5	9	8	10	9	8
Store #2	22	25	23	19	24	26	21	23	25	27	20	22
Store #3	11	14	10	9	12	14	13	15	12	10	11	14

Example of a wide and shallow table

When tables become too long for easy analysis or to fit in the space available, they are sometimes divided into two or more smaller tables and placed above or beside one another.

	1988	1989	1990	1991	1992	1993
Store #1	8	6	5	8	4	7
Store #2	22	25	23	19	24	26
Store #3	11	14	10	9	12	14

	1994	1995	1996	1997	1998	1999
Store #1	5	9	8	10	9	8
Store #2	21	23	25	27	20	22
Store #3	13	15	12	10	11	14

Example of same table as above, except the horizontal axis has been divided into two pieces and the vertical heading repeated

Year	Product	Value
1990	A	2
1990	B	10
1990	C	8
1990	D	22
1991	A	4
1991	B	12
1991	C	7
1991	D	29
1992	A	6
1992	B	17
1992	C	6
1992	D	20

Example of a long and narrow table

Year	Product	Value
1990	A	2
1990	B	10
1990	C	8
1990	D	22
1991	A	4
1991	B	12

Year	Product	Value
1991	C	7
1991	D	29
1992	A	6
1992	B	17
1992	C	6
1992	D	20

Example of the same data as shown in the example at the left, except the vertical axis has been divided into two pieces and the horizontal headings repeated

The examples shown here divide (sometimes referred to as double up) the tables into two pieces. If desired, they could have been divided into three, four, or more pieces.

Multiple values in cells

In most tables a single piece of information is included in each cell. There is; however, no limitation on the amount of data that can be included in a given cell and in some cases multiple entries per cell can be extremely useful. For example, if average or median values are shown, it might also be helpful to note the maximum and minimum values. If end-of-the-month membership counts are shown, it might be helpful to also show comparable figures for the previous year or the five-year average. In some cases combining data in cells can avoid generating multiple tables.

	Number of employees	
	Weekdays	Saturday
1st shift	**320** 287 Full-time 33 Part-time	**135** 82 Full-time 53 Part-time
2nd shift	**215** 198 Full-time 17 Part-time	**57** 52 Full-time 5 Part-time
Total	**535** 485 Full-time 50 Part-time	**192** 134 Full-time 58 Part-time

In this example, the total number of employees is shown (in bold type) as well as a breakdown of how many of the total are full-time and part-time.

	Income in thousands of dollars			
	Source 1	Source 2	Source 3	Total
Income A	**22** (34%) (24%)	**25** (39%) (45%)	**17** (27%) (26%)	**64** (100%) (30%)
Income B	**7** (18%) (8%)	**19** (50%) (34%)	**12** (32%) (18%)	**38** (100%) (18%)
Income C	**62** (56%) (68%)	**12** (11%) (21%)	**37** (33%) (56%)	**111**(100%) (52%)
Total	**91** (43%) (100%)	**56** (26%) (100%)	**66** (31%) (100%)	**213** (100%) (100%)

Value — Percent the value represents of total for the row — Percent the value represents of total for the column

In this example, in addition to displaying the actual values in the cells (in bold type), the percents that the values represent of the column and row totals are also included.

Table heading options

Words, numbers, and symbols

Words, numbers, and symbols might be used alone or in conjunction with one another. Examples are shown below.

Words used as headings

Example of a table with words used as row and column headings

	Over-all Rating	Health Factors Value			Housing Value	Social Value
		Method A	Method B	Method C		
Hospital A	295	125	100	75	50	0
Hospital B	305	150	120	90	0	1
Hospital C	179	75	60	45	30	6
Hospital D	245	50	40	30	25	25
Hospital E	263	100	80	60	0	0
Hospital F	223	0	0	0	10	20

Numbers used as headings

Example of a table with numbers used as row and column headings. In some cases, such as this one, the numbers represent quantitative values. In other cases the numbers are used for identification purposes and have no quantitative significance.

Air Temperature (°F) / Heat Index

110	105	112	123	137	150				
105	100	105	113	123	135	149			
100	95	99	104	110	120	132	144		
95	90	93	96	101	107	114	124	136	
90	85	87	90	93	96	100	106	113	122
85	80	82	84	86	88	90	93	97	102	108
80	75	77	78	79	81	82	85	86	88	91
75	70	72	73	74	75	76	77	78	79	80
70	65	66	67	68	69	70	70	71	71	72
	10	20	30	40	50	60	70	80	90	100

Percent Relative Humidity

Symbols used as headings

Example of a table with symbols used as row and column headings. Some symbols are called icons, images, pictures, etc. Symbols sometimes:
- Convey more information in less space
- Are faster for orienting the viewer
- Overcome language barriers

Accompanying text is sometimes required for a full understanding of the meaning of a symbol.

						Total
 (female)	55	61	28	32	15	**191**
 (male)	42	27	72	39	63	**243**
Total	**97**	**88**	**100**	**71**	**78**	**434**

Location of headings

Row headings are most frequently located on the left of the table and column headings at the top of the table. It is not uncommon, however, to see headings located elsewhere, as in the example of the heat index table shown above, in which the column headings are located at the bottom. In some cases when the table is very wide, the row headings are repeated on the right.

Order in which headings appear in tables

- If the headings represent sequential (e.g., time series, order of occurrence, etc.) or quantitative variables, the individual headings are arranged in their proper sequential or numerical order progressing from left to right along the horizontal axis. On the vertical axis they might progress either up or down.
- Headings representing categories can be arranged in any order. Some typical methods of arranging them are:
    - Alphabetical
    - Ascending or descending order ranked by some row or column of data in the body of the table.
    - Chronological, even if it is not a true sequential series. For example, the thing that happened first, second, third, etc.
    - Geographical
    - Qualitatively – for example, the thing assumed to be best, second best, etc.
    - Frequently compared columns and rows placed closet together when possible
    - Related columns and rows located close to one another, for example columns with actual and cumulative values, or a column of values and the column displaying the percents of the whole that each value represents.
    - More important values placed near the top or left side where they are most easily noticed by the viewer.
- Total columns and rows are frequently located at the far right or bottom. Some people consider total columns and rows the most important in the table, and therefore often make them the first row or column after the headings.

Table headings (continued)

Dependent versus independent variable

In tables there is no generally accepted standard as to which axis the independent variable is placed on; therefore, it is sometimes found on the vertical axis and other times on the horizontal axis.

Interchangeability of column and row headings

In many tables the row and column heading can be transposed with no degradation of the information. To a large extent, the decision as to which variable goes on which axis depends on such things as:

 – Which combination makes the data most meaningful
 – The number of variables
 – The number of entries. Those variables with large numbers of entries are often are placed on the vertical axis.
 – The lengths of the headings. Long headings can be accommodated better on the vertical axis.

Examples of the same data arranged on tables with the headings transposed are shown below.

	1990	1991	1992
Product A	2	4	6
Product B	10	12	17
Product C	8	7	6
Product D	22	29	20

Products listed on vertical axis

	Product A	Product B	Product C	Product D
1990	2	10	8	22
1991	4	12	7	29
1992	6	17	6	20

Products listed on horizontal axis

Illustration of how headings can be transposed with no degradation of information

Positioning of headings

Shown here are some of the practices used for positioning row and column headings. Although the same method is generally used throughout a given table, it is not uncommon to see two or more methods used in the same table.

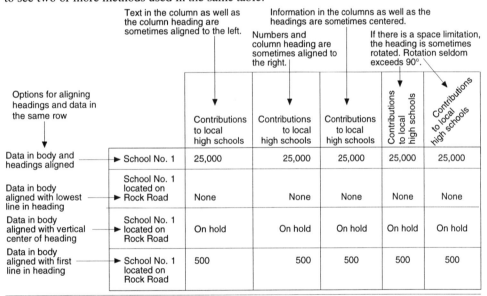

Positioning of material in body of table

Numbers

If the numbers have decimal points, they are almost always aligned on the decimal points. If the column contains no decimal places, the numbers might be aligned to the right or centered. Numbers are almost never aligned to the left except occasionally as scales on graphs.

Centered	Aligned right	Aligned on decimal
4,255	4,255	4,255.1
2,978	2,978	2,978.500
1,029	1,029	1,276.
1,276	1,276	1,029.5
119	119	119.73
127	127	12.4

Words

Words in a column are usually aligned to the left, although they may be centered or aligned to the right. When the column has a mixture of words, numbers, and/or symbols, all of them are often centered.

Aligned left	Centered	Aligned right
Suspended	Suspended	Suspended
Void	Void	Void
Accepted	Accepted	Accepted
Requested	Requested	Requested
Rejected	Rejected	Rejected
N/A	N/A	N/A

Body/field of table

Options for encoding information into the body of tables

There are many options as to how information might be encoded into tables. Shown here are some of the major variations used. When graphics are used instead of words or numbers, the table is sometimes referred to as a graphic table.

Numbers

Most tables that are used for analysis purposes use numbers in the body.

	1990	1991	1992	1993	1994	Total
Product A	2	4	6	3	5	20
Product B	10	12	17	15	13	67
Product C	8	7	6	9	9	39
Product D	22	29	20	24	21	116

Words

Sometimes referred to as word tables. The number of words used might vary from a single word to a sentence or even a paragraph.

	Product introduced	Product growing	Product maturing
Sales	Negligible	Rapid increase	Stable
Profit	Negative	Break even	Positive
Cash flow	Negative	Stable	Positive

Symbols

In its simplest form, a geometric symbol is used to indicate yes or no, available or not available, done or not done, etc. Color or fill can be used to encode additional information.

○ Available in 110 volts only
● Available in 110 & 220 volts

	Brand A	Brand B	Brand C	Brand D	Brand E
Feature 1	○	○			○
Feature 2	●			●	
Feature 3		○	○	●	●
Feature 4	○	●		○	○

A number of techniques can be used to indicate such things as varying degrees of something, a phasing in and out, etc. For example, the responsibility for a task may gradually shift from one department to another or one product might be phasing out and another in.

■ Total responsibility
▨ Partial responsibility

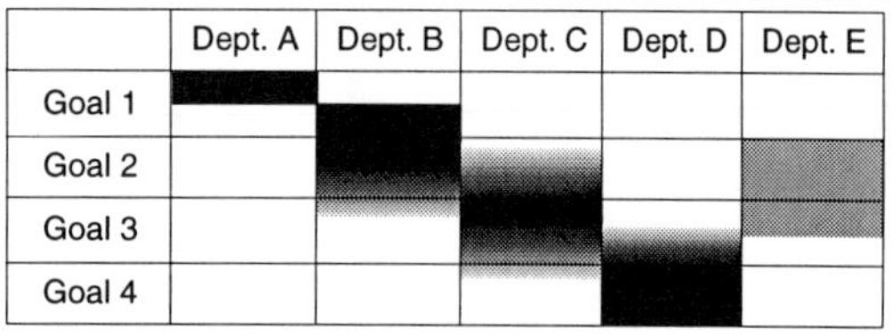

	Dept. A	Dept. B	Dept. C	Dept. D	Dept. E
Goal 1	■				
Goal 2		■	▨		▨
Goal 3		■	▨	▨	▨
Goal 4				■	

The size of the symbol can be shown in proportion to some feature such as price, rating, volume, etc.

Size of circle is proportional to price

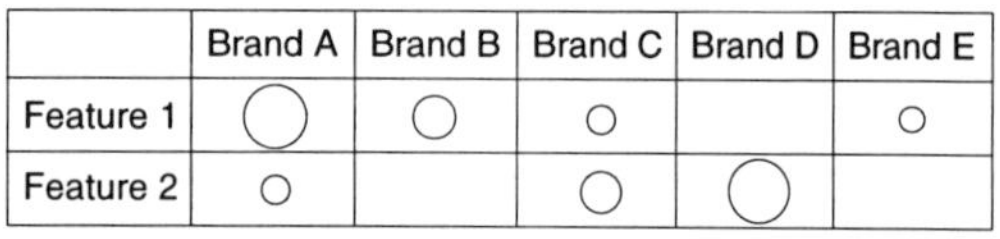

	Brand A	Brand B	Brand C	Brand D	Brand E
Feature 1	◯	○	∘		∘
Feature 2	∘		○	◯	

Various shapes can be used to indicate yes or no to multiple questions: Is a given feature available? Is it available at a given store? etc.

○ Black and white
□ 16 colors
△ 256 colors

	Brand A	Brand B	Brand C	Brand D	Brand E
Product 1	○□△	○△	○	□△	○□
Product 2	○	○□	○□△		○

Qualitative or ordinal information can be communicated by using different color, size, or numbers of symbols.

⬚⬚ Very good
⬚▪ Good
▪▪ Fair
▪ Poor

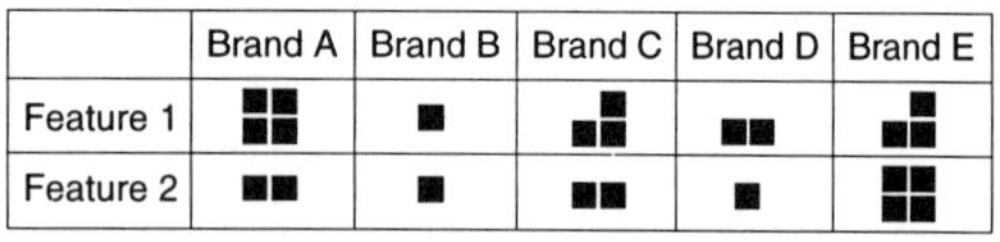

	Brand A	Brand B	Brand C	Brand D	Brand E
Feature 1	⬚⬚	▪	⬚▪	▪▪	⬚▪
Feature 2	▪▪	▪	▪▪	▪	⬚⬚

Quantitative information can be encoded by the use of symbols such as small pie charts, framed rectangles, etc.

○ No backup
◐ 50% backup
● 100% backup

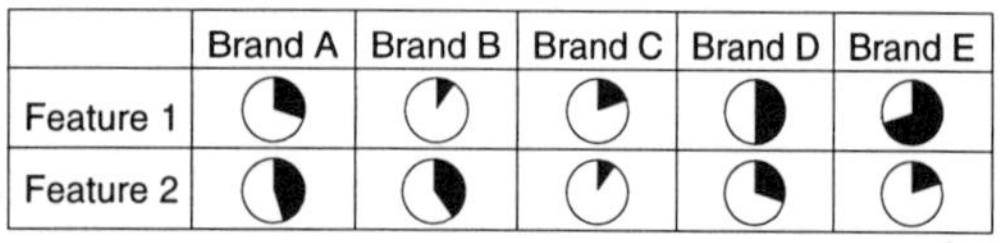

	Brand A	Brand B	Brand C	Brand D	Brand E
Feature 1	◔	◔	◔	◑	◑
Feature 2	◑	◔	◔	◔	◔

Trends, directions, political positions, etc., can be indicated by symbols such as arrows.

↑ Market share increasing dramatically
→ Market share stable
↓ Market share decreasing dramatically

	1990	1991	1992	1993	1994
Company 1	↗	↓	→	→	↑
Company 2	↘	→	↘	↗	↗

Icons and images

Icons and images are sometimes included in the category of symbols. When used in the body of a table, pictures, images, or icons become quantitative if their size or quantity is proportional to some value associated with the entity they represent. Accompanying text is normally required for full understanding of a table with images or icons. Sometimes called pictorial table.

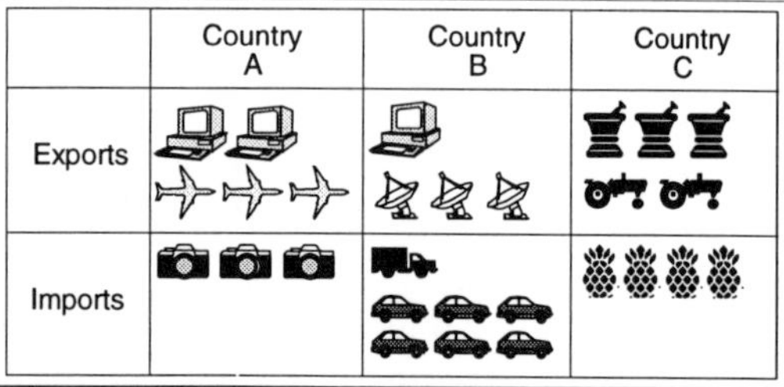

	Country A	Country B	Country C
Exports	💻💻 ✈✈✈	💻 📡📡📡	♟♟♟ 🚜🚜
Imports	📷📷📷	🚚 🚗🚗🚗 🚗🚗	🍍🍍🍍🍍

Body/field of table (continued)

───**Spreadsheet**───

A type of table in which the body of the table contains predominately numeric values. Spreadsheets are used for many different applications including the systematic recording, organizing, and manipulation of large quantities of data. It is common for many of the values on a spreadsheet to be mathematically interrelated such that a change in one value or series of values affects all or many of the other values in the table. • Spreadsheets are frequently used with computers to conduct what-if analyses. In a what-if analysis, hypothetical, predicted, alternative, worst case, etc., values are entered into a matrix of information and their effect on the other values noted. For example, sales values that are 25% below forecast might be entered into a spreadsheet of key financial information for a company to see what would happen to profit, cash flow, return on assets, employment levels, etc., if there was a major down-turn in sales.

───**Graphs arranged in a tabular format**───

This format, sometimes referred to as a matrix display, capitalizes on the advantages offered by graphs as well as the benefits of a tabular display. (See Matrix Display.) Although the example uses scatter graphs, the same concept is applicable to other types of graphs. In addition to displaying graphs in a tabular format, maps, diagram, and even small tables can be arranged in a similar fashion.

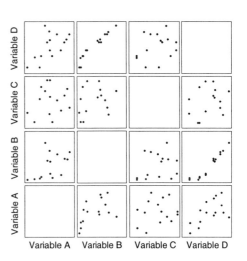

───**Combinations of words, numbers, and other graphics**───

Sometimes tables that display data using words, numbers, and other graphic images can be particularly informative. The example below uses words, numbers, graphs, maps, and symbols. There are no restrictions as to the combination of methods that can be used.

	Company A	Company B	Company C
Headquarters	Dallas, Texas	Paris, France	Tokyo, Japan
Ownership	Public	Public	Private
1996 Sales	$56,000,000	$76,000,000	$87,000,000
1996 Net profit	$4,000,000	$5,000,000	$6,000,000
Percent administrative expenses			
Type sales force	Manufacturers' Representatives	Direct and agents	All direct
Distributed to retailers by	Wholesalers and distributors	Wholesalers	Direct from factory
Percent marketing expenses			
Product strategy	Emphasize new products	Emphasize service	Emphasize improved existing products
Area where major sales efforts are focused			
Trend in market share			

Grid lines, rules, guidelines, etc.

Methods involving grid lines (sometimes called rules), guidelines, bold text, etc., have been developed to do such things as:

- Help organize information in tables
- Assist the viewer in visually tracking data
- Emphasize key information
- Improve the appearance of tables

Examples of several of the more widely used techniques are shown below. Many organizations have standards for how they would like tables constructed to achieve uniformity. There are no industry standards.

Basic table for reference

For reference, this example has no grid lines, rules, shading, or boxes to emphasize particular information or to assist the viewer in visually tracking information.

	1992	1993	1994	1995	1996	Total
Product A	23.1	23.7	24.2	24.9	25.6	121.5
Product B	2.7	3.1	2.5	1.8	0.9	11.0
Product C	10.7	11.2	11.5	11.9	12.5	57.8
Product D	5.9	7.2	9.8	12.4	15.7	51.0
Total	42.4	45.2	48.0	51.0	54.7	241.3

Horizontal and vertical grid lines

This example has grid lines separating all rows and columns.

	1992	1993	1994	1995	1996	Total
Product A	23.1	23.7	24.2	24.9	25.6	121.5
Product B	2.7	3.1	2.5	1.8	0.9	11.0
Product C	10.7	11.2	11.5	11.9	12.5	57.8
Product D	5.9	7.2	9.8	12.4	15.7	51.0
Total	42.4	45.2	48.0	51.0	54.7	241.3

Horizontal grid lines

When the key flow of information is from left to right, vertical lines are sometimes omitted so they do not interfere with the horizontal flow of information.

	1992	1993	1994	1995	1996	Total
Product A	23.1	23.7	24.2	24.9	25.6	121.5
Product B	2.7	3.1	2.5	1.8	0.9	11.0
Product C	10.7	11.2	11.5	11.9	12.5	57.8
Product D	5.9	7.2	9.8	12.4	15.7	51.0
Total	42.4	45.2	48.0	51.0	54.7	241.3

Vertical grid lines

When the key flow of information is vertical, horizontal lines are sometimes omitted so as not to interfere with the vertical flow of information. Lines are generally continuous.

	1992	1993	1994	1995	1996	Total
Product A	23.1	23.7	24.2	24.9	25.6	121.5
Product B	2.7	3.1	2.5	1.8	0.9	11.0
Product C	10.7	11.2	11.5	11.9	12.5	57.8
Product D	5.9	7.2	9.8	12.4	15.7	51.0
Total	42.4	45.2	48.0	51.0	54.7	241.3

Shading or color

In place of or in addition to grid lines, shading or coloring might be used to help the viewer track data, particularly in wide tables.

	1992	1993	1994	1995	1996	Total
Product A	23.1	23.7	24.2	24.9	25.6	121.5
Product B	2.7	3.1	2.5	1.8	0.9	11.0
Product C	10.7	11.2	11.5	11.9	12.5	57.8
Product D	5.9	7.2	9.8	12.4	15.7	51.0
Total	42.4	45.2	48.0	51.0	54.7	241.3

Bolding of selected information

Using bold type in selected rows or columns highlights key data and assists in the visual tracking of data.

	1992	1993	1994	1995	1996	Total
Product A	**23.1**	**23.7**	**24.2**	**24.9**	**25.6**	**121.5**
Product B	2.7	3.1	2.5	1.8	0.9	11.0
Product C	**10.7**	**11.2**	**11.5**	**11.9**	**12.5**	**57.8**
Product D	5.9	7.2	9.8	12.4	15.7	51.0
Total	**42.4**	**45.2**	**48.0**	**51.0**	**54.7**	**241.3**

Boxes around related data

If the material in the table consists of distinctly different groups of information, boxes are sometimes drawn around the groupings. The boxes might run horizontal or vertical.

	1992	1993	1994	1995	1996
Sales	23.1	23.7	24.2	24.9	25.6
Profit	2.7	3.1	2.5	1.8	0.9
Expense A	10.7	11.2	11.5	11.9	12.5
Expense B	5.9	7.2	9.8	12.4	15.7

Boxes around summary data

When the major focus is on particular data, grid lines or boxes might be used selectively to emphasize that information.

	1992	1993	1994	1995	1996	Total
Product A	23.1	23.7	24.2	24.9	25.6	121.5
Product B	2.7	3.1	2.5	1.8	0.9	11.0
Product C	10.7	11.2	11.5	11.9	12.5	57.8
Product D	5.9	7.2	9.8	12.4	15.7	51.0
Total	42.4	45.2	48.0	51.0	54.7	241.3

Guide lines used with symbols

Guide lines aligned with headings and symbols can sometimes be more effective than grid lines between them. Guide lines always pass behind whatever data graphics are used.

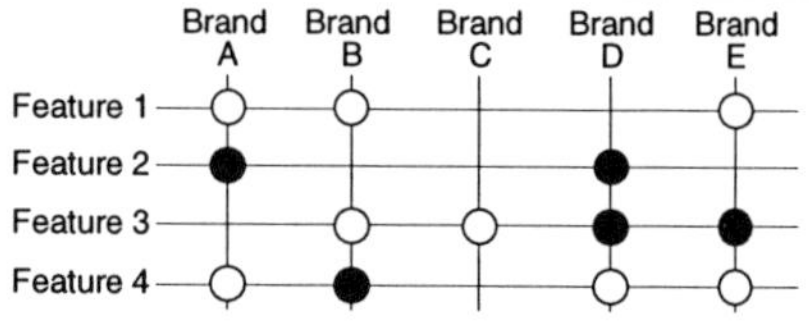

Guide lines used with text

Guide lines can be used with numbers; however, they are generally not as effective as when used with symbols.

	1992	1993	1994	1995	1996
Product A	23.1	23.7	24.2	24.9	25.6
Product B	2.7	3.1	2.5	1.8	0.9
Product C	10.7	11.2	11.5	11.9	12.5
Product D	5.9	7.2	9.8	12.4	15.7

Tally Chart

A type of a frequency distribution chart in which tally or hatch marks are used to record and graphically show the distribution of data elements in data sets. In constructing a tally chart, values or class intervals are shown on one axis, either the horizontal or vertical. Each time a value occurs in the data set, a tally is added to the chart in the appropriate column or row. When all the data elements are accounted for, the resulting series of tally marks forms the equivalent of a simple histogram, as shown at the right. This technique

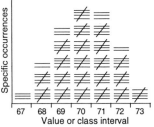

Tally chart used for analyzing the distribution of data elements in a data set

has the unique feature that a histogram is developed as the date is collected, as opposed to collecting all the data

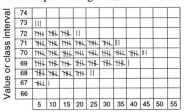

Frequency
Tally chart generated using a preprinted form

and then developing the histogram. If one is only looking for such things as the value that occurs most frequently (mode), the spread, and general pattern of the data, the tally marks do not have to be counted. If exact values are desired, the hatch marks are counted for each column or row. In cases where a task is repetitive, preprinted forms can be used similar to the example shown at the left. With preprinted forms, frequencies can generally be read directly from the form.

Template Chart /Graph

Sometimes referred to as a master chart or graph. A template chart or graph contains all or most of the common elements of a series of charts or graphs. Detailed information is then added to copies of the template to form specific variations. For example, if the sales performance of three different salespersons were to be looked at individually, a template might be generated and the sales data for the individual salespersons added to three copies of the template as shown below. In the example, the three specific charts generated from the template are all column graphs. It is not required that they all be the same, nor is there a restriction as to the type of chart or graph that can be generated using templates. Templates allow more rapid construction and revision of repetitive types of charts and graphs and generally improve accuracy and consistency in a series of charts. Templates are frequently used as the background for presentation charts to provide a uniform appearance throughout the presentation.

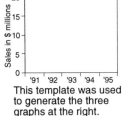

This template was used to generate the three graphs at the right.

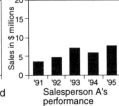

Salesperson A's performance

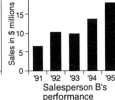

Salesperson B's performance

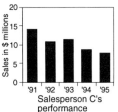

Salesperson C's performance

These three graphs were generated by adding specific information to the template at the left.

Ternary Graph

Sometimes called a triangle, triangular, or trilinear graph. The ternary graph is the only triangular-shaped graph in popular use today. It is used to plot information that has three variables, the sum of which always equals the same amount. That amount can be any value; however, it is typically 1 or 100% since the data is generally plotted in terms of percent of the whole or its equivalent in decimal form (0 to 1). The ternary graph consists of an equilateral triangle with lines from the vertexes to the bases representing each of the three axes. Scales are placed on the axes with zero at the base and 100% at the vertex. To prevent scale labels from interfering with the data points and with one another, the scales are frequently projected onto the sides of the triangle. See Trilinear Graph.

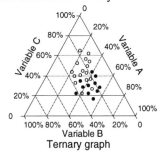

Ternary graph

Tether

Sometimes referred to as a drop line or drop grid. Tethers are thin lines drawn from data points to a reference point, line, or plane to assist the viewer in determining the value or location of the data point. See Drop Line.

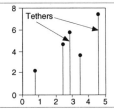

Text

The text used with charts and graphs can sometimes have a significant impact on the appearance and clarity of the material. The purpose of this section is to define some of the major terms used with text and give examples.

Typeface

Sometimes referred to as font. When a set of letters, numbers, and characters have the same design features they are sometimes called a typeface. There are hundreds of typefaces. They have such names as Helvetica, Times, Geneva, etc. Three major categories of typefaces are serif, sans (without) serif, and decorative. Examples of the three types are shown below.

Serif typeface	This is an example of a serif typeface named Times. Serif letters have small lines projecting from the ends of each of their main lines or strokes.	With serifs **T**
Sans serif typeface	This is an example of a sans serif typeface named Helvetica. Sans serif letters do not have the small lines projecting as the serif typefaces do.	Without serifs **T**
Decorative typeface	*This is an example of a decorative typeface named Zapf Chancery. Decorative typefaces are many times used for headings, special effects, and/or to improve the appearance of charts.*	*T*

Type size

Type size is frequently designated in terms of points, based on a standard of 72 points to the inch. Thus, if a type is designated as 12-point, it means it is 12/72 or 1/6 of an inch tall, and 36-point is 36/72 or 1/2 inch tall. The height is measured from the highest point (highest ascender) on any letter of the typeface to the lowest point (lowest descender) of any letter of the typeface. Examples of various type sizes are shown below.

Size 4	A comparison of type sizes	Highest ascender
Size 6	A comparison of type sizes	Type size
Size 8	A comparison of type sizes	Lowest descender **Type lined** Baseline
Size 10	A comparison of type sizes	**the pages.** Leading
Size 12	A comparison of type sizes	Baseline
Size 14	A comparison of type sizes	
Size 16	A comparison of type sizes	
Size 18	A comparison of type sizes	
Size 20	A comparison of type sizes	

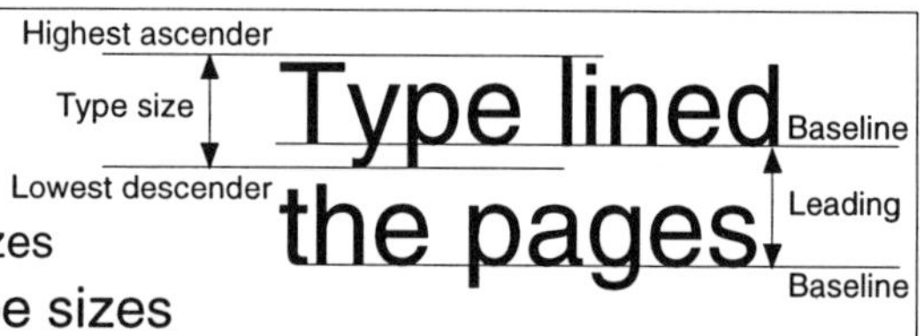

Leading

Sometimes referred to as line space. Leading (pronounced "ledding") is the vertical distance between lines of type. It is measured from the baseline of one line of type to the baseline of the next line of type. (See example above.) Leading is generally specified in the same point system used to specify type size (i.e., 72 points to the inch). The examples below show three different leadings using the same number nine-point type size in each.

# 9 size type and #9 leading. Sometimes designated as 9/9	#9 size type and #11 leading. Sometimes designated as 9/11	#9 size type and #13 leading. Sometimes designated as 9/13
Brief paragraphs sometimes appear below charts and graphs to explain the contents of the chart. Brief paragraphs sometimes appear below charts and graphs to explain the contents of the chart. 9-point leading	Brief paragraphs sometimes appear below charts and graphs to explain the contents of the chart. Brief paragraphs sometimes appear below charts and graphs to 11-point leading	Brief paragraphs sometimes appear below charts and graphs to explain the contents of the chart. Brief paragraphs sometimes appear below 13-point leading

Tracking

Tracking changes the horizontal spacing between letters and words, as shown in the examples below. Number nine size type is used in each example.

Loose spacing	——	This is an example of changing horizontal spacing
Typical spacing	——	This is an example of changing horizontal spacing
Tight spacing	——	This is an example of changing horizontal spacing

Text (continued)

Kerning

Tracking changes the horizontal spacing of whole words, lines, or paragraphs. Kerning changes the horizontal spacing between pairs of letters, one of which is generally a capital. Its major function is to improve appearance in those cases where the normal spacing between letters would seem too large.

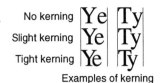

No kerning

Slight kerning

Tight kerning

Examples of kerning

Type style

Different styles of text are used for many reasons, including placing special emphasis on certain text, indicating book titles, denoting totals, highlighting names, etc. Shown below are some of the more widely used styles. All examples are in nine-point type except the superscript and subscript.

Bold	**An example of a particular style**
Italics	*An example of a particular style*
Underlined	<u>An example of a particular style</u>
Shaded or colored	An example of a particular style
Full size capitals	AN EXAMPLE OF A PARTICULAR STYLE.
Small capitals	AN EXAMPLE OF A PARTICULAR STYLE
Reverse type	An example of a particular style
Outlined	An example of a particular style
Superscript	An example of ^{superscript} or $325^{(2x-4)}$
Subscript	An example of $_{subscript}$ or $325_{j,k,l}$

Alignment

A term that is generally used to describe how text and/or numbers are positioned. There are five major types of alignment, as noted below. Examples of how each applies to stand-alone text, tables, columns of numbers, and graphs are shown.

Type alignment	Stand alone text	Table	Column of numbers	Graph
Aligned flush left/ragged right	Sometimes referred to as justified left and/or ragged right. When text is aligned left, the left edges of all lines are even. The right edge of each line is determined by the full words, partial words, and spacings in the line.	Sales 4,255 / Costs 2,978 / Gross profit 1,276 / Expenses 1,029 / Taxes 119 / Income 127	18,250 / 5,750 / 7 / 0 / 60,563 / 145	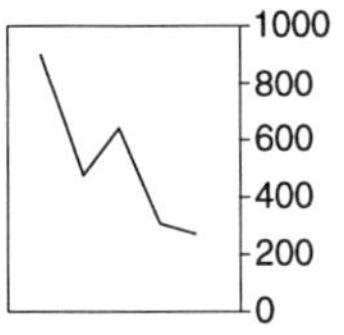 1000 800 600 400 200 0
Aligned flush right/ragged left	Sometimes referred to as justified right and/or ragged left. When text is aligned right, the right edges of all lines are even. The left edge of each line is determined by the full words, partial words, and spacings in the line.	Sales 4,255 / Costs 2,978 / Gross profit 1,276 / Expenses 1,029 / Taxes 119 / Income 127	18,250 / 5,750 / 7 / 0 / 60,563 / 145	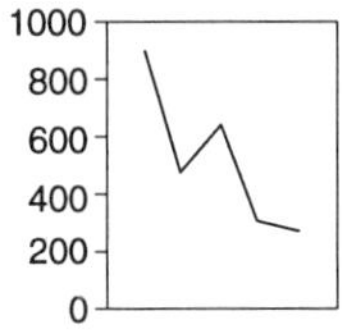 1000 800 600 400 200 0
Centered	Sometimes referred to as ragged right and left. Uniform spacings are used between words and letters. Excess space is distributed equally at the two ends of each line. Generally used only with a few lines of text.	Sales 4,255 / Costs 2,978 / Gross profit 1,276 / Expenses 1,029 / Taxes 119 / Income 127	18,250 / 5,750 / 7 / 0 / 60,563 / 145	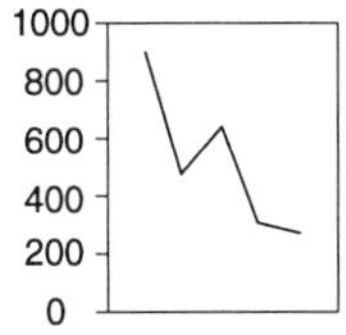 1000 800 600 400 200 0
Justified (aligned flush right and left)	Sometimes referred to as flush right and left. Text that is justified normally has the right and left edges aligned. This is accomplished by varying the space sizes between words and letters and the frequent use of hyphens.	Not applicable	Not applicable	Not applicable
Aligned on decimal points	Not applicable	Sales 4,255 / Costs 2,978.5 / Gross profit 1,276.5 / Expenses 1,029.73 / Taxes 119.37 / Income 127.4	12,365.87 / (658.14) / 3.7985 / 0.0 / 65,88.05 / 0.872	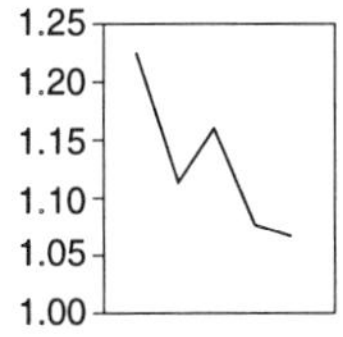 1.25 1.20 1.15 1.10 1.05 1.00

Text Chart

Sometimes referred to as a word chart. A text chart is a single document, page, slide, etc., made up solely or mainly of words and used primarily in presentations. Although a text chart is generally one of a series of charts, each chart tends to stand on its own. Text charts normally contain only the key words or phrases of a subject. Typical guidelines recommend that the number of lines per chart and the number of words per line be limited to between five and eight. Most text charts have no graphics associated with them except for decorative purposes. Two examples are shown at the right. Shown below are five of the major variations of text charts

Improved benefits

– More life insurance

– Bonus program

– 401K improvements

Improved Features
• Speed
• Memory
• Price

Examples of text charts with
and without graphics

Title Text Chart

Typically used as the introductory page or slide in a presentation. In addition to the title, this chart might include such information as purpose of meeting, outline of meeting, key points to be covered, etc.

ABC Company
Annual Meeting

☐ 1996 Results
☐ 1997 Outlook
☐ Questions

Paragraph Text Chart

A text chart in which a complete paragraph is used as opposed to just a few key words as in the other types of text charts.

"Last year was a good year for our company. This year promises to be an even better year thanks to all of our employees."

Mary Evans

Bullet Text Chart

A text chart where symbols (bullets) highlight the key thoughts or phrases. Dots are most widely used; however, any symbol will work. Different symbols can be used to indicate different types of material or level of indentation.

• Adult activities
• Youth activities
• Table games
• Races
• Cookout

Column Text Chart

A chart where the text is arranged in two or three columns as in a newspaper. Sometimes used for making comparisons, organizing related information, or displaying questions and answers.

Product Comparison

	Ours	Theirs
	– 5 pounds	– 6.5 pounds
	– 8 outlets	– 6 outlets
	– 5 colors	– 3 colors
	– 2 speeds	– 1 speeds
	– $37.50	– $39.25

Build Text Chart

Sometimes referred to as a reveal chart. A build chart is one of a series of charts used to develop/ present a broad message, idea, or concept. The first chart of the series would have only the first point on it. The next chart would have points one and two on it. The next chart would have points one, two, and three on it. This would continue until the subject has been fully explained and all points had appeared.

Sales Strategy
✓ More sales calls

Sales Strategy
✓ More sales calls
✓ Bigger incentives

Sales Strategy
✓ More sales calls
✓ Bigger incentives
✓ More backup

Chart #1   Chart #2   Chart #3

Variations of text charts

Textured Map

Sometimes referred to as a choropleth, crosshatched, or shaded map. A textured map is a variation of a statistical map that displays data with regards to areas through shading, color, or patterns. Data is often organized into class intervals, as in the example at the right. See Statistical Map.

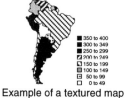

■ 350 to 400
■ 300 to 349
■ 250 to 299
▨ 200 to 249
☐ 150 to 199
☐ 100 to 149
⊞ 50 to 99
☐ 0 to 49

Example of a textured map

Thematic Map

A thematic map addresses one or at most a few themes, ideas, characteristics, attributes, types of data, etc., over a given geographic area. The map might display quantitative and/or qualitative information in relationship to areas such as geographic, administrative, political, trading zones, sales areas, trade routes, networks, etc. Typical themes displayed on this type of map include population density by state; crops grown by region of the world; people who voted for a given candidate by ward; market potential by county; traffic by highway; resources by location, etc. The purpose of thematic maps is to allow visual overviews or impressions as well as to convey detailed information about a specific subject. Topographic maps are generally not included in the category of thematic maps.

Six-year sales
by state

Thematic map

Thermometer Graph

Sometimes referred to as a a segmented line, fever, broken line, or zigzag graph. Thermometer graph is a term occasionally used to describe a line graph in which the data points are connected by straight lines.

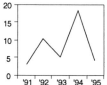

Thermometer graph

'91 '92 '93 '94 '95

Three-Dimensional (3-D) Graph

Sometimes referred to as a stereogram or three-dimensional map. See Three-Dimensional Map. General aspects of three-dimensional graphs are discussed in this section. Details for three-dimensional variations of specific types of graphs are discussed under the headings for those graphs. • The phrase three-dimensional (sometimes abbreviated 3-D) graph is used in two significantly different ways, which sometimes causes confusion. In one case the phrase is used to mean that a graph has the appearance of having depth which makes it look three-dimensional. This frequently is done for cosmetic purposes. In the other case, the phrase three-dimensional graph refers to a graph that has three axes. This may be done for cosmetic purposes or to display the data in a way that simplifies an overall analysis or provides insights not possible in other formats. In this book, three-dimensional graph generally refers to a graph with three axes. The majority of this section is devoted to three-dimensional, three-axis graphs.

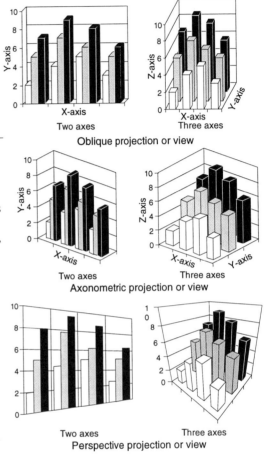

Projection or view

Almost all three-dimensional graphs, whether generated for cosmetic or technical purposes, are drawn with one of three techniques, sometimes referred to as projections or views. These are oblique, axonometric, and perspective. One-, two-, or three- axis graphs can be drawn with either technique. Generally the oblique view is used with one- and two- axis graphs and the axonometric view with three-axis graphs. The perspective view is generally used for aesthetic purposes and works with one, two, or three axes. Examples of two- and three-axis graphs using each of the three different projections are shown at the right. See Oblique Projection (View), Axonometric Projection (View), and Perspective Projection (View). • Three-dimensional graphs are seldom used for monitoring or controlling. They are almost never used where estimating relatively exact values from the graph itself is required, since it is difficult or impossible to determine exact values from such graphs. If it is necessary or desirable for the viewer to read exact values, the values are many times noted on the graph or a reference table is provided. When used for analytical purposes, the inability to read exact values is often not a problem since many such graphs are used to look at the general shape or nature of the data as opposed to determining specific values.

Terminology

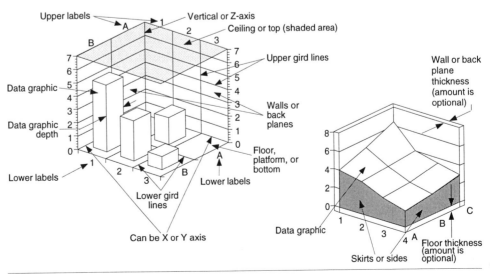

Graph type

One of the most basic options with three-dimensional graphs is the type of graph that will be used. The six examples shown at the right are simplified representations of the major choices available in three-dimensional rectangular graphs. For illustrative purposes, all six graphs were plotted using the same three data series, which, altogether have a total of fifteen data points. Each graph type has its advantages, disadvantages, and individualized features which are discussed under the specific graph headings. Which type of graph to use is based on such things as whether the data is discrete or continuous; if continuous, is it continuous along one, two, or three axes; is the objective of the graph to show specific values or the overall pattern of the data; are there many data points or just a few; are distinct values used or will the graph describe an equation; is the graph for analytical or presentation purposes; etc.

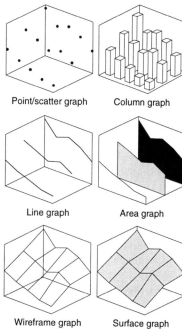

Point/scatter graph   Column graph

Line graph   Area graph

Wireframe graph   Surface graph

Six basic types of three-dimensional graphs. For comparison purposes, they are shown here displaying the same three data series (fifteen data points).

Drop lines

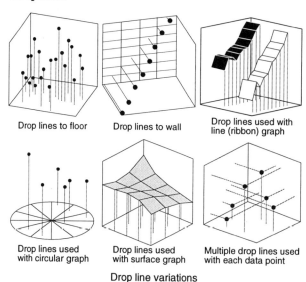

Drop lines to floor   Drop lines to wall   Drop lines used with line (ribbon) graph

Drop lines used with circular graph   Drop lines used with surface graph   Multiple drop lines used with each data point

Drop line variations

When specific data points are plotted, drop lines are sometimes used to help more accurately determine the location and/or value of the points. In other cases drop lines might be used to call the viewer's attention to specific data points or values. The drop lines generally extend from the data point to the floor of the graph. On occasion they extend to a wall. When there are only a few data points, drop lines can extended to two or more surfaces of the frame. See Drop Line.

Patterns of lines used on graph surfaces

When data is continuous along all three axes, surface type graphs are often utilized with patterns of lines describing the surfaces of the data graphics. The examples at the right illustrate three of the major types of line patterns used for this purpose. See Surface Graph.

Contour lines

This type of line connects points with equal values on the vertical (Z) axis. See Contour Graph.

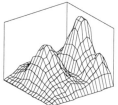

Isolines along the X and Y axes

This type of line connects points of equal X and Y values. They are sometimes called fishnet graphs. See Surface Graph.

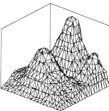

Random lines

The types of lines used in this variation of graph are optional. Sometimes a combination of isolines, contour lines, and diagonal lines are used. Many times called a mesh graph.

Line patterns used to identify surfaces on three-dimensional data graphics

Number of lines used to form the surface pattern

The number of lines used in the pattern of surface graphs has to do with the amount of data available, the nature of the data, the amount of detail desired, the purpose of the graph, and its desired appearance. When selected data points are plotted, often just those data points will be connected with lines. When data is input by means of an equation or when a continuous series of data points is interpolated between isolated points, there is a fair amount of flexibility as to the number of lines used. The only limitations are that at one extreme, if too many lines are shown, the resulting data graphic is a solid silhouette, while at the other extreme, if too few lines are shown, the data graphic becomes distorted. The illustration below shows the two extremes and two intermediate examples. The illustration uses graphs with isolines. The same principle applies with any type of lines. See Surface Graph.

 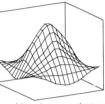 

Plotting too many lines results in a solid silhouette.

There is a wide range of number of lines between the two extremes that produce acceptable data graphics.

Plotting too few lines results in distortion of the data graphic.

Illustration of the effect the number of surface lines can have on the appearance of a data graphic

Opaque versus transparent for the spaces between lines

The areas between lines on three-dimensional graphs may be transparent or opaque. Sometimes they are left transparent so the viewer can see the nature of the data behind the front surfaces. In other cases the graph becomes too confusing with all of the lines exposed and the areas are made opaque so the front surfaces stand out clearly and the relationships between multiple surfaces are easier to see. Three comparisons are shown here.

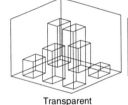 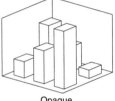

Transparent   Opaque

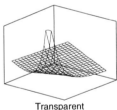 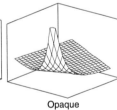 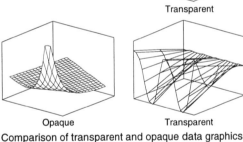

Transparent   Opaque   Transparent   Opaque

Comparison of transparent and opaque data graphics

Color and shading

Four major reasons for using color or shading in three-dimensional graphs are:
– To identify and group data graphics
– To designate areas or data graphics that have equal values
– To make the data graphic more understandable and easier to interpret
– To improve the appearance of the graph.
Examples of each are shown here.

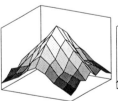  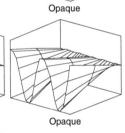

Color or shading used to identify elements belonging to a common data series

Shading used to indicate areas of equal values between isolines

Shading used to indicate areas of equal values between contour lines

Shading or colors used to indicate areas of equal values or ranges of values

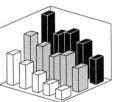 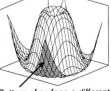 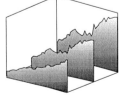

Top and bottom of surface the same shade

Bottom of surface a different color than top

Shading or colors used to make data graphic more understandable

Color or shading used to improve the appearance of the graph

Three-Dimensional (3-D) Graph (continued)

Cross sections of data

With some three-dimensional information it is helpful to have a more detailed graphical representation of hidden surfaces or to see what the data looks like at various points along an axis. Shown below are three techniques for achieving these goals.

Section of data graphic removed

With this method a section of the data graphic is removed or separated exposing the inside/underside. The section might be removed along any of the axes. The plane at which the section is taken might be flat or curved, perpendicular to the axis or slanted. With complex data, multiple sections may be removed or separated.

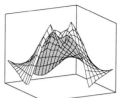
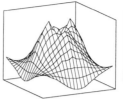

Without section removed With section removed

Example of a section of a graph removed to expose the inside or underside of the data graphic

Data condensed into selected planes

With three-dimensional scatter graphs, the data points for uniform intervals along a given axis are sometimes condensed into selected planes, as shown at the right. In this way patterns within the larger cloud of data points can be noted. See Slice Graph.

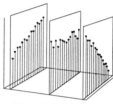
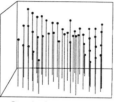

Standard scatter graph Data points condensed onto three planes

Example of data points condensed into selected planes along one axis

Profiles at selected intervals

Three-dimensional graphs that represent a volume as opposed to specific data points or a surface are sometimes shown as a series of area graphs called profiles. The profiles might be spaced regularly or irregularly along the X or Y axis. See Profile Graph [1].

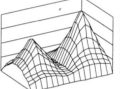

Standard three-dimensional graph with isolines Multiple cross sections displayed at selected points

Example of planes passed through data graphic to depict profiles of the data graphic at selected points

Projections

Projecting the details of a data graphic onto the walls, ceiling, or floor of the graph can provide assistance in its interpretation. The amount of detail shown in the projection varies depending on the type of graph and the purpose of the projection. In some cases only the outline or profile of the data graphic (sometimes called shadow) is projected on to the walls, as shown on the left below. When more detail is desired, isolines, contour lines, outlines, etc., might be projected, as shown in the two center examples. Another use of projections is to identify hidden data graphics such as columns that would normally not be seen in the standard three-dimensional view. An example of this is shown on the right below. The lines, areas, surfaces, etc., displayed in projections may or may not be labeled or color-coded to convey quantitative information.

Shadows of 3-D graph projected onto the walls Surfaces of 3-D graph projected onto the floor Contour lines of 3-D graph projected onto the ceiling Projections on the ceiling help identify hidden columns

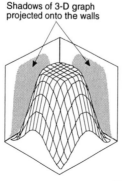

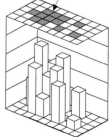

Examples of various types of information projected onto the floor, ceiling, or walls of three-dimensional graphs

Three-Dimensional (3-D) Graph (continued)

Two- and three-dimensional graphs used together

Two-dimensional graphs are good at conveying specific values. Three-dimensional graphs are good for depicting the general nature of the data. Using the two in conjunction with one another provides the advantages of both. In some situations one two-dimensional graph in conjunction with a three-dimensional graph is adequate. In other cases, two or more of one type or the other is advantageous and in some cases multiples of both types are ideal. See Draftsman's Display.

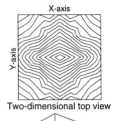

Two-dimensional top view

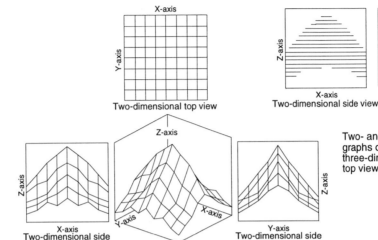

Two- and three-dimensional graphs of the same data. The three-dimensional graph and the top view are contour type graphs.

Two- and three-dimensional isoline graphs used in combination to analyze a single set of data

Tilt and rotation

The amount and direction of tilt and rotation of a three-dimensional graph displayed in an axonometric view can sometimes improve the viewer's ability to discern key features of the data graphics. The example at the right illustrates how the direction of rotation can effect the graph. The examples below illustrate how the degree of tilt and rotation can sometimes make the graph easier or harder to interpret. See Axonometric Projection (View).

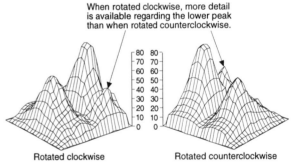

When rotated clockwise, more detail is available regarding the lower peak than when rotated counterclockwise.

Rotated clockwise   Rotated counterclockwise

The same data shown on two graphs with different orientations. In this illustration, more detail is available when the graph is rotated clockwise. With other graphs, the opposite might be true.

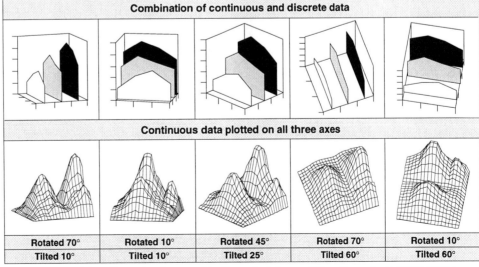

Combination of continuous and discrete data

Continuous data plotted on all three axes

Rotated 70°	Rotated 10°	Rotated 45°	Rotated 70°	Rotated 10°
Tilted 10°	Tilted 10°	Tilted 25°	Tilted 60°	Tilted 60°

Illustration of the effects of various combinations of rotation and tilt on three-dimensional graphs

Width and depth of data graphics

With some graphs, such as wireframe and surface, the width and depth of the data graphics are dictated by the scale and the data itself. In other cases, such as column and bar type graphs, width and depth of the data graphics are optional and can sometimes have an effect on one's ability to read the graph accurately. With oblique views, the width and depth of data graphics have an effect on appearance but generally do not affect one's ability to interpret the data graphics. See examples at right. When three-dimensional graphs are displayed in axonometric view, the width and depth of the data graphics become more of a factor, as shown below. The combination of width and depth that is best for any particular graph is sometimes unique to that graph.

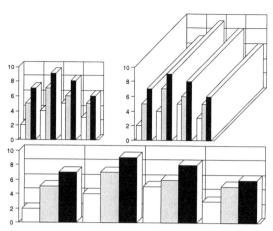

Examples illustrating that wide variations in width and depth of data graphics in the most common type of oblique view have little effect on one's ability to read the graph. The same data is plotted on all three graphs.

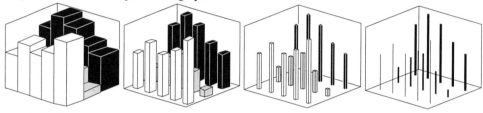

In this illustration the only things varied are the width and depth of the columns. Three columns are totally hidden in the graph on the left. All columns are visible in the two graphs on the right.

Spin or continuous rotation

This is a technique available only on computer screens. With the spin method, the entire graph can be rotated so the viewer can look at it from almost any angle. Depending on the specific program, the graph might be viewed from any of 360° around any of the three axes. In this way patterns, trends, relationships, distributions, outliers, anomalies, etc., can many times be observed. Shown below are nine views (including the top view) of the same three-dimensional floating column graph to crudely illustration what one might see with this process. In an actual application, many more views would be available, however, only one view is available at a time.

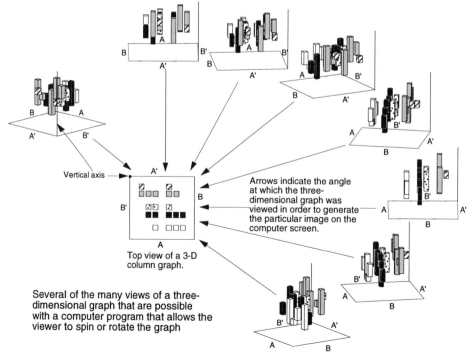

Vertical axis

Arrows indicate the angle at which the three-dimensional graph was viewed in order to generate the particular image on the computer screen.

Top view of a 3-D column graph.

Several of the many views of a three-dimensional graph that are possible with a computer program that allows the viewer to spin or rotate the graph

Three-Dimensional Map

Sometimes referred to as a perspective map whether or not it is truly drawn in perspective. See Perspective Projection (View).

Three major types of three-dimensional maps

There are three major categories of what are sometimes called three-dimensional maps.

- One type of map that is sometimes referred to as three-dimensional is a two-axis map that has been tilted and rotated with depth added for cosmetic purposes, as shown at the right. This type of map yields the same information as a conventional two-dimensional map.

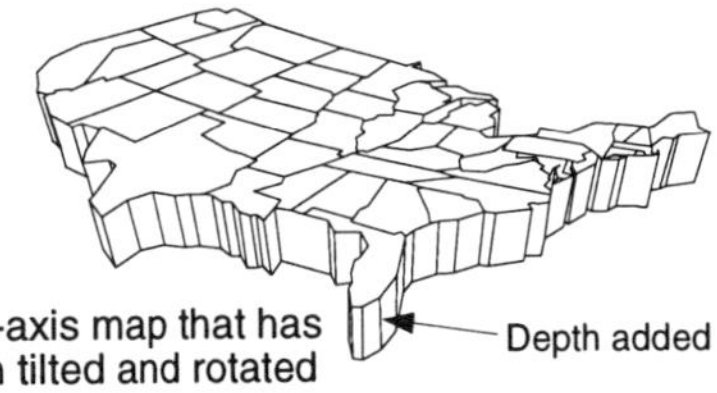

Two-axis map that has been tilted and rotated — Depth added

- A second type of three-dimensional map is often called a shaded relief map. These are two-dimensional maps; however, they have have a third variable, elevation, encoded by means of contour lines. Shading is used to simulate shadows which gives the maps the appearance of being three-dimensional, as shown at the right. The primary function of the shading is to make it easier to interpret the map. Shaded relief maps are a variation of topographic map.

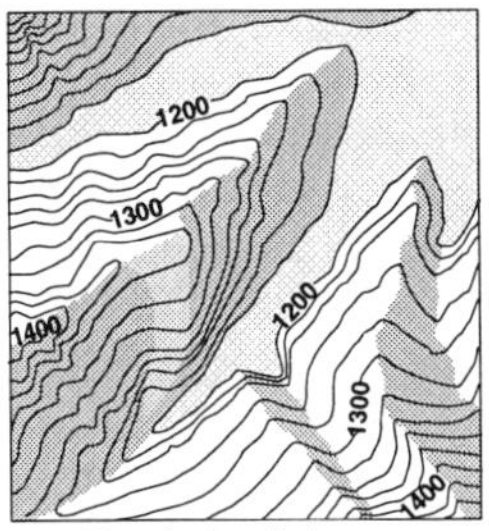

Shaded relief map

- This variation has a third axis along which elevations or quantitative Z-axis values are plotted. They are sometimes referred to as three-axis, three-dimensional maps. When statistical information is plotted on a three-axis map, it frequently resembles and functions like a three-dimensional graph (example at right). The coordinates to locate each point are plotted along the X- and Y-axes (e.g., latitude and longitude). The elevation, or some other quantitative variable, is plotted along the Z- (vertical) axis.

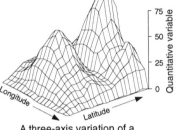

A three-axis variation of a three-dimensional map

Variations of three-axis, three-dimensional maps

Although vertical scales on maps with three-axes are generally quantitative, in most cases it is difficult to accurately determine actual values from the maps. For this reason, three-axis maps are generally used to indicate the general nature of the data or terrain as opposed to supplying detailed information. For example, they might be used for such things as displaying population figures to see where major markets are located, plotting pollution levels to see where major efforts should be concentrated, or showing the strength of a magnetic field around an electrical device to see where shielding should be provided. Since it is so difficult to determine exact values from three-dimensional maps, the values are sometimes noted on the map, provided by means of a reference table or document, or in the case of some computer programs, the data are displayed on the screen by pointing at a specific point with the cursor. Examples of five types of three-axis, three-dimensional maps are shown here.

Discrete data plotted such that the height of each area is proportional to the value it represents. Sometimes called stepped statistical map.

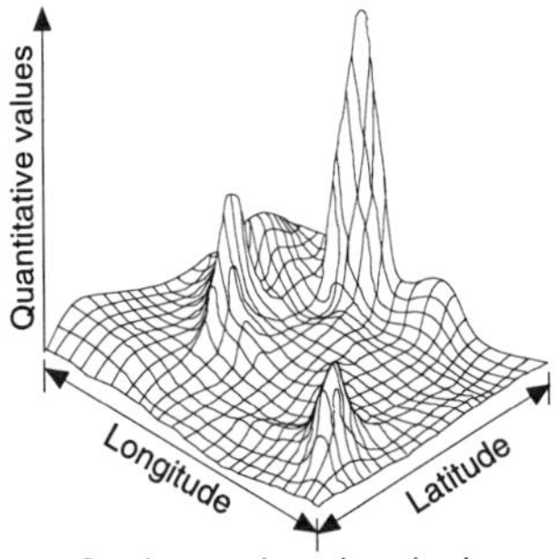

Continuous data plotted using isolines. Sometimes called a smooth statistical map.

Examples of three-dimensional statistical maps

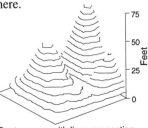

Contour map with lines connecting points of equal elevation

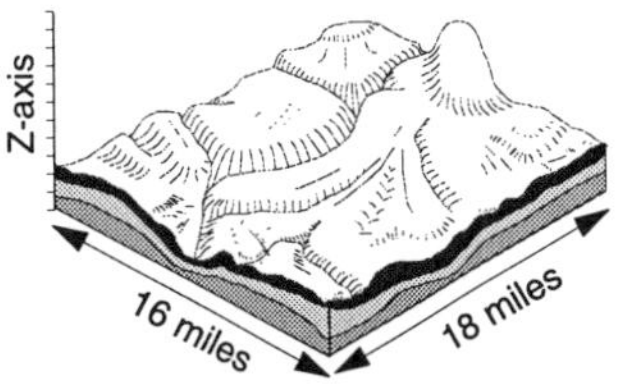

Block map showing general features of the terrain

Examples of three-dimensional topographic maps

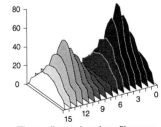

Three-dimensional profile map displaying cross-sections

Direction of rotation

With three-axis maps the direction of rotation can have significance if, for example, the higher elevations are concentrated on one side or if it is important for the viewer to see a particular side. The examples at the right show maps of the same data rotated in two different directions. It can be noted that there is more detail available regarding the shorter peak on the map rotated clockwise than the one rotated counterclockwise.

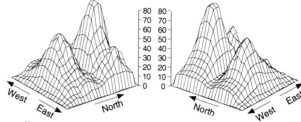

Examples of two maps of the same data with two different orientations. The one on the left is most frequently used.

Amount of tilt and rotation of three-axis maps

With three-axis maps the amount of tilt and rotation can be instrumental in making the map more readable and meaningful. For example, when the values on the Z-axis are relatively small, a slight amount of tilt might be adequate. When there are many high peaks relatively close together, a greater amount of tilt might be advisable so the peaks in the front do not obscure the ones in the back. Examples of several different combinations of tilt and rotation are shown below. Values close to those of the example in the middle are frequently used when there are no special requirements that suggest some other combination. When the degree of rotation and/or tilt is close to zero or a multiple of 90°, some of the advantages of a three-dimensional map are lost since it begins resembling a two-dimensional map.

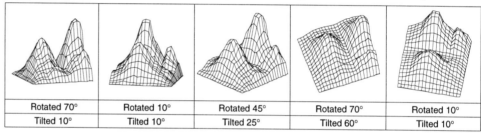

| Rotated 70° | Rotated 10° | Rotated 45° | Rotated 70° | Rotated 10° |
| Tilted 10° | Tilted 10° | Tilted 25° | Tilted 60° | Tilted 10° |

Examples of the same three-dimensional map with varying amounts of rotation and tilt
(Degrees of rotation and tilt are based on an arbitrary set of reference angles. A different set of reference angles would result in different numbers of degrees, but the appearance of the maps would be the same.)

Amount of tilt and rotation on two-axis maps

With two-axis maps the amount of rotation and tilt are less critical; however, orientations close to multiples of 90° can be a problem, the same as with three-axis maps. The concern with extreme degrees of either tilt or rotation is that the large amounts of distortion it introduces can mislead the viewer.

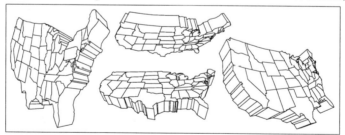

Examples of the same two-axis map with various amounts of tilt and rotation

Box diagram maps

When three-dimensional maps represent portions of a larger entity, they are sometimes referred to as block diagram maps (examples below). Originally hachures and outlines were used to generate such maps. Today most are of the fishnet or shaded relief type. When a block diagram map is used as a topographic map, the sides of the block sometimes display geological information (e.g., soil layers, rock layers) or man-made structures (e.g., tunnels, sewers).

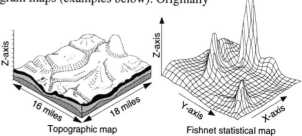

Topographic map   Fishnet statistical map
Examples of box diagram maps

Tick Mark
or
Tick
or
Tic

Sometimes referred to as a stub, stub mark, or scale point. Tick marks are short lines generally drawn perpendicular to a larger line and used to mark off uniform increments along that larger line. Frequently that larger line is an axis or scale line on a graph. The uniform increments might represent intervals of values on a quantitative scale, intervals of time on a sequence scale, distance on a map scale, or simply mark off uniform segments on a category or ordinal scale. Tick marks are optional and frequently are not used, particularly on maps. How many tick marks to use, what type, and where generally depends on the purpose of the chart, the nature of the data, whether or not grid lines are used, and the degree of accuracy desired in decoding the information. Tick marks are used on several different types of charts but are used most extensively on graphs.

General characteristics

Types of tick marks

There are three types of tick marks. They are called major (the longest), minor (the shortest), and intermediate (between the major and minor tick marks in length). Axis labels are normally associated with major tick marks. Unless it is a very large graph, axis labels many times do not accompany minor or intermediated tick marks.

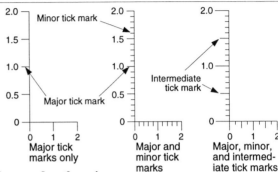
Major tick marks only • Major and minor tick marks • Major, minor, and intermediate tick marks

Orientation of tick marks with regard to the axis

Tick marks might be located external or internal to the body of the graph, across the axis (portions internal and external), or in some combination of the three. Which method to use is largely a matter of preference. One potential hazard with using internal tick marks is their possible interference with the data being plotted.

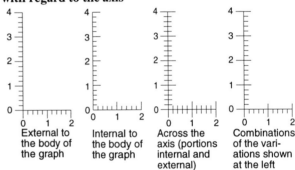
External to the body of the graph • Internal to the body of the graph • Across the axis (portions internal and external) • Combinations of the variations shown at the left

Lengths of tick marks

Tick marks are normally kept short; however, there are no restrictions as to how long they can be. If too long they can detract from the key focus of the chart. The examples at the right illustrate four different lengths of the same set of tick marks. The length of tick marks are many times specified in terms of inches, millimeters, or points (72 points per inch).

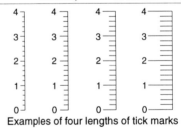
Examples of four lengths of tick marks

Axes on which tick marks are located

The axes on which the tick marks are shown is optional. The most widely used practice is to place the tick marks on the axes with the labels. They sometimes are also placed on the second vertical or horizontal axis, a reference axis, or combinations thereof. Labels may or may not be used with the additional tick marks.

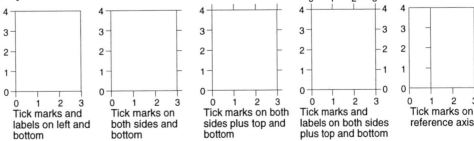
Tick marks and labels on left and bottom • Tick marks on both sides and bottom • Tick marks on both sides plus top and bottom • Tick marks and labels on both sides plus top and bottom • Tick marks on reference axis

Tick marks in the body of the chart

Tick marks are typically located on the axes. Occasionally additional tick marks are located in the body of the chart to improve the accuracy of decoding the information. Such tick marks are more common on maps than on graphs.

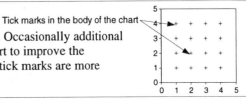
Tick marks in the body of the chart

409

Tick marks and grid lines

To a large extent, tick marks and grid lines serve the same function: helping the viewer decode information in a chart, rapidly and accurately. Thus, many times, if grid lines are used, tick marks are not. Combinations of tick marks and grid lines are also widely used. Several of the more common combinations are shown below. When tick marks and grid lines are used on the same graph they are aligned at the axis.

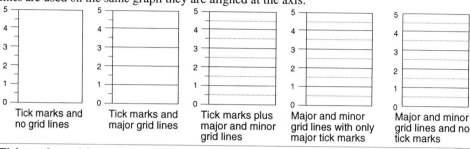

Tick marks and no grid lines Tick marks and major grid lines Tick marks plus major and minor grid lines Major and minor grid lines with only major tick marks Major and minor grid lines and no tick marks

Tick marks and frames

Tick marks can be used with or without graph frames. Variations ranging from full frame to no frame are shown at the right.

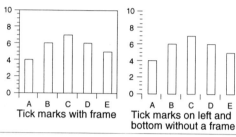
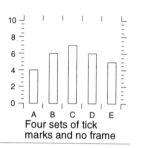

Tick marks with frame Tick marks on left and bottom without a frame Four sets of tick marks and no frame

Tick marks used with other types of graphs

Three-dimensional rectangular graphs

In three-dimensional rectangular graphs, the tick marks on the outside edges might be parallel or perpendicular to their respective planes. When tick marks are drawn internally on this type of graph, there is a hazard that they may interfere with the data on the graph.

When tick marks are drawn external to the graph, the ones on the back sides are hidden by the walls of the graph.

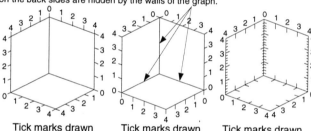

Tick marks drawn parallel to the planes they apply to Tick marks drawn perpendicular to the planes they apply to Tick marks drawn internal to the graph

Circular graphs

Examples of tick marks on three types of circular graphs.

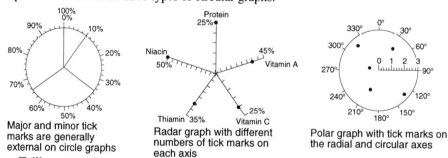

Major and minor tick marks are generally external on circle graphs Radar graph with different numbers of tick marks on each axis Polar graph with tick marks on the radial and circular axes

Trilinear graphs

When used with trilinear graphs, tick marks serve the additional function of helping to orient the viewer as to which grid line a label refers. As shown at the right, tick marks on trilinear graphs typically appear as continuations of the grid line to which they apply.

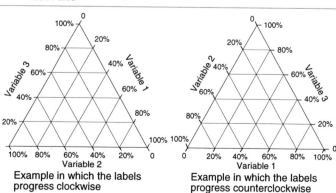

Example in which the labels progress clockwise Example in which the labels progress counterclockwise

Tick Mark
or
Tick
or
Tic
(continued)

Positioning along axis

Where the tick marks are positioned along the axis depends on the type of data plotted on the graph. Examples of some of the key variations are shown below.

———**Quantitative data**———

When quantitative data is plotted using specific data points, the major tick marks are generally aligned with the labels.

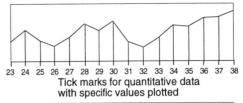

Tick marks for quantitative data
with specific values plotted

If class intervals are used, the tick marks are frequently located at the boundaries of the intervals.

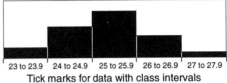

Tick marks for data with class intervals

———**Sequential (e.g., time) data**———

With sequential data, the tick marks may or may not be aligned with the labels. Factors sometimes influencing the decision as to which method to use include:

– When tick marks are not aligned with labels, it is sometimes easier to partition time series information into subgroups such as months or years;

– When period data (e.g., average number of employees for the month) are plotted, tick marks between labels are often used;

– When point data (e.g., number of employees the last day of the month) are plotted, tick marks are sometimes aligned with labels.

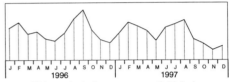

Tick marks between sequential labels

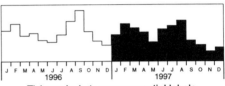

Tick marks between sequential labels

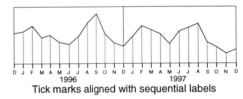

Tick marks aligned with sequential labels

———**Category data**———

Tick marks are frequently not used with category type data. When they are used, they generally are located between the labels and serve as sort of dividers separating one category from another. Minor and intermediate tick marks are seldom used with category type data.

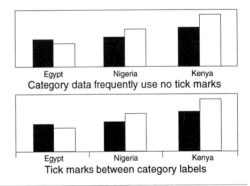

Category data frequently use no tick marks

Tick marks between category labels

Number of minor and intermediate tick marks

Minor and intermediate tick marks divide the space between major tick marks into logical numbers of smaller intervals. For example, if major intervals represent years, minor tick marks might divide the years into 12 smaller intervals to represent months. On a quantitative scale, if the major ticks represent units of ten, the minor tick marks might divide the intervals into units of 1, 2, or 5. Minor tick marks seldom generate intervals that make it difficult for the viewer to relate to. For example, minor tick marks would seldom form intervals such as 1.67, 3.33, 14.3, etc. An exception is when the prescribed parameters require it. For example, if one sets the upper value on the class interval scale of a histogram at 20 and the lower value at 13 and requires there be 10 class intervals, each interval will be 0.7 units wide and the values on the scale will be 13, 13.7, 14.4, 15.1, 15.8, etc.

Tier   Sometimes called bank, cycle, deck, or phase. The major interval on a logarithmic scale.

Time and Activity Bar Chart The bar chart is the most widely used type of time and activity chart. There are many different variations and applications. Most have the following things in common:

- The major purpose is to relate events, activities, actions, etc., to time.
- Time is displayed on the horizontal axis.
- Either the activities or the people, places, or things involved with the activities are displayed on the vertical axis
- Horizontal bars, colors, symbols, etc., are used to designate blocks of time.

Major functions of time and activity bar charts

Five of the major functions of this type of bar chart are scheduling, loading, project planning, monitoring/managing of activities, and communicating. Almost all charts accomplish the function of communicating, and the majority accomplish two or more of the other four functions. Below are examples of charts used for the first four functions.

Charts for scheduling

Shown below are two simple examples of time and activity bar charts used for scheduling. The chart on the left has people's names on the vertical axis. The other has activities on the vertical axis. Key objectives of this type of chart include indicating when things will happen, assuring that things do not overlap, coordinating multiple activities or individuals, etc. There generally is no attempt to fill all of the cells. Additional information can be encoded into the bars such as priorities, required or optional, status, etc.

Vacation Schedule												
Week	1	2	3	4	5	6	7	8	9	10	11	12
Avery, T.		▓	▓									
Chandler, A								▓	▓			
Dorsey, J.				▓	▓							
Goodman, E.						▓	▓					
Hewell, P.										▓	▓	
Leahy, F.	▓	▓										
Mansfield, G.			▓	▓								
Scott, H.								▓	▓			

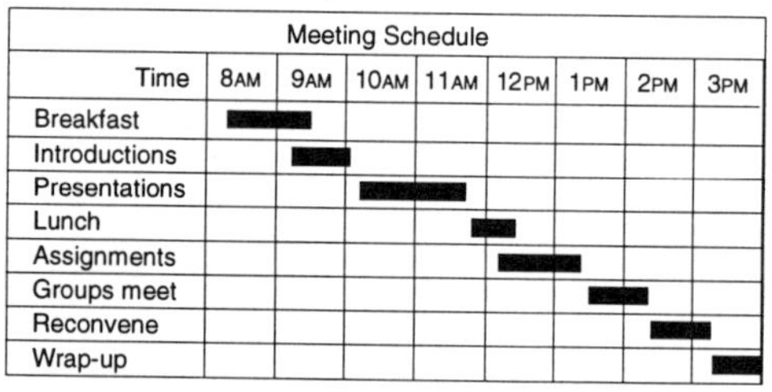

Meeting Schedule								
Time	8AM	9AM	10AM	11AM	12PM	1PM	2PM	3PM
Breakfast	▓							
Introductions		▓						
Presentations			▓					
Lunch					▓			
Assignments						▓		
Groups meet							▓	
Reconvene								▓
Wrap-up								▓

Examples of time and activity bar charts used for scheduling.

Charts for loading

When time and activity bar charts are used for assigning work to individuals, machines, facilities, etc. (sometimes referred to as loading), the charts are laid out basically the same as above; however, they are filled in differently, as shown in the example below. When the cell is blank it generally means that nothing is scheduled for that time period. When an assignment has been made, the cell is filled. In this type of chart, the objective frequently is to have every cell filled since that would indicate maximum utilization of the resource. Bars are often drawn so they fill the entire cell. Additional information is sometimes included in cells such as operator, job number, priority, etc.

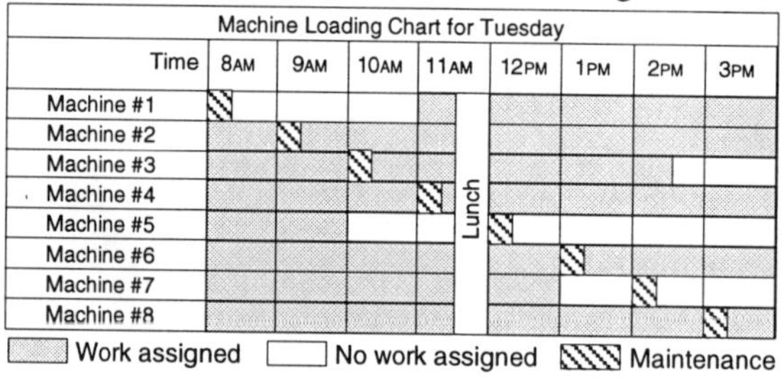

Machine Loading Chart for Tuesday								
Time	8AM	9AM	10AM	11AM	12PM	1PM	2PM	3PM
Machine #1	▨							
Machine #2		▨						
Machine #3			▨					
Machine #4				▨				
Machine #5					▨			
Machine #6						▨		
Machine #7							▨	
Machine #8								▨

▒ Work assigned   ☐ No work assigned   ▨ Maintenance

Example of a bar chart used for loading
(e.g., assigning work)

Charts for project planning

Time and activity bar charts are frequently used to plan activities such as projects that have a distinct beginning and end. Such charts are sometimes referred to as Gantt charts. In this type of chart, each subprogram or activity that is involved in the completion of an overall program is represented by a horizontal bar. There may be one or more horizontal bars per line. The ends of the bar represent the start and finish of an activity. In addition to indicating the time intervals, bars can include considerable additional information. In the example at the right, whether the work is to be done internally or externally is encoded. Other examples are shown on the next page.

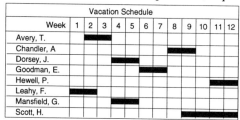

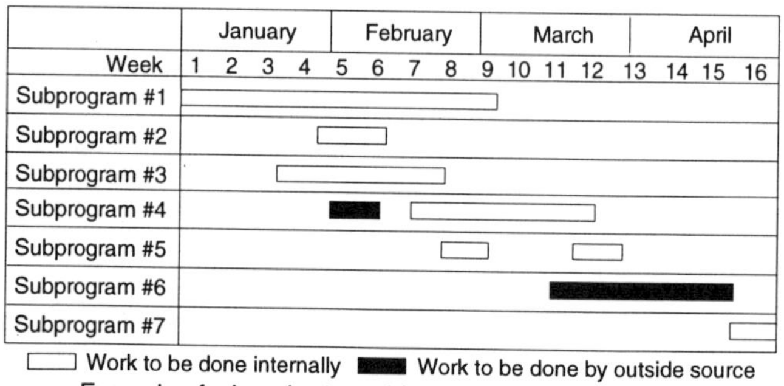

	January				February				March				April			
Week	1	2	3	4	5	6	7	8	9	10	11	12	13	14	15	16
Subprogram #1																
Subprogram #2																
Subprogram #3																
Subprogram #4																
Subprogram #5																
Subprogram #6																
Subprogram #7																

☐ Work to be done internally   ■ Work to be done by outside source
Example of a bar chart used for planning major programs.
Sometimes referred to as a Gantt chart.

Major functions of time and activity bar charts (continued)

──── **Charts for project planning** (continued) ────

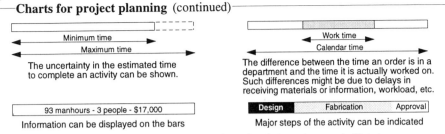

The uncertainty in the estimated time to complete an activity can be shown.

The difference between the time an order is in a department and the time it is actually worked on. Such differences might be due to delays in receiving materials or information, workload, etc.

Information can be displayed on the bars

Major steps of the activity can be indicated

Examples of the additional types of information that can be encoded into bars

──── **Charts for monitoring and managing** ────

Charts used for monitoring and managing purposes generally are expansions of the same charts used for planning purposes. Key information about the tasks, such as the name of the activity, accounting numbers, departments or individuals involved and/or responsible, dollars budgeted and spent, man-hours budgeted and used, etc., are sometimes included on the chart. Color, shading, or patterns can be used to identify certain types of activities. Reference lines and markers can be used to identify key external events, such as when the chart was last updated or when a competitor is expected to introduce a new product that will compete with the one being developed by the project. In some cases bars are darkened or a second set of bars is added to indicate the amount of an activity that has been completed. Examples are shown below.

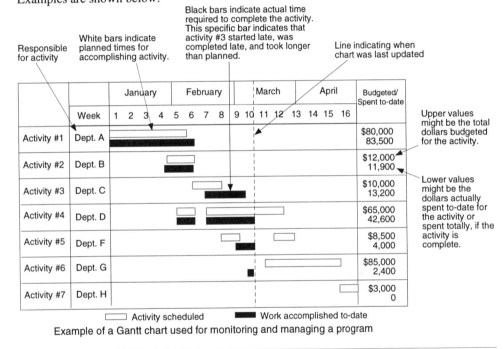

Example of a Gantt chart used for monitoring and managing a program

Interdependencies between activities

Bar charts typically do not indicate interdependencies between the various bars or subprograms; thus, there typically is no way to know which tasks must be completed before which other tasks can be started. This problem can be partially overcome by including arrows on the chart. For example, if an arrow leads from one bar to another, it indicates that the task represented by the bar at the base of the arrow must be completed before the task at the tip of the arrow can be started. If bars are one above the other without interconnecting arrows, it indicates that the tasks represented by the bars can proceed independently. An abbreviated example is shown at the right. For time and activity charts that display all interdependencies, see PERT chart and Critical Path Method (CPM).

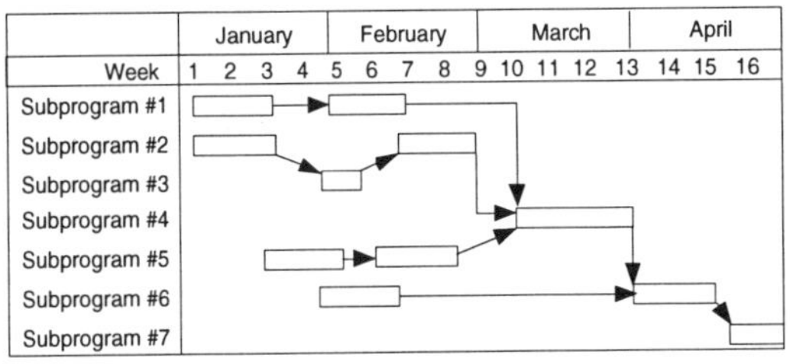

Bar chart where dependencies between activities are indicated by arrows.

Milestone chart

When a subprogram or activity is scheduled to take a fair amount of time, it is sometimes broken into smaller pieces and the completion of each smaller piece recorded. This provides closer control of the project and allows difficulties to be detected earlier. Some sort of symbol or notation is located at the points where the smaller tasks are to be started and completed. These are called milestones and the chart itself is sometimes referred to as a milestone chart. Shown below are four different ways milestones are sometimes denoted.

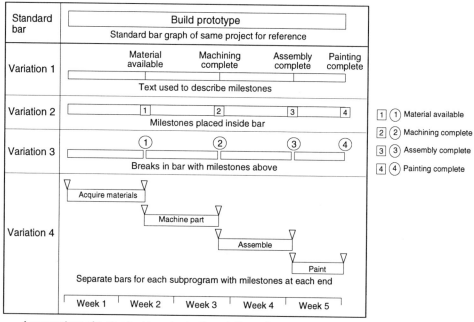

A comparison of a standard bar chart with variations of milestone charts

Milestone symbols for tracking

Once a program is begun there are several methods for tracking the various activities against the proposed schedule. One of the most widely used methods is to darken the bars and/or milestones (see examples below) to indicate what portion has been completed to date. By means of variations in milestone symbol shapes, colors, locations, etc., the status of each aspect of a project can also be tracked. See examples at the right.

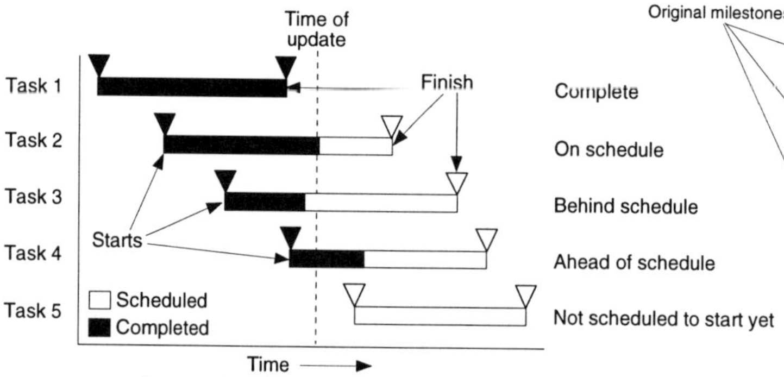

One method for recording progress on a bar chart using milestones

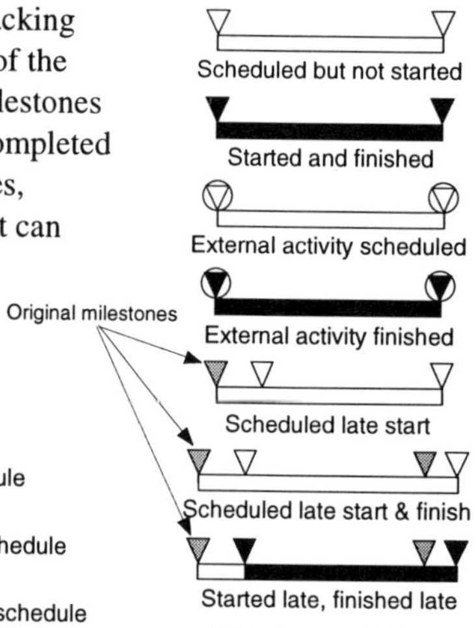

Milestone variations

General

The lengths of the bars in Gantt and milestone charts are generally proportional to the times activities will take. • In the case of scheduling and loading charts, activities are often deleted as they are completed. With Gantt and milestone charts, subprograms are generally not deleted as they are completed. • The only limitation on the number of activities or length of time the activities might take in a time and activity bar chart is the physical size of the chart. An alternative to one large chart is to have a master chart plus several small charts covering major segments of the main program.

Time and Activity Charts

These charts relate activities and time using combinations of words, numbers, and graphics to represent activities and events. In some cases the sizes of the graphic elements are proportional to the time allotted or required to perform the given activities (e.g., Gantt and milestone charts). In other cases, such as PERT and CPM charts, the graphic elements are not proportional. The size and complexity of these charts ranges from small scheduling charts used in most offices to very complicated charts that require a computer to update them. A few of the basic types of time and activity charts have names associated with them. Many do not. Even those that have acquired names over the years (and in some cases have had books written about them) are generally modified by individual users to fit their unique requirements. The major purposes for time and activity charts include documenting, scheduling, analysis, planning, monitoring, controlling and communicating. Shown below are examples of the major types of time and activity charts. See chart types for more detail.

Time line chart

Time line charts show the chronological sequence of activities.

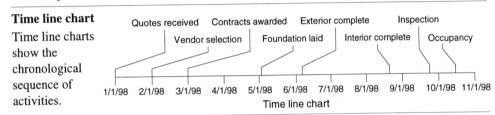

Time line chart

Time and activity bar chart

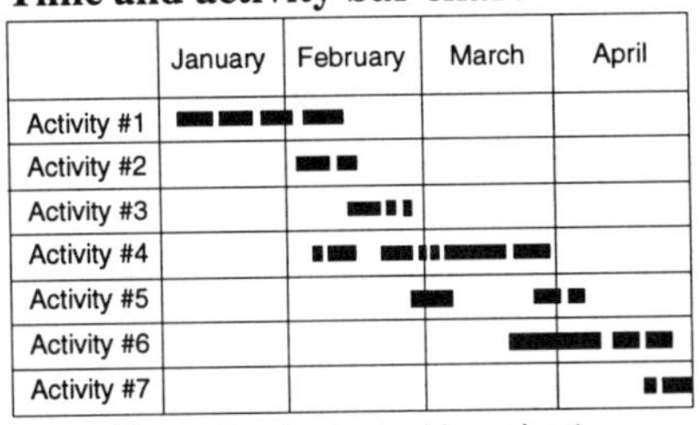

	January	February	March	April
Activity #1	▬▬ ▬▬ ▬▬			
Activity #2		▬▬ ▬		
Activity #3			▬▬ ▮▮	
Activity #4		▮▮▮	▬▬ ▬▬▬	
Activity #5			▬▬	▬ ▮
Activity #6			▬▬▬▬▬	▬▬▬
Activity #7				▬▬

Example of a typical bar chart used to relate time and activities

Bar charts that are used for relating time and activities are referred to by a number of different names including:
- Gantt chart
- Milestone chart
- Scheduling chart
- Loading chart

Many bar charts used for these purposes are customized for specific applications.

Process chart

There are several types of time and activity charts used to document and analyze processes in detail. Included among them are:
- Operation process charts
- Flow process charts
- Operator process charts
- Machine use charts
- Multiple activity process charts

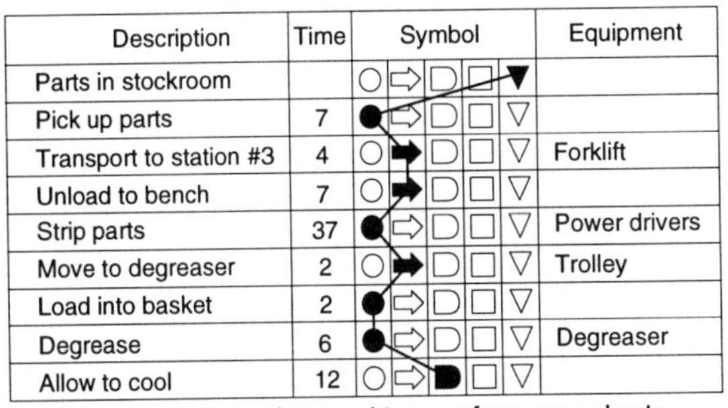

Description	Time	Symbol	Equipment
Parts in stockroom		○ ▷ □ ▽	
Pick up parts	7	● ▷ □ ▽	
Transport to station #3	4	○ ▶ □ ▽	Forklift
Unload to bench	7	○ ▶ □ ▽	
Strip parts	37	● ▷ □ ▽	Power drivers
Move to degreaser	2	○ ▶ □ ▽	Trolley
Load into basket	2	● ▷ □ ▽	
Degrease	6	● ▷ □ ▽	Degreaser
Allow to cool	12	○ ▷ ■ ▽	

Example of one of several types of process charts

Critical path method (CPM) chart

Critical Path Method (CPM) charts are used to plan, analyze, and monitor large programs. In addition to showing the time required to accomplish each activity, they also display the interrelationships of the various activities.

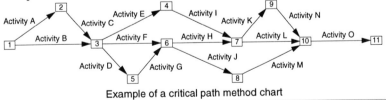

Example of a critical path method chart

PERT chart

PERT charts accomplish the same functions as CPM charts. The major difference is that instead of noting how long it will take to accomplish each task or activity, as shown in the CPM chart, the PERT chart denotes when events (shown in the ovals) will happen.

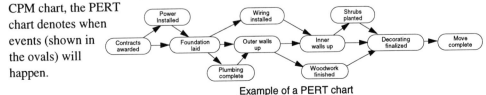

Example of a PERT chart

Time Line Chart

Sometimes referred to as a sequence chart. Generally a time line chart is a one-axis chart used to display past and/or future events, activities, requirements, etc., in the order they occurred or are expected to occur. The major function of time line charts is to consolidate and graphically display time-related information for purposes of analysis and communication.

In addition to enabling the viewer to see graphically when things occurred or are

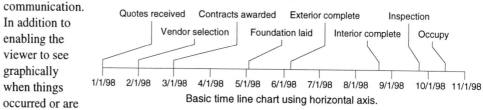
Basic time line chart using horizontal axis.

to occur, a time line lets the viewer assess the time intervals between events. For example, in addition to seeing when each customer placed orders, one can graphically observe whether there is a pattern to the intervals between orders. If the chart is historical in nature and covers very long periods of time, it is sometimes referred to as a chronology chart. The

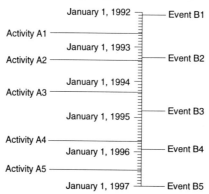

Vertical time line chart with two series of data, major time intervals of years and minor intervals of months.

axis of a time line chart can run horizontally (above) or vertically (left). Time almost always progresses from left to right when the axis is horizontal. On a vertical axis, time might progress up or down. Progression from top to bottom is frequently used.

Scales might be in terms of specific times (e.g., three o'clock, four o'clock, five o'clock), specific dates (e.g., June, July, Aug) or in terms of elapsed time such as first year, second year, third year; ten minutes, twenty minutes, thirty minutes; etc. • Time scales can be linear or logarithmic, but are generally linear. • Multiple data series can be shown on the same chart (above).

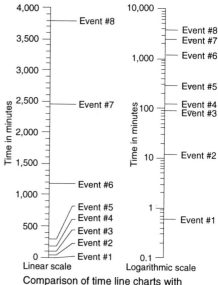
Comparison of time line charts with linear and logarithmic scales

• For repetitive activities, circular time lines are occasionally used, as shown at the left. With this configuration, additional information can be noted or encoded into the segments of the circle. • In some cases time intervals are recorded as opposed to specific times. In these cases, the line is

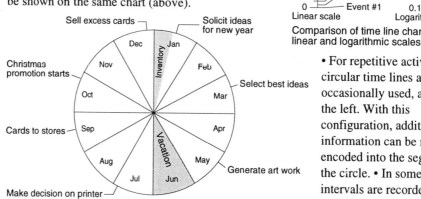
Circular time lines are sometimes used for activities that are repetitive

generally thickened to become a bar (example below) and increments of the bar used to identify the various time periods. Alternate increments are many times colored or shaded so the transitions from one period to another stand out. Different colors or shades can be used for the increments to encode additional information.

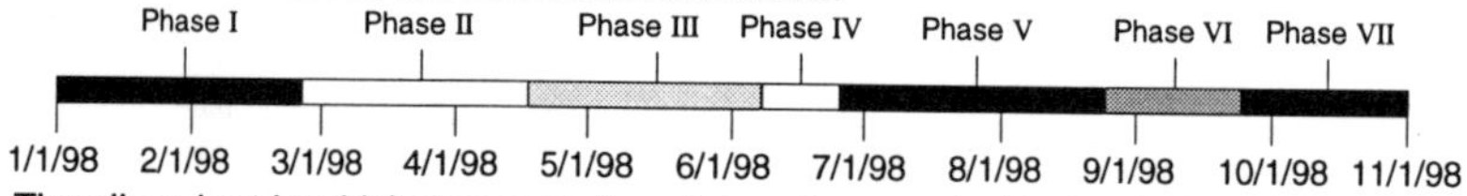

Time line chart in which time periods or intervals are noted instead of specific events

Time Series Axis   An axis with a time series scale.

**Time Series Graph
and
Time Series Scale
or
Time Scale**

Sometimes called a chronological scale or graph. A time series graph is a variation of a sequence graph that has a time series (frequently abbreviated to time) scale on the horizontal axis and a quantitative scale on the vertical axis. Time progresses from left to right. In a three-dimensional time series graph, the time scale is displayed on either the X or Y axis and a quantitative scale on the vertical axis. In circular graphs the time series scale is almost always on the circular axis. Three widely used time series formats are shown below.

Time series formats

Simple time series

This is the most widely used type of time series graph. This format is probably the easiest to read and makes it easier to spot long term trends. It has the advantage over the repeated time scale format, shown below, in that it does not have the discontinuities associated with one time period ending on the right side of the graph and the next time period beginning on the left.

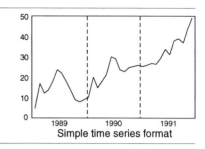
Simple time series format

Repeated time scale

In a repeated time scale graph, a single data series is broken into pieces and each of those pieces represented by a different curve on the same graph. For example, a three-month data series might be broken into three one-month periods, or a three-year data series into three one-year periods, as shown at the right. The same quantitative scale applies to all the curves. This type of graph makes it easier to compare a given point in time (e.g., March, June, etc.) from time period to time period (e.g., year to year) and sometimes makes repetitive patterns stand out more clearly.

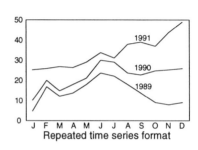
Repeated time series format

Multiple time scale

Multiple time scales are used on graphs that have multiple data series covering different time periods (e.g., two post-war periods, the start-up years for several different stores) plotted on them. Since the data series cover different time periods, each must have its own time scale on the horizontal axis. The data series may or may not use the same quantitative scale on the vertical axis.

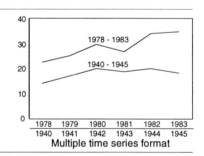
Multiple time series format

Basic types of graphs most frequently used with time series

Line and column type graphs frequently incorporate time series scales. Time scales are sometimes used with point and area graphs, but seldom with bar graphs. With circular graphs, the line format is most frequently used with the time series scale on the circumference. Several examples of time series graphs are shown here.

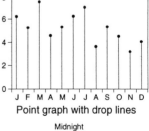
Point graph with drop lines

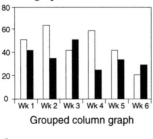
Grouped column graph

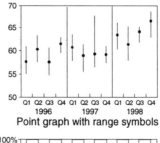
Point graph with range symbols

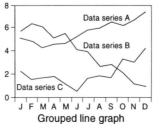
Grouped line graph

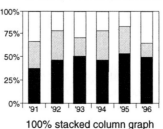
100% stacked column graph

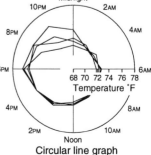
Circular line graph

417

**Time Series Graph
and
Time Series Scale
or
Time Scale**
(continued)

Period, point, and cumulative data

Period data

Period data is data that applies to some interval of time – for example the total cash collected for a month, the average temperature for a week, or the average membership for the year.

Point data

Point data is data that applies to a specific point in time – for example, the cash on hand the last day of the month, the temperature at midnight on Sunday, or the membership as of the last day of the year.

Cumulative data

Cumulative data is period data that is accumulated over an interval of time – for example cumulative sales are the total sales from the beginning of the year to any given point during the year. They are sometimes referred to as year-to-date sales. • Point data is not cumulated.

Equal scale intervals

With few exceptions, the time intervals and the distances between the labels representing the intervals are equal on a given time scale. Exceptions are:

– When very long times are involved and the emphasis is on the most recent times. In these cases the spacings occasionally resemble those of a logarithmic scale.
– When the most recent time period is divided into smaller increments to provide greater detail. For example, if annual data is presented on a graph, sometimes the current year will be broken into quarters or months. The example at the right illustrates one of the hazards of doing this and offers a partial solution to the problem.

An illustration of the effect that changing scale intervals can have on a graph when period data is being plotted. The dashed line indicates a partial solution to the problem.

Missing or irregular data

It is generally recommended that there be no breaks or irregularities in a time scale since breaks can distort the pattern or trend of the data. Even though data might be irregular and in some cases missing, the time scale should be continuous and uniform. An exception is the common practice of eliminating intervals that are immaterial to the graph. For example, if a company never does business on Sunday, that day may not be shown on the graph of daily sales. The key is how the scale intervals are defined. If they are defined as all calendar days, Sundays will normally be shown. If the intervals are defined as all business or working days, Sundays will frequently not be shown. • When data is irregular or missing there are several alternatives for addressing the situation. In column or vertical line graphs, a gap is generally left where there is no data. A value can be interpolated and a dashed column or vertical line drawn in the gap, but this is generally not done. With standard line or area graphs an alternative to leaving a gap is to bridge the gap with a solid or dashed line. The dashed line is sometimes recommended because it calls the viewer's attention to the missing data. When the data is very irregular, a point graph, with or without drop lines, is sometimes used so there are no decisions as to whether or not to bridge gaps. When a line graph is used, plotting symbols are sometimes shown in conjunction with the line so the viewer can see the exact data from which the curve was drawn.

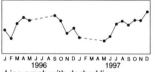

Point graph showing irregular data

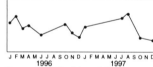

Line graph with data points of
irregular data connected by solid lines

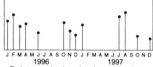

Line graph with gaps
where data is missing

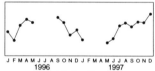

Line graph with dashed lines
bridging gaps where data is missing

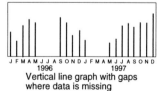

Vertical line graph with gaps
where data is missing

Methods for dealing with gaps in data

Title Text Chart

A variation of a text chart. A title text chart is frequently used as the introductory page or slide in a presentation. In addition to the title or subject of the presentation, meeting, event, etc., this chart might include such information as the purpose of the meeting, an outline of meeting, or key points to be covered. See Text Chart.

<table>
<tr><td>

Topographic Map
or
Topo Map

</td><td>

Topographic maps portray the shape and elevation of sections of the earth's surface. Included on such maps are natural features (mountains, valleys, rivers, lakes, etc.) as well as a limited number of man-made features (roads, quarries, reservoirs, political boundaries, etc.). Topographic maps are many times used as base maps over which buildings, power lines, recreational facilities, airports, statistical data, etc., are superimposed. Two-dimensional types are the most widely used; however, three-dimensional variations are becoming increasingly popular. Years ago, short lines called hachures (example at right) were used to indicate changes in elevations (relief). Today, contour lines (a form of isoline) are used almost exclusively on two-dimensional topographic maps. Topographic maps are sometimes enhanced by filling the areas between contour lines with different shades or colors (called hypsometric tints) or

</td></tr>
</table>

Example of hachures which were used to indicate changes in elevation before contour lines became widely used

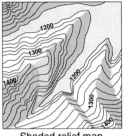

Shaded relief map

by applying shading to represent shadows cast by an imaginary light source (called shaded relief). The shading gives the map the appearance of being three-dimensional as shown at the left. One additional variation of a topographic map simply shows selected elevations (called spot elevations) on a standard two-dimensional map. • An example of a two-dimensional topographic map with contour lines is shown below. An elevation (sometimes called section or profile) map is included to assist in the interpretation of the contour map.

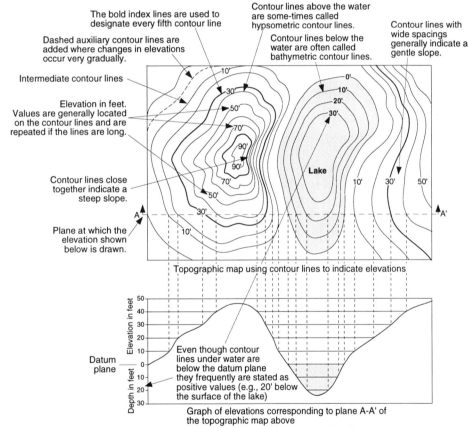

The bold index lines are used to designate every fifth contour line

Dashed auxiliary contour lines are added where changes in elevations occur very gradually.

Intermediate contour lines

Elevation in feet. Values are generally located on the contour lines and are repeated if the lines are long.

Contour lines close together indicate a steep slope.

Plane at which the elevation shown below is drawn.

Contour lines above the water are some-times called hypsometric contour lines.

Contour lines below the water are often called bathymetric contour lines.

Contour lines with wide spacings generally indicate a gentle slope.

Lake

Topographic map using contour lines to indicate elevations

Datum plane

Even though contour lines under water are below the datum plane they frequently are stated as positive values (e.g., 20' below the surface of the lake)

Graph of elevations corresponding to plane A-A' of the topographic map above

Example of contour map and a map of elevations at one cross-section of the map

Contours are shown as solid lines on maps and represent imaginary lines on the terrain along which all points have the same elevation. For example, a contour labeled 50 feet represents every point on the land surface that is 50 feet above a reference plane (referred to as a datum plane). That reference plane frequently is sea level. The numbers indicating elevations of the contour lines are normally shown directly on top of the line to avoid misunderstanding. The change in elevation represented from one contour line to another (called contour interval) depends on the scale of the map and the elevations in the area being plotted (i.e., mountains have larger intervals than plains). Whatever interval is selected, it is generally used consistently throughout a given map. When the changes in elevation are very gradual, contour lines are so far apart that additional lines must be added.

These additional contour lines, called auxiliary contour lines, are shown as dashed lines and placed at one-half the interval between regular contour lines. • For identification purposes every fifth contour line is sometimes drawn with a bolder line. These bolder lines are referred to as index contours, and the dashed lines in-between as intermediate contours. In addition to quantitatively designating elevations, the spacing and shape of contour lines also give graphic indications of the nature of the terrain, as shown in the examples below.

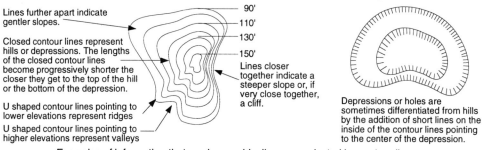

Lines further apart indicate gentler slopes.

Closed contour lines represent hills or depressions. The lengths of the closed contour lines become progressively shorter the closer they get to the top of the hill or the bottom of the depression.

U shaped contour lines pointing to lower elevations represent ridges

U shaped contour lines pointing to higher elevations represent valleys

90'
110'
130'
150'

Lines closer together indicate a steeper slope or, if very close together, a cliff.

Depressions or holes are sometimes differentiated from hills by the addition of short lines on the inside of the contour lines pointing to the center of the depression.

Examples of information that can be graphically communicated by contour lines

Three-dimensional topographic maps

Topographic maps are sometimes drawn as three-dimensional. The illustration below shows the relationship between two- and three-dimensional types. An elevation map is included for reference.

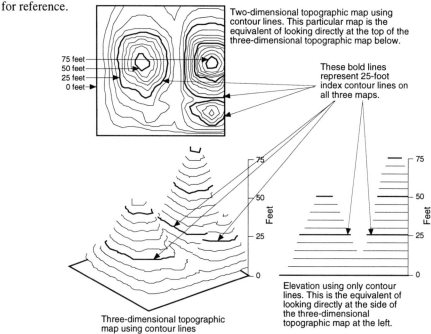

Two-dimensional topographic map using contour lines. This particular map is the equivalent of looking directly at the top of the three-dimensional topographic map below.

75 feet
50 feet
25 feet
0 feet

These bold lines represent 25-foot index contour lines on all three maps.

Three-dimensional topographic map using contour lines

Elevation using only contour lines. This is the equivalent of looking directly at the side of the three-dimensional topographic map at the left.

A comparison of two- and three-dimensional topographic maps using the same data in both. The elevation is included for general information.

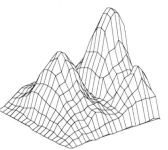

There are a number of possible variations of three-dimensional topographic maps. One variation uses contour lines as shown above. In some cases the areas between the contour lines are filled with different colors, tints, and/or shading. Another variation is to draw isolines along the X and Y axes, as shown at the left. This variation is many times referred to as fishnet. In still other cases the topographic map appears as though drawn by hand, as shown at the right. All three types are occasionally referred to as

Example of a topographic map using isolines along the X and Y axes. This map was generated with the same data used in the three-dimensional contour map shown above.

block diagram maps. Quantitative information is difficult to determine from either the isoline or hand-drawn variations; however, both types frequently make it easier for the viewer to visualize the overall terrain than with contour lines.

Example of a hand-drawn topographic map. Geological information can be shown on the sides of the block.

Tracking	Tracking refers to the horizontal spacing between letters and words as shown in the examples below. Number nine size type is used in each example.

Loose tracking —— This is an example of changing horizontal spacing
Typical tracking —— This is an example of changing horizontal spacing
Tight tracking —— This is an example of changing horizontal spacing

Trajectory Line Graph	A three-dimensional graph containing one or more smooth three-dimensional line curves, as shown at the right. Typically such a graph has three quantitative axes.

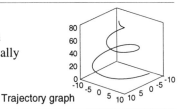

Trajectory graph

Transpose	To transpose is to interchange the rows and columns in a table or matrix. For example, when a table is transposed, the information that originally was shown in the columns is then shown in the rows, and the data originally in the rows is shown in the columns. The data itself remains unchanged. In some computer programs, when the data in a table is transposed, the graphs associated with that table are correspondingly affected. An example of a transposed table is shown at the right. Also shown is an example of the effect such a change sometimes has on a graph of the data.

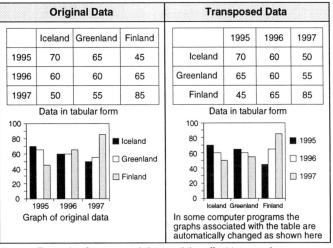

Example of transposed data and the effect transposing sometimes has on graphs generated from the data.

Tree Diagram[1] **or Tree Chart**	Sometimes referred to as a dendrite diagram. A tree diagram typically starts or finishes with a single entry or junction. If it begins with a single entry, that entry has one or more paths leading from it, some or all of which subdivide into multiple paths. Some or all of these subpaths, in turn, subdivide into additional subpaths. This process can be repeated any number of times. If the diagram ends with a single entity or junction, the process works in reverse. Two of the best-known examples of a tree diagrams are organization charts and family or genealogical trees. The point where a single path subdivides into multiple paths is sometimes referred to as a node or junction. These points typically represent such things as people, actions, events, decisions, accomplishments, positions, options, etc. The paths between nodes are sometimes called branches, arms, arcs, connecting lines, or links. Paths are generally used to indicate how the nodes are connected or interrelated. Nodes are sometimes organized into levels, such as levels of management or levels in a food chain.

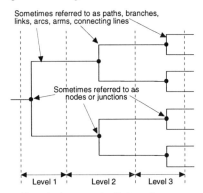

Terminology used with tree diagrams

Orientation of tree and how it is read

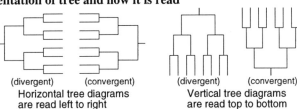

(divergent)   (convergent)
Horizontal tree diagrams are read left to right

(divergent)   (convergent)
Vertical tree diagrams are read top to bottom

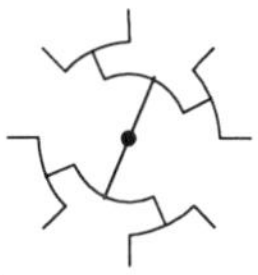

Circular tree diagrams are read from the center out

Tree Diagram[1]
or
Tree Chart
(continued)

Variations in connecting lines and symbols

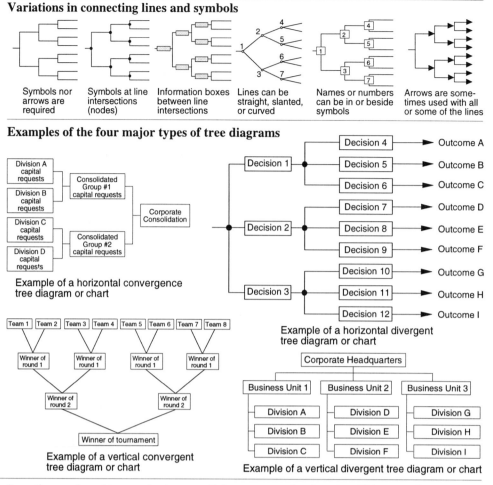

Symbols nor arrows are required

Symbols at line intersections (nodes)

Information boxes between line intersections

Lines can be straight, slanted, or curved

Names or numbers can be in or beside symbols

Arrows are sometimes used with all or some of the lines

Examples of the four major types of tree diagrams

Example of a horizontal convergence tree diagram or chart

Example of a horizontal divergent tree diagram or chart

Example of a vertical convergent tree diagram or chart

Example of a vertical divergent tree diagram or chart

Tree Diagram[2]

Sometimes referred to as a cluster map, linkage tree, or dendrogram. This type of tree diagram is a graphical means of organizing information for the purpose of establishing groupings and/or categorizing individual elements. See Dendrogram.

Tree Diagram[3]

A variation of a decision diagram. See Decision Diagram.

Trend Line
and
Trend Chart/Graph

The term trend line is used in several different applications. In each case it refers to a line or curve that indicates the general direction of the data over time. For example, Are costs trending up and at what rate? Is attendance trending down? Are interests rate leveling off? Is the process stable? The graph on which the trend line is displayed is sometimes referred to as a trend chart or graph. Examples are shown below. In some cases a line is drawn that connects all the actual data points. In other cases a line that smooths or approximates the general pattern or trend of the actual data points is drawn. Such lines might be drawn based on visual observations or, as is frequently the case, arrived at mathematically. The data points through which the trend line is drawn, or that it approximates, might represent individual data elements or groups of data elements such as averages or medians. Trend lines may or may not be accompanied by additional lines, such as upper and lower statistical limits, specification values, goals, targets, confidence limits, etc. In some cases the data points alone are referred to as the trend line, with no additional lines superimposed.

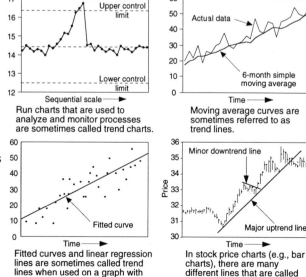

Run charts that are used to analyze and monitor processes are sometimes called trend charts.

Moving average curves are sometimes referred to as trend lines.

Fitted curves and linear regression lines are sometimes called trend lines when used on a graph with time on the horizontal axis.

In stock price charts (e.g., bar charts), there are many different lines that are called trend lines.

Examples of curves that are sometimes referred to as trend lines

Trilinear Graph
or
Triangle Graph
or
Triangular Graph

Sometimes referred to as a ternary graph. Trilinear graphs are the only triangular-shaped graphs in popular use today. They are used to plot information that has three variables, the total of which always equals the same amount. That amount can be any value; however, it is generally 1 or 100%. A typical example is the percent of material, labor, and overhead in the total cost of a product. Different products have different percentages of the three elements, but the percentages of the three elements for most products add up to 100%.

Description of a trilinear graph

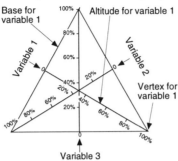

Basic layout of a trilinear graph with scales on each of the three altitudes.

A trilinear graph consists of an equilateral triangle (all 3 sides of equal length) with each line from a vertex (apex) to the opposite base representing one of the three variables. (The line from the vertex to the opposite base is called an altitude.) Scales are distributed along the altitudes with zero at the base and 100% at the vertex, as shown at the left. To prevent the scales and labels from interfering with the data points and with one another, the scales are sometimes projected to the sides of the triangle. The process for projecting the scales to the sides is outlined in the illustration at the right and examples of the graphs after all three sets of labels have been transferred are

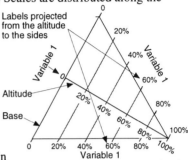

An illustration of how the scale of an altitude is projected to the sides by drawing grid lines parallel to the base. Each altitude scale can be displayed on one of two sides. The chosen side is then used with all three altitudes.

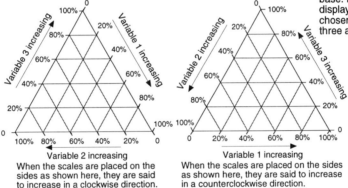

When the scales are placed on the sides as shown here, they are said to increase in a clockwise direction.

When the scales are placed on the sides as shown here, they are said to increase in a counterclockwise direction.

Examples of trilinear graphs after all three scales have been transferred to the sides.

shown at the left. As the illustration above indicates, each altitude scale can be transferred to one of two sides. Depending on the sides selected, the values increase in the clockwise or counterclockwise direction. Where the labels are located has no effect on the location of data points; however, it does have an effect on the location of the scale for a particular variable. For instance, in the example where the scales proceed clockwise, the scale for variable 1 is on the right side of the graph. When the scales proceed counterclockwise, variable 1 is on the bottom.

Labels, tick marks, and grid lines

Whether to display the scales on the altitudes or sides is not an easy decision. Graphs tend to be easier to interpret when the scales are on the altitudes, but they tend to be less visually confusing when scales are located on the sides of the graph. When scales are displayed on the sides of the triangle, care must be taken to make it clear as to which grid line each label applies to, since two grid lines converge at each label. To address this issue, projected tick marks are sometimes added, or the labels are tilted to orient the viewer as to which set of grid lines the labels apply to. In still other cases the label is placed on the grid line just inside the triangle. Examples are shown at the right. • All three scales on trilinear graphs typically range from zero to 1 or zero to 100%. No two zeros are at the same vertex of the triangle. • Major grid lines and tick marks may or may not be shown, though both generally are. Depending on the particular data and the degree of accuracy desired, minor grid lines and tick marks may or may not be used.

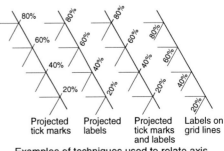

Examples of techniques used to relate axis labels to the appropriated grid lines

Axis terms and titles

The axes of trilinear graphs are referred to in three different ways: by their position such as left, right, or bottom (also referred to as horizontal or base); by the specific thing measured along the axis such as material, labor, ingredient A, ingredient B, etc.; or by the letters X, Y, and Z. There are no guidelines as to which letter should be assigned to which axis. Scale titles are placed either beside the axis to which they apply or at the vertexes. Both are illustrated on the same graph at the right. When the titles are placed at the vertexes, it is sometimes difficult to determine which axis they apply to.

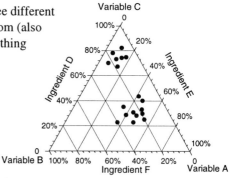

Examples of two locations for scale titles. When titles are located at vertexes, it is sometimes difficult to determine which axis they apply to.

Drop lines

When there are just a few data points, drop lines are sometimes used instead of grid lines. A comparison of the two variations is shown below. Drop lines are drawn perpendicular to the altitudes (parallel to the bases), not perpendicular to the sides of the triangle. Drop lines have the advantage of more easily directing the viewer's eye to the appropriate scale, thus making the estimating of values more accurate.

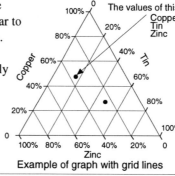

The values of this data point are:
Copper = 48%
Tin = 16%
Zinc = 36%

Example of graph with grid lines

Example of graph with drop lines

Multiple data series plotted on the same graph

Multiple data series can be plotted on the same graph using different symbols to differentiate them. The multiple data series can be for the same set of things at two points in time, or for two different sets of things at the same point in time.

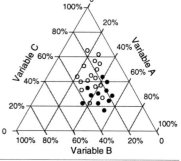

Multiple variables plotted on the same graph using different symbols to differentiate them

Data points connected by lines

When changes over time are to be noted, the data points are sometimes connected by lines or arrows in sequential order as shown at the right. In this way, trends or drifts can easily be noted.

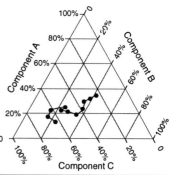

Example of a trilinear graph used to track a set of values over time. The lines connect successive readings.

Isolines

In some cases trilinear graphs are used to establish or record the various combinations of three main ingredients that yield a common result. For example, the combinations of virgin aluminum, recycled aluminum, and scrap aluminum that yield the same compression strength of the resulting metal. Such lines are referred to as isolines. The example at the right illustrates such a graph.

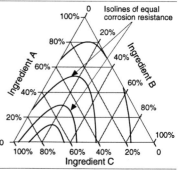

Example of a graph with isolines of some property that is a function of the relationship of the three variables

**Trilinear Graph
or
Triangle Graph
or
Triangular Graph**
(continued)

Trilinear graphs as monitoring devices

As a monitoring device, repetitive data can be recorded on a trilinear graph along with a tolerance or specification area. Periodic data can be added to the chart as it becomes available. In the example below, the circle indicates the specification limits for a given product and the dots record actual test data. It can be seen that the data is concentrated in the upper left portion of the specification, with several samples being out of specification. • When it is known that all of the data points will be concentrated in a small area, sections of the graph can be separated and enlarged, as shown at the right. When this is done the scale labels are the same as in the larger graph and do not run from 0% to 100%.

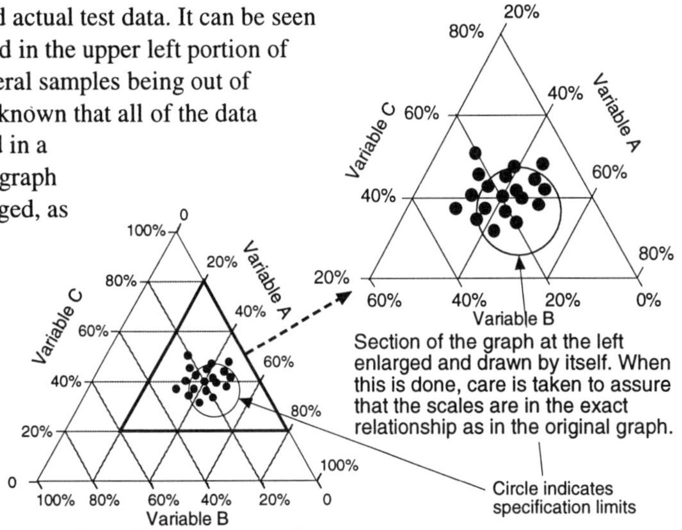

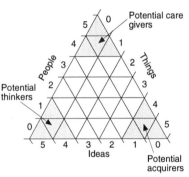

Section of the graph at the left enlarged and drawn by itself. When this is done, care is taken to assure that the scales are in the exact relationship as in the original graph.

Circle indicates specification limits

Trilinear graph used for quality control purposes where each dot represents a different sample and the circle represents the specification limits

Other uses of trilinear graphs

Trilinear graphs can be used in applications where the scales are not in percents. For example, in evaluating personalities, candidates are sometimes rated on the relative importance they place on people, things, and ideas. Any number between 0 and 5 can be assigned to each of the three measurements, but the total of the three must add up to 5. After a group of scores are plotted on a graph similar to the one at the right, those people with similar interests form clusters. Any values can be used on the axes as long as they are all the same and no two zeros are at the same vertex.

Trilinear graph with unique scales

General

• There are no three-dimensional trilinear graphs in general use.

• Although each data point represents three values, only two are required to plot the data point. For example, if a data element has the values of 20%, 30%, and 50%, the 20% and 30% values will accurately locate the point and the 50% occurs automatically. Or, the 30% and 50% values might be used and the 20% occurs automatically.

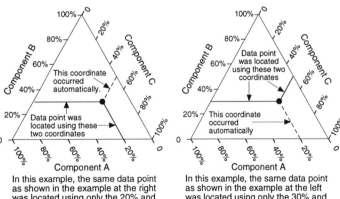

In this example, the same data point as shown in the example at the right was located using only the 20% and 30% values. The 50% value occurred automatically.

In this example, the same data point as shown in the example at the left was located using only the 30% and 50% values. The 20% value occurred automatically.

Examples illustrating that only two of the three values of a data element are required to locate the data point on a trilinear graph

Trivariate	A subcategory of multivariate. A chart, graph, map, group of data, etc., with three variables.
**Two-Axis Graph
or
Two-Dimensional Graph**	A graph with data on two axes, usually the horizontal and vertical axes. This is the most widely used graph format; most of the examples in this book are of the two-axis type. They are frequently referred to as two-dimensional graphs. The companions to the two-axis graph are the one-axis and three-axis graphs.

Two-by-Two Chart

Sometimes referred to as a four-fold chart. See Four-Fold Chart/Graph.

Two-Way Bar Graph

Sometimes referred to as a bilateral, opposed, paired, or sliding bar graph. This type of graph is a variation of a bar graph in which positive values are measured both right and left from a zero on the horizontal axis. The major purpose of such a graph is to compare two or more data series with particular attention to correlations or other meaningful relationships. See Bar Graph.

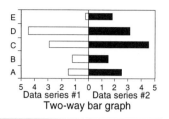

Two-way bar graph

Two-Way Column Graph

Sometimes referred to as a floating column graph. This type of graph is a variation of a column graph in which positive values are measured both up and down from a zero on the vertical axis. The major purpose of such a graph is to compare two or more data series with particular attention to meaningful relationships and changes over time. See Column Graph.

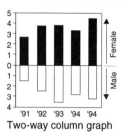

Two-way column graph

Two-Way Histogram

Sometimes referred to as a pyramid graph. A two-way histogram is a variation of a paired bar graph in which spaces between bars are eliminated and the values of the data get smaller as they get higher on the vertical axis. One of the best-known applications of a two-way histogram is a population pyramid, an example of which is shown at the right. See Bar Graph and Population Pyramid.

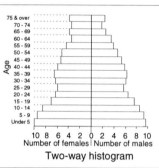

Two-way histogram

Two-Way Table

Year	Product	Value
1990	A	2
1990	B	10
1990	C	8
1990	D	22
1991	A	4
1991	B	12
1991	C	7
1991	D	29
1992	A	6
1992	B	17
1992	C	6
1992	D	20

A one-way table of the same data shown in the two-way tables at the right

A table displaying three variables, one along each of the two axes and the third in the body of the table, as shown at the right. The source of the data for the two-way tables at the right is the one-way table at the left . A comparison of the two formats indicates how much easier it is to analyze data in the two-way format than in the one-way.

	1990	1991	1992
Product A	2	4	6
Product B	10	12	17
Product C	8	7	6
Product D	22	29	20

Third variable in body of table (shaded area)

One variable along vertical axis

Second variable along horizontal axis

	Product A	Product B	Product C	Product D
1990	2	10	8	22
1991	4	12	7	29
1992	6	17	6	20

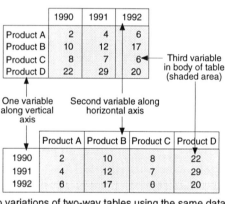

Two variations of two-way tables using the same data

Typeface

Sometimes referred to as a font. When a set of letters, numbers, and characters have the same design features, they are sometimes called a typeface. There are hundreds of typefaces. They have such names as Helvetica, Times, Geneva, etc. Three major categories of typefaces are serif, sans serif, and decorative. Examples of each are shown below. The size of the type in each example is nine-point.

Serif typeface	This is an example of a serif typeface called Times. Serif letters have small lines projecting from the ends of each of their main lines or strokes.	With serifs T
Sans serif typeface	This is an example of a sans serif typeface called Helvetica. Sans (without) serif letters do not have the small lines projecting, as the serif typefaces do.	Without serifs T
Decorative typeface	*This is an example of a decorative typeface named Zapf Chancery. Decorative typefaces are many times used for headings, special effects, and/or to improve the appearance of charts.*	*T*

Unclassed Map

Sometimes referred to as a no class or classless map. A map that does not use class intervals to encode statistical information. See Statistical Map.

Unit Chart

A chart used to communicate quantities of things by making the number of symbols on the chart proportional to the quantity of things being represented. For example, if one symbol represents ten cars and five symbols are shown, the viewer mentally multiplies ten times five and concludes that the group of symbols represented 50 actual cars. Simple geometric shapes or irregular shapes such as pictures and icons are generally used. Each provides basically the same degree of accuracy. When the symbols are geometric shapes, the chart is occasionally called a block chart. When pictures, sketches, or icons are used, the chart is often referred to as a pictorial unit chart. Unit charts are used almost exclusively in presentations and publications such as newspapers, magazines, and advertisements. Legends are generally required with unit charts. Examples of unit charts are shown below.

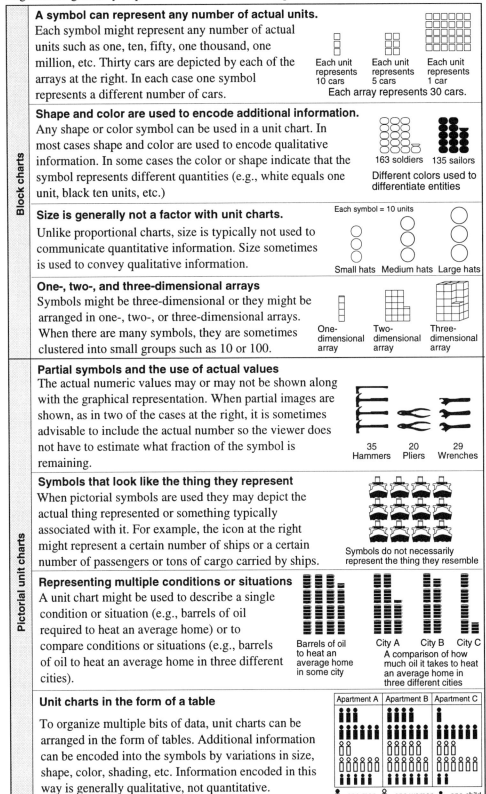

Block charts

A symbol can represent any number of actual units.
Each symbol might represent any number of actual units such as one, ten, fifty, one thousand, one million, etc. Thirty cars are depicted by each of the arrays at the right. In each case one symbol represents a different number of cars.

Each unit represents 10 cars   Each unit represents 5 cars   Each unit represents 1 car
Each array represents 30 cars.

Shape and color are used to encode additional information.
Any shape or color symbol can be used in a unit chart. In most cases shape and color are used to encode qualitative information. In some cases the color or shape indicate that the symbol represents different quantities (e.g., white equals one unit, black ten units, etc.)

163 soldiers   135 sailors
Different colors used to differentiate entities

Size is generally not a factor with unit charts.
Unlike proportional charts, size is typically not used to communicate quantitative information. Size sometimes is used to convey qualitative information.

Each symbol = 10 units
Small hats   Medium hats   Large hats

One-, two-, and three-dimensional arrays
Symbols might be three-dimensional or they might be arranged in one-, two-, or three-dimensional arrays. When there are many symbols, they are sometimes clustered into small groups such as 10 or 100.

One-dimensional array   Two-dimensional array   Three-dimensional array

Pictorial unit charts

Partial symbols and the use of actual values
The actual numeric values may or may not be shown along with the graphical representation. When partial images are shown, as in two of the cases at the right, it is sometimes advisable to include the actual number so the viewer does not have to estimate what fraction of the symbol is remaining.

35 Hammers   20 Pliers   29 Wrenches

Symbols that look like the thing they represent
When pictorial symbols are used they may depict the actual thing represented or something typically associated with it. For example, the icon at the right might represent a certain number of ships or a certain number of passengers or tons of cargo carried by ships.

Symbols do not necessarily represent the thing they resemble

Representing multiple conditions or situations
A unit chart might be used to describe a single condition or situation (e.g., barrels of oil required to heat an average home) or to compare conditions or situations (e.g., barrels of oil to heat an average home in three different cities).

Barrels of oil to heat an average home in some city

City A   City B   City C
A comparison of how much oil it takes to heat an average home in three different cities

Unit charts in the form of a table

To organize multiple bits of data, unit charts can be arranged in the form of tables. Additional information can be encoded into the symbols by variations in size, shape, color, shading, etc. Information encoded in this way is generally qualitative, not quantitative.

Apartment A	Apartment B	Apartment C

= one man   = one woman   = one child

427

Univariate Graph — Sometimes referred to as a one-axis data distribution graph, number axis, or number line. See One-Axis Data Distribution Graph.

Value Axis — An axis with a value scale on it.

Value-by-Area Map — Sometimes referred to as a distorted or proportional map. See Statistical Map.

Value Scale — Sometimes referred to as quantitative, numeric, amount, or interval scale. A value scale consists of numbers organized in an ordered sequence with meaningful and uniform spacings between them. The numbers are used for their quantitative value as opposed to being used to identify entities. See Scale.

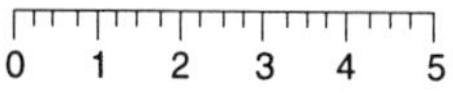

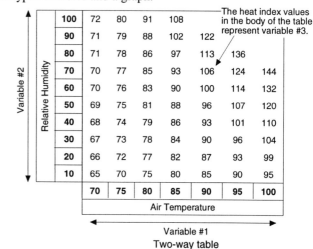

Example of a linear value scale. Other types of value scales include logarithmic, probability, and power.

Variable — A variable is something that changes or has the ability to change. For example, the outside temperature is a variable because it can, and generally does, change from hour to hour. The heights of students in a classroom is a variable because height differs from individual to individual. When reporting election results, the names of the states are variables. Variables are many times referred to by terms such as dimensions, characteristics, attributes, responses, results, etc. Other than the fact that it must have the ability to change, there are few restrictions as to what might be considered a variable. Variables might be tangible, intangible, hypothetical, measurable, theoretical, historical, projected, etc. • The information that constitutes a variable is frequently organized and displayed in tables, graphs, maps, and diagrams. An example of the data for three related variables is shown below in the form of two different types of tables and a graph.

Air Temperature (Variable #1)	Relative Humidity (Variable #2)	Heat Index (Variable #3)
70	10	65
70	20	66
70	30	67
70	40	68
70	50	69
70	60	70
70	70	70
70	80	71
70	90	71
70	100	72
75	10	70
75	20	72
75	30	73
75	40	74
75	50	75
75	60	76
75	70	77
75	80	78
75	90	79
75	100	80
80	10	75
80	20	77
80	30	78
80	40	79
80	50	81
80	60	83
80	70	85
80	80	86
80	90	88
80	100	91
85	10	80
85	20	82
85	30	84
85	40	86
85	50	88
85	60	90
85	70	93
85	80	97
85	90	102
85	100	108
90	10	85
90	20	87
90	30	90
90	40	93
90	50	96
90	60	100
90	70	106
90	80	113
90	90	122
95	10	90
95	20	93
95	30	96
95	40	101
95	50	107
95	60	114
95	70	124
95	80	136
100	10	95
100	20	99
100	30	104
100	40	110
100	50	120
100	60	132
100	70	144

One-way table

Relative Humidity (Variable #2)	70	75	80	85	90	95	100
100	72	80	91	108			
90	71	79	88	102	122		
80	71	78	86	97	113	136	
70	70	77	85	93	106	124	144
60	70	76	83	90	100	114	132
50	69	75	81	88	96	107	120
40	68	74	79	86	93	101	110
30	67	73	78	84	90	96	104
20	66	72	77	82	87	93	99
10	65	70	75	80	85	90	95

Air Temperature (Variable #1)

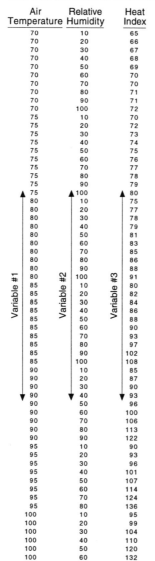

The heat index values in the body of the table represent variable #3.

Two-way table

This illustration shows how variables are sometimes displayed in tables and graphs. For illustrative purposes the same data is used in all three examples. In these examples, numbers are used to express the variables. In other cases, words or symbols might be used.

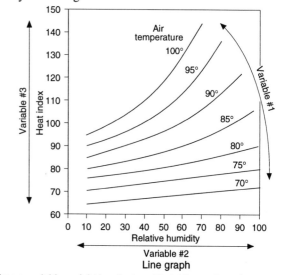

Line graph

Examples of how the data that make up variables might be displayed in tables and graphs

Variable (continued)

The illustration on the previous page has all quantitative variables. In the illustration below, one variable is qualitative (territory) and the other quantitative (sales). The purpose of these two illustrations is to show examples of what might be considered as variables, to demonstrate that the same variable can be displayed in different ways, and to illustrate that the way a variable is displayed can have a significant effect on the types of observations that might easily be made about the variable or groups of variables.

Territory (Variable #1)	Sales (Variable #2)
Boone	$11,587
Branch	9,471
Calhoun	14,128
Cass	12,332
Clinton	19,147
Cook	4,158
Dodge	11,482
Ford	965
Fulton	6,879
Hancock	10,547
Hillside	1,025
Jackson	7,314
Kent	18,511
Mason	7,814
McLean	2,687
Oakland	2,567
Rush	13,210
Shelby	19,387
St. Clair	13,542
Tipton	3,256
Vernon	6,152
Wayne	8,832

One-way table

Bar graph

Shaded map

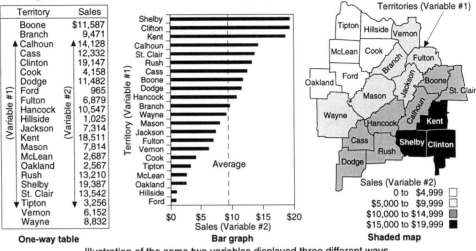

Illustration of the same two variables displayed three different ways

Classification of variables – Variables are classified in a number of different ways. Several of the more widely used classification systems are described on the following pages.

Quantitative, category, and sequence

Quantitative

Sometimes referred to as a value, numeric, or interval variable. Quantitative variables are typically made up of entities that have specific numeric values such as heights, weights, ages, etc. Because they have specific values, they can be plotted on graduated, numbered scales. The elements of a quantitative variable can be ranked and/or mathematically manipulated, such as averaging, subtracting to determine differences, taking logarithms, etc.

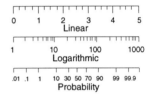

Examples of scales along which quantitative variables might be plotted

Category

Sometimes referred to as a nominal, nonquantitative or qualitative variable. Categorical variables are typically made up of word descriptions of entities such as people, places, things, events, etc. When numbers are used to describe things, they are used only for identification purposes and have no quantitative significance. Each word or number defines a distinct category which contains one or more entities. The elements of category variables can be ordered (e.g., alphabetized, sorted by families, etc.) but not ranked or mathematically manipulated. Reordering of categorical data may make it easier or harder to read a graph or table, but it generally does not degrade the integrity of the data.

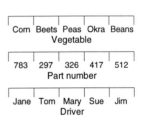

Examples of scales along which categorical variables might be plotted

Sequence

Included in this category are time series, order of occurrence, and ordinal variables. Sequence variables are typically made up of words or numbers that have an ordered, chronological, or nonquantitative numeric progression. The progression might be time, the order in which events occur, the ranking of importance, etc. Time series are the best known examples of sequence variables. When numbers are used they are only for identification and/or ordering purposes and have no quantitative significance (e.g., cannot be mathematically manipulated).

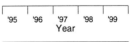

Examples of scales along which sequence variables might be plotted

Independent and dependent variables

Two related variables are many times designated as independent and dependent with respect to one another. There is no single or simple definition as to which variable should receive which designation or whether the terms are even applicable to a given situation. In some cases the decisions are easy. For example, category and sequence variables are almost always independent variables. In other cases it is more difficult. For example, if one is plotting the concentrations of two unknown substances, there may or may not be any interdependency and therefore it would be difficult to decide which, if either, is the independent and dependent variable. Dependent variables are generally plotted on the vertical axis of most graphs except bar. On bar graphs, the dependent variable is typically plotted on the horizontal axis. On three-dimensional maps, the dependent variable is typically plotted on the Z-axis. There is no comparable, generally accepted practice for tables and diagrams. Shown below are several examples of how independent and dependent variables are sometimes defined or described .

Independent variable	Dependent variable
Cause	Effect (the variable that is affected)
Known variable	The variable that one is trying to predict
Things that are manipulated	Things that are measured
Subject to independent change	Affected by changes in independent variable
Explanatory	Response
Under experimental control	Response
Case	Variable
Invariant	Component

Variable described by type of data/measurement

• **Interval**

Sometimes referred to as quantitative or value. With this type of variable, each entity has a distinct quantitative value, such as temperatures in degrees, currency in dollars, weight in kilograms, etc. Data for interval variables typically consists of numbers. Interval data can be ranked, compared, and mathematically manipulated.

• **Ratio**

Sometimes considered a subcategory of interval. With this type of variable, entities are mathematically compared to one another or some reference value. For example, the data might indicate that A is twice as big as B, or B is 3.7 times as big as C. Data for ratio variables are typically expressed as numbers.

• **Ordinal**

Sometimes considered sequential. This type of variable allows the ranking of entities in terms of some nonquantitative criteria such as which entity has more or less of a particular quality or characteristic. For example, things might be ranked as having the most, average, least; first, second, third; upper, middle, lower; or best, average, worst. With this type of variable the viewer can tell whether one thing is bigger, better, faster, etc., than another, but not by how much.

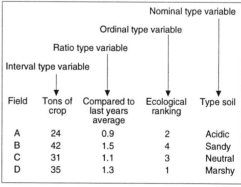

Examples of the four types of variable data

• **Nominal**

Sometimes referred to as categorical or qualitative information. Nominal values are typically made up of words constituting the names or descriptions of people, places, things, or events. Nominal data can be sorted alphabetically but cannot be ranked or mathematically manipulated. Sometimes numbers are assigned to nominal data as identifiers or labels, but the numbers do not change the nature or classification of the data. It still remains nominal data.

Variable (continued)

Continuous and discrete
• Continuous variable

Sometimes call indiscrete. A continuous variable is a quantitative variable that can take on any value within a given range. For example, the outdoor temperature can be any fraction of a degree between its lowest and highest temperatures. The amount of rainfall can be any fraction of an inch from zero to its maximum. Values for continuous variables are typically determined by measurement.

• Discrete variable

Sometimes referred to as discontinuous. A discrete variable might be quantitative or qualitative. When quantitative, the elements take on only specific values and no values in between. For example, when counting people, there are only whole numbers and

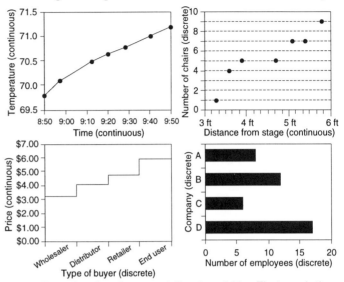

Examples of continuous and discrete variables. The terms in the parentheses behind titles apply only to that particular variable.

nothing in between. There are either ten or eleven people, not ten and a fourth, etc. In many cases the elements of a discrete variable have no relationship to one another, especially if they are categorical. For example, companies A, B, C, and D are discrete entities with no gradations in between and perhaps nothing more in common than that they are all companies. • When a discrete variable is quantitative, the values are frequently determined by counting.

Univariate, bivariate, multivariate, etc.

If a chart, graph, table, or map displays only one variable, it is sometimes referred to as univariate. If it displays two variables, it is sometimes referred to as bivariate, three variables as trivariate, etc. Any number of variables above one may be referred to as multivariate or multidimensional. • There are several different interpretations as to what qualifies as a variable when using these terms. Some people only take into account quantitative variables and therefore they would say a graph with one quantitative and one categorical variable is a univariate graph. Others would call it a bivariate graph. • A similar situation exists with maps. Some people consider the longitude and latitude of a location as two variables. In this case, if a single variable such as population was encoded into each location, the map might be called a trivariate map. Others ignore the latitude and longitude and call the same map a univariate map.

Dummy variable

Sometimes a variable is designated to get more insight into a data series, as opposed to seeing how it varies with respect to some other variable. For example, if a study is done to see how reaction time varies with age, a group of people of all ages might be tested and the data plotted on a scatter graph, as shown at the right. In this example, age and reaction times are the independent and dependent variables, respectively. As an extension of the study,

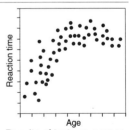

Results of tests on a cross-section of a population

one might identify those data points that represent individuals with some particular characteristic, such as those who are active in sports. This would be called a dummy variable. If circles were used to identify those individuals active in sports, the graph might look like the one at the left, which indicates that those individuals active in sports tend to score higher than those who are not. • There is no limit as to how many dummy variables can be designated for a given data series; however, generally only one is shown per graph.

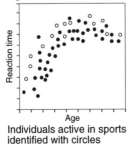

Individuals active in sports identified with circles

Variable Data Chart		Certain quality control charts are categorized as variable data charts. In terms of these quality control charts, variable data is defined as being continuous data that can be measured and expressed numerically. Typical examples would be lengths, diameters, weights, temperatures, strengths, etc. • The counterpart of variable data charts are attribute data charts, which display discrete data that can be counted but not measured. Typical examples would be the quantity of acceptable parts, the number of parts that pass a no-go gauge, the number of scratches, the defects per unit, etc. See Control Charts.

Vector

A measurable entity that has magnitude and direction. Vectors are generally shown as lines or arrows. The length of the line or arrow is proportional to the magnitude of the vector, and the direction of the line or arrow corresponds to the direction of the vector. An example of a vector would be an arrow indicating wind speed and direction as shown at the right. See Vector Graph.

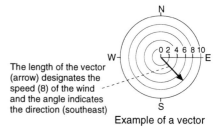

The length of the vector (arrow) designates the speed (8) of the wind and the angle indicates the direction (southeast)

Example of a vector

Vector Chart

A vector chart uses the angle and/or size of arrows to communicate qualitative information. For example, the angle of an arrow might describe the nature of a market. Straight up might mean the market is growing rapidly. Pointing slightly down might mean it is shrinking slightly, etc. In a situation dealing with legislators and their positions on different legislation, an arrow tilted to the left might mean the lawmaker is taking a liberal approach, and an arrow tilted to the right, a conservative approach. Vector charts are often constructed in the form of a matrix or table. The items being reviewed are distributed in rows and columns in some meaningful fashion. The only guideline regarding placement of headings in rows and columns is that similar things be handled uniformly. For instance, if time periods are one of the variables and it is decided to make them into column headings, all time periods should be displayed as column headings and distributed in chronological order. If diseases in various countries are being compared, all diseases might, for example, be placed in rows and all countries in columns. Specific information is encoded into the matrix by placing a vector (arrow) in each of the cells, as shown below.

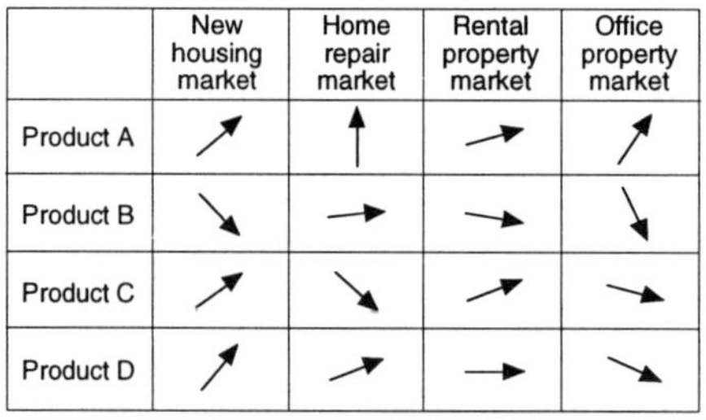

	New housing market	Home repair market	Rental property market	Office property market
Product A	↗	↑	↗	↗
Product B	↘	→	→	↘
Product C	↗	↘	↗	→
Product D	↗	↗	→	↘

Vector chart relating products and markets. An arrow straight up indicates the market is doing very well, straight down means it is doing poorly, and level means the market is static.

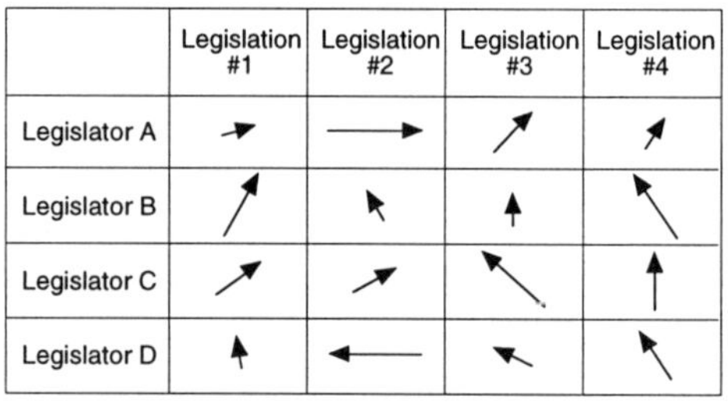

	Legislation #1	Legislation #2	Legislation #3	Legislation #4
Legislator A	↘	→	↗	↗
Legislator B	↗	↖	↑	↖
Legislator C	↗	↗	↘	↑
Legislator D	↑	←	↖	↖

Vector chart on legislators and legislation. Arrows tilted to left indicate the degree of liberalism shown by the legislator, tilted to the right the degree conservativeism. The size of arrow is proportional to the importance the legislator places on the legislation.

The length of the arrow is sometimes used to convey additional information. In the case of the legislators, the angle of the arrow might indicate a conservative or liberal position, while the length of the arrow might indicate how important the lawmaker considers the issue. For example, if the arrow is short, it might indicate the legislator considers the legislation to be of little importance. If the arrow is long, it might indicate the lawmaker considers the legislation of prime importance. Multiple arrows might be used in each cell to address multiple subjects if the arrows are coded in some way so the viewer can tell them apart. • Occasionally any chart that has symbols, groups of words, enclosures, etc., connected by arrows is referred to as a vector chart. • Vector charts are generally qualitative while vector graphs are generally quantitative.

Charts that have symbols, groups of words, enclosures, etc., connected by arrows are sometimes referred to as vector charts.

Vector Graph

Vector graphs are used to plot things that have a value (magnitude) and a direction associated with them. For example, vector graphs are used to plot wind speed and direction, a force and its direction, a vehicle's acceleration and direction, a projectile's velocity and direction, or the distance an object is displaced and the direction of the displacement. The data graphic frequently used with this type of graph is a straight line or arrow called a vector. The length of the line or arrow is proportional to the magnitude of the thing being represented, and the direction of the line or arrow is the same as that of the thing it represents. Arrows are generally used when the thing being represented is unidirectional. Vector graphs can be constructed on two- or three-dimensional, rectangular or circular grids as shown below. Arrows are sometimes coded to indicate multiple data series or different families of vectors.

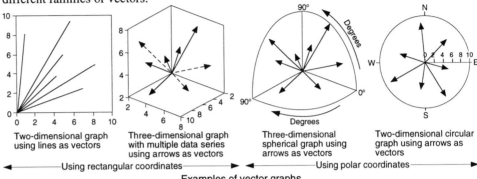

Two-dimensional graph using lines as vectors | Three-dimensional graph with multiple data series using arrows as vectors | Three-dimensional spherical graph using arrows as vectors | Two-dimensional circular graph using arrows as vectors

◄—————Using rectangular coordinates—————► ◄—————Using polar coordinates—————►

Examples of vector graphs.

Starting points of vectors

In many cases, the vectors all originate from the same point. That point might be the origin of the graph or some other meaningful point. The examples at the right show vectors originating from locations other than the origin. The upper two examples show all vectors emanating from a common point. The lower two examples show each vector starting from a different point. In one case, the beginning of each successive vector coincides with the end of the preceding vector, as it would when an airplane flies a zigzag course (e.g., fifty miles east, thirty miles northeast, fifteen miles southeast, etc.). In this type of graph only one pair of coordinates is required for each vector, since the coordinates for the tip of each arrow become the coordinates for the base of the next arrow. In the other example, the start and finish of each vector is independent of all other vectors (e.g., the speed and direction of a group of boats at a given point in time). In this case two pairs of coordinates are required for each vector.

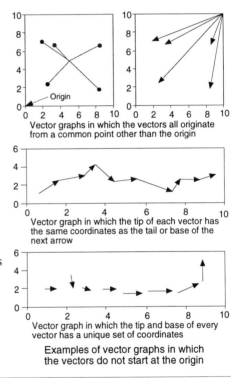

Vector graphs in which the vectors all originate from a common point other than the origin

Vector graph in which the tip of each vector has the same coordinates as the tail or base of the next arrow

Vector graph in which the tip and base of every vector has a unique set of coordinates

Examples of vector graphs in which the vectors do not start at the origin

Vector graphs to display flow or fields

In some situations vectors are used to depict the flow or field of things such as air, water, magnetism, etc. With this type of graph, each arrow is independent of all others; however, it is the composite of all the arrows that conveys the full pattern of the thing being plotted. In this type of display, the scales on the graph often locate the point at which the arrow will be placed. A scale in the legend is used to relate the length of the arrows to the actual values they represent. If directions other than north, south, east, and west are used for orienting the vectors, this too is explained in the legend.

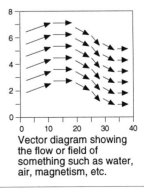

Vector diagram showing the flow or field of something such as water, air, magnetism, etc.

Vector Graph (continued)

Vector diagrams

Vector graphs are many times used to analyze systems of forces. When used in this way they are generally referred to as vector diagrams. When vector diagrams are drawn, the length of the vector (arrow) represents the magnitude of the force and the position of the arrow indicates the location and direction of the force. The resultant of two forces can be determined by drawing the diagonal of the parallelogram formed by the two forces. Vector diagrams can be constructed using either a circular or rectilinear type grid. The example at the right uses a circular or polar type grid.

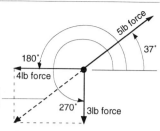

Example of vector diagram analyzing three forces at a point. The dashed arrow represents the resultant (equivalent) of the 3 and 4 pound forces.

Methods for specifying vectors

There are two major methods for specifying vectors. One is called the polar coordinate method and the other the rectangular coordinate method.

Polar coordinate method

In this method the length of the vector and an angle associated with it are specified. The length of the vector (sometimes referred to as radius), is typically positive but can be either positive or negative. The base of the vector may or may not be located at the origin of the graph. In two-dimensional graphs (example at right) or diagrams, the angle of the vector (angular coordinate) is generally measured from the X-axis, but can be referenced from any line. Degrees can be measured in either the clockwise or counterclockwise direction and therefore must be stated or agreed to by convention. With this method, the angle is not limited to 360° and consequently the same data point on a graph can have a

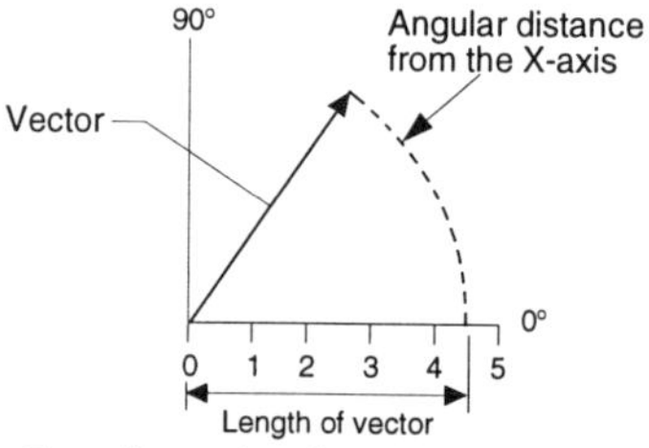

Two-dimensional vector described by the length of the vector and an angular coordinate.

limitless number of angles describing it based on the number of revolutions involved. With three-dimensional vectors, one additional angle or linear coordinate is required for each vector, as shown in the examples at the left.

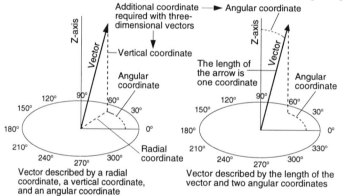

Vector described by a radial coordinate, a vertical coordinate, and an angular coordinate

Vector described by the length of the vector and two angular coordinates

Examples of how three-dimensional vectors might be specified

Rectangular coordinate method

The second method for specifying vectors is to describe the distances along the axes that the tip of the vector is from its base. With two-dimensional graphs, this distance is a certain number of units along both the X- and Y-axes. With three-dimensional graphs, coordinates along all three axes are required, as shown at the right. With this method, each point on the graph has only one set of coordinates that describe it.

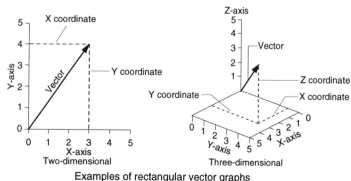

Two-dimensional

Three-dimensional

Examples of rectangular vector graphs

General

Polar and rectangular coordinates can be converted from one to the other graphically or mathematically. • In both the polar and rectangular methods, the point at which the vector starts must be known. When not specifically indicated, the starting point for the base of the vector is generally assumed to be at the origin of the graph.

Venn Diagram

Sometimes referred to as a set diagram or Ballantine diagram. Venn diagrams are graphical tools used in many different fields including mathematics, psychology, education, advertising, and sociology. They are used for such diverse purposes as the study of complex concepts, the introduction of young children to mathematics, the analysis of interrelationships between groups of people, the illustration of ideas in presentations and promotions, etc. Venn diagrams are typically used to describe a relationship between two or more sets of things or information. They accomplish this by the relative positioning of geometric shapes (normally circles) representing the sets of information. For example, two circles might be used to qualitatively depict two sets of buyers. One circle might indicate the set of buyers that bought brand X cars from a given dealer and the other circle the set of

Circles used to represent one set of buyers that purchased brand X cars and another set of buyers that bought red cars

buyers that bought red cars from the same dealer, as shown at the left. If one wanted to convey the idea that some of the same buyers appear in both circles (i.e., they bought red brand X cars), the two circles might be drawn over-lapped, as shown at the right.

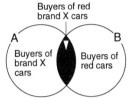

Venn diagram indicating that some of the people in the circle of buyers of red cars are also in the circle of buyers of brand X cars

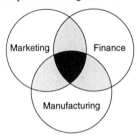

A Venn diagram with two levels of overlap

• An example of a Venn diagram with three symbols and two levels of overlapping is shown at the left. One level of overlap indicates that each pair of departments shares responsibility in some areas (gray). The other level indicates that all three departments share responsibility in other areas (black). Nonquantitative Venn diagrams, such as these, are sometimes considered variations of conceptual diagrams.

Quantifying Venn diagrams

Venn diagrams can be quantified by placing values directly on the data graphics or by means of identification numbers or letters and a legend. Generally no attempt is made to have the size of the symbols proportional to the values they represent. • Shown at the right is a repeat of the car example above, except values have been added and a rectangle drawn around the circles. Translating this diagram into words, it says that out of 500 people who bought cars, 175 bought brand X cars; 100 bought red cars; 25 bought red, brand X cars; and 250 people bought cars that were neither brand X nor red. The illustration below explains how this information is encoded into the diagram.

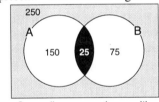

Same diagram as above, with values and a box added

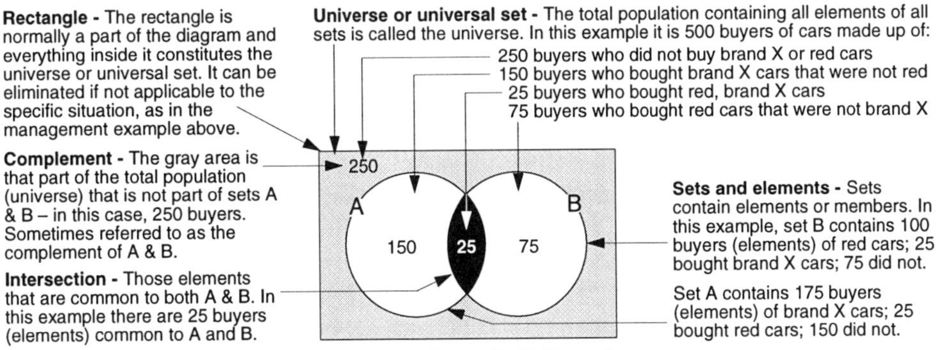

Rectangle - The rectangle is normally a part of the diagram and everything inside it constitutes the universe or universal set. It can be eliminated if not applicable to the specific situation, as in the management example above.

Complement - The gray area is that part of the total population (universe) that is not part of sets A & B – in this case, 250 buyers. Sometimes referred to as the complement of A & B.

Intersection - Those elements that are common to both A & B. In this example there are 25 buyers (elements) common to A and B.

Universe or universal set - The total population containing all elements of all sets is called the universe. In this example it is 500 buyers of cars made up of:
— 250 buyers who did not buy brand X or red cars
— 150 buyers who bought brand X cars that were not red
— 25 buyers who bought red, brand X cars
— 75 buyers who bought red cars that were not brand X

Sets and elements - Sets contain elements or members. In this example, set B contains 100 buyers (elements) of red cars; 25 bought brand X cars; 75 did not.

Set A contains 175 buyers (elements) of brand X cars; 25 bought red cars; 150 did not.

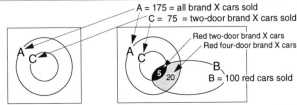

Terminology generally used with Venn diagrams

Variations of Venn diagrams
Subsets

When a set is totally inside another set, it is referred to as a subset. For example, two-door brand X cars form a subset of all brand X cars. The example at the right incorporates a subset into the example from above.

A = 175 = all brand X cars sold
C = 75 = two-door brand X cars sold
Red two-door brand X cars
Red four-door brand X cars
B = 100 red cars sold

C is a subset of A   Subsets can sometimes provide more detail   435

Variations of Venn diagrams (continued)

———————**Disjointed sets**———————

When two sets have no common elements or members, they are referred to as disjointed sets. For instance, in the example at the right the A set represents all the buyers of brand X cars and the B set represents all the buyers of gray cars. Since brand X does not product a gray car, there are no common members and thus no overlap of A and B.

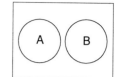

Sets that do not touch are said to be disjointed

———————**Union**———————

When two sets combine it is called a union. Unions function the same as a single set and can intersect with other sets.

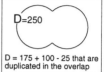

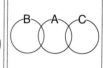

A union occurs when two sets combine into one such as A and B on the left into D on the right.

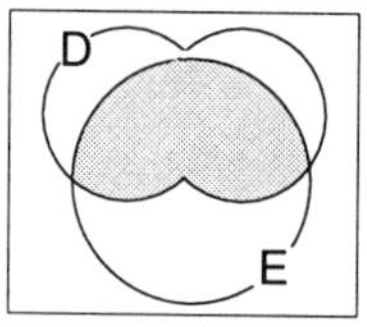

A union functions like a single set and can intersect with other sets.

General

———————**Number and arrangement of sets**———————

Most diagrams use between two and four sets. Technically, there is no restriction on the number that can be used or the way they might be arranged.

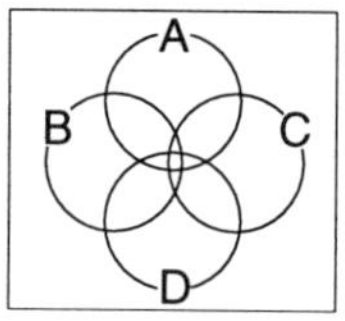

Four sets 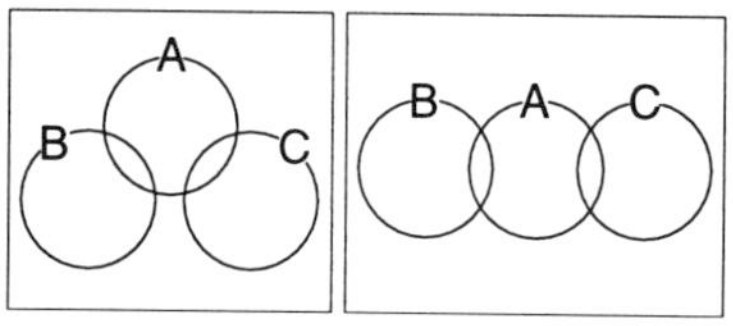   Three sets arranged two different ways

———————**Shape of symbols**———————

Although circles are typically used with Venn diagrams, other shapes are sometimes substituted. The symbols representing each of the sets in a given diagram do not have to be the same shape.

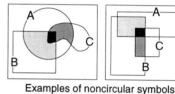

Examples of noncircular symbols

———————**Color, shading, and patterns**———————

Shading, coloring, and/or patterns are used extensively to designate specific areas or to highlight certain features.

Variations in the use of shading or color

———————**Labels**———————

The letters or numbers identifying the sets can be placed inside the circles or on the circumference. On the circumference is sometimes preferred because it reduces the potential confusion as to whether the label applies to the entire set or just a portion of the set.

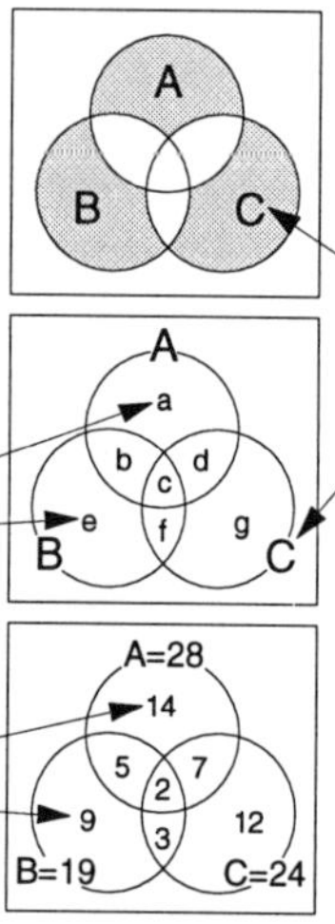

Labels identifying sets might be located inside the set or on the circumference.

Sometimes the individual segments of the intersecting circles are identified by lowercase letters and uppercase letters used to identify entire sets.

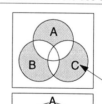

Other times the actual values for various segments are shown, including the totals for the sets.

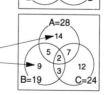

Vertical Axis

In both two- and three-dimensional graphs, the vertical axis is the one that runs up and down. In two-dimensional graphs, it is frequently called the Y-axis. In three-dimensional graphs the vertical axis is generally called the Z-axis. Positive values typically progress upward.

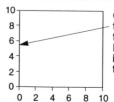

Generally referred to as the vertical or Y-axis in a two-dimensional graph. Positive values typically increase from bottom to top.

Vertical Bar Graph

Sometimes referred to as a bar graph, rotated bar graph, or column graph. See Column Graph.

Vertical bar graph. Sometimes called a column graph.

Vertical Line Chart

Sometimes referred to as an open-high-low-close (OHLC) graph, high-low-close-open (HLCO) graph, high-low-close (HLC) graph, high-low graph, bar chart, or price chart. This type of chart is used extensively to record and track the selling prices of securities, commodities, markets, etc. Several prices are typically recorded for each time period with highest, lowest, close, and open being the most common. See Bar Chart [2].

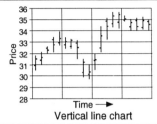

Vertical line chart

Vertical Line Graph

Sometimes referred to as a needle or spike graph. On this type of graph a separate vertical line is used to designate each individual data point. The top of the vertical line designates the actual data point. Such graphs are frequently used when many data points are to be plotted at uniform intervals, as in many sequence type graphs. Vertical line graphs might be two- or three-dimensional. See Line Graph.

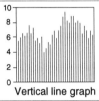

Vertical line graph

Volume Chart

A variation of proportional charts, volume charts use the volume of data graphics to convey information about the relative sizes of the data elements they represent. For example, if data element B is twice as big as data element A, the volume of the data graphic representing element B will be twice as large as the volume of the data graphic for A. Volume charts are difficult to interpret and therefore are generally used only for aesthetic purposes or to give a general impression when values vary significantly. The data graphics can be displayed several different ways, including side-by-side, stacked, and nested. Two examples are shown at the left. • Even though each data graphic on a proportional volume chart appears three-dimensional, it

Two examples of proportional volume charts.

typically represents a single value. On rare occasions an effort is made to quantify variables on all three sides of the data graphics, as shown at the right. • When "depth" is added to the data graphics on an area proportional chart for aesthetic purposes, it does not reclassify it as a proportional volume chart since the depth is typically uniform on all data graphics and included for cosmetic purposes only.

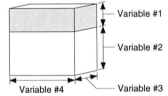

Example of a seldom-used technique for quantifying multiple dimensions on a data graphic of a proportional volume chart

Volume Graph

Volume graphs are often used for recording the number of shares of a stock sold in a given interval of time (i.e., day, week, month, etc.). The shares might be of an individual security or commodity, a family of securities, an exchange, a market, etc. Volume charts are also used for recording such things as number of contracts executed, number of contracts open at the end of trading, number of price reversals, etc. Volume figures are recorded on the vertical axis and time on the horizontal axis. Vertical line graphs are generally used. Segmented line graphs are sometimes used. Examples of both are shown at the left.

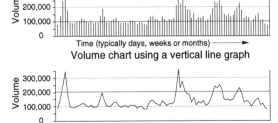

Volume chart using a vertical line graph

Volume chart using a segmented line graph.

Volume charts recording shares sold are generally used in conjunction with price charts, since the price-volume relationships are sometimes considered as important as the volume chart alone. This combination is sometimes referred to as a price-volume chart. An example is shown at the right.

Example of typical price and volume chart

Voronoi Diagram

A graph consisting of a series of reference points and a web of lines. The lines form cells or polygons around each reference point such that any point within a given cell is closer to the reference point in that cell than it is to any other reference point on the graph. For instance, in the example at the right, every point in the cell shaded gray is closer to reference point D than to any other reference point on the graph (reference points are the black dots with letters alongside). If this were a map instead of a graph and every reference point represented a restaurant, the people whose homes were in the gray area would be closest to the restaurant designated by the letter D. Some companies use this technique to locate their nearest outlet for potential customers who phone in. In such an application, the scales on the graph might be latitude and longitude, miles from a known reference point, or some arbitrary scale that can be referenced back to a map.

Voronoi diagram in which each point in a cell is closer to the reference point in that cell than to any other reference point on the diagram.

Wage and Salary Graph

A wage and salary graph summarizes wage and salary information in such a way that key information and relationships are graphically displayed. Rate of pay is plotted on the vertical axis. The scale is typically graduated in values such as dollars per hour, dollars per week, dollars per month, etc. Wage and salary grades are shown along the horizontal axis with the lowest grade on the left. For each grade, a box is shown with the top of the box representing the maximum specified for the grade, the bottom the lowest, and a line across the middle representing the midpoint. A dot is entered above each grade for every employee in that grade. If the employee's pay is within specification, the dot for that employee lies within the box. If it is above or below specification,

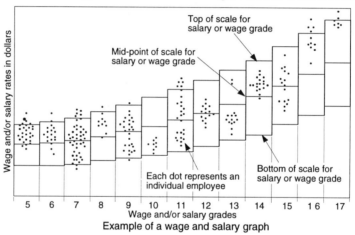

Example of a wage and salary graph

the dot will lie outside the box. • The following are typical obser-vations that might be made from such a graph.

– The grades in which there are high concentrations of employees
– The distribution of employees along the entire spectrum of grades
– Whether employees are clustered in particular sections within a grade
– How many employees are above the maximum or below the minimum in each grade
– The relationships between the midpoints of each of the grades
– The spreads between maximum and minimum for the various grades
– The overlapping of pay levels from grade to grade
– Whether or not salary grades form a smooth progression

By superimposing competitive or industry standards on top of such a graph, the relative position of the company against an outside benchmark is possible.

Wall

Sometimes referred to as a back plane. Walls are the vertical planes on three-dimensional graphs formed by the vertical or Z-axis and the X and Y axes. Grid lines for one or more of the axes are many times displayed on the walls. For aesthetic purposes, walls sometimes are given the appearance of having thickness.

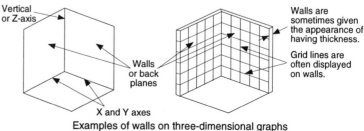

Examples of walls on three-dimensional graphs

Weather Map

Weather maps are a blend of descriptive, statistical, and flow maps. Below are examples of some of the more widely used weather maps and the graphics used to encode information.

Temperature

Colored bands are used to indicate areas of equal temperature ranges. This usage is sometimes referred to as an isopleth (bands of equal values) map, or an isotherm (bands of equal temperature) map. The range of values that apply are generally shown in the bands. For example, a band labeled 40s means that the temperatures for the areas included in the band range from 40° to 49°.

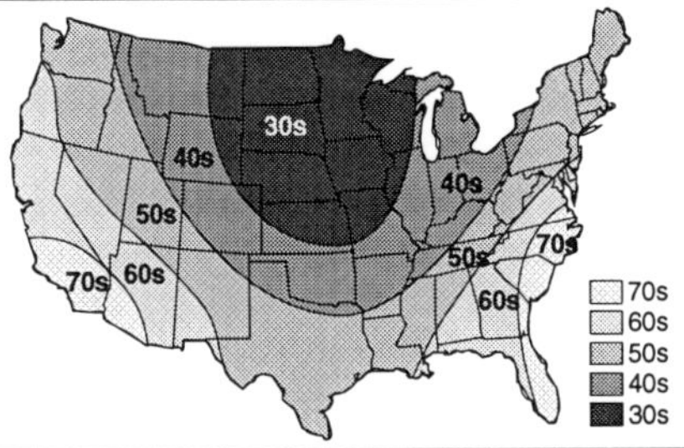

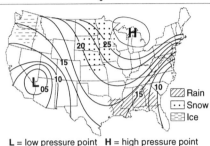

Pressure

When lines connect points of equal pressure, the map is sometimes called an isoline or isobar map. Values are shown on the lines and are generally stated in terms of millibars (mb), using only the last two or three digits. For example, 20 on the map equals a value of 1020 mb; 207 equals 1020.7 mb, and 996 equals 999.6 mb. The letters H and L designate the points of highest pressure in the region. Precipitation is sometimes also shown.

Rain
Snow
Ice

L = low pressure point   **H** = high pressure point

Warm and cold fronts

Symbols for fronts convey information about what type of front it is, where it is located, and the directions it is moving. Six of the more commonly used symbols for fronts are shown below.

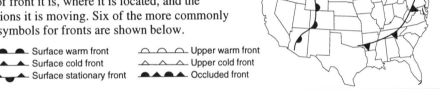

Surface warm front
Surface cold front
Surface stationary front
Upper warm front
Upper cold front
Occluded front

Local weather information

Certain maps give very detailed data for specific locations using a combination of meteorological symbols and numerical data. The example at the right shows how symbols are sometimes used to designate the weather at specific points. Variations of the information shown below are generally encoded into or included with the symbol.

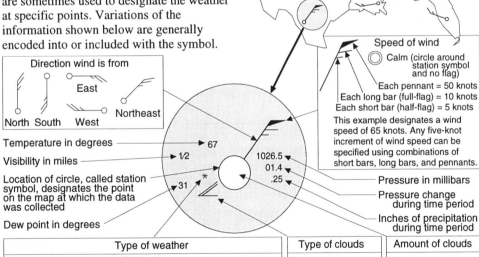

Direction wind is from

North   South   West
East
Northeast

Temperature in degrees ———→ 67
Visibility in miles ———→ ½
Location of circle, called station symbol, designates the point on the map at which the data was collected ———→ 31
Dew point in degrees

1026.5
01.4
.25

Pressure in millibars
Pressure change during time period
Inches of precipitation during time period

Speed of wind

Calm (circle around station symbol and no flag)

Each pennant = 50 knots
Each long bar (full-flag) = 10 knots
Each short bar (half-flag) = 5 knots

This example designates a wind speed of 65 knots. Any five-knot increment of wind speed can be specified using combinations of short bars, long bars, and pennants.

Type of weather			Type of clouds	Amount of clouds
⸴ Intermittent drizzle	Snow shower	═ Mist	Altocumulus	Clear sky
⸴⸴ Continuous drizzle	✳ Intermittent snow	≡ Fog	Altostratus	A few clouds
Rain showers	✳✳ Continuous snow		Cirrocumulus	Scattered clouds
• Intermittent rain	Slightly drifting snow	Lightning	Cirrostratus	Partly cloudy
•• Continuous rain	Heavy drifting snow		Cirrus	Mostly cloudy
Thunderstorm			Cumulonimbus	Cloudy/overcast
Heavy thunderstorm	△ Sleet		Cumulus	Sky obscured
Squall	▲ Hail shower		Nimbostratus	
Tropical storm	Freezing rain		Stratocumulus	
Hurricane			Stratus	

Wedge

Sometimes referred to as a slice, sector, or segment. The term wedge is frequently used to refer to one portion of a pie chart or circle graph. Each wedge typically represents one data element of a data set. The example at the right shows a pie chart divided into four wedges.

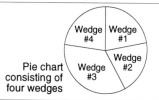

Pie chart consisting of four wedges

Wind Rose Graph

A wind rose graph summarizes wind information over a period of time such as a day, week, month, or year. A series of roses are frequently used to compare wind patterns in multiple locations or in the same location over different time periods. Wind rose graphs can be used as stand alone graphs or as symbols on maps. The graphs generally show actual wind values for eight to sixteen different directions, equally spaced around the 360°. Each value or set of values might be shown separately (below left) or connected by lines to form a polygon or series of polygons (below right), which may be filled or unfilled. The angle denoted by the symbol or data point designates the direction from which the wind was blowing. The total length of the symbol or the distance of the furthest data point from the center of the circle indicates the percent of time the wind came from that direction. Different widths, colors, or patterns of the data graphics can be used to give an indication as to what proportion of the winds from that direction were of various speeds. For example, the graphs below indicate the wind was from the east 6% of the time. There were no winds from that direction above 38 miles per hour. One third of the time the winds were in the range of 16 to 38 mph. Another third of the time the winds were between 9 and 15 mph and the rest of the time they were less than 9 mph.

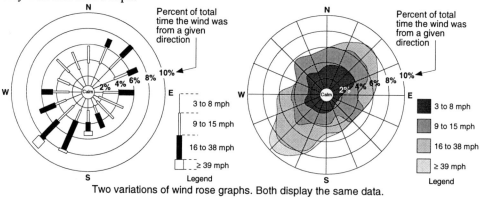

Two variations of wind rose graphs. Both display the same data.

Wireframe

A wireframe is a line drawing consisting of a series of lines that simulate the surface of an object or data graphic. The areas between the lines are transparent. If the areas between the lines are made opaque, the result is generally called a surface or lined surface. The purposes of wireframes in graphs include:

– Serving as a mechanism for graphically displaying complex information
– Better visualization of the arrangement and patterns of data points
– Providing a framework from which surfaces can be generated

In some cases wireframes are generated by connecting all adjacent data point with lines. This many times results in very irregular patterns but does help define the relationships of the data points. Another method is to draw lines connecting points of equal value along one or more axes. If equal values along the vertical axis are used, the graph is often called a contour graph. If equal values along the X and Y axes are connected, the pattern that results is many times call a fishnet. Examples of both are shown below. (See Surface Graph)

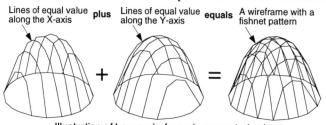

Lines of equal value along the X-axis **plus** Lines of equal value along the Y-axis **equals** A wireframe with a fishnet pattern

Illustration of how a wireframe is generated using lines of equal values along the X and Y axes

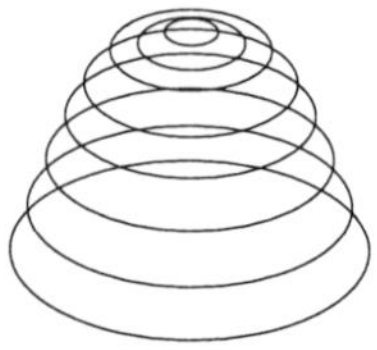

A wireframe of the same data shown at the left, except lines of equal value along the Z-axis are used

In some cases the fishnet and contour lines are combined on the same drawing. In still other cases, the lines are drawn randomly, choosing only those lines that best portray the surface. There are no requirements as to what or how many lines should be used. Some computer programs give the operator the option as to how many lines they would like and in some cases which lines. Wireframes can be two- or three-dimensional, but are generally three-dimensional.

Wireframe (continued)

The examples below compare data displayed using wireframes and other techniques. The column on the left shows the data using standard scatter, column, and line graphs. The second column shows the same data using wireframes. The two columns on the right show what the graph looks like when the areas between the lines of a wireframe are made opaque (called surface or lined surface) and then filled with a color, shading, or pattern (called filled surface).

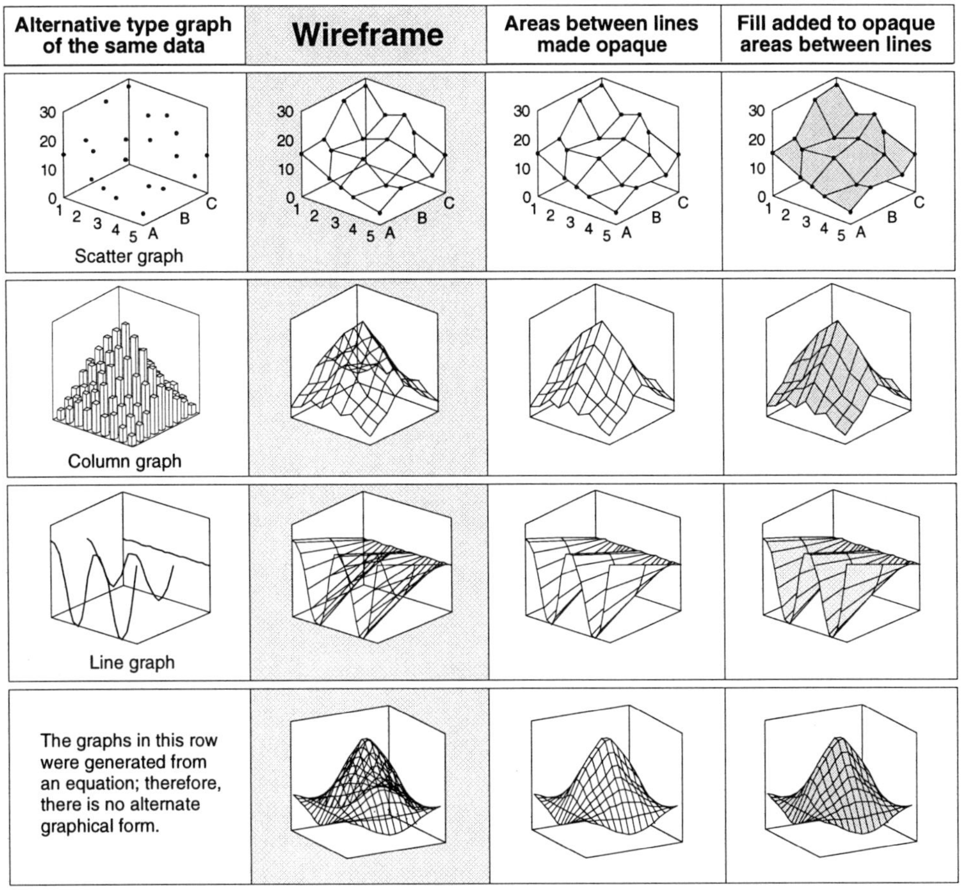

Alternative type graph of the same data	**Wireframe**	Areas between lines made opaque	Fill added to opaque areas between lines
Scatter graph			
Column graph			
Line graph			
The graphs in this row were generated from an equation; therefore, there is no alternate graphical form.			

Word Chart

Sometimes referred to as a text chart. A word chart is a single document, page, slide, etc., consisting solely or mainly of words and used primarily in presentations. Although a word chart is generally one of a series of charts, each chart tends to stand on its own. Word charts normally contain only the key words or phrases of a subject. Typical guidelines recommend that the number of lines per chart and the number of words per line be limited to between five and eight. Most word charts have no graphics associated with them except for decorative purposes or general orientation. Two examples are shown at the right. See Text Chart.

Word charts with and without graphics

Word Table

A table consisting largely or entirely of words, as shown at the right. The number of words in each cell might vary from a single word to a sentence or even a paragraph.

	Product introduced	Product growing	Product maturing
Sales	Negligible	Rapid increase	Stable
Profit	Negative	Break even	Positive
Cash flow	Negative	Stable	Positive

Word table

X-Axis
or
X Axis

In a two-dimensional graph, the X-axis is traditionally the horizontal axis. Positive values generally increase from left to right and negative values become more negative from right to left. In three-dimensional graphs the X-axis might be either of the nonvertical axes. See Axis, Graph.

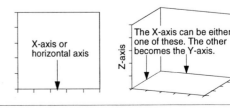

X-Bar/R Chart	X-bar (sample average) and R (range) type charts are used extensively in the field of quality control. When used together they are often referred to as an X-Bar/R chart. The two charts are sometimes shown separately, with the R chart below the X-Bar chart. In other cases the two are merged to appear as a single chart with a common horizontal scale and two different quantitative scales on the vertical axis, as shown at the right. The X-Bar chart generally displays sample averages of consecutive lots. The R chart displays the range of values for the same data. Based on the combination of these two charts, observations can be made about the process being monitored. See Control Charts.	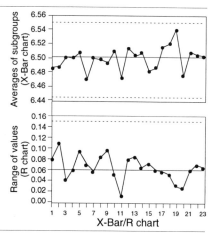

XY Graph

The letters XY are many times used in front of the name of a graph to indicate that it is a two-axis, two-dimensional graph with quantitative scales on both axes. The XY prefix is almost never used when the graph has a category scale. It is occasionally used when one of the axes has a sequence scale. Examples are: XY scatter graph, XY scatter line graph, and XY area graph. Occasionally the term XY graph refers to a two-axis scatter graph.

XY Plane
XZ Plane
YZ Plane

The two-letter prefix designates the two axes that form the plane being referred to. For example, the XY plane is formed by the X and Y axes, as shown below.

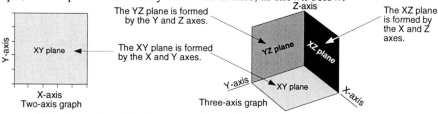

Major planes formed by the axes on two- and three-axis graphs

XYZ Graph

The letters XYZ are many times used in front of the name of a graph to indicate that it is a three-axis, three-dimensional graph with quantitative scales on all three axes. The XYZ prefix is almost never used when the graph has a category scale. It is occasionally used when one of the axes has a sequence scale. Examples are: XYZ scatter graph, XYZ scatter line graph, XYZ wireframe graph, and XYZ surface graph. Occasionally the term XYZ graph refers to a three-axis scatter graph.

Y-Axis
or
Y Axis

In a two-axis graph, the Y-axis is traditionally the vertical axis. Positive values generally increase in an upward direction and negative values become more negative in a downward direction. In three-axis graphs the Y-axis might be either of the nonvertical axes. See Axis, Graph.

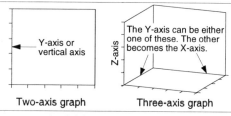

Yield Curve

Sometimes referred to as term structure of interest rates. A yield curve displays the relationship of interest rates on fixed-income securities (e.g., bonds), and their time to maturity. Length of time to maturity is shown along the horizontal axis and percent yield along the vertical axis.

Scales

The vertical scale is always linear with units of percent. The lower and upper values are typically slightly below and above the lowest and highest values plotted, respectively. The lower and upper values on the horizontal axis are usually three months and 30 years, respectively. The placement of values on the horizontal axis varies widely. Sometimes they are placed uniformly, other times there is no apparent pattern. As a result, curves of the same data can look quite different, as shown in the examples below.

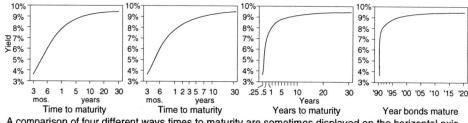

A comparison of four different ways times to maturity are sometimes displayed on the horizontal axis

Yield Curve (continued)	**Positive and negative yield curves**

When the percent yield is higher on bonds with longer term maturities, the curve is called a positive yield curve. When the percent yield is lower on bonds with longer term maturities, the curve is referred to as a negative or inverted yield curve.

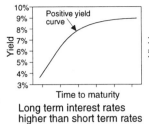

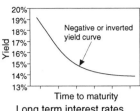

Z-Axis or Z Axis

The term Z-axis generally refers to the vertical axis on a three-axis, three-dimensional graph. Positive values generally increase in an upward direction and negative values become more negative in a downward direction. In some two-dimensional graphs, such as contour graphs, the Z-axis is considered perpendicular to the paper.

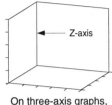

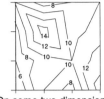

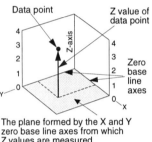

On three-axis graphs, the Z-axis is traditionally the vertical axis.

On some two-dimensional graphs, such as contour graphs, the Z-axis is perpendicular to the paper.

Zero Base Line Axis

A zero base line axis serves as a reference from which quantitative values are measured in order to establish the proper location of data points. In addition, they define the origin, quadrants or octrants, and basic planes of a graph. A plane serves the same function in a three-dimensional graph that the zero base line axis does in the two-dimensional graph. Examples are shown below. See Axis, Graph.

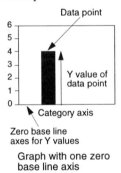

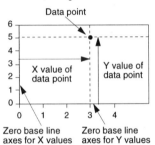

Graph with one zero base line axis

Graph with two zero base line axis

Graph with a plane formed by two zero base line axes

Z Graph

A Z graph is a special application of a line graph. Its major attribute is its ability to graphically display three different aspects of a given variable on the same graph. The three aspects are monthly values, cumulative values, and twelve-month total values. This type of graph is generally started at the beginning of a twelve-month period, at which point the graph consists of three points, two of which are on top on one another (monthly and cumulative). As the year progresses, the Z configuration begins to form, until at the end of the twelve-month period the lines for the three different values converge to form a Z shape.

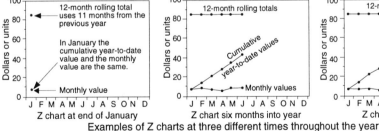

Examples of Z charts at three different times throughout the year

After about six months of data, the cumulative and the twelve-month total lines can be projected to get an estimate of what the year-end total will be, assuming no major seasonal element. In some cases a dual value scale is required to make all three curves legible. Graphs can be plotted in terms of dollars or units. Z charts are frequently used for tracking sales type information.

Zigzag Graph

Sometimes referred to as fever, thermometer, broken line, or segmented line graph. A zigzag graph is a variation of a line graph in which the data points are connected by straight lines, as shown at the right.

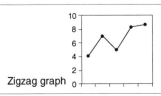

Zigzag graph

Bibliography

Articles, General

Cleveland, William S. and Robert McGill. "Graphical Perception: Theory, Experimentation, and Application to the Development of Graphical Methods." *Journal of the American Statistical Association* 79 (1984): 531-554.

McCullar, Robert L. "Improving the Image of Numbers." *Journal of Accountancy* (August 1995): 37-40.

Snee, Ronald D. and Charles G. Pfeifer. "Graphical Representation of Data." *Encyclopedia of Statistical Sciences* 3 (1983): 488-511.

Basic Information Including How To Draw Charts and Graphs by Hand

American National Standards Committee Y15. *Time-Series Charts, A Manual of Design and Construction.* New York: American Society of Mechanical Engineers, 1979.

Cardamone, Tom. *How to Plot & Construct Charts & Graphs For Reproduction.* New York: Art Direction Book Company, 1988.

Coburn, Edward J. *Business Graphics: Concepts and Applications.* Boston, Massachusetts: Boyd & Fraser Publishing Company, 1991.

Lefferts, Robert. *How to Prepare Charts and Graphs for Effective Reports.* New York: Barnes & Noble Books, 1981.

Robertson, Bruce. *How to Draw Charts & Graphs.* Cincinnati, Ohio: North Light Books, 1988.

United States General Accounting Office. *Visual Communication Standards.* Washington, D.C.: US Government.

Warner, Jack. *Graph Attack! Understanding Charts and Graphs.* Englewood Cliffs, New Jersey: Regents/Prentice-Hall, 1993.

Washington Training and Development Services. *Presenting Statistics Using Graphs, Charts, and Tables.* Washington, D.C.: US Office of Personnel Management, 1988.

Charts and Graphs, General

Breckon, C.J., L.J. Jones and C.E. Moorhouse. *Visual Messages: An Introduction to Graphics.* England: David & Charles Publishers, 1987.

Brinton, Willard C. *Graphic Presentation.* New York: Brinton Associates, 1939.

Carlsen, Robert D. and Donald L. Vest. *Encyclopedia of Business Charts.* Englewood Cliffs, New Jersey: Prentice-Hall, 1977.

Costigan-Eaves, Patricia James. *Data Graphics in the 20th Century: A Comparative and Analytic Survey.* New Brunswick: Rutgers University The State University of New Jersey, 1984.

Enrick, Norbert Lloyd. *Effective Graphic Communication.* New York: Auerbach Publishers, 1972.

Henry, Gary T. *Graphing Data: Techniques for Display and Analysis.* Thousand Oaks, California: SAGE Publications, 1995.

Herdeg, Walter. *Graphis Diagrams: The Graphic Visualization of Abstract Data.* Zurich: Graphis Press, 1983.

Karsten, Karl G. *Charts and Graphs.* New York: Prentice-Hall, 1923.

Lockwood, Arthur. *Diagrams.* New York: Watson-Guptill, 1969.

Charts and Graphs, General
(continued)

Lutz, Rufus R. *Graphic Presentation Simplified.* New York: Funk & Wagnalls, 1949.

Meyers, Cecil H. *Handbook of Basic Graphs: A Modern Approach.* Belmont, California: Dickenson Publishing Co., 1970.

Murgio, Matthew P. *Communications Graphics.* New York: Van Nostrand Reinhold, 1969.

Rogers, Anna C. *Graphic Charts Handbook.* Public Affairs Press, 1961.

Schmid, Calvin. *Statistical Graphics, Design Principles and Practices.* New York: John Wiley & Sons, 1983.

Schmid, Calvin F. and Stanton E. Schmid. *Handbook of Graphic Presentation.* New York: John Wiley & Sons, 1979.

Selby, Peter H. *Using Graphs and Tables.* New York: John Wiley & Sons, 1979.

Spear, Mary Eleanor. *Practical Charting Techniques.* New York: McGraw-Hill Book Co., 1969.

Spear, Mary Eleanor. *Charting Statistics.* New York: McGraw-Hill Book Co., 1952.

Swindle, Robert E. and Elizabeth M. Swindle. *The Business Communicator.* Englewood Cliffs, New Jersey: Prentice-Hall, 1985.

du Toit, S.H.C. et al. *Graphical Exploratory Data Analysis.* New York: Springer-Verlag, 1986.

Computer Software Manuals

Most information graphics can be drawn by hand; however, more and more are being originated on computers. The majority of computer programs tend to focus on particular types of information graphics such as basic graphs, maps, flow charts, investment charts, scientific graphs, or statistical graphs. Generally the manuals or books dealing specifically with particular software are the best source of information for each application. Some of the computer manuals provide excellent general information as well as describing the details of the graphics included in the application. Because of the number of different manuals and the speed with which new ones are issued and older ones out-dated, these types of reference books have not been included in this bibliography.

Format and Appearance

Bowman, William J. *Graphic Communication.* New York: John Wiley & Sons, 1968.

Holmes, Nigel. *Designer's Guide to Creating Charts and Graphs.* New York: Watson-Guptill, 1984.

Lotus Development Corp. *Looking Good With Lotus 1-2-3.* Cambridge, Massachusetts: Lotus Development Corp., 1991.

Meilach, Dona Z. *Better Business Presentations.* Torrance, California: Ashton-Tate, 1988.

Talman, Michael. *Understanding Presentation Graphics.* San Francisco, California: SYBEX Inc., 1992.

Tufte, Edward R. *The Visual Display of Quantitative Information.* Cheshire, Connecticut: Graphics Press, 1983.

White, Jan V. *Using Charts and Graphs.* New York: R.R. Bowker, 1984.

White, Jan V. *Graphic Idea Notebook.* Rockport, Massachusetts: Rockport Publishers, 1991.

Zelazny, Gene. *Say it with Charts.* Homewood, Illinois: Business One Irwin, 1991.

Material Focused on a Specific Type of Information Graphic

Andersen, Anker V. *Graphing Financial Information: How Accountants Can Use Graphs to Communicate.* New York: National Association of Accountants, 1983.

Anson, R. W. *Basic Cartography for Students and Technicians, Volume 2* . New York: Elsevier Applied Science Publishers, 1984.

Dorsey, Thomas J. *Point and Figure Charting: The Essential Application for Forecasting and Tracking Market Prices* . New York: John Wiley & Sons, 1995.

Edwards, Robert D. and John Magee. *Technical Analysis of Stock Trends* . Boston: John Magee Inc., 1966.

Evarts, Harry F. *Introduction to PERT.* Boston: Qallyn and Bacon, 1964.

Hardy, C. Colburn. *The Investor's Guide to Technical Analysis* . New York: McGraw-Hill Book Co., 1978.

Hiam, Alexander. *The Vest-Pocket CEO* . Englewood Cliffs, New Jersey: Prentice-Hall, 1990.

International Labour Office. *Introduction to Work Study* . Geneva. 1957.

Ishikawa, Kaoru. *Guide to Quality Control.* White Plains, New York: Quality Resources, 1991.

Jarett, Irwin M. *Financial Reporting Using Computer Graphics.* New York: John Wiley & Sons, 1987.

Laseau, Paul. *Graphic Problem Solving for Architects & Builders.* Boston, Massachusetts: CBI Publishing Co., 1975.

Leebov, Wendy and Clara Jean Ersoz. *The Health Care Manager's Guide to Continuous Quality Improvement.* Chicago, Illinois: American Hospital Publishing, Inc., 1991.

Lobeck, Armin Kohl. *Block Diagrams and Other Graphic Methods Used in Geology and Geography*, Amherst, Massachusetts: Emerson-Trussell Book Company, 1958.

McGill, Robert et al. " Variations of Box Plots." *The American Statistician* 32 (February 1978): 12-16.

Modley, Rudolf and Dyno Lowenstein. *Pictographs and Graphs.* New York: Harper & Brothers, 1952.

Molnar, John. *Nomographs: What They Are and How to Use Them.* Ann Arbor, Michigan: Ann Arbor Science Publishers, 1981.

Montgomery, Douglas C. *Introduction to Statistical Quality Control* , New York: John Wiley & Sons, 1990.

Muehrcke, Phillip C. and Juliana O. Muehrcke. *Map Use, Reading, Analysis, and Interpretation.* Madison, Wisconsin: JP Publications, 1978.

Mundel, Marvin E. *Motion and Time Study, Principles and Practices.* Englewood Cliffs, New Jersey: Prentice-Hall, 1970.

Nelson, Wayne. *How to Analyze Data With Simple Plots.* Milwaukee, Wisconsin: American Society for Quality Control, 1986.

Nison, Steve. *Japanese Candlestick Charting Techniques: A Contemporary Guide to the Ancient Investment Techniques of the Far East* , New York: New York Institute of Finance, 1991.

Nixon, Judith M. *Organization Charts.* Detroit, Michigan: Gale Research Inc., 1992.

Bibliography (continued)

Material Focused on a Specific Type of Information Graphic (continued)

Oakland, John S. *Statistical Process Control: A Practical Guide* . Great Britain: Billing & Sons Ltd., 1990.

Reynaud, C. B. *The Critical Path, Network Analysis & Resource Scheduling* . London: George Godwin Limited, 1970.

Schiller, Bradley R. *The Economy Today.* New York: Random House, 1983.

Thompson, Allen R. *Economics.* Reading, Massachusetts: Addison-Wesley Publishing Co., 1985.

U.S. Geological Survey. *Map Projections.* Reston, Virginia: U.S. Department of the Interior, 1994.

Witzling, Lawrence and Greenstreet, Robert. *Presenting Statistics.* New York: John Wiley & Sons, 1989.

Theoretical and Statistical

Bertin, Jacques (Translated by William J. Berg). *Semiology of Graphics: Diagrams, Networks, Maps.* Madison, Wisconsin: University of Wisconsin Press, 1983.

Chambers, John M. et al. *Graphical Methods for Data Analysis.* New York: Chapman & Hall, 1983.

Cleveland, William S. *The Elements of Graphing Data. Pacific Grove, California:* Wadsworth Advance Book Program, 1985.

Pettersson, Rune. *Visuals for Information Research and Practice.* Englewood Cliffs, New Jersey: Educational Technology Publications, 1989.

Pfaffenberger, Roger and James H. Patterson. *Statistical Methods for Business and Economics.* Homewood, Illinois: Irwin, 1987.

Tukey, J. W. *Exploratory Data Analysis.* Reading, Massachusetts: Addison-Wesley Publishing Co., 1977.

Material on factors that confuse or mislead viewers

Huff, Darell. *How to Lie with Statistics.* New York: W.W. Norton and Co., 1954.

Kosslyn, Stephen M. *Elements of Graph Design.* New York: W.H. Freeman and Company, 1994.

Zelazny, Gene. "Grappling With Graphics." *Management Review* (October 1975).